*For my Darling Rebecca*

*Your precious heart and wonderful spirit mean everything to me.*
*Without you, this book would not have been possible.*

*"There cannot be any conflict between truth revealed from heaven and truth revealed through the research of man; for truth is a unit and never is found in conflict with itself."*

Joseph Fielding Smith
(Man... His Origin and Destiny)

*"When a person receives intelligence from the LORD and is willing to communicate that for the benefit of the people, he will receive addition to that intelligence."*

Lorenzo Snow

*"Some people say, "How can you live without knowing?" I do not know what they mean— I always live without knowing. That is easy. How you get to know, is what I want to know!"*

Richard Feynman

# IN THE BEGINNING

The Complete Telling of the Three Creations of God
(A Latter-Day Perspective)

Mark Campbell

Ron & Vi

May God always bless you with knowledge and wisdom – May He fill your hearts with the desire to discover –

You friend always
Mark Campbell
Aug 2020

Thank you both for being important to both Lynda and I and our family

*Creation Cubed Publications*



Library of Congress Control Number: 2017951479

ISBN-10: 0-692-89617-4
ISBN-13: 978-0-692-89617-4

First Printing: November, 2017
First Edition: November, 2017
Second Edition: August, 2018

SECOND PRINTING

Printed in the United States of America

## Note from the Author

This treatise represents the culmination of a lifetime of work, one filled with great heights of joy and the deep sadness of loss. Through it all, I have always stood fast in my faith in God.

Though some of the ideas discussed here are not new and to a few may sound familiar, their application to a framework of a tiered creation, in three distinct parts, is entirely unique. There is much to be found within these pages that has never been printed or discussed previously, indeed even formulated or conceptualized before. If it has, I am unaware of it.

I present here both a theology and a cosmology, unique and never before disclosed. That this book includes both is intentional, for if either is to be true and complete, it must do so with the other; the two cannot stand alone.

I am not speaking here on behalf of the Church; I am not authorized to do so. Except where noted, in a few places, this book does not pretend to convey authorized church doctrine. This work was not submitted to the church correlation committee for approval.

What this work does represent, is a thorough study of the Scriptures, and the creations of God, in truth the *ultimate* source. The understandings developed and published here are the culmination of almost 60 years of living in the light of our common Father, constantly seeking His guidance and knowledge in prayer, ever learning and always standing in awe.

True understanding of the truth of the things of God is available to ALL that humbly seek after it. It is revealed to all sons and daughters of God; it is not reserved for a precious few. The keys of knowledge are not handed out from a collection of keys held tightly in a small box. Knowledge does not come through appointment of position to those who never seek it. Neither is it dispensed through dogma however oft it is reiterated, but true knowledge comes through inspiration by diligently seeking with an open mind and a dedicated heart.

My resource materials were for the most part confined to the certain truths of the Scriptures (being those identified as the Standard Works of the Church of Jesus Christ of Latter-day Saints) and the application of the best commentary on their interpretation, both ancient and modern in origin.

I also include much from the Temple Endowment and its ordinances (the narrative and other sacred ordinances spoken and performed within Latter-day Saint Temples) as they are applied, as a source of divinely inspired knowledge, drawing upon them to expand our understanding of the things of God.

The science used here remains distant from that which is speculative, relying only on the most respected, thoughtful, understandings of the workings of the Universe.

Not much will be found here in the way of commentary from Latter-day Church leadership, beyond that of Joseph Smith and a smattering of other prophets and apostles. Most readers will be pleasantly surprised to learn, as they will here, that the deepest meanings of the truth that is revealed by God had been revealed before, to those who simply sought it, long before the opening of this, the Last Dispensation and the birth of those that were to be called to lead it.

There is nothing revealed here that is not backed by the best of science and the religious truths revealed by God.

I have been blessed to have received the truths I have shared here. I am humbled by such a gift from God. Though I have labored all my life to learn what I have, I am aware that the most precious words contained herein are not mine, but they are those of the Spirit. Uncountable passages seemed to just appear in front of me as I wrote them with a clarity I never held before. I would commit great sin if I did not acknowledge this divine assistance from beyond the veil. *(Proverbs 5:18)* → Lynda ♡

It is my wish that as you read these things, the Spirit will testify of their truthfulness as it has to me. Read and enjoy – a new vision of the Cosmos and its Creator awaits you.

*Mark Campbell*

*November 2017*

# TABLE OF CONTENTS

## List of Tables

## List of Figures

## PROLOGUE

# A LESSON TO BE LEARNED

Complexity it seems, like beauty, is in the eye of the beholder. Enormously intricate systems at the highest level of examination can seem so simple. The details of how a thing works can completely belie its designed purpose and even contradict its end function. Life is like that, the Universe as a whole speaks to the truthfulness of this observation.

It is estimated that the diameter of the observable Universe is about 28 billion parsecs (93 billion light-years), putting the edge of all that we can see at about 46–47 billion light-years away. Within this ever expanding sphere exists everything we will ever know; from our perspective everything there is or ever has been, all that 15 billion years could produce: 10 million super clusters, 25 billion galaxy groups, 350 billion large galaxies, 7 trillion dwarf galaxies, and some thousand, billion, trillion stars.[1] All of this, and perhaps even more, teems with diversity, individuality and wonder, yet in aggregate none of it is as complex and varied as the geology and water driven biosphere of our home planet and the life that inhabits it.

Imagine for a moment the Earth, its oceans, continents, all of their complexities, depths, shallows, mountains, canyons, rivers, lakes, glaciers and valleys. Think of all the flora and fauna that ever was, all of history, all of humanity's struggles and triumphs, all of our loves and dreams, all any of us ever were or will be. All of this is, has been, or has taken place on what, from a distance, appears as nothing more than, as Dr. Carl Sagan said so eloquently, "*a pale blue dot... a mote of dust suspended in a sunbeam*" (*Figure 1*).[2]

Each of us individually is a remarkable wonder of complexity. Our bodies are made of some 74 trillion cells, each producing 2000 proteins a second. A single cell can completely reproduce itself in precise, molecular-level detail in only ten hours.

---

[1] These large numbers represent uncountable quantities and unfathomable realities – they are the best estimates available at the time the first edition of this book was printed. In fact, some of them were revised twice before the first writers draft was completed and perhaps stand in need of yet another revision at the time of this second printing, the values always upwards ever enlarging the volume of what is knowable. The point made here is manifest; these are enormously large numbers representing a vast and expansive universe.

[2] From, The Pale Blue Dot: A Vision of the Human Future in Space, Carl Sagan, Random House, 1994, pg. xv–xvi.

Figure 1 *The Earth as Seen from a Distance. The Earth and the Moon photographed by the Cassini spacecraft in orbit around Saturn some 870,000 miles distant; across the bottom, in the full image photographed by Cassini, at the upper middle, the dot shown below the rings is our home; the upper left is yet another enlargement of that same bottom picture showing the Earth on the left and our moon to its right. The fourth photograph, on the upper right, is a picture of the Earth taken from the Voyager spacecraft nearly 4 billion miles distant; an almost imperceptible dot almost obscured by what is actually an artifact of diffracted sunlight within the lens of the camera. Our home is indeed a very small place in the vastness of space.*

Within us, our heart beats about 100,000 times in one day and about 35 million times in a year. During an average lifetime, the human heart will beat more than 2.5 billion times. With each beat, your heart muscles contract with the same intensity your hand would use to squeeze a stiff, new tennis ball.

Miracles hardly ever happen, or so we say, but in reality, they go on all around us every moment of every day. Consider all of our bones, muscles and sinews, our senses and their receptors; our thoughts and joys come together in perfect unison to kick a ball across the fresh cut lawn on a beautiful spring day. As we run along the field on a small part of our Earth, under the warmth of a star we call the Sun, spinning away in a galaxy which is itself hidden in an obscure corner of the immense Universe, we give the Cosmos no thought at all. But in that moment in time, in that brief instance, everything is aligned. We have no consciousness of the complexity of the moment; to us it is just a simple game of kickball.

In the perspective of how things actually are, having an understanding of the Universe as it is, our home the Earth, and the wonder of the dynamics that make up the lives of each of us, it is not hard at all to fathom a creation in three parts. We can understand that three separate events are all part of one grand plan to bring to pass the immortality and eternal life of man.

Science, a hallmark of modern day thinking, seeks for the simplest theory as the gold standard for truth.[3] This, however, is a human construct, born in no small part by the absolute need of scientists to simplify a complex problem so that our restricted human understanding can grasp it. Our greatest minds, with our greatest advances, made no such attempt to simplify the world to meet their needs. Indeed they constructed complex theories requiring huge advances in the way we saw ourselves and the world. Many times the worlds they discovered were far more complex than what they even imagined. In many cases, centuries later, the Universe their minds understood is still without an explanation simple enough to be comprehended by more than just a few. The rest of us believe in what they say based on faith, faith in them, faith in the process, and the results we have achieved in controlling the Universe by the principles and laws they established as fact.

This same application of faith and understanding, with all of its rigors and experience can, and should, be applied to our understanding of all that has been revealed to us in the Scriptures about how we all came to be here in this amazing Universe around us. At first, just on the surface, it may seem much more complex for our liking, more complex than necessary, certainly more complex than we thought. In reality, however, this complexity disguises the simple beauty of the whole Plan of Salvation and the intent of God as He created us and our Universe.

---

[3] *Ockham's Razor* (Latin, *lex parsimoniae*) is a problem-solving principle devised by the English Franciscan Friar, *William of Ockham* (c. 1287–1347). William was a scholastic philosopher and theologian. His principle states that among competing hypotheses, the one with the fewest assumptions should be selected. Other more complicated solutions may ultimately prove correct, but in the absence of certainty, the fewer assumptions that are made, the better.

# CHAPTER 1

# A CREATION IN THREE PARTS

## Creation Accounts

In this book, we will work our way back through the three separate and distinct creations following in order from the last (most recent), to the first (most ancient). Peeling back in layers from what we see today, through the eons to the way things are eternally, the way they always have been, in fact are now, and always will be. In so doing, we must not lose track of the fact that these three creations are part of a single Plan of God, not as much separate events as they are elements of one single process of "*bring[ing] to pass the immortality and eternal life of man*" (Moses 1:39).

For each creative event, we will examine the Scriptures (both ancient and modern), Latter-day, teachings and what science has to tell us, to formulate a complete and descriptive understanding. These creative processes are pivotal in the history of each of us, the sons and daughters of God. They are necessary in the formative process that moves each of us forward, towards our ultimate potential and destiny, becoming like our Father in Heaven. Because of this, as can be expected, information can be gleaned all over the Scriptures regarding them. However, the events are detailed most fully in seven separate accounts found in nine separate tellings, eight in the pages of Scripture and one spoken of in the Temple. We will reference and focus here, on the passages/teaching listed:

- Genesis 1-2:3
- Genesis 2:4- (Genesis 5:2)(D&C 30-50 [31,32])
- Proverbs 8
- St. John 1:1-5
- Moses 2,3
- Abraham 3:22 through 5:21
- Temple Endowment

It is common to talk of the *Spiritual World* and the *Physical World*. These two phrases and derivations thereof dominate the landscape of religious thought. By them we define the *World/Universe* we live in mortally today and differentiate it from the *Eternal Sphere* that God inhabits, the Heaven we all hope to attain and live in forever. However, these two concepts (one mortal and tangible and the other eternal and spiritual) are insufficient for our discussion here and should be considered inadequate in any discussion of the creations of God.

It is clear that we live in and inhabit a mortal *World/Universe*, within which everything is dying, all living beings, including man, and indeed all matter itself, deteriorates and passes away. Our theology also bespeaks of the existence of another much different world, an *eternal* world, inhabited by immortal Physical beings and constructs as tangible as that we see about us today. We are told that this *immortal* world had a beginning in the distant past and that its existence will stretch into the *eternal* future in the presence of God.

So far so good, we comprehend *two* spheres, a mortal one and an eternal one, except any examination of what we know to be true will lead to the realization that there was something that existed *before* the creation of the immortal world. We will discuss the details of this further on in this chapter; but suffice it to be said for now, that before all things were created *spiritually*, there existed God, the Father and the Son and "*intelligences*" that were eternal with them. All things proceed from the Father and are created by Him, so we have need of defining this world as well and the separate elements, creative processes and attributes associated with it.

Here, we will define these three distinct spheres, their contents, and the creative processes associated with bringing them into being as: Spiritual, Physical and Temporal (*Figure 1-1*).

-The Spiritual sphere encompasses all things created and existent without body or construct, including eternal entities identified in the Scriptures as "*intelligences,*" the existence of which long preceded the creation of eternal Physical bodies and worlds. We will discover that all things of this Sphere are eternal (for they will have no end), but we will learn that some of these eternal things have a beginning, a time when they were created.

-The Physical sphere is that which includes what we will call our eternal physical bodies and the eternal worlds, kingdoms and constructs created to exist along with them throughout eternity. As we shall see, it is the creation of this sphere that is spoken of in Genesis 2:5 and the Endowment narrative. This is the sphere that has been traditionally associated with things termed *spiritual* (the Spirit World, Eternal Life and the multiple Kingdoms of God) but doing so leads to the concealment of the fact that this world is *tangible* (it exists physically), though immortal. Identifying it as solely *spiritual* also shrouds the fact that there is a recorded *beginning* for this eternal sphere.

-The Temporal sphere contains everything that is mortal or Temporary in nature; it is the construct the Scriptures most frequently term "*natural*." The Temporal is the Universe we inhabit today constrained and governed by the laws of physics. It contains all that we can "*sense.*" It is a sphere that is identical in most all respects to the *physical world* of tradition.

**Spirit Creation (Abraham 3:22-26)**

- God (Father) Organizes/Creates *Intelligences*
- Son created (first)
- Father and Son create all else (Proverbs 8; John 1: 1-5)
- Eons pass

**Physical (Immortal) Creation (Temple Endowment)**

- Son, at direction of the Father (with Michael) Create the Physical (Immortal) World. Father and Son Create *Embodied Spirits*
  - All things (Earth, Stars, Plants, Water, Animals (Man, Woman) (Christ is the Father of these Bodies)

    *All Spirit is matter; it is just more refined and we cannot see it. Adam is created from the (dust) substance of this spiritual matter.*

    ***Note:** This is a progressive, tiered creation (organized from building blocks previously in existence).*
- Father and Son introduce the Neshama into Adam *(Man **and** Woman together are named Adam)(Genesis 5:2)*
- This creation is eternal not mortal, NOT the Universe around us

**History of this World begins (Moses 1:35 and beyond)**

**(Genesis 2:4 – 3:24 still in the Physical [Immortal] World)**

*Adam and Eve introduced into the Garden of Eden, Flesh and Bone Bodies, Satan allowed to test Adam and Eve, Adam passes the test (The Fall is precursor to the Temporal World)(Flesh and Bone and Blood)*

**Temporal Creation (Genesis Chapter 1, Moses 1:35, 40; 2:1-3:3, Abraham 4:1-5:3)**

- Big Bang
- Six days of creation mirrors Physical Creation (Physical Laws)
- Six days of creation lasts 13+ billion years as measured in current Earth time frame

  *Adam (Sons of God?) introduced late in the history of the Earth (Genesis 1:26-28 /Moses 2:26-28)*
- Spiritual World governs the Temporal World (Complete Free Agency)
- *Zero Point Field*
- *Quantum Mechanics*
- *Cosmology (Dark Matter)(Dark Energy)(CMBR)(Relativity)*
- **Tzimtzum** – creation by subtraction (Temporal World *less* than the more perfect whole)

Figure 1-1 *The Three Creations*

# The Three Creations

## *The Temporal Creation*

Temporal dies, this is the most significant property of this sphere. In this aspect, the Temporal sphere is fundamentally distinct from the other two we will discuss, which are both eternal and immortal. Death in its most *inclusive* definition is not an exclusive characteristic of life; at its most basic level, death affects all that there is in the Universe. The inanimate does not escape the effects of death. Death is degradation, disorder and decay.

Death represents a progression from a higher state of order to a lower state of disorder. Scientifically, death is defined within the concept of the Second Law of Thermodynamics,[1] which is most commonly expressed in the following way, "*energy and all systems run downhill.*" For living systems, this most fundamental law means aging, death and eventually total decay of structure down to the most basic elements. But the Universe itself is also dying by this definition, and it has been from the very moment it began. Every atom and subatomic particle will lose energy, order, and pass away as the energy of the Universe becomes empty space.[2] This degeneration is what is known scientifically as entropy.

---

[1] The second law of thermodynamics states that in a natural thermodynamic process, there is an increase in the sum of the entropies of the participating systems. The second law is an empirical finding that has been accepted as an axiom of thermodynamic theory. The law defines the concept of thermodynamic entropy for a thermodynamic system in its own state of internal thermodynamic equilibrium. It considers a process in which that state changes, with increases in entropy due to dissipation of energy and to dispersal of matter and energy (see Appendix, Entropy – the Great Equalizer, pg. 538).

[2] This is a true principle and it escapes me as to how this fundamental relationship has not been codified into a law of physics. Energy and Space are inseparably bound in the same way that matter and energy is. Matter is just a special form of energy, energy that is constrained, fixed in location, as in a standing wave. Energy and Space, however, are inversely proportional, ***ES=k*** *where k is a fixed constant representing the total content (volume and energy/mass) of the Universe (the constraints of the Zero Point Field).* At the beginning of the Universe, there was infinite energy (all of it potential energy) and all of space was drawn up in the Primeval Singularity with a value of absolute zero. At the moment of the Big Bang, energy began to go to zero (changing from potential to kinetic and thermal energy and eventually matter, then dissipating) and space began to increase. As the Universe plays out, constrained energy in matter will diminish and go to zero, finally all of the energy of the *Zero Point Field* will drop to zero as space itself will increase to infinity. It is the drawing up of space by the fluctuations in the *Zero Point Field*, as those fluctuations tug on the fabric of space, which gives space its curvature. It is the flattening of those fluctuations, as they diminish in scale, which is increasing the expansion of space today (Dark Energy) as the rate of dissipation of energy accelerates in the field. Dark energy cannot be seen because anything that moves freely through the *Zero Point Field* or does not vary from it in amplitude remains undetectable to any observer within the field. Since matter and energy are interchangeable and energy and space are interchangeable, nothing is lost when matter falls into a black hole. Beyond the *Event Horizon*, as matter is decomposed, the energy from which the matter was comprised continues to hold the waveform, joining it, like a symphony, to the other waveforms that fell onto the surface of the black hole. The resultant black hole has a waveform comprised of all of those it consumed. The proposed equivalency also accounts for the apparent loss of energy by photons due to the Cosmological Redshift. These concepts and the direct correlation between energy and space seem most obvious, and I submit them here as my contribution to Theoretical Physics.

With death (Entropy) comes time. The arrow of time, what we observe as entropy, irreversibly moves events forward. Within the Universe, we observe complexity to be increasing and growing. Today large galaxies stretch out in strands around the surface of large vast voids; in the distant past, the galaxies were more uniformly clumped together. We observe change over periods exhibiting an irreversible growth of complexity which provides us with a perception of direction in time; our sense of time is a result of Entropy.

The Temporal sphere is a Fallen Universe because of entropy, but because it is *Fallen*, agency exists as it never did before. We will throughout this work, on multiple levels, examine *agency* in its various facets. For now we posit that the Temporal Universe, in composition and structure, was designed and exists, to allow for an agency that is truly "*free*," permitting the expression of unrestricted will in ways that are significantly different than had been possible within the Physical sphere.

When Einstein put forth the theories of General and Special Relativity and explained the photoelectric effect by positing the existence of quanta of light, our understanding of this Universe changed in an all-encompassing way. *Quantum Uncertainty* was discovered, resulting in what Einstein termed "*spooky*" actions for matter.

Uncertainty means that, on the most fundamental level, matter in its most basic form, responds variably to the actions of a conscious observer. Matter is both particle and waveform; it can be at all places at the same time in a manner that is entirely unknown until an attempt is made to observe it. This action is not *smoke and mirrors*; it is fundamental reality. Einstein reeled at the thought; in the world he preferred, all actions could be known and defined. Such a world, however, would be ordered, and any one action, anywhere, would determine all actions everywhere at every time. Everything would be interlinked and predetermined, no *Free Will* could exist.

As the principles of Quantum Uncertainty were being discovered in the microscopic world, great advances were being made in Cosmology. As a result, the Universe could no longer be seen as eternal and unchanging, but it now had a definite beginning and would probably have an end. This discovery, for the first time, aligned the scientific world with the scriptural accounts. These understandings and advancements define a *Temporal Universe*.

The Temporal sphere cannot support the requirements of immortality; it is designed to provide the conditions necessary for the "*free*" agency of Man. At no place, at no time, could an immortal existence be sustained in the environs of this Universe. Death is the price of a passport to this Universe and always has been for everyone who enters. This fact is so eloquently expressed in the incarnation of the Creator of this Universe Himself, Jehovah, Christ the LORD, who also had to suffer death for His sojourn here. Within this work, we will explore this point in detail.

## *The Physical Creation*

In contrast to the Temporality and death of the sphere we live in today, stands the immortality of the Physical Creation. The Scriptures clearly state that all things were created *spiritually* before they were "*in the Earth*" (naturally) (Moses 3:5, Genesis 2:5). Associated with this statement, we have the story of the Garden of Eden where death was not known. In Eden, God walked freely alongside Adam; in Eden, the Tree of Life could be freely consumed. Both of these facts identify the Garden as an eternal abode. Where was it? It was *not* in this Universe and as a necessity not on the Earth we inhabit today.

In the Temple, we are told of Adam's participation in the creation that eventually led to his placement in a Garden somewhere Eastward in Eden. In the three scriptural accounts of the creation of the world we live on, Adam's (Michael's) creative participation is not mentioned at all, it is, however, integral to the story told in the Endowment narrative. This is because the creation story told in the Temple Endowment is of a different creative event than the one told in the first chapter of Genesis and the two matching accounts of the books of the Pearl of Great Price.

We know that the creation Michael participated in was that of an eternal (immortal) sphere. Yet we are told that Adam, within the Garden, had a Physical body of *Flesh and Bone*. The immortal creation Adam assisted with then resulted in a Physical (tangible) sphere. Again we can turn to the events attendant to the Savior's advent for an example of this. After our LORD's resurrection, He purposefully appeared instructing those He attended to that His "*Spiritual*" (Physical) body is of *Flesh and Bone* and is as tangible as it had been when He walked among them, fully Man.

Another important observation that can be made of the Endowment narrative are the critical differences in the sequencing of the creative period events from those detailed in the scriptural accounts. The sequence changes speak truth about the distinctive individuality of the Physical creation as separate from the Temporal creation. The Physical sphere is separate and distinct from the Temporal sphere we inhabit today, it always has been. The Garden and the placing of Adam within it, are not early components in the history of this Universe, nor could they ever have been. But they are to be understood as things attendant to another immortal yet Physical sphere.

## *The Spiritual Creation*

The Spiritual sphere is the most enigmatic of the three we will discuss here; this is so because we are told so little of it and it is, in fact, a realm of our habitation so very distant in the past. Despite this, however, there are some very direct conclusions that can be made about this important sphere. We are told:

> *"There is no such thing as immaterial matter. All spirit is matter, but it is more fine or pure, and can only be discerned by purer eyes; We cannot see it; but when our bodies are purified we shall see that it is all matter." (D&C 131:7, 8)*

This takes on more meaning through the full context provided through the Endowment narrative. In it, at the outset of the Physical Creation, we are made aware that matter, in an unorganized form had an existence prior to the creative periods of the immortal Physical sphere. We can conclude then, that matter, what D&C 131 identifies as "*Spirit*," pre-dated the Physical Creation.[3]

We witness Michael (Adam) interfacing with the Father and the Son throughout the creative processes of the Physical Creation. This informs us of several things. First, Michael pre-existed and was a distinct created being long before the Physical sphere was fashioned. Second, he existed in such a way then that allowed him to distinguish himself to the extent that he was chosen to act as co-creator of the Physical sphere and to be the progenitor of all Mankind. This required his differentiation in some way from the other Intelligences. It cannot be said for certain, but may be properly inferred, that this differentiation resulted from Michael's actions in this sphere and may have been a process of the organization of the many "*Intelligences*" there (as witnessed by Abraham both through actual participation and in subsequent vision) (Abr. 3:22, 23). From this we know that there are beings (*Intelligences*) dating from a period *before* the eternal Physical world was formed.

Our theology informs us that Christ is "*the image of the invisible God, the firstborn of every creature*" (Col. 1:15). Further we know, "*All things were made by him, and without him was not any thing made that was made*" (John 1:3). Christ therefore existed in the Spiritual sphere and He predates all of the individual Intelligences there.

Let us be clear, we are told, "*Intelligence, or the light of truth, was not created or made, neither can be*" (D&C 93:29). Of this we hold no dispute; however, *we believe that eternal intelligence is the Character of God*, and that the "*Intelligences*" (distinct individual beings) were created and composed of individual quanta of intelligence given by God, *directly from Him*. We were created by our Father in an intimate way, not unlike the sharing of genetic material from Temporal parents to their Earthly offspring today. Indeed we are *literally* sons and daughters of God with the "*stuff*" of His existence deep within us.

## *Fitting Them Together*

With our initial understandings of the three creations established, let us discuss their interrelationships (*Figure 1-2*). The top of the diagram shows the three creative *principles*, the Father, the Son and Michael. The middle portion of the diagram shows the creative sequence; though, as we will discuss later in this book, it is improper to speak of time in any environment other than the Temporal Universe we

---

[3] We will thoroughly explore the term "*Spirit*" in yet another context later in this work as a proper name for a type of being, embodied and comprised of two distinct parts, *intelligence* and organized physical, yet *immortal body*.

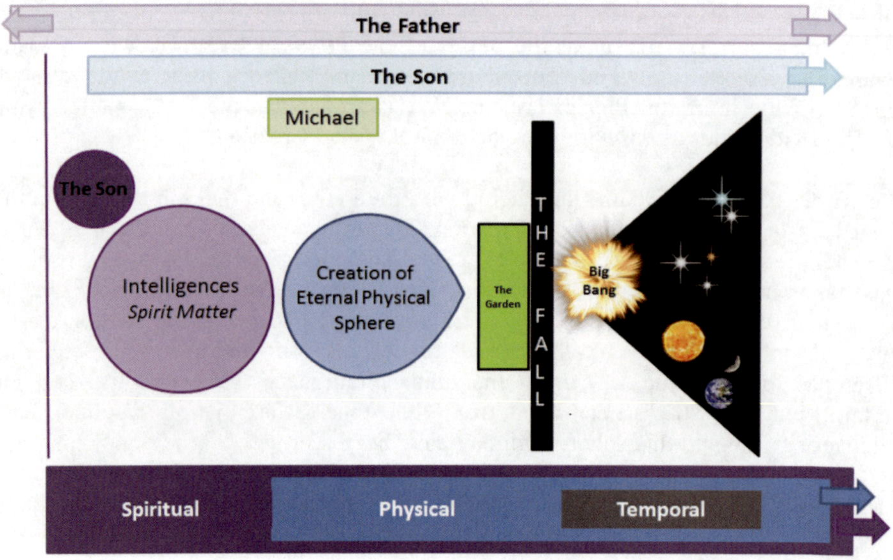


Figure 1-2 *Diagram of the Three Creations. The chronology and interrelationships of the creative events, players and the periods of their creative activity are shown here. Note that along the bottom of this illustration, it is indicated that the Temporal sphere is entirely enveloped within the Physical sphere, which is in turn entirely enveloped by the Spiritual sphere.*

inhabit today, and this would properly include any discussion of the concepts of *before* and *after*. (For this reason, I believe, the LORD identifies Himself as Alpha and Omega, both the beginning and the end, in a way that includes them both simultaneously.) In the diagram, "*time*" flows from left to right. The bottom of the diagram indicates the Temporal (in grey) is entirely included as a subset of the Physical, and the Physical in turn is a subset of the Spiritual. Additionally, it can be seen the Temporal has a terminus and as such ends; the Physical and Spiritual, however, are eternal without end.

At the top of the diagram, we see the Father to have had no beginning, but for the Son there was a beginning, correlated with a sphere representing His creation predating the sphere representing the creation of the *Intelligences* in the Spiritual sphere. At the top of the diagram, the creative existence of the Son and the Father (as shown by the horizontal bars) has no end. Though the process of Exaltation makes the same to be true for Michael, he is shown here as a short green bar, correlating with his only known creative period recorded in the Scriptures.

The middle of the diagram shows the progression of the creative process in sequence: Intelligences first, the Physical realm next, followed by the Garden, the Fall and resultant Temporal Creation.

## Creation By Subtraction

Creation is a divine act of tzimtzum (Hebrew, *ṣimṣūm*, צמצום, meaning a contraction or withdrawal). The Kabbalah[4] describes tzimtzum as a spiritual contraction by which God withdraws His *Eyn Sof* (אֵין סוֹף) (infinite light) from the Universe in order to allow for a "*conceptual space*" in which the finite and *seemingly* independent realm we inhabit exists. The primordial initial contraction (*tzimtzum*) formed a *chalal hapanui* (חָלל הפנוי) (vacant space) into which new creative light could beam. Though these concepts from the Kabalah are metaphysical, it will be seen (*as we explore the Temporal Creation and its initial moments later in this work*) that the concepts of tzimtzum closely parallel actual physical events leading to the Universe we have today.

Tzimtzum (*creation by subtraction*) describes a process that runs parallel to the scientific principle of entropy of which we will also look at in more detail later in this work. For now it will suffice to say that entropy can be understood as a subtraction of order in which complexity later appears where there had been the undifferentiated simplicity of one. Scripturally we read, "*Hear O Israel the Eternal our God the Eternal is One*" (שְׁמַע יִשְׂרָאֵל יְהוָה אֱלֹהֵינוּ יְהוָה אֶחָד׃) (Deut. 6:4).[5] As God's withdrawal (*tzimtzum*) increases, so do complexity and the potential for imperfection. Because of Tzimtzum (its creation by subtraction), the Temporal world is less than the perfect whole; *in a very real sense*, it is a Fallen state.

To illustrate how a subtraction can be a form of creation as tzimtzum and entropy imply, I would like to draw on a conceptualization shared with me through the works of Bernard Haisch.[6] It is important that we have this fundamental understanding in our minds as we move forward.

In optics, the process of creating an image involves taking something away. The pure white light emanating from a projector and filling the bright white screen is capable of depicting and communicating nothing. With just white light filling the room everything is effectively invisible, bathed in the same white light, everything appears exactly the same.

---

[4] Kabbalah is an assimilation of inferred ancient wisdom detailing how the universe and life work as revealed through interpretations of the Hebrew Bible. The word Kabbalah (קַבָּלָה) means "*to receive.*" Kabbalah is an esoteric method, discipline, and school of thought that originated in Judaism. Kabbalah's religious origin is an integral part of Judaism, and it means to explain the relationship between an unchanging, eternal, and mysterious *infinite light,* the *Eyn Sof* (אֵין סוֹף) and the mortal and finite Universe (God's creation).

[5] Throughout this work, we will refer to the Authorized King James version for Biblical passages. The English text rendered here for Deut. 6:4 is a more literal transliteration of the Hebrew text than that rendered in the KJV.

[6] *The God Theory, Universes, Zero-Point Fields, And What's Behind It All*, Bernard Haisch, Weiser Books, 2006.

Intervene in the light beam, block its path to the screen, and you subtract something from the wholeness of the light; a shadow appears and an image forms. This is the most basic understanding of creation by subtraction and brings to mind Plato's story of Socrates' prisoners, viewing the world as mere *shadows on a cave wall.*[7]

If we return to the projector, the white light contains the potential to create every image imaginable, precisely because it is whole, and includes all the frequencies of visible light. Place a developed frame of film in front of it and a miraculous image will appear, in full and vivid color. What happened? The film is a carefully constructed filter that blocks (subtracts) out specific frequencies of light from the beam. The image and any infinite imaginable number of others like it are as Bernard Haisch said, "*contained* ***in potentia*** *in the formless white light*" that beams from the projector.

Place the images in rapid succession in front of the beam and we can create the sensation of motion. A motion picture, though just a series of still shots, taken together, provide a sum effect that becomes greater than its parts. Such a film, however detailed, is merely a *replica* of our real world. Through the constraint of the filtering effects of the film (by the subtraction from the full potential of the pure white light behind it), a less than perfect image of the world around us is rendered. However real and beautiful the image may appear on the screen, it is something less than what the white light once was. We are as incapable of comprehending a complete picture of the beauty of God's multiple creations around us as the film is in displaying the full potential of the pure light it was created from. What we observe today of the things of God are as incomplete a picture of the fullness of the world as the one viewed by Socrates' prisoners in the shadows moving on the cave wall.

In a very real sense, the white light with its infinite potential to create any desired image through subtraction represents the perfect potential of God, who through *intelligent subtraction* causes the tangible (Temporal) imperfect Universe to emerge from all that was and is possible. "*By limiting the infinitely possible, you can create the finitely real.*"

This is the power of tzimtzum – creation by subtraction. This also bespeaks how limited the Temporal Creation is when measured against the totality of the full potential of the *Three Creations of God*.

---

[7] In his work, <u>*The Republic*</u> (514–520 BC), Plato tells of Socrates describing a gathering of people who have lived chained to the wall of a cave all of their lives, facing a blank wall. The people watch shadows projected on the wall by things passing in front of a fire behind them and begin to designate names to these shadows. The shadows are as close as the prisoners get to viewing reality. He then explains how the philosopher is like a prisoner who is freed from the cave and comes to understand that the shadows on the wall do not make up reality at all, as he can perceive the true form of reality rather than the mere shadows seen by the prisoners.

# PART 1

# THE TEMPORAL CREATION

*"Mine hand also hath laid the foundation of the earth,*
*and my right hand hath spanned the heavens:*
*when I call unto them, they stand up together."*

*ISAIAH 48:13*

# CHAPTER 2

# THE TEMPORAL WORLD

## Familiar Surroundings

We are most familiar with this creation; actually we are most familiar with the *conditions* this creation established. This creation entails all that is tangible and known to us, the whole of the observable Universe, all of its laws and much of what is still unknown to us. Before we talk about this creation, it will be helpful to lay some groundwork to help define what this creation is.

### *God of Our Fathers*

We, as the Christian faithful, own our theology. We cannot turn the discussion over to the non-believing who posit their definitions of God then go about using those definitions to methodically show that God does not and cannot exist. They cannot control the discussion in that way; they don't have the right to do so. We own the definitions of whom and what we believe in. We know the theology; we are those inspired by the same Spirit the un-believers work so hard to deny and drive away. We have the answers through the grace of God.

> ***We cannot turn the discussion over to the non-believing who posit their definitions of God then go about methodically using them to show that God does not and cannot exist.***

Those who do not know any better envision a God that is all powerful; this, as it turns out, is a correct understanding, but one that is woefully incomplete. The world we live in evidences a God whom, through his *own* choosing, is restricted. Such can be seen in varied ways. It is evidenced from and through the reality that many choose not to believe in God and it is evidenced in the horrible death of innocents from very evil causes. It is seen in disease and death from infection, in the mutation of microbes and genes, and in a Universe *full* of deadly occurrences. Black holes, supernova, careening asteroids and deadly radiation, all capable of destroying life in an instant, testify of a Universe which runs on its own, apart from God being in every moment in every way.

A proper understanding of our Christian faith and the Scriptures reveals that God, though all powerful, chooses to limit Himself in a way that allows for agency and free will. This is more than an intellectual construct; it *literally* involves the creation of a Universe in which EVERY individual part has an undetermined reality. How can this be so? The answer to that lies in the fact that it *has* to be so. Any alternative would preclude free choice. This is not just a game of logic; it is a reality that is verifiable at all times and in all places.

At very well near the beginning of the last century, Albert Einstein wrestled with the results of experiments that verified the very concepts that sprung from his brilliant mind. The Universe unveiled by Einstein was ultimately counterintuitive to him. To Einstein *and to most of his contemporaries*, the Universe was a place of order, something that was on the whole eternally static and unchanging. His theories (the math that sprung from them) and the results of experiments conducted as a direct result of those theories, proved otherwise. The Universe was shown to have a beginning, to be constantly changing towards a definite future much different than its past. More troubling than that, it was found that it was impossible to determine the precise actions of anything. We could come close to knowing how subatomic particles would react, but we could never be sure. Incredibly this uncertainty includes such varied probabilities, that in fact it can be said, individual particles both do and do not exist at the exact same time; particles can be both here and at the completely distant, far side of the Universe at the same moment.

Einstein rebelled against this at first, in no small part because ingrained into his very being were fundamental notions that came from Judeo/Christian concepts of the Universe and God. His God, and as an extension the creation that sprung from Him, was perfect, unchanging, and eternal. Protesting the uncertainty of the quantum world he uncovered, Einstein was famously quoted as saying, "*God does NOT play dice with the Universe.*" To Einstein, however, and others looking at what the science was telling them, it looked precisely like He in fact did.

Quantum uncertainty and the misunderstanding of both believers and unbelievers as to just who God is, what His creations are all about, and their incomplete and disparate understandings of the Scriptures themselves, developed a rift between those of faith and science that widened and deepened until now it seems beyond crossing.

The Bible, however, does not speak of a God who is always directing events in a supernatural way. Consider the following passages, just three of many like them:

> *"I will send my fear before thee, and will destroy all the people to whom thou shalt come, and I will make all thine enemies turn their backs unto thee. And I will send hornets before thee, which shall drive out the Hivite, the Canaanite,*

> *and the Hittite, from before thee. I will not drive them out from before thee in one year; lest the land become desolate, and the beast of the field multiply against thee. By little and little I will drive them out from before thee, until thou be increased, and inherit the land." (Exodus 23:27-30)*

In the preceding passage, God details His using natural processes to secure the land for the Israelites; in this instance, He used hornets to do his bidding. The underlined sentence is most telling of His action. The God of this Biblical passage is aware of the natural order, the constraints placed upon the wildlife by the delicate ecological balance of the land, flora and fauna. The common, false understanding of God, would expect God to accomplish this task paranormally, without any concern for the laws of nature, providing a perfect environment *at will* for His chosen people. Here, however, we see Nature working independently, outside of the direct control of the Creator; more properly, we observe God allowing nature to run its due course.

> *"When thou goest out to battle against thine enemies..., be not afraid of them: for the Lord thy God is with thee..., ...let not your hearts faint, fear not, and do not tremble, neither be ye terrified because of them; For the Lord your God is he that goeth with you, to fight for you against your enemies, to save you. ¶And the officers shall speak unto the people, saying, What man is there that hath built a new house, and hath not dedicated it? let him go and return to his house, lest he die in the battle, and another man dedicate it. And what man is he that hath planted a vineyard, and hath not yet eaten of it? let him also go and return unto his house, lest he die in the battle, and another man eat of it. And what man is there that hath betrothed a wife, and hath not taken her? let him go and return unto his house, lest he die in the battle, and another man take her." (Deut. 20:1-7)*

This passage is interesting in that it depicts a God who is confident of His power in the macroscopic view (as in winning the battle) but appears far less confident in the microscopic view (saving the individual lives of individual soldiers). Here the LORD God promises to fight for and in behalf of Israel, to save them. He tells them to *fear not* because they (Israel) are assured victory. Such power is expected of God in the common understanding. Yet, in the same passage, God orders a release for some from a call to battle beside Him, lest they die! How can this be so?

Could not the God of the Universe extend complete protection *at will* when and to whom He chose? How is it possible that He would order exemptions for some "*lest he die*"? Would He not be able to protect them and ensure those faithful enough to fight for His cause, beside Him, survive to return home? The common understanding of God teaches that this must be true. Here, however, the Scripture tells a different story, one in which the actions of men and their free agency allow for detailed results independent of the will of God.

> *"And he cometh to Bethsaida; and they bring a blind man unto him, and besought him to touch him. And he took the blind man by the hand, and led him out of the town; and when he had spit on his eyes, and put his hands upon him, he asked him if he saw ought. And he looked up, and said, I see men as trees, walking. After that he put his hands again upon his eyes, and made him look up: and he was restored, and saw every man clearly." (Mark 8:22-25)*

The event written of in this passage has been referred to as the imperfect miracle. That description is misleading and incorrect. In actuality, the result was perfect; it is the getting-to-the-result that appears not to align with the common concept of God. It is interesting to note that this healing is different from many that were recorded as happening seamlessly, immediately upon command, some even without direct activity from the Savior. The sacred record reveals immediate healings through just a touch, the spoken word, and in one instance a woman with an issue of blood was healed just by touching His garment. No, it is clear the healing recorded here is out of the norm, but the details (*intentionally included in the scriptures with purpose of instruction*) bespeak some independence of the Universe to the will of God.

The last sentence of the previous paragraph discloses the truth. The God of the Bible, of the Scriptures, is not at odds with the Universe we know. He is not at odds with the discoveries of science, the laws of physics nor the artistry of mathematics that attend them. Any understanding of God, as not being in complete harmony with these things, themselves creations of Him, is incomplete and incorrect.

So with all of this in mind, let us press forward. The Universe we live in is imperfect; it was created so, for our own benefit to fully empower us with "*free*" agency. God, who is all powerful, can at any time restore it all to perfection, making it immediately responsive to His every desire, but in doing so, He would also take away our ability to freely choose, to fully function as His literal children; that is not His intent or plan.

This is exactly why the Atonement is necessary, why it was preplanned and foreordained. Through the Atonement and *only* through the restorative effects of the Atonement, can the imperfection of this sphere and the sin that accompanies it be expunged and healed. God is in complete control through the Atonement. He, in a very literal way, created this Universe, all of its complexity and the uncontrollable negative consequences. He placed us here in its midst to experience an imperfect world where there is joy and pain, evil and grace, love and loss, so that we can grow and learn. He is not the creator of evil, we are.

Our free agency, and God's desire to allow us to exercise it, limits His ability to control the Universe in its Temporal state. Because of this, our loving Father does not send us here alone. The Holy Ghost was commissioned to be our companion, to ease the pain and influence us for good. The Scriptures were delivered to provide us with guidance and a familiar kinship with other sons and daughters of God who have gone through the same trials. The commandments, if followed, are not to constrain or affect a test, but as intended, provide us with relationships and understandings that are strong and active, bolstering us in our times of need. Finally in the end, God speaks to us in an assuring way, if we but listen; He (in effect) says, *I have a plan, I will restore you to perfection and rescue you from evil.* He impresses on our Soul a feeling of warmth, *Trust me my child, it will be all right in the end, I love you.*

Against this backdrop, we find the whole of the Temporal Creation played out.

## Temporal and Physical Creations Delineated

Why do we feel the need to discuss a Temporal Creation as a separate and distinct event from the Physical Creation? The answer to that question is both simple and direct; *because it is true*. To begin with, let us identify the accounts and passages that define The Temporal sphere. The account is specifically addressed in the Bible within the passages of Genesis 1:1-2:3 and, St. John 1:1-3. The same story is detailed in the Pearl of Great Price (POGP) in Moses, Chapters 2 and 3 and in Abraham, Chapters 4 through 5. To a lesser extent, in more general terms, the verses of Proverbs 8 tell the same story as well. We will go through each in detail here, pointing out the elements therein in a new way. Additionally, we will examine significant details of the Temple Endowment that point to the Temporal Creation as separate and distinct from the Physical Creation.[1]

### *Five Evidences*

The first evidence we have detailing the Temporal Creation as separate and distinct from that which is told in both the Genesis and POGP accounts is that the creation monologue is immediately followed by a disjointed second account that tells of another creation, one we will call the *Physical Creation*. This is not by chance, nor is it a detail that can be overlooked.

This "*second*" account has been troubling to non-LDS Biblical scholars for centuries. Many explanations have been offered up, including but not limited to considering this account to have separate authorship from the one provided in Genesis, Chapter 1 and by their association, the corresponding POGP chapters. The most widely held view is that this second account is older, more original than the first account, rising from the early days of Israel before the captivity. This view attributes the fuller account of Genesis 1, as a later addition to the record drawing heavily on Babylonian mythology; the fuller account then was created and inserted at the front of the Torah by Deutero Priests after the Babylonian Captivity.[2] These

---

[1] In this treatise, we will be respectful of the sacred nature of the blessings given in the Temple Endowment. I will not provide details of the eternal truths revealed therein. We will only discuss in general terms the elements of the endowment that are crucial to understanding the concepts of this treatment. I will be careful not to reveal anything not already detailed in publications approved by the church. Though complete details can easily be found online, published by forces both friendly and unfriendly to Latter-day Saints. I encourage those seeking to more fully understand the details of the endowment to seek out that information prayerfully within Temple walls.

[2] Keith E. Norman in his Essay Adam's Navel, states: "*The first step in getting past misconceptions about the Biblical creation story is to recognize the setting in which the first chapters of Genesis were written. Chapters 1 and 2 offer two differing accounts of the creation story with conflicting details. Further analysis of style, vocabulary, theology, and purpose has allowed scholars to date Genesis 1 through 2:4a to the time during or immediately after the exile in Babylon. The writer is concerned with preserving and restoring the religion of Israel in a time of great uncertainty, and his themes are stability, legitimacy, and the proper performance of ritual. He is known anonymously as the priestly (P) writer or writers. P's magnum opus was the book of Leviticus, but he contributed to other parts of the Bible, including the genealogical lists of Genesis. It is not surprising that P ends his account of creation with the summary statement, 'These are the generations of the heavens and of the earth' (Gen. 2:4). Even the ground we stand on has a pedigree. The King James version of Genesis, following the manuscripts available at the*

attempts to explain away the apparent discrepancies of these two accounts stem from a misunderstanding of the creative process recorded. The second account is far less problematic to Latter-day Saints because we have an understanding that includes a spiritual creation before the creation described in Genesis 1 and the parallel accounts in the POGP.

*(Note: for simplicity and to avoid unnecessary confusion we will refer to the Genesis 1-2:3 account and the parallel POGP accounts of Moses 1:35, 40 and all of Chapter 2, and Abraham 4-5:3 as the* ***Genesis 1 Account****)*

Though a second spiritual creation is known to Latter-day Saints, most, if not all, fail to understand it completely and therefore do not wholly separate it from the Universe we inhabit today. Almost without exception, all hold that the spiritual creation wholly encompasses this world and *is* this world prior to its Fallen state. This is an incorrect interpretation and has led to many false concepts and troubling divisions;[3] a correct view eliminates these. *Our understanding here, in this work, will be of a truly eternal, corporeal, wholly Physical Creation, which exists in the place of what has been called spiritual in the Scriptures and tradition. We will unabashedly substitute Physical Creation in these places for many reasons, all of which will be detailed later in this work.*

---

*time, obscures the transition to the second account by running P's conclusion into the new statement, which is now numbered 2:4b. A better translation of this opening might read, 'In the day that the Lord made the earth and the heavens, before there were any plants in the earth or herbs growing in the fields ...' This is another way of saying, 'In the beginning,' but this dry, barren world awaiting the Creator is a very different scenario from the chaotic deep P describes. We note right off that this second writer calls the deity 'Lord,' while P referred to him as 'God' (Hebrew Eloheim). Lord is the English equivalent of Adonai, the Hebrew circumlocution used to avoid pronouncing the sacred name of God, YHWH, from which we derive the name Jehovah (scholarly consensus now vocalizes this as Yahweh). According to P's version, this name was first revealed to Moses (Ex. 6:3), and any use before then would be an anachronism. The liberal use of the name Yahweh (or Jehovah) has earned for the author of Genesis 2:4b-4:27 the designation of J. He is a masterful storyteller but a less careful scholar. J's is actually the earlier of the two creation accounts by several hundred years. He reflects the national self-confidence and less restrictive theology of David's and Solomon's reigns."* (*The Search for Harmony;* pgs. 231 and 232, edited by Gene A. Sessions and Craig J. Oberg; Signature Books)(Norman K. Gottwald, *The Hebrew Bible: A Socio-Literary Introduction* Philadelphia: Fortress Press, 1985, pgs.140, 170) (E. A. Speiser, Genesis (1967; rprt., New York: Doubleday, 1982, pgs. 14-15) (Gottwald, Hebrew Bible, pg. 137; Speiser, Genesis, xxvii.)

[3] See *The B.H Roberts/Joseph Fielding Smith/James E. Talmage Affair*; by Richard Shelock and Jeffrey Keller; *The Search for Harmony* (pgs. 92 -110; edited by Gene A. Sessions and Craig J. Oberg; Signature Books) and *Agreeing to Disagree: Henry Eyring and Joseph Fielding Smith;* by Steven Heath; (Ibid, pgs. 136-153).

| Genesis 2 |
|---|
| 4 ¶These are the generations of the heavens and of the earth when they were created, in the day that the Lord God made the earth and the heavens,<br>5 And every plant of the field before it was in the earth, and every herb of the field before it grew: for the Lord God had not caused it to rain upon the earth, and there was not a man to till the ground. |
| **Moses 3** |
| 4 And now, behold, I say unto you, that these are the generations of the heaven and of the earth, when they were created, in the day that I, the Lord God, made the heaven and the earth,<br>5 And every plant of the field before it was in the earth, and every herb of the field before it grew. For I, the Lord God, created all things, of which I have spoken, spiritually, before they were naturally upon the face of the earth. For I, the Lord God, had not caused it to rain upon the face of the earth. And I, the Lord God, had created all the children of men; and not yet a man to till the ground; for in heaven created I them; and there was not yet flesh upon the earth, neither in the water, neither in the air; |
| **Abraham 5** |
| 4 And the Gods came down and formed these the generations of the heavens and of the earth, when they were formed in the day that the Gods formed the earth and the heavens,<br>5 According to all that which they had said concerning every plant of the field before it was in the earth, and every herb of the field before it grew; for the Gods had not caused it to rain upon the earth when they counseled to do them, and had not formed a man to till the ground. |

Table 2-1 *The Second Creation. I have intentionally segregated the first part of Genesis Chapter 2, Verse 4 from the rest of that verse because Verse 4 should properly start at "in the day," and the highlighted words from the beginning of Verse 4 actually belong to Verse 3. How the verse got to be assembled the way it has been is a lengthy story involving many factors all well documented. For a very good treatment of this issue, refer to the commentary of the Jewish Study Bible, Featuring the Jewish Publication Society, Tanakh Translation, Berlin. Adele and Brettler, Marc Zvi, editors, Oxford University Press 2004.*

Even the most cursory examination of the passages in *Table 2-1* will reveal that they are very similar and clearly describe the same set of events. Undoubtedly, Moses 3:4-5 provides the most clarity, but all three passages contain identical details; they are:

- Plants and herbs created before the Earth was capable of supporting their life.
- There was not yet man on Earth.

To these the Moses account adds:

- All the children of men created.
- Not yet flesh on the Earth, in the air or in the water.
- All things created spiritually before they were naturally on the Earth.

Among these specifics are items that are most interesting. We are told, plants and herbs are created before they can survive on earth. How can this be so? How can a living thing be fully created in the sense that it meets its purpose if the conditions for it to live do not yet exist? The Moses account provides us with an answer; we are told, "*For I, the Lord God, created all things, of which I have spoken, spiritually, before they were naturally upon the face of the earth.*"

The creation here is of truly *physical* things (*though termed spiritual, because they are eternal and of a different realm than the temporal we inhabit today*) formed in their entirety at a time *before* the Temporal (Natural) Creation. Plants, animals and all the children of men, created in Heaven not on the Earth. This is **Evidence Number 1.**

The second Evidence is to be found yet again in the book of Moses, this time in passages which precede the Genesis 1 account in historical chronology. In Moses Chapter 1, Moses (after seeing in grand vision every particle of the Earth and the inhabitants thereof) called upon the LORD saying, "*...Tell me, I pray thee why these things are so, and by what thou madest them.*" In reply, the LORD responds:

> *"But only an account of this earth, and the inhabitants thereof, give I unto you. For behold, there are many worlds that have passed away by the word of my power. And there are many that now stand, and innumerable are they unto man; but all things are numbered unto me, for they are mine and I know them." (Moses 1:35)*
>
> *"And now, Moses, my son, I will speak unto thee concerning this earth upon which thou standest; and thou shalt write the things which I shall speak." (Moses 1:40)*
>
> *"And it came to pass that the Lord spake unto Moses, saying: Behold, I reveal unto you concerning this heaven, and this earth; write the words which I speak. I am the Beginning and the End, the Almighty God; by mine Only Begotten I created these things; yea, in the beginning I created the heaven, and the earth upon which thou standest."(Moses 2:1)*

Immediately following these defining words, the LORD then recounts the familiar words of Genesis 1. *Note this important detail*, the creation account of Genesis 1 is *specifically* related to the very place where Moses was standing, on the Temporal Earth within this Temporal Universe.

Previously in Chapter 1 of Moses, the great prophet was shown and told of many lands, each land was called earth and each land was inhabited. In Verse 35, the LORD reveals that many worlds had already passed away. If Adam had indeed introduced the Universe to death and destruction through the Fall a mere 26 generations and but a few thousand years previous to the prophet's receipt of this great vision, these inhabited planets would have had but a very brief existence; more on this point later in this work.

Of significance to our discussion here, is the important point that the creation account Moses records at the commandment of God (our Genesis 1 account and the parallel Abraham account) pertains ***only*** to *this* temporal Earth and the surrounding temporal universe. This important (but most often overlooked) fact is clearly stated and emphasized by God, for He speaks it, not once, but *three times*. This is **Evidence Number 2.**

The third, fourth and fifth evidences that the Temporal Creation is distinct from the Physical Creation are to be found in the Temple Endowment. First of these three is the sequence of the creation events themselves, as described in the Endowment (*Table 2-2*). This impactful change of sequence in the events of creation has many levels of significance, yet it is so often entirely un-noticed by members of the church. We will probe deeper into the revelatory meanings of the change in the event sequence later in our discussion of the Physical Creation. But for now, we just need to concern ourselves with the fact that the sequence is different and why that difference is there.

There can only be two explanations for the Endowment sequence to differ from the account in Genesis; they are:

- *LDS originators were ignorant of the Biblical account AND their own POGP accounts, and they simply made a mistake in drafting the Endowment.*
- *The sequence of creation events in the Temple Endowment is intentional. The difference is of significance and it clearly identifies the creation depicted as a different event from the one detailed in Genesis and the POGP.*

The first of these two explanations, though the most often drawn conclusion by detractors of the Latter-day Saints, [4] is entirely implausible. Early Saints, especially

---

[4] Consider this observation recorded on the website antagonistic to Latter-day Saint teaching. "*A while back, a friend called my attention to the fact that the order in the creation account given in the*

| DAY | Temple Endowment (Physical) | Genesis/ Moses/Abraham (Temporal) |
|---|---|---|
| ONE | World Organized | Earth Organized, Light Created, Light and Dark Separated |
| TWO | Water and Land Separated, Earth Formed | Firmament Created |
| THREE | Light and Darkness Divided -Lights in the Firmament-Night and Day Separated, Greater and Lesser Lights | Land and Water Separated – Plant Life Created |
| FOUR | Plant Life Created | Lights in the Firmament-Night and Day Separated |
| FIVE | Creation of Animal Life - Land, Air and Sea | Sea Life and Birds Created |
| SIX | Creation of Man | Land Animals and Man Created |

Table 2-2 *Differing Sequence of Creation Events.*

their leaders and Joseph Smith were keenly aware of the Bible and its details; the Prophet Joseph had in fact personally completed the inspired Joseph Smith Translation of the Bible (JST). Under inspiration, he also revealed the words in the Book of Abraham. Throughout the early days of the church, Joseph Smith and others in church leadership exposited often on the details of the creation story of Genesis, including specific discussions of the meanings of individual Hebrew words. It is clear the originators of the Endowment in this dispensation were well versed in the Genesis 1 account.

---

*endowment ceremony doesn't match up with the order given either in Genesis or the Book of Moses (part of Smith's supposedly inspired restoration of Genesis.) The order of events in Genesis and the Book of Moses do match up with each other, as well as with Joseph Smith's 'translation' of the Book of Abraham, but not with the Endowment. The creation account in the Book of Abraham and the Book of Moses are actually nearly word-for-word identical to Genesis, with the exception that The Book of Moses account is written in the first person ('I, God, created the heavens and the earth....' etc.) and the Book of Abraham's account is in the third person plural ('In the beginning they, the gods, created the heavens and the earth...' etc.) What does this actually mean? Frankly, at this point, I have no clue, other than perhaps Joseph Smith just couldn't be bothered to keep things straight.*" (lds.temple.net, a Christian apologetic site dedicated to exposing so called errors and fallacies in LDS teachings, the site has since been taken down and can no longer be accessed.)

The second explanation then must be correct: The sequence of events in the Endowment *is* intentional and therefore of extreme significance.

First, let us put aside all explanations that would describe this sequence as something offered to 'correct" an altered (*mistranslated*) Genesis 1 account. Though a corrupted Biblical account requiring correction is the all-too-frequent go-to for Latter-day Saints when resolving any apparent conflict with modern day revelation and Biblical renditions, it cannot be the case here. The Prophet Joseph reiterated the Genesis 1 account in *exact* sequence not once, but twice; Moses and Abraham both confirm the Genesis sequence in precise detail. Yet, still the same man, the Prophet Joseph, gives us a completely different sequence in the Endowment; *this is telling*. Additionally, the change in order of the creative events does not constitute a more complete understanding as may be expected in the Temple. Other components of the Endowment indeed do that; no, this is not additional information; it is just a change in sequence.

I propose here that the change in order is due to the fact that the Endowment creation story is an account of the Physical Creation. There are extremely important details in the Endowment account that clearly identify the creation it details as one that is fundamentally much different than the Temporal Creation we live in today. We will discuss this more completely later. For now, however, the intentional change in sequence of the Temple Endowment creation story points to a separate and distinct event than that of the Temporal Creation, this is **Evidence Number 3.**

The next evidence involves one of those fuller understandings in the Endowment I previously spoke of. Michael (Adam) assists in the creation events detailed in the Endowment. It is understandable that this truth is missing from the account in Genesis, but where is he in the accounts of the POGP? Particularly in the account of Abraham, for it follows directly after Abraham saw in vision:

> *"... the intelligences that were organized before the world was; and among all these there were many of the noble and great ones;...These I will make my rulers; for he stood among those that were spirits, and he saw that they were good; and he said unto me: Abraham, thou art one of them; thou wast chosen before thou wast born." (Abraham 3:22, 23)*

Surely Michael (Adam) would have been pre-eminent in the gathering of the noble and great. Michael's presence in the Endowment narrative as co-creator with Jesus Christ and his absence in the other creation accounts is significant and is meant to inform us of a great truth. Additionally, pay attention to when Michael quits repeating back his commands to the Son in the Temple narrative; enough said about that for now; we will discuss this important point later in this book. These details are there in the Endowment for a reason. I strongly disagree with those who so readily say these accounts are mythological; they are analogous; that they are meant to just direct our thoughts to God; that the accounts are meant to be descriptive of each of our individual lives and because of that should be forgiven *minor historical inaccuracies*. We must proclaim here with full understanding that these accounts are not inaccurate. They are fully detailed and full of the very fine and revelatory things discussed here and so much more.

That Michael is in the Endowment account and absent from the inspired Genesis 1 accounts of the POGP is **Evidence Number 4**.

The last evidence is subtle but important as well. The Endowment account proceeds through the six days of Creation and onward; it details the Garden and the Fall, then at that point, and precisely at that point, when Adam and Eve are cast out into the lone and dreary world, the participants don robes signifying our Temporal bodies.[5]

Why then? Why not before, perhaps when Adam was created from the dust? No, it is clear that point represents the vivid dividing line from immortality to mortality. What does that tell us about the creation account of the Endowment and the location of the Garden? What happens at that juncture? Remember back to our previous discussion of the worldly views of the two creation accounts in Genesis. We disclosed that the second account was most usually agreed to be the first, an older account than the one it follows. It is significant to note that the Garden story is tied to that *second*, earlier account, not the more detailed and previously related six day creation sequence told in Genesis 1.

We don robes[6] (our Temporal bodies) after the Fall in the Endowment narrative to signify a change from an immortal existence in the Physical sphere to a Fallen mortal state in Temporality. In the Endowment, we symbolically move from a higher state to a fallen state; this is **Evidence Number 5**.

**Evidence 1** – Genesis, Moses, and Abraham each record that life existed in Heaven before the creation of the Earth.

**Evidence 2** – Moses and Abraham are shown the Temporal Creation, our Earth; Moses is also shown other earths, some of which had already passed away through the Word of God's power by the time of Moses.

**Evidence 3** – The order of creation detailed in the Endowment differs from the Genesis, Moses, and Abraham accounts; this is not an error, but an inspired view of a separate creation.

**Evidence 4** – Michael (who is Adam) does not participate in the creation accounts of the inspired Genesis 1 or POGP.

**Evidence 5** – Adam and Eve receive their "*coats of skins*" from God, after the Fall, signifying their passage from immortality to mortality.

Table 2-3 *The Five Evidences of a Separate Physical Creation.*

---

[5] See our discussion, The Temple, a Bridge Between the Spheres, later in this work, beginning on pg. 367.

[6] In many temples this point is also accompanied by a change of rooms, from a room signifying the Garden to one signifying the Telestial world.

## Time Is A Relative Thing

Time is local and temporal; in the simplest of terms, it depends entirely on the location of the observer. My location on the Earth relative to your location may make my day coincide with your night. The latitude of any location and the tilt of the Earth's axis relative to the location of the Sun determine when the Sun rises and sets and dictates how many hours there are in the local day or night. A day is determined by a single rotation of the Earth[7] and a year by a single course of the Earth as it orbits the Sun. This is our understanding today, but the concept of day and year and month were established well before any human knew of such astronomical truths.

Further complications come in the way time is reckoned. Calendars have changed throughout history and within them months are measured differently, in a varying number of days on the Julian calendar and by phases of the moon on Jewish and Islamic calendars. However, all of these things become insignificant when we change our perspective to that of someone beyond the bounds of the Earth or this particular point in its history.

The Earth's rotation is slowing and has been doing so for a very long time. This makes the day ever longer by any measurement. A single circuit around the Sun today takes less time than it once did as the Earth's orbit shrinks; steadily (almost imperceptibly now) over the eons drawing closer to the Sun with each pass. Other planets in our Solar System have different lengths of time for their day and different lengths of time for their years. Our Sun orbits the Milky Way at a speed that has varied over time. Any *big-picture* understanding of these courses and rounds will reveal that the reckoning of time is completely dependent upon one's individual perspective.

The ultimate complication comes when we recognize that not only the way one *reckons* time is fleeting, but that time itself is relative to the way one moves through the Universe. Today we know that time itself, at the most fundamental level, is relative to the effects of speed and the effects of gravity. These are proven facts that have very real consequences for life. Accelerators speed subatomic particles along at substantial percentages of the speed of light increasing the lifespans of each by immense factors. When you consider that most of what we know the Universe to be moves around freely in space, at or very near the speed of light, it is clear that we are bound to a conception of time that is out of the norm.

We confidently move about in our world today guided by GPS phones and devices without ever giving thought to the Theory of Relativity. However, these now common devices would not work at all if not for adjustments made to the clocks of the interlinked satellite network which allow for the relativistic changes made to time due to orbital speeds and the effects of gravity. If the effects on time predicted

---

[7] Sidereal time aside.

by relativity were false and incorrect, each of us may never be able to find that new restaurant we are seeking.

As one would expect, the truth of these things can be found in Scripture as well. Though a complete understanding may have been elusive to whom these truths came and to those who recorded them, the Scriptures do tell a story we can put in context today and understand. The Apostle Peter pleaded with us to not be ignorant of such truths when he said:

> "*...Where is the promise of his coming? for since the fathers fell asleep, all things continue as* they were *from the beginning of the creation. For this they willingly are ignorant of, that by the word of God the heavens were of old, ...But the heavens and the earth, which are now, by the same word are kept in store, ...But, beloved, be not ignorant of this one thing, that one day* is *with the Lord as a thousand years, and a thousand years as one day.*" *(2 Peter 3:4-8)*

Though there is a context to these verses that is metaphorical, they do demand we recognize that there is more than one reference point for time as it refers to man and his relationship with God. This concept is inferred throughout the entire Biblical account.

The inspired account in Abraham provides a more detailed representation and an understanding of the relative nature of time that is much closer to what we know today:

> *"And the Lord said unto me, by the Urim and Thummim, that Kolob was after the manner of the Lord, according to its times and seasons in the revolutions thereof; that one revolution was a day unto the Lord, after his manner of reckoning, it being one thousand years according to the time appointed unto that whereon thou standest. This is the reckoning of the Lord's time, according to the reckoning of Kolob. And the Lord said unto me: The planet which is the lesser light, lesser than that which is to rule the day, even the night, is above or greater than that upon which thou standest in point of reckoning, for it moveth in order more slow; this is in order because it standeth above the earth upon which thou standest, therefore the reckoning of its time is not so many as to its number of days, and of months, and of years. And the Lord said unto me: Now, Abraham, these two facts exist, behold thine eyes see it; it is given unto thee to know the times of reckoning, and the set time, yea, the set time of the earth upon which thou standest, and the set time of the greater light which is set to rule the day, and the set time of the lesser light which is set to rule the night. Now the set time of the lesser light is a longer time as to its reckoning than the reckoning of the time of the earth upon which thou standest. And where these two facts exist, there shall be another fact above them, that is, there shall be another planet whose reckoning of time shall be longer still; And thus there shall be the reckoning of the time of one planet above another, until thou come nigh unto Kolob, which Kolob is after the reckoning of the Lord's time; which Kolob is set nigh unto the throne of God, to govern all those planets which belong to the same order as that upon which thou standest. And it is given unto thee to know the set time of all the stars that are set to give light, until thou come near unto the throne of God." (Abraham 3:4-10)*

These Scriptures from both the New Testament and Latter-day records clearly describe an understanding of time that is relativistic and fluid, dependent upon locality and position. This is a concept that aligns with Einstein's General Field Theories. God, who has an eternal perspective and a universal view, has a conception of time that includes all spoken of here; His is an eternal sphere. With this in mind, we can unshackle from errant views that confine God and things grand in scope to our commonly narrow view of space and time. The truth of all that has been revealed can only be understood fully when we do so. With this freedom achieved through a broader perception, we will see that the Scriptures have, all along, provided details that have eluded many.

How old is the Earth? When was Adam placed here? When did the Temporal Creation, as it is described in the Genesis 1 account, take place? How long did it take to accomplish, *just 6 Earth Days*?[8] These questions have been ardently debated for centuries. Both sides of the discussion have argued with certainty. The debate has been waged both within the circles of those who believe in God and in circles without. In recent decades, answers to these questions have defined a well-marked line and opened a dividing gulf between those who believe in a creator and those who do not. Here we will present a view that is wholly supported by Scripture *and* in harmony with the most current discoveries of science. This view will no doubt be surprising to all but a very few.

The Temporal Creation followed the Fall, which occurred while Adam and Eve were in Eden. Eden was in an immortal creation (sphere). It is true that death did not precede the Fall in any way. It is important, however, to fully comprehend what "death" entails. Of course it means the cessation of life in living organisms of every imaginable kind, but in the sense of the story told in the Genesis 1 accounts, it also means any decay and movement from order to disorder. In scientific terms, this process is entropy. Everything in the Universe, every subatomic particle and the "*stuff*" they are made up of, muons and gluons and other like particles, are decaying, in fact dying. As a course of fact, this means *nothing* that exists in this Universe (not one *speck* of it) could have predated the Fall of Adam. This fact alone necessitates the distinction between the Physical Creation and the Temporal one; the ramifications of this realization are far reaching.

The Garden of Eden was not located on this planet. There is no sense in trying to identify any of the geographical features associated with the Garden account with

---

[8] The actual meaning of the Hebrew word *yō·wm* (יוֹם) translated as "*day*" in the KJV has been debated for millennia. Augustine, writing in the fifth century, expressed his doubt about a literal meaning of day as it is used in the Genesis 1 accounts: "*What kind of days these were it is extremely difficult, or perhaps impossible, to determine*" (City of God, 11.7). We will not make any attempt in this work to *uncouple* the creation account from 6 actual day long periods. We will resolve this debate within these pages as we show that the relative nature of time allows us to overlay almost 15 Billion years of Universal history onto just six days as measured in God's reference frame for time.

any known feature on the Earth today.[9] Adam and Eve did not exist in this Universe until they were cast out of the Garden and entered the lone and dreary world.

Other ramifications are exciting and compelling. The Earth we inhabit and this Universe were created from the outset in a Temporal state with all of the laws of physics in place, including entropy and quantum uncertainty,[10] which are the scientific terms for death, time and free agency. These terms should sound very familiar for their religious overtones. Because they are embedded into the entire Universe at every level, death, procreation and evolution did not have to wait until Adam was introduced into the lone and dreary world. The history of the Universe can be, and is, much older than the history of man on this Earth, this is spoken of in Moses (Moses 1:29-38). In these verses, Moses was shown that there were many inhabited Earths predating ours, "*that had been and had already passed away.*" Remember this vision of Moses was a mere 3000 years (just 18 generations had been born) since Adam had stood upon this Earth at Adam-Ondi-Ahman; a very short lifecycle for innumerable inhabited planets.

With the antiquity of the Universe established, it can be applied to the Earth as well. All of the geology, paleontology and anthropology can be accepted for their value and truths. Adam can be a late addition to the Earth's overall history, as indeed he was. Death and procreation can predate Adam on this planet with no theological consequence. Additionally, since the Fall of Adam predates the Universe in its entirety, life (*all of it*) anywhere in the Universe, if ever it is found, is experiencing the effects of that Fall. The world that Adam brought death to, through the Fall, is not just this planet Earth but the Universe as a whole. Indeed, the effects of the Fall are universal in consequence. No matter where we look, how far into the distant past we see, entropy is at play; *everything* is dying.

We will return again to this subject of time and entropy in more detail later in this work.

---

[9] Though it is my belief that things of the Temporal sphere closely mirror things of the eternal Physical sphere, and this would include the Earth, they are *not exact copies*. Much like those born with deformities are not imagined to have spiritual bodies with the same defects, the Earth with its changing temporal geographies and desolate places is different from its spiritual twin. Continental drift and other temporal realities are not expected to be at play there, though they may be. The point to be understood here is that the Physical (spiritual) Earth that housed the Garden is not and should not be expected to be identical to the Earth we inhabit today.

[10] Quantum uncertainty (its attendant randomness) is NOT the *source* of free agency; it does not *emerge* from it, but is empowered by it, unshackled and set free. The uncertainty of the Temporal allows for the emergence of novelty, the creation of the previously unknown. Daniel Dennett gets it wrong when he points out: "An indeterministic spark occurring at the moment we make our most important decisions couldn't make us more flexible, give us more opportunities, make us more self-made or autonomous in any way that could be discerned from the inside or the outside, so why should it matter to us? How could it be difference that makes a difference?" Though he is right that immersed as we are within the uncertain Temporal, discerning the uncertainty is difficult and near impossible. For this reason, we instinctively assume a deterministic worldview. However, it is the spark of uncertainty that frees us to act with our will, exploring paths not taken before or previously encouraged by any fixed influence. We are flexible to choose one way over another, best over better, good over evil; by that we *establish* where nothing once was. How could it be difference that makes a difference? It doesn't. It is the sentient, purposeful action of consciousness that makes a difference in a sphere where a difference of course is possible.

## Incorrect Assumptions about the Fall

There is NO MECHANISM for transitioning the Universe, or any part of it, from an immortal state to a mortal one. If the fall was 6000 years ago or even a generous million years ago, or 4.5 billion, such an event would have left a form of shock wave (in scientific terms, a phase transition) pulsing through the Universe as the change moved outward from this planet. In fact, everywhere we look, in any direction, we see objects and the matter that compose them, 13+ billion years old, all of it mortal, initially composed mortal and dying that way. Do not take this point lightly; this understanding of the Creation in two parts, a Temporal one and a Physical (spiritual) one, is accepted without debate by Latter-day Saints. However, the concept that the Temporal and Physical Creation are *fully* disconnected, separate events as I propose here (and as I will show the Scriptures and our sacred teachings detail precisely) is not commonly understood; though it should be. In full realization, this theological component is the only one that is compatible with other life in the Universe, any life at all. All other theologies will be left in crisis at any such discovery. This theology, and only this theology, takes the wind completely out of the sails of Atheists who depend on disbelief and science to support their claims.

***There is NO MECHANISM for transitioning the Universe, or any part of it, from an immortal state to a mortal one.***

It is the "transitioning" from the immortal to the mortal and the associated timing of that event that has been the bane of theologians for centuries. A lack of understanding that the Temporal Creation produced the Universe in its entirety has led to many false assumptions and teachings.[11] Attempts to extend the qualifications of immortality and an eternal state to the whole or any small part of the Universe that we can perceive are incorrect and result in troubling theologies. These errant attempts stem from incomplete and dated information, from worldviews that developed in times since past, not incorporating the many truths God has allowed us to know today. Though in centuries past theology was ahead, *and at times way out ahead*, of what was known by science, today theology still bases its worldview and Biblical interpretations on what was known about the Universe over 120 years ago. Conversely, science has made

[11] *"Both the things created and the order of their production must have been many times repeated in the multitudinous worlds of the universe, where creations in some manner have been going on eternally."* The Truth, The Way, The Life (B.H. Roberts, edited by Stan Larson; Smith Research Associates, 1994, pg. 266). Here Elder Roberts details his view that the creation event of the Genesis 1 account occurred wholly within the Universe we inhabit today and that the Universe is eternal. This worldview predates the concepts of the Big Bang and the discoveries of Fred Hoyle and Edwin Hubble.

gigantic strides in those same 120 years, much of it in just the past three decades, dramatically changing what we thought we knew about our Universe. Many of these gains point directly to a creator and intelligent design, but theology with its failure to catch up and provide religious interpretation of these new revelations, has left a dangerous vacuum all too easily filled by those who do not know and refuse to know the things of God.

As we have discussed previously, in 1910, the Universe was thought to be unchanging, in an eternally static and stable state. The Earth was thought to be perhaps a few hundred million years old and the fossil remains of man and animals from strata in the 100 of millions of years in age were straining beliefs and perplexing many.

Almost 100 years ago, Elder B.H. Roberts wrote:

> *"... we hold the universe to be self-existent and eternal. Duration, space, matter, force, mind – each infinite after its kind – are its prime and included factors. The universe is comprised of these.*
>
> *Such a universe can have no first cause, since that would imply a time when there was no cause, and there has been and can be no such time. Causation is eternal, and in the eternal and infinite universe there has been operating always Eternal Cause."*[12]

In association with the controversies that continued to surround the book from which the previous quote of Elder Roberts was taken and by extension the debate carried on by its supporters and detractors, Joseph Fielding Smith responded to Henry Eyring some years later in the following way:

> *"'...The scriptures teach that Adam was the first flesh on the earth. This is the doctrine in the Bible, Book of Mormon, and Pearl of Great Price, but it is rejected by the advocates of organic evolution.'*

---

[12] The Truth, The Way, The Life (B.H. Roberts, edited by Stan Larson; Smith Research Associates, 1994, pg. 260). That this great work of Elder Roberts was not published until 1994 is an incredible story of the discord caused by the lack of understanding of the true nature of the three creations discussed in this book. As we will see in this treatise, Elder Roberts made many false assumptions in his book correctly identified by the Brethren (most associated with Roberts' belief that the fossil record told of a creation of mortal life on Earth *including human life forms* which were all wiped out in some great cataclysm before the creation described in the Genesis 1 accounts occurred).We will also address in this work the fact that there existed on the brethren's side of the issue an equally abundant set of false assumptions and preconceived notions that perpetuated errant theologies as well. In addition to the excellent treatment of this discord and the very interesting processes involved, a comprehensive summary of information regarding this matter and subsequent associated matters can be found in the pages of a book that is out of print but available online.

See *The B.H Roberts/Joseph Fielding Smith/James E. Talmage Affair*; by Richard Shelock and Jeffrey Keller; *The Search for Harmony* (Edited by Gene A. Sessions and Craig J. Oberg; Signature Books, pgs. 92-110) and *Agreeing to Disagree: Henry Eyring and Joseph Fielding Smith;* (by Steven Heath; Ibid, pgs. 136 -153).

*'...These scriptures teach that Adam was not subject to the mortal and spiritual death before the Fall, and that the fall brought these deaths into the world...'*

*'... This earth is passing through seven days of temporal existence of one thousand of our years for a day, and that it was not temporal before the fall. This is clearly stated in the Bible, the Doctrine and Covenants and the Pearl of Great Price, but it is definitely and positively denied by organic evolutionists.'*

*'These divine records promise us that the earth on which we dwell will be renewed and restored to its primitive beauty for one thousand years and be cleansed of all its iniquity. This is denied by most scientists.'*

*'These divine records declare that the earth shall die, for it is a living body, and will rise again in the resurrection through the redemption of Jesus Christ, to become a celestial globe and the abode of the righteous. Scientist[s] preach a far different doctrine.'"*[13]

The preceding quotes frame nicely the incorrect worldviews that were attendant to the best minds of the early and mid-last century. Elder Roberts clearly expresses his belief in an eternal universe, a universe to which the Fall *happened*. He is correct in stating that just such an eternal universe cannot have a "*first cause*." However, he is incorrect in that he has missed two great truths because of his predetermined worldview. First, there was a "*first cause*," the Fall. Second, the Genesis 1 creation account is a definitive statement that the Universe did have a beginning. The Fall did not occur *within* the Universe we inhabit today; the Universe was created by God, not *ex nihilo*, but organized from spiritual matter in direct response to the "*first cause*," the Fall of Adam *and because of the needed introduction of the full effects of free agency*. Bear in mind, we propose here that the Universe was created in its entirety within an eternal continuum that is the dwelling place of our Eternal Father and that as Elder Roberts stated, there is an *"Eternal Cause"* that has been in "operation always." We propose this Eternal Cause is the work of God as revealed to Moses:

> *"For behold, this is my work and my glory—to bring to pass the immortality and eternal life of man..." (Moses 1:39)*

Joseph Fielding Smith's doctrinal points are in error because, among many things, they discount entirely the fact that the Scriptures speak of many Earths, all inhabited, that preceded and passed away before this Earth was created. If Adam was the first flesh on Earth, as understood by President Smith, then to which of these multiple Earths would he have come?

The President goes on to incorrectly conclude (as does Elder Roberts) that the Fall was an event that happened in *this* Universe in a Garden located on *this* Earth.[14] He

[13] From a letter written by Joseph Fielding Smith to Henry Eyring: Agreeing to Disagree: Henry Eyring and Joseph Fielding Smith; Steven Heath; *The Search for Harmony* (; Edited by Gene A. Sessions and Craig J. Oberg; Signature Books, pgs. 136-153).

[14] For those made uncomfortable by my seemingly critical remarks regarding a man who became a Latter-day Prophet subsequent to his making the statements in discussion here, let me remind you of the following: In early 1961, Smith gave a talk at a Stake Conference in Hawaii, he said there: "*We will*

speaks of the return of the Earth you and I stand on today to an immortal and paradisiacal state from which it changed as a result of the Fall. These beliefs are driving factors for the extrapolations he makes of the Scriptures to justify his views. In his mind, these doctrines are clearly subtended by the Scriptures. In the President's view, the Scriptures clearly speak of the Earth we stand upon as precluding the death of any living thing prior to the arrival and Fall of Adam, a mere 6000 years ago. He spoke many times about fossil evidence to the contrary of that claim as being false and a ruse. As we know today, this can in no way be true.

Without a doubt, the Scriptures are the Word of God and as such are correct. If our reading of them contradicts what we know from other sources are to be true, then it is our reading that is in error and we must re-read them with an open mind that allows our knowledge to expand. It does no good at all and is an affront to our God, who gave us reasonable minds, to dig-in and stubbornly ignore the truth around us; none of our worldviews are so valuable to justify that.

---

*never get a man into space. This earth is man's sphere and it was never intended that he should get away from it. The moon is a superior planet to the earth and it was never intended that man should go there. You can write it down in your books that this will never happen.*"[A] Earlier, Smith had written that "*it is doubtful that man will ever be permitted to make any instrument or ship to travel through space and visit the moon or any distant planet.*"[B] Ironically just 6 months and three days after the astronauts of Apollo 11 first landed on the moon a press conference, where Smith was introduced as President of the LDS Church, was held and he, of course, was asked about these statements. Smith reportedly responded, "*Well, I was wrong, wasn't I?*" [C] If he was faced with what we know to be true about the latest scientific discoveries, I am convinced Smith would reply in a similar way about his comments made in this matter. *Some question this mea culpa actually occurred, but it is certain the prophet knew he was wrong; he was in fact presented with a Utah state flag on the 14th of September 1971 by the astronauts of Apollo 15 who had personally carried it with them as they traversed the lunar surface, some 234,000 miles above his head.*

[A] D. Michael Quinn, *Elder statesman: A Biography of J. Reuben Clark* (Salt Lake City, Utah: Signature Books, 2002, pg. 498).
[B] Joseph Fielding Smith, *Answers to Gospel Questions* (Salt Lake City. Utah: Deseret Book, 1957, 2:191)
[C] *"Mormonism and science/Joseph Fielding Smith claimed that man would never walk on the Moon"*, FAIR Mormon, Foundation for Apologetic Information and Research.

# CHAPTER 3

# THE SIX DAYS OF CREATION

*The heavens declare the glory of God; and the firmament sheweth his handywork. Day unto day uttereth speech, and night unto night sheweth knowledge. There is no speech nor language, where their voice is not heard. Their line is gone out through all the earth, and their words to the end of the world.*

*Psalm 19:1-4*

## Creation Begins

**H**ow was the Temporal Creation accomplished? Who created it? And what can we learn about it from the scriptural accounts[1] and the truths we know from science?

> *"In the beginning God created the heaven and the earth. And the earth was without form, and void; and darkness* was *upon the face of the deep. And the Spirit of God moved upon the face of the waters." (Genesis 1:1,2)*
>
> *"In the beginning was the Word, and the Word was with God, and the Word was God. The same was in the beginning with God. All things were made by him; and without him was not any thing made that was made." (John 1:1-3)*

The opening words of the Hebrew Bible, *Beresheet bara Elohiym*, literally means, at the head, God created. The phrase, "*at the head*," *Beresheet* (בראשית) refers directly to the beginning of time. This entire phrase is restated again by the Apostle John in the opening words of his Gospel, this time in Greek, *en arche en ho logos* (Ἐν ἀρχῇ ἦν ὁ λόγος). John, in his use of *en arche* is speaking of the most ancient of times. It is impossible to get a time more ancient than that of the beginning. It must be noted that both the author of Genesis and John chose to begin their stories

---

[1] In this chapter, we will refer to the Genesis 1 account as we previously defined it using the words of the Book of Moses. The words found there faithfully follow the Biblical account with minor but important additions and no omissions. When necessary, we will include important insights from the account in Abraham.

at the point when time began. They both chose their words carefully. In separate revelations, spanning thousands of years, differing cultures, and languages, they both speak of the beginning of time. Time is a principle of the Temporal sphere alone. It is a manifestation of entropy, the physical law that provides us with free agency. Either of the writers could have spoken of a time before time existed, they both chose not to do so. They both have marked out a point in time, literally as it were, that could not have existed before the Fall of Adam.

**PREFACE**

1 And it came to pass that the Lord spake unto Moses, saying: Behold, I reveal unto you concerning this heaven, and this earth; write the words which I speak. I am the Beginning and the End, the Almighty God; by mine Only Begotten I created these things; yea, in the beginning I created the heaven, and the earth upon which thou standest.

The second chapter of Moses begins with a preface to the familiar Genesis 1 account. The preface reiterates that the story to follow pertains to, "*this heaven and this earth; ... upon which thou standest,*" and nothing more. Though the terms the speaker uses to delineate himself are often used by the Father, the one actually speaking here is the Logos[2] and Word declared by John, the great "*I Am*" of Sinai, even Jehovah[3] of the Old Testament and Christ of the New. For a more detailed discussion of the identification of the Son as the Creator of the Temporal sphere, turn to Chapter 10 of this work.

---

[2] The Greek *Logos* means *Word*. We will return to it time-and-time again in this work, each time visiting yet another of the many understandings associated with this extremely important and nuanced appellation for the Son of God.

[3] The name Jehovah only appears 28 times in the standard works, 13 times in the Old Testament, once in the New; 3 times in the Book of Mormon, 7 in the D&C and 4 in the POGP. It is used as the appellation for the Son in the Endowment narrative. Later in this work, we will detail how this name, as it is spelled and ultimately pronounced today, may be in error due to efforts taken by the pious to prevent the taking of the name of God in vain; see -The Unutterable Name, pg. 178. The reason Scripture and the Endowment use Jehovah today, is probably due to its established familiarity and not because it is the Son's proper name.

*Day 1 (15 Billion to 7.75 Billion years ago)*

| **DAY 1 — Moses 2** |
|---|
| 2 And the earth was without form, and void; and I caused darkness to come up upon the face of the deep; and my Spirit moved upon the face of the water; for I am God.<br>3 And I, *God, said*: Let there be light; and there was light.<br>4 And I, God, saw the light; and that light was good. And I, God, divided the light from the darkness.<br>5 And I, God, called the light Day; and the darkness, I called Night; *and this I did by the word of my power, and it was done as I spake*; and the evening and the morning were the first day. |

The narration of the first creative era begins with something identified as earth. It is not yet the object we stand on today; it is something fundamental, yet without form and open in architecture. In the early stages of the Universe, the density was enormously high and space itself was very compact. Today, in our era, "*space*" can be accurately described as at least appearing to be largely void, an expanse.

The Revelator here is God Himself. He had the knowledge then (like we do now), that the void we see today, came from the structure of the initial fireball of the Big Bang, and in fact, is the same thing. This fact is key to understanding the complex meaning, a vital characteristic, of the single Hebrew word translated "*void*" in Verse two.

Inserting Hebrew into the opening phrase of Verse two, we read, "And the earth was tohu (תֹּהוּ) and bohu (בֹּהוּ):" *Without form* (or chaotic) is a reasonable translation for *tohu.*[4] The same cannot be said for the English rendering of *bohu*, as it means so much more than void. Both the Talmud[5] and the great Jewish scholar Nahmanides[6] (1194-1240 AD) state that *bohu* means, *filled with the building blocks of matter*. Thus a more complete and accurate rendering of this phrase is, *And the earth was in*

---

[4] We will explore the greater complexities of the word *tohu* as it pertains to the Spiritual Creation later in this treatise.

[5] The "*Talmud*," as I use it here, is a collection of Jewish civil and ceremonial law and legend comprising the Mishnah and the Gemara. The Jerusalem Talmud (*Talmud Yerushalmi*), also known as the Palestinian Talmud, is thought to have been redacted as early as the year 350AD. The Babylonian Talmud (*Talmud Bavli*) was compiled later, throughout perhaps the entirety of the 3rd to 5th centuries of this era. The writings of the Talmud are central to Rabbinic Judaism today. The Talmud is also traditionally referred to as the "*six orders*," a reference to the six orders of the Mishnah.

[6] Also known as *Rabbi Moses ben Naḥman Girondi, Bonastruc ça (de) Porta*; Nahmanides was a leading medieval Jewish scholar and Biblical commentator. He was raised, studied, and lived for most of his life in Girona, Catalonia, Spain. He is also commonly known by his Hebrew acronym (RaMBaN, for Rabbi Moshe ben Naḥman). He was born in Girona in 1194AD and died in the Land of Israel about 1270AD.

*a state of chaos within the developing structure of the void, and the void was filled with the building blocks of matter.*

In the Moses translation, we find that God Himself caused darkness to come upon the face of the deep. In Genesis, the darkness appears to be just an initial state that pervaded the deep. Both are true, because God created the conditions that darkened the deep in its initial state.

What is the deep? It is something much like the seas, more correctly the vast oceans, we find on Earth today, only much more immense. The deep has a face and is synonymous with a vast body of water. It is significant that the Temporal account begins here with a great sea and darkness.

The initial state of the Universe we know today began with an event commonly known as the Big Bang. It is often imagined as a fiery explosion; however, this is not entirely descriptive of what it was like. To an observer, witnessing the explosion, in its initial stages (just about one second old) the Universe would have looked like an enormous sphere of water, a "*water-world*" in overall scale about the size of our Solar System and expanding fast. Close up it appeared chaotic and turbulent, but overall, from a distance, it was homogenous and remarkably uniform. The footprint and signature of this remarkable great sea can be seen in the remarkably uniform patterns of the cosmic microwave background radiation (CMBR) that reach us today from the very edges of the Universe in every direction.

Within the depths of this great sea, the initial building blocks of atoms, the matter (the earth, of the Scriptures[7]) teemed as the order of the cosmos began to form. As the principle building blocks (*bohu*) of everything there is within the Temporal sphere began to coalesce, so too were forming the very laws of physics that would control them. Today we know that the most fundamental element in the Universe is information itself. Without information, the laws of physics, chemistry, and biochemistry would not exist, indeed nothing would. Everything there is, has been infused with this information. All things are filled with this knowledge, allowing them to "*filleth the measure of its* [their] *creation*" (D&C 88:25).

In Verse two we read, the Spirit of God "*moved*" upon the face of the water. The Hebrew word translated as *moved* in the KJV is *merahepet* (מְרַחֶפֶת) and can be more properly translated as *brooded* or, *was brooding*. This is the same Hebrew word that would be used to describe a hen as she anxiously tended to, gathered-up

---

[7] We will detail in this work that the "*earth*" of the scriptures takes on multiple meanings and is used to represent material, fundamental in form. The word "*earth*" is used by the scriptures to identify the building blocks of matter within the Physical sphere; in this form, *earth* is the matter spoken of by the Prophet Joseph when he declared that "*all spirit is matter.*" In a similar fashion, earth is used by the scriptures to denote the primary particles and elements we define as matter in the Temporal Universe today; this is the earth that was without form, as Genesis, chapter one begins. Earth is also the name given to the planet we inhabit today and that name given to many other inhabited celestial orbs, both within this Temporal Universe and it seems perhaps many orbs in the eternal Physical sphere as well.

and protected her chicks from harm. *Merahepet* is an expression of determined will and active and resolute care.

> ***"And I God said: Let there be light…" Within the execution of this simple statement, lies the primary connection between all that is known by science and all that has been revealed by God…***

The concern shown by the mother hen towards her offspring accurately describes the actions of God as He carefully tended to the infant Universe imbedding within it all the knowledge it would need to carry out its purpose over the billions of years that would follow. This detail is not insignificant. Here we see the initial input of Intelligent Design – the loading of the critical knowledge into the "*stuff*" of the Universe that will allow it to grow, organize and evolve in accordance with the wishes and the will of God. As we progress through our telling of the six creative days, we will see that in many instances the elements of the creation bring forth life and creatures on their own; they do so at the command of God as He "*spake.*" In a very real sense, those elements have power within themselves to do so. That power was at first instilled within the matter of the Universe at this very point on the first day.

## LIGHT, THE TIMEKEEPER OF THE UNIVERSE

"*And I God said: Let there be light…*" Within the execution of this simple statement, lies the primary connection between all that is known by science and all that has been revealed by God pertaining to our existence today. Just four words, given as a command, tell a tale that is over 14 billion years in the making, one that provides an explanation for all that is.[8] It is revealing that this event, recorded over and over again by those who could not begin to fathom its ultimate importance, is even here, in texts recorded so long ago, stands as testimony to the divine nature of the Genesis 1 creation account, and its accuracy.

The Genesis 1 account is ancient in origin; first recorded by Moses at the time of the Exodus, approximately1500 BC and before that by Abraham 400 hundred years earlier. It would have been counterintuitive for anyone of that age to associate light in a tangible sense as it is spoken of here.

---

[8] The creation of light (energy) itself is the foci for the Holy Grail of science, the Grand Unification Theory and super-symmetry breaking.

Today we understand light as being energy, and because of Einstein we know energy to be matter. So it is precisely correct to record the creation of light as the initial work of God in this Universe. For it is truly energy that makes up all there is and drives time forward. God was pleased with the way this creation performed.

As the infant Temporal Universe continued to mature, it experienced a phase change which would leave a mark that remains visible to this day. The "*mark*" is a result of a fundamental conversion of the young materials of the Universe (*Figure 3-1*). During what is known as the "*decoupling era*," some 377 thousand years after the initial Big Bang, the Universe had cooled to the point that allowed what had previously been free protons and electrons to merge and form atoms. This is critical, for prior to this era, the energy of light was *coupled* to the particles, being reabsorbed by them as fast as it could be unbound, never able to travel freely in the ever expanding space.

Space was opaque and dark; the manifestation of energy we most closely associate with light was not yet possible. With the formation of atoms from the free protons and electrons, the energetic photons could *decouple* from the particles that formed the atoms and move freely for the first time. By this process, literally, "*light*" was "*separated*" from what had been "*the darkness of the deep.*"

The detailed description of this primordial event provided through inspiration could not possibly have been imagined or understood by any man living in the days of Abraham and Moses, yet here in ancient Scripture we find the incredible detail.

The "*mark*" decoupling left on the Universe is the Cosmic Microwave Background Radiation (CMBR) of which we previously referred. This remnant of the early stages of the Universe is critical and extremely important. Here in the ancient text God has revealed it to us, at this precise point in the creation story. He points us to a *marker* that would be a key to opening a whole series of truths; a key that is at once readily apparent, no matter where one looks and yet would remain entirely hidden from view for millennia after its presence was first revealed by God to Moses.

Time starts here at the beginning, in Day 1. The insight shown in the scriptural record (that time *begins* with the appearance of matter) is quite extraordinary, for it details processes only most recently known by science. When matter forms from energy, disorder is ordered and entropy is established. From that instant, the arrow of time points back to the direction of disorder and decay. Entropy literally is the "*stuff*" of time and begins with the first formation of matter, *exactly* when God says it does.

The Theory of Relativity began as one very elaborate *Gedankenexperiment*. These "*thought experiments*" were used to perfection by Albert Einstein. He conducted conceptual rather than actual experiments to imagine the complex settings needed for his intricate theories to play out. We too will conduct one of these *thought experiments* here to help us understand how the CMBR is the key to understanding the true nature of time; let us tell a parable of sorts.

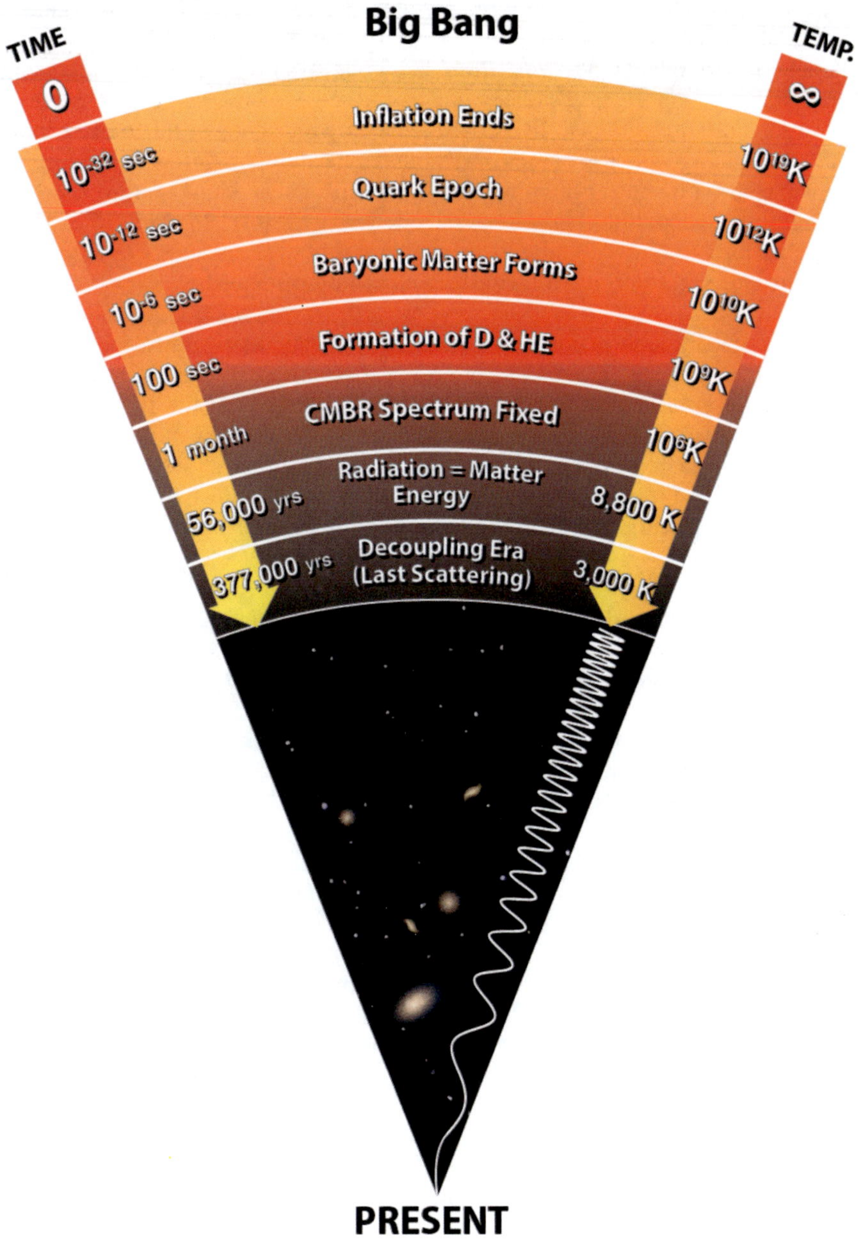


Figure 3-1 *CMBR and the Early Epochs of the Universe. Measureable Entropy (Time) begins just 0.0000000000001 seconds after the Big Bang, when the transition from pure energy to stable matter occurred at the beginning of the Quark Epoc.*

Imagine sitting in a large darkened theater, actually living out your whole existence within its closed walls. Your only reference to the passage of time, dimension, and structure comes from the image played out on the screen in front of you and the light and sound that bathes you from the images shown. The film plays in an endless, repeating loop and has been unperceptively slowing in front of generations of viewers like you, who also lived out their entire existence in this closed room, in front of the image on the screen. Without any external reference for time, it is impossible for you and for them to measure the film's slowing,[9] and in short order, all context of what the original speed once was is lost. In fact, it is sensibly impossible for you to even know that the rate of time was changing; to any observer within the closed theater, it isn't happening, nothing is noticed.

Over time, generations of viewers each live out their lives entirely encompassed within segments of the film that have grown shorter and shorter. There are ancient records of those in the past who chronicled seeing lengths of the film much longer than the snippets of the film played out in front of the viewers in the theater today. Once-upon-a-time, it is told, audiences saw much of the film, perhaps even all of a single loop and a bit more; now an individual lives to see just a few scenes, never viewing the full picture or knowing the complete story it has to tell.

On the screen, in the film, there is a piano being played by a musician keeping rhythm with a metronome in a room that has a clock on the wall. The setting on the metronome is clearly visible to you, *72 beats per minute.* You watch the pendulum swing side-to-side in time, listening to the player as she perfectly plays in time with the clock on the wall in the film, *72 beats per minute*. Unknown to you and to everyone else who has ever seen the film, the projector in the projection room has been forever slowing, the player, the clock, the metronome, the song, and the film, has forever been slowing in unison.

In time, in the theater, on a wall, someone discovers an object that looks just like the clock in the film. It has been, until now, unseen, hidden in the darkness at the edge of the room. This new found object is carefully observed and discovered to be keeping time just like the clock in the film, but strangely at a rate much faster, a rate that, upon observation, actually seems to be ever increasing in speed.

When the "*clock*" in the theater was first discovered on the wall in the darkness, it wasn't clear what had actually been found. No one knew what to make of it, or how to apply any of the information it disclosed. At some point, within the theater, some of the viewers (the occupants of the closed universe) thought to apply the metronome setting on the screen to the clock they had found on their wall in the darkness and not to the clock in the film.

---

[9] A detail we will discount here is any sense of "*local*" time as measured from the locality of the observer (within his or her seat); that local time is what we observe on Earth as, years, days, hours and minutes. Local time, which changes or does not change relative to the background over period, is what today allows to count an apparent age of the Universe in billions of years (by our reckoning) and not by an age that may be relevant to any time frame intrinsic to space itself. The analogy still holds in that we have no sense that Universal time is changing, for it surrounds and bathes us and everything there is.

When the adjustment was made, *72 beats per minute* occurred at a much faster rate. The result was remarkable, the film was restored to its original rate of speed and the song was heard again in the key and at the rhythm in which it was originally recorded so very long ago. The viewers in the theater, once again, could see all of the film and not just a portion in their lifetimes. What had once been incomprehensible about the clock discovered on the wall now made perfect sense.

The clock was a key that restored to the viewers in the theater much, much they never knew they had lost. The CMBR today is our clock, once unseen, but discovered, at the edge of our Universe in the darkness. It too is a key to understanding time and it's slowing, just like the clock long lost on the wall in the darkness of the theater.

The analogy is not perfect but provides a good description of time, its properties, and our observation of it. The CMBR is not a clock per se, and it does not remain unaffected by the universe symbolized by the realization of the film in our thought experiment. For unlike the discovered clock on the wall, the CMBR is within this Universe (the film) and is transformed by its expansion.

God, in the second verse of the Genesis 1 account, tells us how to find the marker (the clock on the wall, the CMBR); He tells us where to look (within the light), and He tells us when to look (the creation point of matter). Light, the energy that it is, has properties with established relationships one-to-another: frequency, energy, and the temperature associated with motion caused by that energy. Through these relationships, the electromagnetic spectrum (light) establishes the order and parameters mimicked by the metronome in our imagined film. The creation of matter (the *when* of the Scriptures) has a fixed position on that spectrum; it is the point at which we must look. When Penzias and Wilson heard the microwave background "*noise*" on their antenna,[10] the clock in the darkness was discovered on the wall. Realizing that the CMBR actually marks the transition energy, that it is a remnant of the first light of creation, is the moment those in the theater discovered what their clock in the dark meant. When science applied that information to the structure of the electromagnetic spectrum (our metronome), determining the changing rate of time, we made adjustment to the speed of the film, restoring a truth that had long since been lost.

---

[10] The CMBR was accidentally discovered in 1964 by Arno Penzias and Robert Wilson as they experimented with the supersensitive, 6 meter (20 ft.) Holmdel Horn Antenna in New Jersey. The pair had to eliminate all recognizable interference from their receiver, and when they did, they found a low, steady, mysterious "*noise*" that remained and persisted. The residual noise, 100 times more intense than they had expected, was evenly spread over the sky and was present day and night. Through its uniformity, without directional bias, both concluded that this noise was coming from outside our own galaxy. When this information was paired with research being conducted at the time by Princeton University astrophysicists, Robert H. Dicke, Jim Peebles, and David Wilkinson, the "*noise*" was found to be the CMBR we study in detail today.

At the beginning of the Universe around us, with the institution of time, came the need of a method to measure it and clock it in a relative way. From the outset, there could be no single time frame experienced by all of the points within the Universe; Einstein's theorems have proven this fact to be true. What is needed is an understanding of the speed of the clock as it was originally set, before entropy, when the Universe was still little more than a single point in space and time. This time frame, or more precisely its initial setting, is found in the relation of the threshold temperature to the CMBR.

Let us turn to our imaginations once more and picture a fully obscuring cloud cover above us at the peak of day. It conceals from our view anything beyond it, preventing us from seeing the source of the light that filters through and fills the entire atmosphere around us below. When we look up we can see the detail of the bottom of the clouds above, but nothing beyond that. When we examine the light, and adjust for the changes made to it by interactions in the clouds, we can see in its spectrum details of the Sun which lies beyond the clouds, hidden from view.

The period of last scattering, like our imagined cloud cover, effectively provides a barrier which obscures any vision of the Universe beyond. At the point of last scattering, light broke through what is in effect a barrier in time, as if passing through a veil. When it did, light was, for the first time, free to move in the ever burgeoning expanse, unfettered, until it could fall upon our eyes and instruments, releasing the mysteries it held.

This light, from the nascent Universe, holds within it the very details of its beginning, recording those details in its broad spectrum which was put in place at the temperatures of the threshold point, when the particles which make up the atoms of the Universe began to form. This light is the recording of the birth of the Universe. In this second decade of the $21^{st}$ Century, we have the results of years of observation of this light from three ever-more-sensitive space borne observatories that have mapped the bottom surface of the clouds of last scattering. The Cosmic Background Explorer (COBE), the Wilkinson Microwave Anisotropy Probe (WMAP) and the Planck Space Observatory have shown us what that *surface* looks like and have shown us what lies beyond through the spectrum of the light they have received. (*Figure 3-2*)

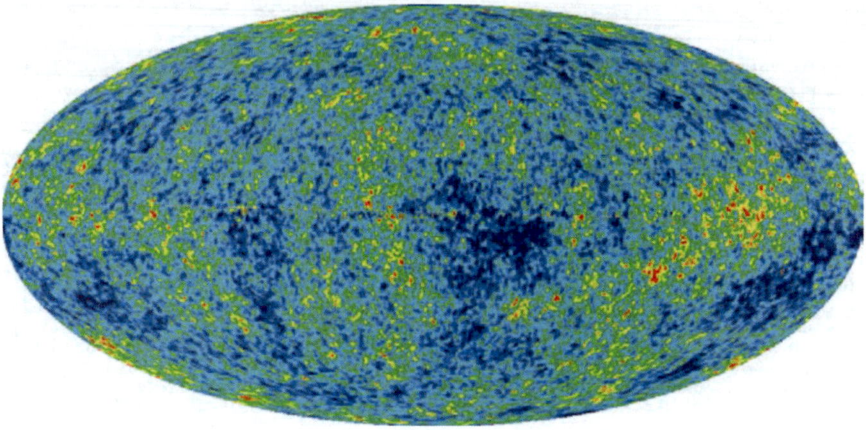

Figure 3-2 *WMAP Image of the CMBR. This 360 degree circumspective image of the Universe shows the bottom surface of the clouds of last scattering. The CMBR variation indicated in the image is exaggerated, with the total range shown by all the colors differing from the nominal by no more than 18μK and with a uniformity that varies by less than 1 part in 100,000.*

Today, everywhere we look, *in any direction*, we see (as the backdrop of everything there is) the base of the "*clouds*" of last scattering and the CMBR that radiates from them. The CMBR is the remnant of the initial temperature and the glow of that first flash of light made at decoupling. From the moment of its creation, the once bright, hot, light has cooled as the Universe has inflated and expanded, now just 2.73 degrees above absolute zero, in the microwave range of frequencies ranging from 0.3 GHz to 630 GHz and peaking at 160.4 GHz. The CMBR, its wavelengths, act as a cosmic metronome, precisely counting out the timing increments of the Universe. The CMBR has locked on to the heartbeat of time and provides the code we need to understand the timepiece of God. Despite the dramatic events of the Big Bang (the entire contents of the cosmos in what appears to be a state of utter chaos) the Universe, at the point of decoupling and before, was much simpler than it is today. Entropy had not yet had 14 billion years to spread disorder and grow complexity. As a result, it is possible to precisely define the parameters of the Universe at these very early stages.

Our modern technology can determine the specific temperature and hence the energy of the Universe at the threshold point, *quark confinement*, when matter began to form from energy.[11] Bear in mind, this value is not theoretical or

[11] The CMBR gives a snapshot of the Universe when, according to standard cosmology, the temperature dropped enough to allow electrons and protons to form hydrogen atoms, thus making the Universe transparent to radiation. When it originated some 380,000 years after the Big Bang (this time is generally known as the "*time of last scattering*" or the period of *recombination* or *decoupling*), the temperature of the Universe was about 3000 K. This corresponds to energies of about 0.25 eV, which is much less than the 13.6 eV ionization energy of hydrogen. Since decoupling, the temperature of the background

extrapolated from estimates of conditions separated from us by vast quantities of space and time; the temperature is measured right here on Earth, *precisely* in the most advanced particle accelerators we have. The threshold point was reached more than 370 million years before the light we see in the CMBR broke free in decoupling (when negatively charged electrons bonded to positively charged ions). The temperature of this threshold point is $10^{12}$K, very nearly a million, million times hotter than the current 2.73K of the CMBR. That radiant energy had a frequency a million, million times longer than that of today's CMBR.

The temperature of the threshold point represents the instant (entropy) Biblical time begins, when the transition from pure energy to stable matter occurred.[12] The radiation from the threshold point has been stretched a million, million fold by the expansion of the Universe since its creation and its release at the time of last scattering. Its redshift (z) as observed today is $10^{12}$. The stretching of light waves has slowed the cosmic clock by a million, million, slowing the perceived swings of our metronome by the same amount.

The threshold temperature (frequency) for the creation of matter we have identified in our particle accelerators, matches the energy state of the Universe just 0.0001 seconds after God said, "*let there be light.*" We know that the threshold temperature is a million, million times hotter than the CMBR today. To measure the age of the Universe, we look outwards to the most distant depths of space effectively looking back in time. From the perspective of our Earth-based clocks, in our time frame, we establish an age for the Universe at approximately 15 billion years.[13] The Scriptures adopt this Earthly time frame, but only for times after the arrival of Adam in the Temporal sphere. With an understanding of relativity that predates Einstein by millennia, the clock of the Scriptures, from an age before Adam, is not tied to any one location. It marks a time frame that looks forward from the Creation, encompassing the entire Universe, a universal timepiece tuned to the cosmic radiation at the moment when matter was formed and entropy began. That timepiece of God, as observed by us today, ticks a million, million times more slowly than it did at its inception. The stretching of space that has occurred since the

---

radiation has dropped by a factor of roughly 1,100 due to the expansion of the Universe. As the Universe expands, the CMBR photons are redshifted, making the radiation's temperature inversely proportional to a parameter called the *scale length* of the Universe. The temperature Tr of the CMBR as a function of redshift (z) can be shown to be proportional to the temperature of the CMBR as observed in the present day *(2.725 K or 0.235 meV): Tr = 2.725(1 + z)*. The concept of time, an ability to measure it, begins, however, at the end of the inflationary era, with the creation of quarks, the building blocks of baryonic matter (Hadrons and Leptons) just $10^{-12}$ seconds after the Big Bang at the threshold temperature of $10^{12}$ K.

[12] According to the Big Bang Theory, in the early Universe, massive particles of matter (photons and fermions) would inter-convert freely. As photon gas expanded and cooled, some fermions would be left over (in extremely small amounts $\sim 10^{-10}$) because low energy photons could no longer break them apart. Those left-over fermions have become all the matter we see today in the Universe around us.

[13] The age of the Universe is most commonly accepted today to be 13.79 billion years. This age is not exact, however, and has been shown by data collected by Norbert Przybilla of the University of Erlangen-Nuremburg, using the 10 meter Keck II telescope on Mauna Kea, Hawaii, to indicate a potential 15 percent error in those data (www.Arxiv.org/astro-ph/060279). In this treatise, we will use the nominal value of 15 billion years.

threshold point was reached has produced a million-million-to-one ratio in the perception of time.

The universal clock continues to advance as it always has, recording the passage of just one minute, while we here on Earth record the passage of a million, million minutes. To put that in perspective, the dinosaurs went extinct 65 million years ago as measured by the decay of radioactive nuclides here on Earth, and that measurement is correct for *our* time frame. But to know the *universal time*, we must divide our perceived time of 65 million years by a million, million. This million-million-to-one ratio results in 65 million Earth years lasting a mere 33 minutes on the cosmic clock. In the eyes of our creator, those leviathans roamed a Mesozoic Earth just a half hour ago.

When we apply all that we have learned here, about the stretching of space and the dilation of time itself to the Genesis 1 accounts, an enlightening correlation appears.

The Universe is 15 billion years old, by the reckoning of our slowed clocks today on Earth; they have slowed by a factor of a million, million from their original setting. If we divide the age of the Universe as counted in days (24 hour periods) by this factor of a million, million, the resultant number is 5.475. Astonishingly, this number places us squarely within the sixth day of the age of the Universe as counted by the universal time piece, precisely where the scriptural record positions the creation of man. Note that this is just six and a half days, *six and a half 24 hour periods as measured by Abraham and Moses and us today……*, not five or seven, not three or twenty two, but <u>six</u> as counted by those who were recording the revelation. Coincidence? We should know better than that by now. (*Figure 3-3*)

God spoke clearly giving us incredible detail. Had the Universe been any other size, by any tiny factor, or had any other mass; had it cooled and stretched at any other rate, had the temperature of quark confinement (the point at which protons and neutrons formed) been different in any way, this precise relationship *would not exist.*

Scientists today speak of six numbers that define the Universe.[14] These important numbers, giving value to critical parameters, appear to be incredibly fine-tuned to one part in $10^{120}$. This and the relation of 15 billion years to just six days, testify of divine design for the Universe. Within the simplicity of the message found in the Day 1 account, lie a wonderful complexity describing relativity, biology, geology, cosmology and all the other disciplines of science as well.

---

[14] See Appendix, <u>Just Six Numbers</u>, pg. 532 and <u>The Impossible Universe</u>, pg. 540.

**Threshold Energy 1MeV** **CMBR Energy 0.235 meV**

1,000,000 000,000.235

The order of magnitude for the energy of the CMBR today is one million, million times smaller ($10^{-12}$) than the Threshold Temperature it is a remnant of.

| **Age of the Universe in Earth Days** |
|---|
| **5,840,000,000,000 Days/16 Billion Years** |
| **5,475,000,000,000 Days/15 Billion Years** |
| **5,019,000,000,000 Days/13.75 Billion Years** |

**Adjusting these ages by $10^{-12}$**

(*the nominal age of the Universe, as measured in the time frame relative to the Threshold Temperature set just 100,000,000,000$^{th}$ second after the initial Big Bang*)

**Reveals that the Universe is just**

**5.475 Earth Days Old**

**That is *Midday* on Day 6 of Creation.**

Figure 3-3 *Relative Time Scale of the Universe.*

As the Universe expanded and cooled, the rate of change for the CMBR frequency slowed, making each of the six periods seem shorter in length than the one that preceded it (when measured in terms relative to our current time frame). This six day cycle accounts for all of the known age of the Universe to this point and beyond. Each of the "*days*" (periods) is of sufficient length and just the correct length to match the creative events of the Scriptures with the cosmological, geological and biological chronologies of science, without any tampering or adjustments (*Table 3-1*).

| Creative Period | Length of Time Period | Years Before Adam | Genesis 1 Account | Scientific Account |
|---|---|---|---|---|
| | | | | |
| **Day 1** | 7.25 Billion Years Duration | 15 Billion | The Temporal Universe is Created; Spirit of God *broods* over the deep, light separated from the darkness. (Gen. 1:1-5) | The Big Bang, Inflation, @ 377 thousand years of age, light literally breaks free as electrons bond to atomic nuclei (Decoupling). Galaxy and first generation stars begin to form. |
| **Day 2** | 4 Billion Years Duration | 7.75 Billion | The heavens are stretched out above. (Gen. 1:6-8) | The Milky Way galaxy forms; our Sun and other main sequence (second generation) stars form. The T-Tauri wind blows. |
| **Day 3** | 2 Billion Years Duration | 3.75 Billion | Oceans and dry land appear; the first life (plants) are *spoken* into existence. (Gen. 1:9-13) | 3.8 billion years ago the Earth cools, large bodies of water form, followed almost immediately by the first forms of single cell life, photosynthesis begins. |
| **Day 4** | 1 Billion Years Duration | 1.75 Billion | Sun, Moon and Stars become visible in the heavens. (Gen. 1:14-19) | Photosynthetic life produces an oxygen rich, transparent atmosphere. Celestial objects appear in the skies. |
| **Day 5** | 500 Million Years Duration | 750 Million | The waters bring forth animal life, which swarm plentifully in the seas, followed by the appearance of *moving* creatures (reptiles and amphibians) and winged animals. (Gen. 1:20-23) | First multi-cellular animals, 600 million years ago in the Precambrian, Ediacaran fauna flourish. ≈530 million years ago oceans and seas suddenly swarm with animal life having the basic body plans of all future animals (the Cambrian Explosion "Biology's Big Bang") as recorded in the Burgess Shale fossil record; massive winged insects appear, the largest creatures ever to fly. |
| **Day 6** | 250 Million Years ***and Counting*** | 250 Million | The earth brings forth land animals, mammals; God creates (*bara*) Man in His own image. (Gen. 1:24-31) | Dinosaurs rule the Jurassic period. Then approx. 65 million years ago, an asteroid strike leads to a massive extinction that destroys over 90% of all life on Earth. Mammals survive and the land is repopulated. Hominids appear and evolve, **then Man**. |

*Table 3-1 The Six Days of Creation. This chart represents the six creative days in relation to the scientific account of the development of the Universe. Information provided here and in this section of the book draws upon work developed by Dr. Gerald Schroeder.*[15]

[15] The time frames listed here represent a simplification of the expansion of the Universe using the most basic equations describing deceleration and gravitational attraction. Most recent information applies a

"*And the evening and the morning were the first day.*" What came first the chicken or the egg? This is an enigmatic question that teaches the futility of a circular progression. Sometimes our traditions, language or concepts take a similar path. Was our first mother named Eve because the word predating her existence meant "*mother of all living*" (*a name that would have had a nonsensical meaning before the birth of her children*), or did the name Eve later come to have that meaning because it was given to her, the *de-facto* mother of all mankind?

Why does the Jewish day start and end at sundown? Because that is what is taught right here in the Genesis 1 account. Or is that actually what *is* taught?

Perhaps a better understanding of this repeated phrase in the Genesis 1 account can be found when we strip away what millennia of culture, tradition and misunderstanding may have heaped upon it.

The now traditional translation of this phrase comes from a false precept that God was trying to express an understanding now familiar with the rising and setting of the Sun marking the daytime period of an Earth day, but this could not be so. At this point in the creation sequence, the Solar System had not yet formed, there were no Sun and Moon to rise and set. The concept of morning and evening could not exist.

It is only later, much later in the history of the Universe, with the forming of the Earth and subsequent history of Man that those understandings could be represented by the words of this sentence. *(In time the true and original meaning of this statement became shrouded in misunderstanding and heaped upon with tradition by men trying to make human sense of it through adoption of their own experience. In-the-end, the repeated phrases were ultimately comprehended in an Earthly sense with the result that today, a whole culture defines a day differently than anyone else, one that begins as the Sun goes down instead of when it comes up.)* This misunderstanding of the words spoken by God was firmly in place as early as the Abraham account and perhaps much earlier.[16]

There is, however, another more profound meaning to be discovered here. The Hebrew words translated in the KJV as evening and morning are, *erev* (עֶרֶב) and

---

repelling force from dark energy (lambda) that provides "*shape*" to an otherwise flat Universe which alters the age and rate of change in the expansion of the Universe. The effect of this force coupled with various lengths of time estimated for the initial inflationary state of the Universe, combine to reduce the overall age of the Universe to 13.75 billion years and shorten the lengths of Days 1 and 2, while increasing the lengths of Days 5 and 6.

[16] In Abraham, the rendering offered for the phrase, "*And the evening and the morning were the first day*" reads: "*And it came to pass that from the evening until morning they called night; and from the morning until the evening they called day; and this was the first, or the beginning, of that which they called day and night.*" This is an obvious extrapolation of the same misconception expressed in the Genesis 1 account.

*boker* (בֹּקֶר). Each of these two words is compounded from root words meaning, disorder (chaos) and order respectively.[17]

Thus in the basest, most *original* form, this phrase in Day 1, in its intended context, reads: *"from disorder came order."* Again we are back to terminology that emulates an understanding of entropy and precisely defines the physical processes taking place in the Universe at its earliest stage, the stage being described in the Day 1 account.

So here, the simple meaning of the revelation as it was originally recorded was lost and replaced with yet another meaning that took form in a later era and in a much different environment. What was the result? A context that was originally intended to express progress in organization (order from chaos, the development of complex structure in the Universe where none had existed before), was replaced with a meaning that resulted in an entire culture turning the most coherent meaning of a day upside down. For the children of Israel, a day starts when everyone else's ends, beginning when the Sun sets, instead of when it rises.

When properly understood, the information provided by revelation at least four thousand years ago provides insightful detail of incredibly complex events that could not have possibly been known by those who recorded it without divine intervention. The sequence of processes of Day 1 was counter intuitive to man of 4000 years ago. The images are remarkably precise in description, and the concepts put forth were much too advanced for the period, given their limited understanding and lack of experiences so common to us today.

SUMMARIZING WHAT WE HAVE LEARNED, *Day 1*

The four Day 1 verses describe the Universe in its infant stages as if seen from without in some way, an immense sea of free particles at first and later a sea of Hydrogen, Deuterium and Helium. The Spirit of God is described as brooding over this vast expanding sphere until all of the nascent matter is infused with all the information (laws of physics) it needs to continue its existence: forming stars, galaxies, planets and eventually life for countless billions of years to follow.

We are told of the very creation of energy, described as light, with an understanding that was not known to science until the mid-1800's. The narrative provides an incredible view of the separation of energy from matter, a process that was not even capable of being theorized until advances in physics and mathematics of the late twentieth century made that possible. And through-it-all, there exists an understanding of time, temporality and entropy that is far beyond the understanding of any man in 4000 BC.

---

[17] The root meaning of *erev*, the Hebrew word for evening, we learn is *mixture, disorder, chaos*—just as vision becomes blurry and chaotic at dusk. The root meaning of *boker*, the Hebrew word for morning, is *orderly, able to be discerned*, as with improved vision upon sunrise.

The verses start with a beginning and end with a statement detailing the formation of order from disorder, which we now know provides the very framework upon which the delicate lacework of time hangs.

It must be remembered, in antiquity as well as in the modern era, the Universe was considered to be eternal, never changing, static and stable. The concept of a beginning to the Universe was there in Scripture all along, but never understood by man. As the 1900's began, science still envisioned a "*Steady State,*" eternally unchanging Universe. All of the advancements of that century told a story that was otherwise, but the greatest minds were often very reluctant to accept. Yet here in these four simple verses, just 107 words, the full story was told, so very long ago.

Subtly, in a way that almost assumes that the fact was commonly known to all, the relative nature of time is addressed as well as the ultimate code book for linking all independent time frames back to the original. The knowledge included in these verses is tremendous in scope and volume, befitting the opening words of God to man regarding the Universe we all live in.

It is my view that Scripture supports an ancient creation of the Universe some 15 billion years ago. That these incredible verses precisely describe creative events that lay over the scientific understanding so completely provides vivid testimony of the truthfulness of this position.

The first day of the Genesis 1 account was 8 billion years in duration as reckoned from our time frame, though it may have been much closer to a 24 hour time period in the reference frame for time established at the outset of the Big Bang.[18] This creative period began approximately 15 billion years before the entry of Adam into the Temporal sphere, give or take a few hundred thousand, but then, who is counting.

*This ends the account of Day 1.*

---

[18] I hesitate here to place God in any time frame, because time really is an attribute of the Temporal Universe and God is an eternal being. It is much more correct and by far less problematic, to speak of the *initial time frame* of the Big Bang; we can do so without tying it in any way to the eternal sphere. It is indeed understandable, proper and expected that any entity dealing with this Universe from a sphere external to its existence would affix a universal time frame for the whole of it. The most universal of all time frames would be the one that represented its initial state, at its point of maximum negative entropy when the Universe could be described as having a single state. This is the time frame I would expect God to relate to our Universe, if He relates to time at all.

## *Day 2 (7.75 Billion to 3.75 Billion years ago)*

| **DAY 2 — Moses 2** |
| --- |
| 6 ¶And again, I, *God, said*: Let there be a firmament in the midst of the water, and it was so, *even as I spake;* and I said: Let it divide the waters from the waters; and it was done;<br>7 And I, God, made the firmament and divided the waters, yea, the great waters under the firmament from the waters which were above the firmament, *and it was so even as I spake.*<br>8 And I, God, called the firmament Heaven; and the evening and the morning were the second day. |

As Day 2 of the Temporal Creation begins, we are introduced to the firmament. This has a complex understanding that we will explore; but before we do that, we need to address something else.

### THE VOICE OF GOD

The phrase "*Even as I spake*" appears five times in the Moses translation, twice in Day 2. It is associated with the phrase "*God said,*" which is found in all three records of the Genesis 1 account, seven of those times in the context of commanding the elements. What is the significance of these repeated statements? How can speaking, be a conveyance of such creative power?[19] Again, the answer demands an understanding of the Cosmos that far exceeds that available until the most recent decades, but here, subtly, it has been nestled in the Genesis 1 account, first penned well over 4000 years ago.

Today we understand the principal structure of the Universe to be what science has identified as the universal vacuum state, which exhibits a ground state energy often associated with the *Zero Point Field*. The *Zero Point Field* is an underlying quantum energy field with fluctuations (Zero Point Energy or *ZPE*) on the tiniest of scales, one hundred million billion times smaller than the electron.[20] Though extremely tiny they may be, the contribution *ZPE* provides to the Universe is immense, for the *Zero Point Field* occupies all of space (the matter we are capable of sensing leaves the space it occupies 99.9999999999999% empty). In totality, the energy attendant to the *ZPE* amounts to more than 75% of everything there is in the Universe.

---

[19] The Book of Mormon provides another testimony of the creative power of the voice of God. "*For behold, by the power of his word man came upon the face of the earth, which earth was created by the power of his word. Wherefore, if God being able to speak and the world was, and to speak and man was created, O then, why not able to command the earth, or the workmanship of his hands upon the face of it, according to his will and pleasure?*" (Jacob 4:9). Here the Nephite Prophet Jacob recognizes both the creative power of God's voice and as the Apostle John would later do as well, Jacob makes an intimate connection with the voice of God and His Word.

[20] See Appendix, A Zero Point Field, pg. 531.

This field carries quantum fluctuations (noise) in the vacuum of space. The average energy value of these fluctuations is called the *Vacuum Expectation Value* (VEV).

These fluctuations bring EVERYTHING into being, communicate EVERYTHING from one place to another, and exist outside of the frame of reference we call time, and control and transmit through their structure and existence EVERY law of physics. Most important to our discussion here is that these fluctuations are most simply expressed as waves, vibrations, like those on the taut surface of a drum. The waveforms, being the structure of the fluctuations in the *Zero Point Field*, conform to, travel and propagate, and transmit information, in the exact same way as the vibrations (waves) do on the surface of a drum. These waves are exactly like those that would be expected if one "*spoke*" towards the drum surface. In this context the references to God *speaking* and the Creation obeying are remarkably insightful and compelling proof of the divine origin of the Genesis 1 account.

The Day 2 verses also offer divine confirmation of the most recent scientific understandings.[21] And it must be remembered that the *Zero Point Field* represents virtually all that there is; it comprises 99.9999999999999% of all of the space each of us identifies as what we are, the volume of our bodies. The whole of our consciousness, the sum of our actions and relationships; our communication, understandings, thoughts, and desires, all play out on a background that is all but entirely the *Zero Point Field*, sent vibrating at the moment of Creation:

> *"...and this I did by the word of my power, and it was done as I spake;" (Moses 2:5).*

We will return to the *Zero Point Field* and its association with the Word of the LORD in our later discussion of the Physical Creation.[22]

THE FIRMAMENT

So what of the firmament? Here, as in the case of *erev* and *boker* changing to mean evening and morning as their original context was forgotten, we have a chain of events that impacted the original meaning of the ancient text. The ancient Hebrew word *raqia* (רָקִיעַ, *which originally meant, to stretch out, or spread*), is translated as firmament in the KJV. *Raqia* was conflated millennia ago through association to mean hammering out metal (because hammering spreads metal as it is worked). Later *raqia* became associated with *hardness* and *firmness*, which hammering will do to iron if done when the metal is heated. *Follow along and we will see how this began a transformation of meaning that led to the word firmament.*

---

[21] Paul Davies wrote: "*These rules look as if they are the product of intelligent design. I do not see how that can be denied. Whether you wish to believe they really have been so designed, and if so by what sort of being, must remain a matter of personal taste. ..... Although many metaphysical and theistic theories seem contrived or childish, they are not obviously more absurd than the belief that the Universe exists, and exists in the form it does, reasonlessly.*" The famed mathematician, Roger Penrose said, "*I would say the Universe has a purpose. It's not there just somehow by chance.*"

[22] See EVERYTHING FROM NOTHING, pg. 159.

The terms hardness and firmness align closely with many ancient misunderstandings of cosmology in which the sky was understood to be a vast solid dome or vault from which the stars are hung. So, by the second century BC, when the Septuagint was translated in Egypt (at Alexandria into Koine Greek), these interpretations of *raqia* were inseparably associated with the cosmologies of the time, especially Egyptian cosmology that held the dome of the sky to have been chiseled out of solid stone. Because *raqia* was understood to be the hammering and stretching of metal to *harden it*, the Septuagint used the Greek word *stereoma* (στερέωμα) (*meaning something made firm or solid*) when it sought to render *raqia* from the original Hebrew text.

Later, despite St. Jerome's insisting on *Hebrew Veritae* (translating from the Hebrew only), rather than using the original meaning of *raqia* to express a heaven that is *spread out* above us, he differed back to the Greek connotation of *stereoma* from the Septuagint to the extent that (when he produced his Latin Vulgate around 400 AD) he used the Latin word *firmamentum* (*meaning a strong or steadfast support*). 1300 years later, in the early 17th Century, the King James translators merely transliterated Jerome's Latin word—and thus was born the firmament. Because of this turn (the KJV use of firmament) and advancements made by cosmology in the centuries that followed, the word *firmament*, took on yet again another new definition as it now became the *Biblical* name for the star filled expanse above us (a definitively diffuse and unbounded, non-solid, vacuous space).

Today, after the work conducted in the early 20th Century by George Hale, Edwin Hubble, Georges Lemaitre, and H.P Robinson, and work near the close of that same century by Alan Guth, we understand that the Universe around us (the heavens we see above) is expanding and spreading out, exactly as *raqia* was originally intended

Figure 3-4 *The Firmament. As the focus of the creative account of Day Two narrows, the firmament that stretches across the sky is that of our own Galaxy, the Milky Way. However, Day Two encompasses the age of the Universe in which Galaxies were forming everywhere. As a galaxy evolves, it transitions through a process that flattens it out forming a spiral shape. This process could have been viewed and understood in terms similar to the hammering of metal into a bowl.*

to express. We know today that there was once a period of immense spreading in the past (known as *Inflation*), and we know that the Universe continues to spread out today because of the diminishing of vibrations (Dark Energy) in the *Zero Point Field*, which were first put there by the very Word of God *speaking* into the *Raqia*.

Close to four thousand years needed to pass before our understanding of what the revelation delivered in these Day 2 verses could come full circle. By returning to the original meaning of the word *raqia,* the correct context of firmament is restored. The linguistic transformations uncovered here and the understandings they invoked are a fine example of divine support for the integrity of the true meaning of the Word of God, in the face of great distortion caused by man through his best efforts to understand.

SUMMARIZING WHAT WE HAVE LEARNED, *Day 2*

Keeping in mind that the Genesis 1 account is of, "*this heaven and this earth ... upon which thou standest*" and nothing more, the focus of Day 2 narrows to that of our home galaxy, *the thing that is stretched out in the sky*[23] (*Figure 3-4*). We are told that the vast expanse of stars formed the vault of the sky, precipitating out of the materials (*waters*) seen in Day 1. The firmament enclosed within it some of those same "*waters*" while the processes at work separated (*raqia*) our galaxy from the rest of the Universe. Our Scientific understanding of this same period tells us that the early Universe was dominated by gravity forming galaxies in every corner. Our galaxy, the Milky Way, was just one of hundreds of billions coalescing at the time in the expanse of space.

From the perspective of one shown this grand scene in vision, as God continues the narrative, the forming and coalescing galaxy presents a visual barrier between the "*universal waters*" of the great deep, seen in Day 1, and constrained primordial waters within the galactic structures (i.e., galactic gas clouds). The viewer is shown and explained the Milky Way as it appears from within, which especially in its infancy housed tremendous expansive formations of (as of yet still calming) unorganized matter. These tremendously beautiful diffuse formations undoubtedly would be very familiar to the observer who had viewed the unorganized, nascent matter revealed in the vision of Day 1.

*This ends the account of Day 2.*

---

[23] The Hebrew noun *raqia* is derived from the verb *raqa*, which means "*to spread abroad, stamp, or stretch.*" The intended understanding was the stretched out nature of the *raqia*. This understanding is consistent with the terminology of many other verses, such as Psalm 104:2 and Isaiah 40:22, which speak of the stretching out of the heavens. The Hebrew word used in these verses for heaven is not *raqia*, but *shamayim* (literally, "*heavens*"). However, in Genesis 1:8, God explicitly calls the expanse "*heaven*," thus equating *raqia* with *shamayim*. If the stretched out nature of the *raqia* is what is intended, then firmament may not be the best translation; expanse is more accurate. The context of Genesis 1:6–8, 14–22 makes it clear that Moses intended his readers to understand *raqia* simply as the sky (atmosphere and heavens or space) above the earth, as even the Sun, Moon, and stars were placed in them. In fact, in Modern Hebrew, *raqia* is the word used for sky.

## *Day 3 (3.75 Billion to 1.75 Billion years ago)*

| **DAY 3 — Moses 2** |
|---|
| 9 ¶And I, *God, said*: Let the waters under the heaven be gathered together unto one place, and it was so; and I, *God, said*: Let there be dry land; and it was so.<br>10 And I, God, called the dry land Earth; and the gathering together of the waters, called I the Sea; and I, God, saw that all things which I had made were good.<br>11 And I, *God, said*: Let the earth bring forth grass, the herb yielding seed, the fruit tree yielding fruit, after his kind, and the tree yielding fruit, whose seed should be in itself upon the earth, and it was so *even as I spake*.<br>12 And the earth brought forth grass, every herb yielding seed after his kind, and the tree yielding fruit, whose seed should be in itself, after his kind; and I, God, saw that all things which I had made were good;<br>13 And the evening and the morning were the third day. |

Day 3 begins with the gathering of waters on the face of the Earth. It is of some worth to note that the latest consensus of scientific thought holds that the water on the surface of the Earth today came through outgassing from molten rock and impacts of water bearing comets and perhaps asteroids in the early Hadean Era approximately 4 billion years ago and earlier.[24] And it is here in Verse 10 that the Hebrew term for water took on its present meaning; prior to this verse, the term referred to the initial building blocks of atoms, *matter* itself (earth, of the Scriptures), the principle building blocks (*bohu*) of the Universe.

By the power of the voice of God the waters are gathered into seas and the dry land appears. From this point forward, it is important to keep track of the times God speaks and how He does so. In Verse 11, we are introduced to a concept that has extreme significance. Here God commands the earth to bring forth grass, herb and trees. Note that *earth* here represents the initial materials called earth back in Day 1.

We can return to the *brooding* of God, over the face of the deep. Remember that event, recorded in the Scriptures as having happened only once? We noted in our discussion of Day 1 that the *brooding* represented the initial input of Intelligent Design; the loading of the knowledge into the building blocks of the Universe (*bohu*) that would allow it to grow, organize, and evolve in accordance with the

---

[24] No doubt a significant fraction of the Earth's surface water came from these extraterrestrial sources though not the majority of it. The majority of the water in our oceans today most probably came from the molten rock of the Earth itself. See our treatment of this subject and process in the subsequent section of this work A MIST GOES UP FROM THE EARTH, beginning on pg. 62.

wishes and the will of God. We said as we moved forward, we would see many instances when the elements of the creation, at the command of God, would bring forth, on their own, life and creatures. This is the first of those instances.

## A RARE EARTH

In Verse 9, God seems directly intervening in the process, why? Perhaps the answer lies in the development of Earth in the vastness of space. The Universe is a very dangerous place for life: stellar objects spewing out high energy plasmas in jets, comets and asteroids hurtling through space with planet killing speed and mass, and black hole behemoths, some large enough to devour huge portions of great galaxies.

Yet our Earth, in its protected location within the galaxy and the Solar System and with its unique Moon, atmosphere and magnetic field, shield us from almost all of the dangers. Intelligent life on this planet would not be possible if these things were not just so. The Earth in its early history and the neighborhood surrounding it were far different from the way they are today. I am speaking of the approximate 400 million year period in which our home planet was still coalescing and the 500 or so million years after the ball we call Earth began to take form. The environs of the Earth in these stages were an extremely violent place and indeed they needed to be volatile to get the final mix of the elements just right.

The process of planet building is much different than the methodical and delicate process of creating life. When the correct mix of life-building, knowledge-infused matter is gathered together, the processes of life will begin on their own. However, collecting that material together from inanimate rocks and raw materials hurtling at enormous speeds along ballistic paths through space is quite another matter altogether.

Today our inner Solar System is very much a stable, largely empty place; most of the materials have long since been swept up by the planets or moved out into the Oort and Hills clouds,[25] but when the Earth was forming, the material was local and in constant collision. Like a massive billiard table with trillions and trillions of balls careening off and impacting one another, the contents of the Solar System, then as it is now, followed the precise laws of physics it was instilled with in Day 1. Those

---

25 The Oort cloud is thought to occupy a vast space from somewhere between 2,000 and 5,000 AU (0.03 and 0.08 ly) to as far as 50,000 AU (0.79 ly) from the Sun. Some estimates place the outer edge at between 100,000 and 200,000 AU (1.58 and 3.16 ly). The region can be subdivided into a spherical outer Oort cloud of 20,000–50,000 AU (0.32–0.79 ly) and a doughnut-shaped inner Oort (Hills) cloud of 2,000–20,000 AU (0.03–0.32 ly). The outer cloud is only weakly bound to the Sun and supplies the long-period comets to inside the orbit of Neptune. The inner Oort cloud is believed to have tens or hundreds of times as many cometary nuclei as the outer halo; it is seen as a possible source of new comets to resupply the tenuous outer cloud as the latter's numbers are gradually depleted. The Hills cloud explains the continued existence of the Oort cloud after billions of years. The outer Oort cloud may have trillions of objects larger than 1 km (0.62 mi) and billions approximately 20-kilometer (12 mi) diameter, with neighboring objects tens of millions of kilometers apart. Its total mass is not known, but assumed to be roughly five times that of Earth.

laws however, would not, could not, spontaneously generate a planet as special as ours on their own, given ten times the age of the Universe.

There have been twenty or more parameters identified as necessary for advanced life to form on a planet, among them are those shown in *Figure 3-5*. If we assign a very generous 1 in 10 probability rate for each of these parameters being associated with any one planet, the odds of just such a planet occurring are 1: 1,000,000,000,000,000, just $10^{-15}$, or as spoken one, one thousandth of one, one trillionth. To put that miniscule number in perspective, if the obvious rarity isn't already apparent from all the zeros, it should be noted that there are only approximately one hundred billion stars in our galaxy (100,000,000,000).[26] That means if you were to assume each and every star had a planetary system surrounding it, you would need to gather up ten thousand (10,000) galaxies identical to our own before you could expect to find *just one planet* like the Earth.

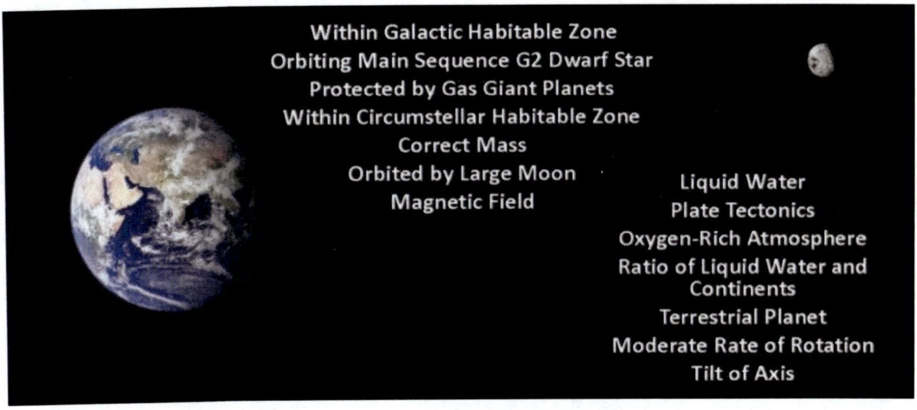


Figure 3-5 *The Rare Earth. Fourteen of the twenty or more critical parameters required for the existence of a planet capable of producing advanced life like our own.*

The Earth is indeed a rare and particular mix of the "*stuff*" of the Universe. It is easy to imagine the gentle hand of God down in the waters, directing the flow with His outstretched fingers until just the right mix coalesced, ensuring life was provided with all it needed to thrive. Then, just as assuredly, with the same gentle touch, He cleared a space for the newly formed Earth to survive. This caring intervention is what is alluded to in Verse 9.[27]

Some may ask and even argue as to why the Universe is so enormous if its sole purpose is the existence of Man on just this one planet. If we are indeed alone in the Universe (and it is a probability that we are not), then the answer may lie in the

---

[26] Some most recent counts attributed to advances in our instrumentation which are now capable of "*seeing*" much less luminous dwarf stars, place the total count of star like objects within our own Milky Way at perhaps 200,000,000,000. Those dwarf stars, however, do not appreciably affect the mathematics of planet probability discussed here.

[27] See <u>A GREAT WIND BLOWS</u> later in this chapter, pg. 69.

numbers we have discussed here. The Temporal Creation is an imperfect sphere and as such it takes an enormous amount of free agent material to bring to pass one perfect place for sons and daughters of God to dwell.

## A MIST GOES UP FROM THE EARTH

*As the Earth cooled down, most of the water vapor condensed from the atmosphere and formed the oceans.* This statement reflects what was assumed by science to have been true for a very long time; however, that assumption proved to be problematic. There are precise constraints placed on what such a cloud of vapor could be like and its size. There are exacting formulas for just how much water it could contain. The parameters are defined by the upper surface of the cloud needing to be hot enough to maintain water as vapor and close enough to the surface of the Earth to prevent the outer layers of the cloud from escaping into space. These parameters precisely set the maximum volume of the cloud. Our oceans today contain 310 million cubic miles of water; the vapor cloud could only contain a tiny fraction of that volume. Our oceans did not form from condensation of a vapor filled atmosphere. We mentioned earlier that water collected on Earth from colliding asteroids. This no doubt was a contributing factor, but analysis of this understanding suggests that this too, even when added to condensation of a vapor cloud, leaves the Earth woefully short of the water it has today. The Scriptures, however, have pointed to an answer all along.

We are told in the Bible: "*there went up a mist from the earth, and watered the whole face of the ground*" (Genesis 2:6). Over 4000 years ago, revelation told that the Earth itself was the source of the water in the Garden; could this also be true of the water on the Temporal Earth? Scientific calculations make it appear so.

Research that began in 1937 has found that water readily dissolves in molten rock (silicates, basalts and granite). These three types of rock are the primary constituents of volcanic lava. The amount of dissolved water molten rock can hold is approximately 5% by weight. As the rock cools and solidifies, the water is pushed out as steam. As it turns out, the quantity of steam expelled in the volcanic period of the Hadean eon by the rock that formed the planet, can account for all the water on the Earth today. This coincides with the scriptural account and would have been completely unknowable to man 4000 years ago. These two points should not come as a surprise to us by now.

## LIFE BEGINS "IN" THE EARTH

"*And the earth brought forth grass, every herb yielding seed after his kind, and the tree yielding fruit, whose seed should be in itself, after his kind…*" (Moses 2-12). Here we have a description of the elements themselves forming into the complex molecules and structures necessary to bring forth life. They form in separate, independent, phyla in a manner unlike the evolutionary tree often described by amateur Darwinians, but self-form they do, in obedience to God as he "*spake.*" This detail is very descriptive in that it identifies a partnership of God with the mortal physical processes and laws He created. It is a common and incorrect understanding that views all creative processes as miraculously occurring. The Universe has, and always has, operated in accordance with these laws and processes just as it does on a regular basis around us today. Remember, God is depicted here in Scripture as creating a Temporal home for his children. This home must include entropy to allow free agency. It must conform to the rules of quantum uncertainty and provide a stable environment for billions of Earth years. It should be expected, given these parameters, that God would instill within the Temporal world an ability to grow, and self-heal when needed and adjust to a changing environment when necessary. In fact, every day we see evidence all around us of the dynamic stability and adaptability of life.

Further evidence that the laws of physics (established at the outset, when God brooded over the face of the deep) play a significant, *if not primary*, role in the creative processes for the Temporal sphere is found within the account of Day 3. You may have noticed that plant life is created here, prior to the arrival of the Sun, the Moon and the stars in the sky. Why is this so, and what does it tell us? It should not be surprising to find that again, we uncover detail that speaks of a factual understanding of science that far exceeds those available to man when the account of Day 3 was first recorded; this remarkable detail provides evidence of the divine origin of the account.

In the mid 1900's, science began to match up fossil evidence of single cell plant life with dated geological strata that positively identified the earliest appearance of life on Earth in the mid Hadean Eon, almost 4 billion years ago[28] (*Figure 3-6*). That life started so early was very problematic to the evolutionary theory. Such an early appearance meant that there was not nearly enough time for it to have evolved on the Earth through the random processes of chance and mutation that had been assumed. It also left huge questions why such life forms existed for 2 billion years without any advancement at all and for nearly 3.5 billion years until the explosion of all phyla existing today occurred in the Cambrian Period. Such an early date for the eruption of life, and the staccato like evolution that followed, led to completely unsupportable theories like Panspermia to explain it.[29]

In contrast to these fantastical theories and in concurrence with the scientific discovery, the Day 3 account correctly places the appearance of life very early on the Earth before the atmosphere becomes transparent, more than 500 million years

---

[28] See studies conducted by Dr. Ruth Blake, Yale University, and by Craig E. Manning, UCLA Department of Earth and Space Sciences; T. Mark Harrison, a UCLA Professor of Geochemistry, Director of UCLA's Institute of Geophysics and Planetary Physics, and University Professor at the Australian National University; and Stephen J. Mojzsis, Assistant Professor of Geological Sciences at the University of Colorado, Boulder.

[29] The evidence for a discontinuous path in the development of life on Earth is so marked that some scientist today embark on a fantastical journey of pure speculation suggesting that life may have started and been completely destroyed many times over on this planet. They imagine *multiple, recurring* panspermic events, driven by the carousel like advance of our solar system around and through the Milky Way. The Earth (and even other Worlds), it is envisioned, through the assistance of asteroid strikes, spreads life throughout interstellar space, frozen and dormant, until the transmitting space debris collides again with our frequently sterilized home planet starting anew where it had left off. See; Neil deGrass-Tyson, Cosmos, A Space-Time Odyssey, 2014, Episode 11, The Immortals.

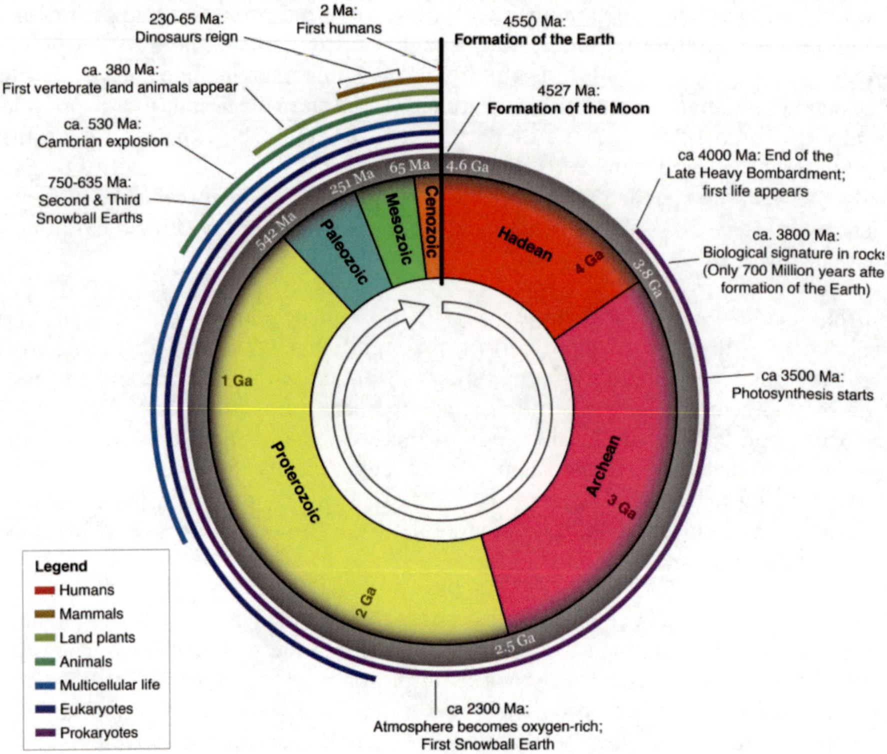


Figure 3-6 *The Eras of the Earth. Life first appears at about two o'clock on this diagram approximately 4 billion years ago, the atmosphere does not clear allowing the full view of the Sun first, then Moon, and then finally stars until about 6 o'clock. Note that life remains unchanged until almost 7 and largely unchanged until 10 when in a period spanning less than about 1 minute, diagrammatically, virtually all of the variation of life we have today springs forth in a single event called the Cambrian Explosion.*

later. The dense early Earth atmosphere was opaque and milky. It acted just like a thick, not very transparent, plastic covering on a greenhouse, letting the light through while obscuring details of anything outside from view. Light through such a covering can be sensed and felt, but its source cannot be clearly seen; shadows of things beyond the screen are vague and undefined. The atmosphere of early Earth was much like this. Though it was radiant during the day, conducting light to the surface where the first life (single cell prokaryotes) could appear and bloom, the Sun and Moon and stars could not yet be seen through the thick heavy blanket of gasses. However, as the prokaryotic life multiplied in the hothouse conditions, it dumped oxygen copiously into the atmosphere in quantities large enough that the atmosphere became oxic and began to clear, allowing the celestial objects to appear. We will provide more on this subject in our discussion of Day 4 that follows.

*I, God, saw that all things which I had made were good* (Moses 2:10, 12); this phrase appears twice in Day 3. We should pay particular attention to it as it serves to identify the coming to a close of an individual creative process and its ultimate acceptance, marking the completion of a particular work of God. It will be seen through both Scripture verses and from what is now known by science, that each of the events concluded with this phrase required great lengths of time and effort to achieve. Here in Day 3, those processes include planetary formation and the beginning of life on Earth itself. In Day 1, a similar phrase marks the close of the creation of the energy of the Universe, the *bohu* from which all other things are composed. As it is used, this phrase attends and describes the patience of the Creator as He waited for the elements to obey;[30] when they did, and they always do, He proclaimed His work and theirs to *be good*.

## SUMMARIZING WHAT WE HAVE LEARNED, *Day 3*

As in Day 2, the focus of Day 3 shifted and narrowed yet again, this time finally coming to rest on the Earth, within the Temporal sphere, "*upon which thou standest,*" where it will remain for the rest of the creative periods. We see the order of the creative events break away from the order of events for the Physical Creation for the second time (*Table 2-2*). This is revealing for it emphasizes that this Creation, from its outset, has necessarily conformed to physical laws established when God infused the basic elements of the Universe with intelligence and information. The creation of the Temporal follows a pattern driven by the natural laws; we will see that the creation of the Physical sphere does not follow that pattern at all.

During this creative period, the seas formed and the continents uplifted. We are informed in the Scriptures that the source of water in the Earth was a mist "*up*" from the ground below, not condensation "*down*" from the sky above. We have learned that this has been proven correct scientifically.

The appearance of grass and herb and trees (plant life) in Day 3, corresponds precisely with the scientific record and bespeaks an advanced understanding of the biological and geo-chemical history of the Earth far beyond what was known to Man 4000 years ago. We are shown that life formed on Earth under the protective shield of an opaque atmosphere, rich in nutrients. The physical laws of the creative process of planet forming are shown to be in place and monitored by a loving and caring God building a habitable environment for Man.

His direct intervention in the early stages of Day 3 are properly understood in the context of a creator bringing disassociated movements of distantly placed, naturally formed, objects into proximity assembling, through perhaps the act of collision, the precise mix of materials (water, minerals, gases) to allow the elements themselves to bring forth life on the Earth as they had been pre-programmed to do early in Day 1.

Again, as before, the incredible detail revealed by modern science is all here in an ancient text penned some 4000 plus years ago.

*This ends the account of Day 3.*

---

30 "And the Gods watched those things which they had ordered until they obeyed." (Abraham 4:18)

| DAY 4 — Moses 2 |
|---|
| 14 ¶And I, *God, said*: Let there be lights in the firmament of the heaven, to divide the day from the night, and let them be for signs, and for seasons, and for days, and for years;<br>15 And let them be for lights in the firmament of the heaven to give light upon the earth; and it was so.<br>16 And I, God, made two great lights; the greater light to rule the day, and the lesser light to rule the night, and the greater light was the sun, and the lesser light was the moon; and the stars also *were made even according to my word.*<br>17 And I, God, set them in the firmament of the heaven to give light upon the earth,<br>18 And the sun to rule over the day, and the moon to rule over the night, and to divide the light from the darkness; and I, God, saw that all things which I had made were good;<br>19 And the evening and the morning were the fourth day. |

Day 4 details the appearance in the sky of the Sun, Moon and Stars, in that order. We are told that they have multiple purposes, giving light to the Earth during both the day and the night and for providing a basis for recording the passage of time. At this point, we will return to why these celestial objects are spoken of here in the Creation account and not earlier, for we know that the Sun and certainly the vast majority of the stars that are visible from Earth, *not to mention the trillions and trillions*[31] *we cannot see from our Earthly vantage point,* date from ages well before the Earth was formed.

An uninformed reading of the words in the Day 4 account in English could lead to some to conclude the Sun and stars are "*created*" here in Day 4. However, the Hebrew *way-ya-as* (וַיַּעַשׂ) (rendered, *made* in the KJV) expresses something much different than creation; in its correct context here, it properly means, *established*.

The Hebrew *me-o-rot* (מְאֹרֹת) translated as "*light(s)*" in these verses can best be rendered "*luminary*." This is significant for two reasons. It is proper for us to understand that God is not creating the objects or the light they provide at this time but that He is referring to them as *providers* of light to the Earth. To this extent, the inclusion of the Moon here is informative and crucial. The Moon does not *produce* its own light like the Sun and stars, it is *not a source of, but merely reflects* light. This provides us with a solid understanding that the events in Day 4 are to be understood to be as observed from an Earthly perspective, because the Earth, and

[31] Some good estimates place the total number of stars in the observable Universe at a Septillion, or $10^{24}$.

only the Earth, is close enough to the Moon to see the Moon as a *luminary* through its reflectivity.

Let us step way back in time for a moment and look at the Solar System in its earliest days of development. Today our nearest neighbor star, Proxima Centauri, is 4.22 light years away, or 271,000 Astronomical Units. A single Astronomical Unit (AU) is the average distance of the Earth from the Sun, so we see that in terms common to our usage, though Proxima Centauri is by far our *nearest* neighbor, it is indeed *very far away*.

Define a sphere 50 light years in radius out from the Sun and you will enclose roughly 1400 star systems containing some 2000 stars. Our Solar System formed from but a fragment of a molecular cloud that was incredibly much larger than that, some 65 light years across. The fragment, our fragment (the pre-solar nebula), was huge in-and-of itself at over 3 light years in diameter. The density was extremely low with a total mass just a little over that of the total mass of the Sun spread evenly throughout this immense region. Over time, gravitation worked to collapse our fragment, forming the planets, moons and the Sun of the system we are familiar with today.

When the process of gravitational collapse had reduced the fragment to a dense core of less than 20,000 AU, its angular momentum was conserved and the nebula began to spin, slowly at first, then faster and faster, eventually flattening the sphere into a disk. Over the next 100,000 years, this collapsing, spinning, and flattening process continued until a protoplanetary disk (about 200 AU in diameter) had formed around a hot, dense embryonic proto star that was about to ignite its nuclear furnace and become the Sun.

Within the protoplanetary disk, the planets began to form through accretion; this process gave rise to the appellation, *accretion disk*. Inside a radius of 4AU, the disk was too warm for molecules like water and methane to condense, so the planets that began forming there would do so from compounds with high melting points like iron, nickel, aluminum, and rocky silicates. Further out, beyond the *frost line* that occurred at about 5AU, at a point between the orbits of Mars and Jupiter, the giant (Jovian) planets accreted.

Jupiter, Saturn, Uranus and Neptune formed from volatile icy compounds; for beyond the frost line (*sometimes called the dew line as well*), the disk was cool enough for those materials to remain as solids. The comparative size of the Jovian giants to their diminutive sister terrestrial planets is a direct result of the relative abundance of the ices that formed them when compared to the rarity of the metals and silicates that formed the rocky planets. The compounds from which the rocky planets are formed are uncommon in the Universe, comprising just 0.6% of all visible matter. Precisely in line with this percentage, we find that in our Solar System the four gas giants together comprise nearly 99% of all the mass orbiting the Sun.

As the planets began to form, the Sun ignited bringing us to a point in time some 4.5 billion years ago, just before the beginning of Day 3. In the last section, we discussed the forming of the Earth in its proper sequence in Day 3, after the processes of the formation of the Solar System we have covered here. With this understanding of the accretion disk and what was occurring in the environs of space which surrounded the infant Earth, we are up-to-date and now prepared to talk about the processes we will discuss in Day 4 of the Temporal Creation account.

## THE MOON IS FORMED

The Sun and Stars predate the formation of the Earth, but the Moon does not. It is thought to have formed through one of three processes occurring early in the Hadean eon, approximately 30 to 40 Million years after the Earth first had a form. Three of the most likely theories for the formation of the Moon are detailed here:

- *The Condensation Theory:* The Moon and the Earth condensed together from the accretion disc that formed the Solar System.
- *The Colliding Planetesimals Theory:* The interaction of Earth-orbiting and Sun-orbiting planetesimals (very large chunks of rocks like asteroids) early in the history of the Solar System led to their breakup. The Moon condensed from this debris.
- *The Ejected Ring Theory:* A planetesimal the size of Mars struck the Earth, ejecting large volumes of matter. A disk of orbiting material was formed, and this matter eventually condensed to form the Moon in orbit around the Earth.

A detailed comparison of the properties of Lunar and Earth rock samples has placed very strong constraints on the possible validity of these hypotheses. For example, if the Moon came from material that once made up the Earth, then Lunar and Terrestrial rocks should be much more similar in composition than if the Moon was formed somewhere else and only later was captured by the Earth.

Analyses indicate that the abundances of elements in Lunar and Terrestrial material are sufficiently different to make it unlikely that the Moon formed directly from the Earth. Generally, work over the last 10 years has essentially shown the condensation theory to be unlikely. At present, the Ejected Ring hypothesis, which states that the Moon was formed from a ring of matter formed of ejecta from the proto Earth as well as the remains of a large object that collided with the Earth, is the favored hypothesis. However, the question is not completely settled and many details remain to be accounted for.

## THE EARTH IS SHROUDED IN DARKNESS

There are two complementary obscuring factors that contributed to the overall effect of rendering celestial objects out of sight from the surface of the Earth during its early history. At first, the accretion disc itself would have completely obscured the Sun and Stars. In the earliest stages of the Earth's existence, the very early Hadean eon, the Earth had little to no atmosphere for various reasons, so the atmosphere itself would have not prevented the viewing of the celestial objects, though the accretion disc would. As the accretion disc dissipated, however, the Earth's atmosphere evolved through several iterations, some more substantive than others and resultant from many processes. By the middle to late Hadean, a dense, opaque atmosphere was in place, effectively (now perhaps more than ever) obscuring the Sun, Moon and stars from view at the planet's surface. These two factors working in concert, one stepping in when the other began to wane, kept the Earth in an effectual shroud as it would be perceived by one who would see it in vision through revelation.

## A GREAT WIND BLOWS

Stars in their early stages (like our Sun was when the Earth was yet infant) lose huge quantities of mass through an explosive developmental period known as the T-Tauri. Our Sun would have undoubtedly undergone such a transitional phase that extended for several million years. The ejecta of the T-Tauri phase produced a solar wind so fierce that it blew all the residual elements of the accretion disc beyond the farthest reaches of the Solar System. The winds, however, had a devastating effect on much more than the accretion disc. The young atmospheres of the terrestrial planets were blown away as well. The Earth has a life supporting atmosphere of 20% oxygen and 80% nitrogen so the timing of the Sun's T-Tauri wind was critical, occurring not too early and not too late, but like in the fabled tale of Goldilocks, its timing was *just right*.

The Sun's T-Tauri period blew away accretion materials totaling two to three times the mass that remains orbiting our star today. If that had happened too soon, the materials in the accretion disc would not have had time to coalesce and the frost and ices in their locales would have been lost into interstellar space before they could be swept up to become part of the still forming inner planets. The result likely would have been an Earth without oceans and seas, perhaps without water at all.

The T-Tauri wind that comes after planet forming is less destructive to a planet, but the timing is still critical in that it cannot come too late.

Most of the water and almost all of the gasses comprising the atmosphere were buried within the Earth's mass early on and as such protected from the T-Tauri wind; this is a crucial requirement and narrows the *Goldilocks-like* timing of the Sun's T-Tauri to a critical period. The source of our oceans and atmosphere are, to a large part, gasses and vapors released during a prolonged period of volcanism occurring in the Hadean eon, some 4 billion years ago. Because the T-Tauri of the Sun occurred in that narrow range, *after* agglomeration of the Earth, but *before* the major volcanic outgassing of the early Hadean, the water that comprises our oceans and the volatile gasses that make up our atmosphere were protected from the solar wind.

If that period of volcanism had occurred earlier or the T-Tauri wind later, after the outgassing had occurred, the Earth would have been stripped of its water and atmospheric gasses. Mars today is perhaps void of water and has a very thin atmosphere because the evolution of the planet did not align well with the T-Tauri phase of the Sun; it's outgassing happened much too soon. Because of its much smaller size, makeup, and distance from the Sun, Mars cooled much quicker than the Earth, ending its volcanic period before the Solar wind began to blow. In this respect, the Goldilocks timing of our home planet is the result of the Earth having just the correct mass and just the correct distance between it and the Sun.

The average distance for the Earth from the Sun is 93,200,000 miles; this varies by only just under 2.5 million miles throughout the course of our year, which is a difference of only 3 percent. This small variation means that the Earth's orbit is practically circular. The Orbit of Mars, the planet next door, is quite elliptical varying in distance from the Sun throughout its annual course by 31 million miles. If the Earth varied this much in its mean orbital path, our crust would pan fry each January when we reach our closest approach to the Sun each year. This is critical for life on Earth. If our distance from the Sun were just 6.2 million miles closer (just 20% of the variance seen in the orbit of Mars, or a 7% change in our mean distance), water vapor would not be able to condense on Earth, resulting in no rain and there would be no oceans.

The science of the early Hadean eon is by far the most speculative we have broached in this treatise. Tremendous variables are at play in understanding the earliest stages of the Earth's development and composition before life appears. Admittedly, much of the variables are completely uncertain. We can be confident about the T-Tauri wind, to a lesser extent our speculation regarding the difference in the timing of volcanism periods on the Earth and Mars. We can assume with some certainty that the Earth did have at least a minimal primordial atmosphere, which was swept away by the same solar wind that swept the accretion disc away. I do not want to speculate too much about how the Earth remained in obscurity through the almost first billion years of its existence, but I feel confident, given the accurate understanding of science shown by the Genesis 1 account so far, that the scriptural record can be trusted to fill in these gaps of scientific information.

We will begin again here with our telling of Day 4 of the Temporal Creation at a much more certain point scientifically, 4-3.8 billion years ago.

The Earth's earliest permanent atmosphere was produced by volcanic gasses, rich in carbon dioxide. Very early on in Earth's history (almost immediately on a geologic time scale), plant life appeared, and photosynthesis began to produce much of the atmospheric oxygen we breathe. As the oxygen built up, UV radiation converted it to ozone. Ozone has two very practical qualities; it is transparent to visible light, and it absorbs UV radiation. The forming ozone layer protected life on the surface from lethal UV radiation while the warmth and energy of the Sun could pass through making it available for photosynthesis. During the Proterozoic eon, photosynthesis by cyanobacteria transformed an opaque atmosphere into one that was transparent very quickly.[32] Fossil abundances of cyanobacteria increase about 2.3 billion years ago, and it is believed this correlates with the time the atmosphere became transparent allowing, first the Sun, then the Moon and finally the stars to appear. Finally "*establishing*" (*way-ya-as*, וַיַּעַשׂ) sunrises and sunsets and the periods of both day and night on Earth as the Day 4 verses originally detailed.

## CELESTIAL CLOCKWORK

Why six days, with a seventh for God to rest? Why is a week even used to measure time? Where did this unit of time come from, if not divinely dictated through the detailing of the creation story itself? Without the creation story driving the period of a week forward, there would have been little to establish the week as a measurement of the passage of time. How would man get to the concept of a week, six days of work and a day of rest, on his own accord? The concept is counterintuitive.

Hunter-gatherers don't perform any tasks based on a seven day cycle; they hunted and gathered every day to survive. Counting the cycles of the Moon from full to new and back to full again, would be helpful and more likely as they attempted to anticipate the seasons and the cycle of fruiting and herd migration, but those are 28 or 14 day periods, not seven.

As civilization began and agriculture became the primary method of feeding the populations, the same would hold true. Farming and the domestication of animals requires work and equal quantities of it every day. The Moon would be helpful in the same way as before in planning, planting and gathering the crops, but as before that would be cycles of 28 and perhaps 14, not seven with a day of rest.

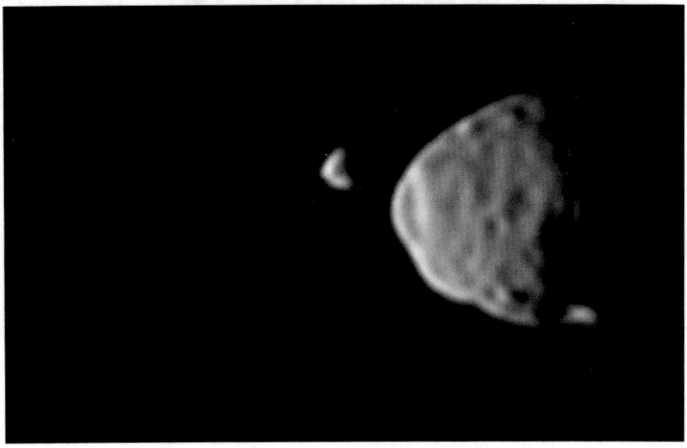

Figure 3-7 *Deimos and Phobos As seen by Curiosity in extreme magnification from the surface of Mars August 1, 2013.* (NASA)

[32] This process of oxygenation of the atmosphere would likely have happened very quickly on the geologic scale. To illustrate this, consider the following. The current rate of photosynthesis for the plant life on the Earth today can, in just a little over 2000 years, produce a quantity of oxygen which exceeds the total found in our atmosphere.

City building and canal work, armies, and the administration of kingdoms all were daily tasks with no need to measure time in weeks. Base ten numbering systems developed because we have 10 fingers, so why not ten day periods of measurement? They don't exist. If the story told in the Genesis 1 account was a fabrication of man, why six days with a day of rest?

We have already seen that the imprint of a clock is embedded into the very fabric of the Universe with the CMBR, but the celestial clockwork described here in Day 4 is something different and independent of the universal timepiece, yet it remains confirming evidence of the 6 day creation account. It is a consequence of a set of remarkably unusual circumstances that the Biblical week of seven days can be tracked by the revolutions of the Moon and the Earth.

The Moon is unique in the Solar System for its size; it is much larger than should be expected when related to the Earth's volume. It is the only true moon in the inner Solar System. In contrast, consider the two "*satellites*" of Mars, Phobos and Deimos (*Figure 3-7*), which are actually captured asteroids and often improperly termed *moons*. The Martian "*moons*" are in fact extremely temporary in nature and for most of the history of the Solar System were not to be found in orbit around the Red planet.

Figure 3-8 *Sun and Moon appear the same size, approximately, as seen from Earth.*

Deimos orbits at just over 5800 miles from the surface of Mars and in less than 50 million years will either break apart forming a ring or crash into the Martian surface. Deimos revolves around Mars in just about 7 hours, meaning it rises in the West three times at differing times each Martian day. Because Phobus is so small, from the surface of Mars, it appears no larger and brighter than a star (like a manmade satellite does from Earth). Phobus' orbit is nearly geosynchronous at 1.25 Martian days, so its movement through the sky is far less noticeable. Both of Mars' moons are forever out of view from substantial areas of the Martian surface. All of these facts, and a few more, render the moons of Mars incapable of providing a means for telling the passage of time on Mars.

In contrast, the Earth's moon is perfectly suited for the task of telling time in many ways. The Moon is of a precise size, its plane of orbit and distance is precisely correct to make a total solar eclipse possible (*Figure 3-8*). Although the Sun's diameter is about 400 times larger than that of the Moon, the Sun is also about 400 times farther away. So the Sun and Moon appear nearly the same size as seen from Earth. This ratio of size and distance is a function of precise timing. The needed parameters are changing with time, the size of the Sun, the Earth-Sun distance, and the distance of the Moon.

With each orbit, the Moon is getting minutely farther from the Earth, by about 4 centimeters every year. This means, in past eons, the Moon completely eclipsed the Sun more often because the Moon in Earth's sky always appeared bigger than the Sun. But as the eons continue to pass, the apparent size of the Moon will continue to grow smaller. Eventually the Moon will not appear large enough to ever be capable of completely obscuring the Sun.

At this point in the history of the Solar System, the Earth orbits the Sun once very nearly every 365 days. Though not precisely divisible into seven day periods, this annual circle is remarkably close.[33] We are all well aware of the famous ¼ day each year which necessitates an added day in February almost every four years. The need for a leap year is not instinctively apparent; however, without it, the calendar year would lag behind the seasons as the Earth's northern-hemisphere-tilt towards the Sun would move from January into December, then on to November, eventually putting the northern winter into the months of summer. It takes the Earth approximately 365.242199 days (365 days, 5 hours, 48 minutes, and 46 seconds) to circle once around the Sun. However, the Gregorian calendar has only 365 days in a year. If a day wasn't added on February 29, nearly every 4 years, we would lose almost six hours off of our calendar every year, taking nearly 24 full days off the calendar every 100 years.

In the end, we find that our position within the Solar System is wonderfully tuned at this moment in time. The Creation account of Genesis 1 details six creative periods called days; the CMBR records the age of the Universe in the original time reference frame as six days. Our unusually large moon, the only true moon in the inner Solar System, at this time in its orbital history, counts out months as fractions of our annual course around the Sun. Each month divides into weeks, 52 in a year. This is a snapshot in time for a Solar System that was once a much different place and will be much different in the future.

The timing of all of these things, relative sizes and distances, are a violation of a golden rule of Physics, the Cosmological Principle. Despite the principle which *insists that there is no preferred location in space for an observer*, we find these alignments. These are not by chance and bespeak truths about the intentional placement of man within this Universe, on this planet, at this point, and at this time.

---

[33] The rotational speed of Earth (today 24 hours but only just over 18 hours some 1.4 billion years ago), the diameter and eccentricity of its orbit, gravitational tugs of the Sun, the Moon, and the other planets, particularly Venus and Jupiter, all ever changing, have an effect on this timing. The most recent studies detail that the eccentricity of Earth's orbit varies periodically with a time scale of about 450,000 years, from almost zero (0.0167) (a nearly circular orbit), as it is now, to about three times its current value (0.050).
Perhaps at a defined point, *perhaps of some eternal significance*, the orbital period was precisely 364 days (or some other number equally divisible by 7), perhaps even marking the precise timing of the fixed orbit of a Celestialized Earth, or the Earth of the Physical sphere on which the Garden is located.

From a perspective clearly taken from the Earth itself, we see the events of Day 4 unfold. This presents a rapid history of the Earth as it went through the most dramatic changes of its geological history. The events told in Day 4 mark the culmination of the creation of the Solar System itself. The vision is grand and epic.

Here in this work, we backtracked just a bit in the chronology of the creation of the Solar System allowing us to put in place a backdrop for the information we subsequently (and have yet) encountered. The backdrop helped us prepare for the rest of the story that follows. The *backtracking* we accomplished was a necessary pause, like the taking of a deep breath between chapters of an unfolding story.

With an understanding of the forming of the Solar System from an extended nebula that gave way to the Sun and planets, we returned to the details given in the Day 4 accounts of Scripture.

As we did in Day 3, we again saw the order of the creative events break away from the order given for the events of the Physical Creation (*Table 2-2*). The divine record reveals (if however counterintuitive) that plant life predates the appearance of the *light* of the Sun, Moon and Stars in the sky. We are aware, as the record of Day 5 will show, that animal life is yet to appear. This information presents a very interesting progression: flora first, then celestial objects in the sky, and then back to fauna. This progression is telling of an advanced understanding the storyteller had of things that were unknowable through any natural or intellectual means.

We are left wanting for a complete scientific theory that covers all of the events detailed in the record of the Day 4 account. However, even with the shortfalls in science, we found that both sides of the gap in reliable theory correlated precisely with the events of Day 4 as told. We feel assured the scriptural record revealed to us by God millennia ago, accurately filled in the things missing in our modern scientific understanding.

*This ends the account of Day 4.*

## *Day 5 (750 Million to 250 Million years ago)*

| DAY 5 — Moses 2 |
| --- |
| 20 And I, *God, said*: Let the waters bring forth abundantly the moving creature that hath life, and fowl which may fly above the earth in the open firmament of heaven.<br>21 And I, God, created great whales, and every living creature that moveth, which the waters brought forth abundantly, after their kind, and every winged fowl after his kind; and I, God, saw that all things which I had created were good.<br>22 And I, God, blessed them, saying: Be fruitful, and multiply, and fill the waters in the sea; and let fowl multiply in the earth;<br>23 And the evening and the morning were the fifth day. |

In Day 5 we are first introduced to animal life, as it appears on the Earth initially developing in the sea. As before with the substance of the Earth itself bringing forth plant life in Day 3 (as if it had within it the very powers of creation and self-generation), we see the waters bringing forth life on their own, by and through processes that were resultant from the information God had instilled into the *bohu* at the outset of Day 1.

Throughout the detailing of the creative events of the Genesis 1 accounts, the vantage point of those recording the revelation was from their personal perspective in place and time as they were moved through the vision by the hand of God. They looked back on events of the past, understanding what they saw in contexts that were familiar to them. They interpreted and understood all that they viewed with an eye and realization that what they saw would progress and become what they were accustomed to in their time. This explains the inclusion of trees and grasses in his recording of the initiation of life in Day 3; the writers had "*seen*" (*no doubt the science was explained to them*) that the earliest form of prokaryotic life sustained itself by the same photosynthetic processes used by the more advanced forms of life familiar to them in the day they lived. (*It must be noted here that the earliest forms of life were microscopic, and though they still exist in similar form today, to Man of 4000 years ago, they would be completely unknown and undetectable individually. That the recorders of the Genesis 1 account could even associate such life with the grass and trees around them, as they so correctly did, is another proof of the divine source of the knowledge they recorded.*) Here in the Day 5 account, we see a grouping of animal life that bespeaks a similar divine insight. Fish and fowl, creatures that were familiar to the recorders of the events of Day 5, were the resultant forms of life that started in the seas long before mammals and other land based life ever took form. This insight could not have been intuitive.

The appearance of *great whales* seems out of place here, but that may be just because of the perspective of the viewer or the definitions imposed by their language. A whale, to the recorders of the Day 5 events, may just have been considered a large fish (we have fossil records of extremely large fish from this era).

The recorders most certainly would not have known such huge creatures to be mammalian, as today we understand them to be. In this sense, any large sea creature would have been understood to be a whale. This same translation and understanding would have been at play in the minds of all who subsequently rendered this verse into the language of their day. The same would be so for all that did so down to the most recent centuries when through our understanding we have learned to segregate sea mammals (whales) from gill breathing fishes.

The word "*created*" is only used in two separate points in the six creative periods. Count them – first, here in Day 5 and later in Day 6. The word created is used in the preamble, and again on day seven, but it is only used at two points in the context of the actual creative periods; this is of extreme significance and importance.

The KJV translators rendered the Hebrew word *Bara* (*Barau*) (בָּרָא), as *created* in the account of Days 5 and 6. *Bara* (*Barau*) is used only 45 times in the whole of the Hebrew Bible. The paucity of its use indicates the special nature of the word.

Although many verbs used to describe God's activity in bringing the Temporal Creation into existence are used elsewhere in the Bible to describe the activities of others, *bara* is distinguished in that it is *exclusively* used to describe an activity of God. By this distinction it is understood that man cannot reach up to the powers inherent in this word, for it describes something only God can do. Throughout the six creative periods, God: *forms*, *broods*, *blesses*, *sets*, *calls*, *makes* and *commands* (all words used in other associations as well), but only twice is *bara* used. This is significant in that it is associated, and *only* associated, with the infusion of the Spirit of God into life. We will detail this concept further in our discussion of Day 6.

In the Day 5 account, life continues to bring forth after its kind and God blesses them that they may be able to do so. Again we see the phrase, "*I, God, saw that all things which I had created were good*;" as before, this brings closure to a creative process that required a great length of time to accomplish and complete. The process of developing life in the sea on Earth began approximately 2.3 billion years ago, with the first Eukaryotic life forms, but the first animals on Earth appeared in the sea precisely on cue when the period of Day 5 begins, 750 million years ago. The earliest ancestors of birds were the raptor Therapods, fully developed by the early Mesozoic, but they evolved from creatures first formed in the sea tens of millions of years earlier.

SUMMARIZING WHAT WE HAVE LEARNED, *Day 5*

With the first true animals appearing in the waters in Scripture on Day 5 and in the fossil record 750 million years ago, we again see that the Genesis 1 account provides a remarkably accurate account of the true record of the Earth that aligns with scientific understandings of today.

*This ends the account of Day 5.*

## *Day 6 (250 Million to the Present and Beyond)*

| **DAY 6 — Moses 2** |
| --- |
| 24 ¶And I, *God, said*: Let the earth bring forth the living creature after his kind, cattle, and creeping things, and beasts of the earth after their kind, and it was so;<br>25 And I, God, made the beasts of the earth after their kind, and cattle after their kind, and everything which creepeth upon the earth after his kind; and I, <u>God, saw that all these things were good</u>.<br>26 ¶And I, God, said unto mine Only Begotten, which was with me from the beginning: Let us make man in our image, after our likeness; and it was so. And I, *God, said*: Let them have dominion over the fishes of the sea, and over the fowl of the air, and over the cattle, and over all the earth, and over every creeping thing that creepeth upon the earth.<br>27 And I, God, created man in mine own image, in the image of mine Only Begotten created I him; male and female created I them.<br>28 And I, God, blessed them, and said unto them: Be fruitful, and multiply, and replenish the earth, and subdue it, and have dominion over the fish of the sea, and over the fowl of the air, and over every living thing that moveth upon the earth.<br>29 ¶And I, God, said unto man: Behold, I have given you every herb bearing seed, which is upon the face of all the earth, and every tree in the which shall be the fruit of a tree yielding seed; to you it shall be for meat.<br>30 And to every beast of the earth, and to every fowl of the air, and to everything that creepeth upon the earth, wherein I grant life, there shall be given every clean herb for meat; and it was so, *even as I spake*.<br>31 And I, God, saw everything that I had made, and, behold, <u>all things which I had made were very good;</u> and the evening and the morning were the sixth day. |

As the narrative of Day 6 unfolds, we are told that animal life is brought forth *from* the earth in the same manner that sea life was brought forth *from* the water. This should not be construed as a separate and distinct process from the appearance of sea life, but an extension of it, just as the evidence we see in nature discloses. The *earth* spoken of in Verse 24, is not so much associated with the surface of dry land (the land masses of the Earth) as it should be understood to be the initial intelligent material, of which the Earth is composed, that *bohu* which was created by God back in Day 1.

What is perhaps most *significant* about the appearance of these animals in Day 6, is that mammals and the advanced forms of life on the Earth today are properly sequenced in the creative periods texts, correlating precisely with the fossil record.

Remember this important detail, the Scriptures (some would say ignorant mythology) records the appearance of life in a tiered order; why is that so? Why not describe the creation of life as one single event occurring on one single day of creation? This was done with purpose and through an understanding that could not have been available to mankind of 4000 years ago by any natural means.

To fully understand that animal life, both within the sea and on land, is a coherent continuous whole, it will be necessary for us to turn back to the creation (*bara*) of animal life spoken of in Day 5. The Genesis 1 account recorded that specific act of God using the Hebrew word *bara*. This word is only used twice in the creative record, once when animals are created within the sea, and again later when man is created; this is a *very* significant observation. We will explore the meaning of the *bara* being used just twice, and when, later in this chapter.[34] But we must note here that by mentioning *bara* only once, and not twice, in relation to animal life, sea and land animals are shown by the ancient text to be understood as part of a singular spectrum of life; this is of *extreme* significance.

To be clear, *creation* is an ongoing process in a very real way. Each new life created, each hatching or birth, in its millions and billions of iterations, is associated with organized, Temporal matter becoming a living Soul. What is spoken of in Day 5 as creation of animals, and continued here in Day 6 for Man, is the inaugural event of each *unbroken* and *continuing* chain.

## LIFE, A SINGLE STRAND OF PEARLS

What we propose here, and contend the Scripture accounts detail, is not a-teleological Darwinian evolution as it has come to be understood by the masses. The Genesis 1 account, in its subtleties, suggesting the water and earth itself brought about life, it's methodical, structured order of what type of life formed and when, through its thoughtful use of the *bara* at specifically critical points along the way, does describe a developmental process for life on Earth. The scriptural account, however, details a process with marked differences from the gradual phyletic evolution accepted as fact by so many of the uninformed today.

As the fossil record continues to expand and fill in, it has become increasingly clear that the gradual evolution of one life form to that of another (*Figure 3-9 A*) is rarely, *if ever*, seen. Though natural selection is a valid and thoroughly descriptive theory for the processes we see operating in the world around us today (genetic modifications in agriculture and medicine, natural immunities, developed resistance to antibiotics, speciation in the plant and animal communities, both natural and induced), it does

---

[34] See our discussion, IN THE IMAGE OF GOD on pg. 84.

Figure 3-9 *Morphologic Chart of Evolutionary Processes. Phyletic gradualism (A) would consist of steady evolutionary change in small steps; in contrast to the much more punctuated process shown in the fossil record (B).*

not, nor has it ever, adequately described the origin of life itself or even phyletic gradualism.[35]

Because of the complete lack of fossils showing a gradual evolutionary process, science today clearly identifies a more staccato like process of evolution called *punctuated equilibrium*. Even this modification to the theory is not truly indicative of what is shown in the fossil record. Theorists advocating punctuated equilibrium, in doing so, are still attempting to indicate a sort of phyletic evolution when no evidence for that process is known. What we do see in the fossil record on the large scale is more closely illustrated in (*Figure 3-9 B*).

The development of life on Earth, as recorded in the fossil record, describes a discontinuous path, with extremely varied rates of change and individual phyla springing forth without a common ancestor in the same period. These independent events are found multiple times in the fossil record with the most widely known and understood being the Burgess Shale formation.

The Burgess Shale records the primary event in the development of animal life on Earth. In a remarkably short period of time (by gradual evolutionary standards), the Cambrian Radiation (or more commonly known as the Cambrian Explosion)

---

[35] Evolutionary Biologist Richard Dawkins argues that such constant-rate gradualism is not present in the professional literature; thereby the term only serves as a straw-man for punctuated equilibrium advocates. In his book The Blind Watchmaker, Dawkins argues against the idea that Charles Darwin himself was a constant-rate gradualist. Dawkins makes the following claim, "*in the first edition of On the Origin of Species, Darwin stated that 'Species of different genera and classes have not changed at the same rate, or in the same degree. In the oldest tertiary beds a few living shells may still be found in the midst of a multitude of extinct forms... The Silurian Lingula differs but little from the living species of this genus.'* "

produced almost all of the animal phyla that have ever been on the Earth. This event was far too short for random mutations to be the source of the development of all of the diversity of life on the planet today; yet the fossil record clearly shows that in a mere 20 million years (a period of time representing an equivalent of what would be just shy of 2 minutes in a 24 hour history of the Earth), the Cambrian Radiation did just that.[36]

***In an average lifetime, the human body will produce proteins numbering in total on an order of magnitude equivalent to the total number of seconds the Universe has been in existence.***

It should come as no surprise to know that the Burgess Shale formations are dated to a time beginning around 542 million years ago, in the Cambrian Period. This placement in time correlates precisely with the *bara* of animals recorded in Day 5. The record of creation of animals in Day 5 and Day 6 clearly shows a development of life on Earth from its origination as single-cell photosynthetic plant forms, which appear on Earth in Day 2.

The fossil and scriptural accounts align perfectly, and they both detail a common heritage for all life on Earth.

The Genesis 1 account clearly describes life, all of it, forming from the intelligent *bohu* infused with knowledge (law) by God in Day 1. We know that all life is composed of the same materials and that advanced life, animals and Man share a common creative experience, that of the *Ruach* of God. Within every cell of our bodies, plant, animal and Man, are the evidences of this common heritage; we all share the same "*stuff*" of life.

Each individual cell in our bodies produces approximately 2000 proteins every second. Multiply that by 74 trillion for the number of cells in a typical human,[37] and we become aware that there is an incredible amount of proteins being produced within us. In an average lifetime, the human body will produce proteins numbering

---

[36] See Appendix, The *Instantaneous* Development of Advanced Life, pg. 542.

[37] This number represents a total count that includes single cell bacteria and other organisms that call our digestive tract home. Some estimates of the total number of these have them accounting for over half of the cells in our body on any given day, though most thorough studies represent the numbers of the symbiotic partners within us to be 20-30 percent of what we are. This fact, not often discussed, casts a different light of the concept of "*me*," what we consider to be ourselves. Our individual identity is actually a collection of organisms; "*I*," therefore, is really "*we*" or "*us*." Life is in fact much more interdependent in many more ways than we think.

in total on an order of magnitude equivalent to the total number of seconds the Universe has been in existence.[38]

Despite these vast numbers in production and the immense probability of variety that could be produced,[39] there are only 100,000 different proteins in use in the human body, and most of these are in use in other animals and plants as well. Protein molecules are often very large and are made up of hundreds to thousands of amino acid units. Notwithstanding this plethora of immense structures, there are only 20 amino acids used by animals and humans to build these complex proteins. Plants use the same 20 amino acids and add a little over 100 more to their mix of the "*stuff*" of life. [40] The point of all of this is that despite the enormous amount of diversity, the tremendous amount of production and the nearly infinite amount of possibilities, life is very similar. This fact is not to be overlooked and is not by chance. The probability that two separate organisms would randomly produce nearly identical proteins is less than one in $10^{103}$.

Let's return to the age of the Universe to put this number in perspective. The total age of the Universe in seconds is in an order of magnitude of $10^{17}$. Random events occurring once each second would require a time frame more than five times longer than the age of the Universe itself to spontaneously generate identical proteins in two distinctly individual organisms. So indeed, the similarity of proteins used by diverse life forms is *very strong statistical evidence* that life on Earth has a common ancestor, but as strong as this evidence is, it is not as compelling as the statistics surrounding the fact that all life forms share vast similarities in DNA and RNA.

Because these structures are so complex, it is statistically impossible that individual life forms from the three separate animal, fungal and plant kingdoms could have developed them separately and individually. It is clear all life shares a common heritage formed of the same materials and built upon one form of life after another in a deliberate and intelligent manner. It is foolishness for Bible believing individuals to turn a blind eye to facts such as these.

That all life on Earth shares a common heritage should pose no problem to a reader of the Genesis 1 account. We have seen that in Day 1 God *brooded* over the primal "*stuff*" of the Universe (*bohu*) and infused it with knowledge to self-organize,

---

[38] The number of cells in the human body that are actually the stuff of the human form (discussed in the preceding footnote) don't actually affect this concept at all; whether the count is 30 trillion or 74 trillion, does not matter. Here we are discussing trillions, times thousands, times billions; the significant numbers don't change the math at all; it is all the zeros that count.

[39] An *extremely* conservative estimation of different protein structures that can be produced (innovability), based on amino acid chains just 100 in length, results in $10^{123}$ unique configurations.

[40] Note here that as the complexity of the structures of life increases, the amount of separate amino acids in use diminishes. This may be attributed to the diversity of plant life and its attendant antiquity in comparison to much more junior animal life.

repair, rebuild, adapt, and progress according to His laws in their own course and in their own time. This is intelligent design as it is properly applied and understood. [41]

The Genesis 1 account details a punctuated description of the development of life that continues to be expressed in the fossil record. The punctuations in development and the appearance of complex, organized life in bursts, far too short to be explained by random mutations, remains problematic to science and in fact unanswerable. The Scriptural account details these events in their correct chronological sequence; that this was done in a record 4000 years old is evidence of the divine nature of the revelation and by association the truthfulness of the correct science that verifies the text.

The appearance of animal life on the Earth is completed and set off from the verses to follow by: "*And I, God, saw that all these things were good.*" This is by now, to us, the familiar close to a creative process that took a great deal of time to accomplish. At this point in the account, it has added significance in that it serves to predate the appearance of animals on the Earth to the appearance of Man, representing a chronological sequence in reverse of the one detailed in the Garden account and introducing a problematic element to those that hold that death did not enter the world until the Fall of Adam.[42] We will address this issue at length elsewhere in this treatise.

## EVOLUTION IS NOT WHAT IT SEEMS

The evolutionary processes described in the Scriptures and within this work are not entirely undirected and random. We do not attempt here to describe a God, who in this sphere has fashioned all the diversity of life in sculptured detail by the work of His hands alone. But we understand the principle building blocks of the Universe were infused with knowledge and obey the laws of a God who creates through that information with those laws. Let me explain.

Build a marvelously ornate and intricate structure out of blocks. Take all the time you need and use as many blocks as you want, mix colors and sizes, make towers and doorways and bridges, pile them high into the sky. No matter how long you take, your structure will pale in comparison to the intricacy inherent in human DNA. The complexity in your structure and in DNA has the same source – *intelligence*. Specifically, the intelligence is external to the structure.

No one pretends to suggest the blocks themselves are intelligent, that they assembled themselves in such a spectacular way, of course not; they do not have that intrinsic ability. The same limitation is true for the basic components of the helical DNA molecule. The nucleic acids, the atoms that make them up, are no more

---

[41] For a brief discussion of the human Genome as it relates here, see The Introduction of the Sons and Daughters of God in the Appendix, pg. 543.

[42] For a detailed discussion of a Pre-Adamic Earth, see LATE ADAM, pg. 92.

sentient and capable than a child's building block. Some may choose to be cute here and add complexity itself to this analogy; adding magnets and interlocking shape to the blocks may give them some capability of self-assembly, but that requires yet another layer of design and external intelligence to the story here. Simply put, when we see complex structure, we KNOW intelligence was at work.

Now comes the next question: How can information be *added* to the complex structure? No sequence of random attacks on the structure (mutation) will affect such a result. The system only decays, becomes shorter and disassembles. The law of Entropy, actually the way we define it and measure it, may view the falling structure as more complex (more difficult to describe, more random in appearance), but the intricacy of the structure, the *information* it once held, has been lost.[43] Mutations, natural processes, only result in information being lost; the structure never gains height, and keystones don't position themselves.

What we have detailed in this analogy (*playing* with our blocks here) is indicative of the real world around us. The information in DNA came from somewhere; the components of that wonderful molecule could not have gathered the immense amount of detailed information they contain through *natural* means. No amount of time, no sequence of natural and random processes, could have loaded information into the double helix. Today, no scientist can describe a *single instance* where a mutation or an evolutionary process added information to the Genome, *not one*. Consider what that means:

We do not deny that billions of years of uncertainty in reproduction of these fabulously complex structures have resulted in uncountable errors in coding, disorganization, mutation and loss of information. Yet the structure still holds a tremendous amount of information, capable of forming atoms numbered on the order of $10^{28}$ (approximately seven billion billion billion) into wondrously sentient human beings still today. Think of the complexity and the information those structures must have originally held. Where did that information come from? Who among us has the knowledge to create such a structure?

Within this work, we affirm the existence of an intelligent designer, our Creator and Father God. We submit that the information to form DNA molecules was loaded into the very elements of the Universe as recorded in Day 1, when the Spirit of God "*brooded*" over the deep, but the *information* contained in DNA was loaded by God through the *bara* events detailed in the Genesis 1 accounts. To the Atheist who asks why we feel compelled to assume that a designer is the prime reality of the Universe and that the Universe is not the prime reality of itself, to those who ask: "*If God is that which does not need to be created, why can't the Universe be that which does not need to be created?*", we respond: We believe we have given an answer here

---

[43] There is a stream of thought in science (an assumption that subtends what has been proposed as a Physical Law in some quarters) that information itself, like matter and energy, is always conserved. This still hypothetical "*law*" itself, supports the contention here that information is external to the physical constructs of the Universe in which we live, and therefore must be sourced from elsewhere, returning to that same source when it is lost, or <u>*NOT*</u> lost as the case may in fact be.

through our analogy and the description of the facts at hand. The blocks, the elements of the Universe, are not inherently endowed (by natural means) with information. Information, complex information, must be sourced from an *intelligent agent.* Our blocks, our DNA, do not assemble themselves out of random chaos. We believe the evidence displayed by the Universe, in both the microscopic and macroscopic scales, insist on an intelligent, agent, and sentient designer.

Once the information was loaded through the actions of the Spirit of God, on Day 1, that information has worked within the very *bohu* to construct, evolve and sustain life in the Universe through all of the processes revealed by legitimate science and testified of by the whole of creation. In the following pages, we will examine another component of our being which works to direct the development and sustainment of life.

## IN THE IMAGE OF GOD

*"Let us make man in our image, after our likeness; and it was so."* The word *image* here in the KJV is a translation of the Hebrew word, *besel'em* (בְּצֶלֶם). The root in this compound word is *tselel*, which literally means *shadow*. A shadow derives its existence as an extension of another tangible; it is separate and distinct from its source object, yet it remains inextricably attached and dependent upon it. So in a real sense, God set out to form Temporal Man as an extension of Himself; we are dependent on His existence for ours. To accomplish this task, God shares His Spirit with Man. How does He accomplish this and what does this mean?

To fully understand the making of Man in the image and likeness of God, we must turn briefly from the account of the Temporal Creation (Genesis 1) to another account of a creative period that preceded it, that of the creation and planting of the Garden and the placing of Adam and Eve within it. We will discuss in depth later in this work the creation of the Garden and the very special set of circumstances that made its existence a bridge from the Physical to the Temporal, from immortality to mortality,[44] but here, to complete our picture of the creation of Man in the image of our Father in Heaven, we need to touch upon a few verses from that account and the meanings they provide for us. First let us consider the very breath of life:

> *"And I, the Lord God, formed man from the dust of the ground, and breathed into his nostrils the breath of life; and man became a living soul," (Moses 3:7)*

The Hebrew word *Ruach* (רוּחַ) has the common meaning of wind. In Hebrew tradition however, as *Ruach* applies to God, the word means very much more. In the Genesis 1 account, it is the *Ruach* that is identified as the breath of God when He bestows *the breath of life* to both animals and man.

---

[44] See Day Six and the Garden (A Special Place in Paradise), pg. 211.

Tradition tells us that every living thing has a *Ruach*. Because of this, the classification of this word slips from a verb (taking a breath) to a noun (the breath) as it is understood that through the act of *breathing* God *bestows* a tangible thing, *the breath of life* (*Ruach*).

Hebrew tradition teaches that the *Ruach* is the life-giving force that animates a living thing, perhaps if *however loosely*, correlating to the Latter-day concept of the Light of Christ (more on why this association is incomplete to follow later in this work). The *relationship* between this life-giving force and the Light of Christ remains significant because the *Ruach* is considered to be the source of emotion (love) within living things, an attribute most often associated with the persona of the Savior.[45]

In Genesis 1, *bara* (*created*) is used in direct association with *Ruach*, the "*breathing*" of God, and *only* with *Ruach*. The *breathing* of God is the infusing of the Spirit of God into Temporal bodies. In the Genesis 1 accounts, we see the same process at play for both Man[46] and animals: 1) Create (*bara*), 2) Breath (*Ruach*) and 3) Temporal bodies becoming living Souls. The detailed recording is as follows:

For Animals:

- *"And I, God, created great whales, and every living creature that moveth,"* (Moses 2:21).
-*"and they were also living souls; for I, God, breathed into them the breath of life,"* (Moses 3:19).

For Man:

-*"And I, God, created man in mine own image, in the image of mine Only Begotten created I him; male and female created I them."* (Moses 2:27).
-*"And I, the Lord God, formed man from the dust of the ground, and breathed into his nostrils the breath of life; and man became a living soul,"* (Moses 3:7).

---

[45] See our later discussion, The Light of Christ, beginning on pg. 443. We will make a strong case there that The Light of Christ is Physical Law, the information that was infused into the matter of temporality on Day 1 when God, The Son, *brooded* over the face of the deep.

[46] Genesis does not specifically discuss God breathing life into Eve. This fact has been used in some odd ways in scriptural arguments. Today, a common popular misconception is that the Council of Nicaea (circa 323-325 AD) debated whether or not women actually had Souls since God did not breathe His essence into them. After 325AD, or so the story goes, the ensouled status of women became accepted as an official part of orthodox doctrine when the Council of Nicaea voted in favor. In actual point of fact, the debate did not take place in an official ecumenical council, but it took place informally in a Church Synod in France in 585 AD, as recounted in the *Historia Francorum* of Gregory of Tours. This debate did not make use of the Genesis text at all, but instead focused on the Latin word *homo* (man) and argued whether Biblical passages referring to *homo* or *homines* were equally applicable to women. This account appears to be the seed from which the later legend grew—a legend that the church decided fairly late that women had Souls. In actual fact, even at the time of the Council of Nicaea, women were already being baptized, taking communion, and taking last rites—all rituals that presuppose in the recipient the presence of an applicable human Soul. That suggests many (and probably most) early Christians believed women had Souls, even if a few unorthodox believers might not have shared that thought. As we have seen here, the breath of life (the *Ruach*) was in fact given to animals (Moses 3:19) and it would be inconsistent for Eve not to have received as well through some process.

Plant life, specifically trees in the Garden, is also identified as a living Soul, "*And it became also a living soul. For it was spiritual in the day that I created it*" (Moses 3:9) though, it must be noted, the sacred record never mentions the *Ruach* as being associated with plant life.

The Temporal Creation is a shadow (*tselel*) of the Physical Creation, simpler in form with less dimensionality and complexity. The Temporal is a creation by subtraction, entirely dependent upon, and derived from, the Physical world in a very literal way. Through the *Ruach* we understand the Physical Creation, the reality of it, to be bonded to the Temporal Creation through God's sharing of his Spirit. The sharing of God's Spirit was conducted at two separate times in the creative process each following the same process (*the breath of life*) but each fulfilling fundamentally different purposes.

***The Temporal Creation is a shadow of the Physical Creation, simpler in form with less dimensionality and complexity.***

From the Scriptural record and Hebrew tradition, we are aware of three forms of the Spirit of God, two of which are associated with the *breath of life* and directly to animals and Man. Each of the three forms of Spirit has its own value and purpose as it pertains to God. The entry of these types of Spirit into Temporal forms along with embodied Spirit is what makes plants, animals and Man a three part being called a *living Soul.*

The several forms of Spirit emanate from the same source, are that same thing, but vary in completeness of delivery, of intimacy, and connection with God. The first form applies to all things in the Temporal sphere both sentient and insentient. It is the carrier of the information that is manifest in the laws of Physics which govern this mortal sphere. The second form adds to the first, things associated with instinct and purpose. The third adds to the first two yet again and is most capable of connecting the receiver with all of the attributes of the Father.

All living things become a living Soul when they spring to life in Temporality. Their Temporal (mortal) forms, mirroring their more complete and immortal Physical form, housing that *Embodied Spirit* entirely, shielding it from the deleterious effect of uncertainty and absorbing the blows of Temporality. The bonding of the Temporal to the Physical is made possible by the first form of the Spirit of God; it is the information which empowers any created thing to seek its purpose, its ability to fulfill the measure of its creation. We will later associate this first form of Spirit to the Light of Christ.

The next form of Spirit is the first of two which is shared through the *Ruach* and is associated by Hebrew tradition with animal life; it is called the *Nefesh* (נֶפֶשׁ). We are to understand that every "*created*" sentient thing in the world has a *Nefes*h. It will be remembered that create (*bara*) was only used twice in the body of the Genesis 1 account, at the two separate points when Animals and Man were formed. *Nefesh* is the *Godly force* that gives them life. *Nefesh* is most often associated with animal instinct[47] and the source of self-identity.

In Day 5 God *created* animals on the Earth, being the initial introduction of advanced life (living Souls) into this Universe. It is proper that this process began in Day 5, for we now know that animal life on Earth originated in the sea. In the Day 6 account, life continues to advance finally developing on the land. This ordered and structured progression of life is in concurrence with the observed fossil record. The initial *bara* event for animal life, with the introduction of the *Nefesh* through the *Ruach* of God, occurred in the period of the history of the Earth covered in the account of Day 5 and was not repeated with the appearance of land animals, for they too are part of the continuous spectrum of animal life.

The second form of Spirit to be shared through the *Ruach* pertains *only* to Man and is considered to be of a higher order than the *Nefesh*. Man has both the *Nefesh* <u>and</u> the *Neshama*. In Hasidic Rabbinical thought, the *Neshama* (נְשָׁמָה) is said to be "*an actual part of God*;" we will adopt this understanding here as well. However, by tradition, it is taught that the *Neshama* is only found in the Jews; this extremely narrow view, that only the chosen people of God receive the *Neshama*, of course is incorrect.

We understand here that it is the *Neshama* that sets *all* of the *Sons of God* (Genesis 6:2, 4) apart from the rest of animal life on Earth,[48] and we agree that it is an actual part of God, an extension of Him. *Neshama* is considered to be intellect; it is congruent to the Christian divine source of reason within us. The *Neshama* is the *logos* within each of us; it is that which emanates from the Divine *Logos,* the Word of God, "*In the beginning was the Word, and the Word was with God, and the Word was God.*" (John 1:1).

The *Nefesh* (Spirit of animals) and *Neshama* (Spirit of men) join with the *Sh'khinah* (a sensible display of the presence of God) in the Hebrew tradition as the three manifestations of Spirit of God in the Temporal world; they are distinct from the Holy Ghost (to these three we add the Light of Christ).

Having laid this groundwork, let us see what the Scripture tells us.

---

[47] We will note here that modern science, with all of its arrogance and insistence of knowing of facts, still has not put forth a single credible explanation for the source of instinctual behavior found in life. Its source is a divine source, the *Nefesh*, the Spirit of God, and the information and knowledge which attend it.

[48] See <u>THE SONS OF GOD AND THE DAUGHTERS OF MEN</u>, pg. 95.

We have seen that all life on Earth shares a common source, in a literal sense. Physiologically all life is the same on the most basic level. What makes humans (The Sons of God) unique is our Soul (*Neshama*), *not the composition of our bodies.*

The Garden creation account explicitly states that the body of Adam, and by extension those of mankind within the Temporal sphere, is formed from the dust of the ground (Moses 3:7). The connection of our Temporal bodies with the "*earth*" is of such intimacy that the name chosen for the first Man was Adam. In Hebrew, the word Adam, *Adamah*, (אדמה) literally means earth or ground, or more correctly clay. In the contexts we have developed in this chapter, Adam is made literally of "*matter*" itself, the very same "*stuff*" of which our Temporal Universe is composed. In the case of Adam and the animals still within the Garden, before they are created Temporally, this clay is the eternal matter of the Physical Creation (more on that later in this work, see Part 2, The Physical Creation).

The scriptures also explicitly state that the bodies of animals were formed from the same material, the ground: "*And out of the ground I, the Lord God, formed every beast of the field, and every fowl of the air*" (Gen. 2:19).

Understanding the Hebrew of the Biblical text, however, provides us with a crucial difference between these two creative events; one that cannot be overlooked. The Hebrew word for "*formed*" is *yat-sar* (יִצֶר). When used for the *forming* of mankind (*yat-sar*) in Verse 7, it is spelled with two Hebrew letters *yud*, (thus rendered יִיצֶר). The structure of the grammar in Verses 7 and 19 of Genesis 2 is identical, because they are both describing the *formation* of something, man and animals respectively. However, when used in Verse 19 to describe the formation of animals, *yat-sar* is spelled with one *yud;* in Verse 7, in describing the formation of man, it is spelled with two. It is important to note that EVERY Torah scroll, whether from Yemen, Jerusalem, or Los Angeles, California, is written this same way; this is not an error or oversight. The double *yud* for man is intentional, what does this mean?

*Yud* (י) is the abbreviation of the Name of God, *YHWH*. In doubling the *yud* for mankind, the Bible is telling us that although mankind and animals may share a common physical origin, there is an extra spiritual input in humanity that is not present in animals or the rest of creation. The *Neshama*, not only binds the Spirit to the Temporal but it also binds the Soul of humankind to God in a most intimate way. It is the *Neshama* which distinguishes the sons and daughters of God (man) from beast.

The introduction of the *Nefesh* (נֶפֶשׁ) and *Neshama* (נְשָׁמָה) to animals and man respectively is represented by the two divine creative (*bara*) acts mentioned in the Genesis 1 account:

> *"And I, God, created great whales, and every living creature that moveth, which the waters brought forth abundantly, after their kind, and every winged fowl after his kind; and I, God, saw that all things which I had created were good." (Gen. 1:21)*

> *"So God, created man in his own image, in the image of God created he him; male and female created he them." (Gen. 1:27)*

In the single creative (*bara*) event described in Verse 21 of the Genesis 1 account, all animals are given the *Nefesh*. This is the point of infusion of the Spirit (*Nefesh*), spoken of in Moses 3:19 and Gen. 2:19 (Gen.1:21); "*and they were also living souls; for I, God, breathed into them the breath of life.*" That thc *Nefesh* of animals is a different, lesser form, of Spirit than that of Man (*Neshama*), is signified by the single *yud* used to describe the formation of their bodies.

## MALE AND FEMALE CREATED I THEM

After the creation of animals of the land, God turns to the creation of man. Of note here is the word *bara* and the introduction of the *Neshama* which we already have addressed. Other significant elements are the timing of the creation of man, the use of the word "*them*" as it refers to man and not the word "*he*" used in the non-Genesis 1 accounts and the creation of *both* male and female.

Adam and Eve are sent here, *to the Temporal Universe*, from the Garden together; that is clear. So we would expect them in this narration to appear together, and they do. The record shows that God calls "*them*" Adam in Genesis 5:2. In contrast, in the Physical Creation (The Garden), Adam comes first and Eve is created *from* him. It is significant that Adam and Eve appear simultaneously together on the Temporal Earth. *Note that Eve does not receive her name from the LORD God, but from Adam after the Fall and its attendant consequences are proclaimed and in place.* In this account, Eve is not fashioned out of the body of Adam. These specific and intentional details do not represent a summary or abridgement of a more detailed or alternately described account provided elsewhere, but they are in fact an intentional, detailed, representation of a complete and separate event that stands entirely on its own.

## THE BLESSING OF MORTALITY

Fundamental to Latter-day Saint theology is the understanding that our Temporal existence, mortality, is a progression, a step forward in our eternal existence, not a regression or step back. Placing the verses of the Genesis 1 account and the proclamations of God told therein in their proper chronological sequence, after the Fall and not before, we see that we are indeed *blessed* to be here and not cursed by God; our Fall was indeed a *fall forward*.

> *"And I, God, blessed them, and God said unto them: Be fruitful, and multiply, and replenish the earth, and subdue it: and have dominion over the fish of the sea, and over the fowl of the air, and over every living thing that moveth upon the earth." (Gen. 1:28)*

Without our Temporal bodies, procreation would not be possible.

The account of the Temporal Creation draws to a close with two verses which describe the diet of man and that of the beasts. An examination and proper understanding of these verses reveal truths and again provide testimony of the divine source of the revelation.

> "¶*And I, God, said unto man: Behold, I have given you every herb bearing seed, which is upon the face of all the earth, and every tree in the which shall be the fruit of a tree yielding seed; to you it shall be for meat. And to every beast of the earth, and to every fowl of the air, and to everything that creepeth upon the earth, wherein I grant life, there shall be given every clean herb for meat; and it was so, even as I spake." (Moses 2:29,30)*

Photosynthesis is the fundamental energy source for life on Earth. This process is the *only* means life has of capturing the energy provided by the nuclear processes of the Sun. Regardless of its final classification, herbivore or carnivore, all animal life on Earth, man and other beasts, mammal, amphibian, reptile or fish, *all life*, is dependent upon the photosynthesis cycle in herbs, trees and all plant life for sustenance.

The photosynthesis process (the chemical details of which are complex and beautifully intricate) was not understood or described until the 1770's by Dutch scientist Jan Ingenhousz. It could be argued that detailing the diet of man and beasts in the manner that it is accomplished here in Genesis 1, is counterintuitive to the mind of man 2000 to 3000 years ago and provides proof yet again that the source of the story told was, and is, divine.

In contrast, it may be and has been weakly argued that these two verses describe an ideal utopian world where death does not exist and carnivorous activities are sinful, but such an argument is not sustainable on multiple counts. First, plant life *is* life, and its consumption as food leads to its death. *We have previously discussed the Scripture* (Moses 3:9) *that states plant life is a living soul.* Second, there is no place to be found in Scripture where carnivorous beasts, fulfilling the "*measure of their creation*" by eating the flesh of other animals, are said to be committing sin. In a traditional Christian worldview that insists on a perfect, non-evolving creation, fully complete from the outset, carnivorous beasts were created as omnivores having no way to survive from day one without the eating of meat.

For those who are convinced the word *meat* used in these verses somehow intends to communicate a vegetarian diet was intended where the consumption of plant life is to be used in place of meat, I will simply refer to the "*meat*" offering described in Leviticus, Chapter 2. Leviticus describes a *meat* offering that is made up of fine flour, firstfruit produce including corn, all seasoned with salt and frankincense,[49] which is to be presented to the sons of Aaron and sacrificed on the Alter at the

---

[49] It is the inclusion of Frankincense in the meat offering that provides the rich symbolic meaning to this precious herb being included in the gifts of the Magi to the infant Christ.

Temple. Proponents of a vegetarian intent in verses of Moses, 2:29 and 30, point to the word "*meat*" used in Leviticus in association with what is clearly a grain offering as proof that the word *meat* in the context of the Moses account conveys God's intent that grain should be the meat of Man.

The Hebrew word translated as meat in Leviticus, Chapter 2, however, is *min-hah* (מִנְחָה), which literally means grain. In Hebrew then, this offering was a grain offering and had nothing to do with flesh or any word like it. As in Leviticus, the Hebrew word translated as *meat* here in Genesis 1 has a different meaning than the English text reveals; it is *le-ak-lah* (לְאָכְלָה) and means *for food*, it does not represent flesh. The early English translations render both *le-ak-lah* in Genesis 1, and *min-hah* of Leviticus, Chapter 2 as *meat*; why was this done?

In medieval Europe and Feudal England, *meat* in the context it is understood today (the flesh of animals) was a rare treat indeed, consumed only infrequently as part of the typical person's diet. The context that was used by Tyndale and others in forming the English words of the first Bibles in that tongue, understood *meat* to mean whatever is eaten, simply food, or specifically *fine food* as in those that would be offered as *firstfruits*. The more extended definition, that of *fine food* came to be associated later with the flesh of animals, because as living conditions improved in the modern eras, a *fine meal* always included "*meat*" as we know the word to be defined today. In time, *meat* eventually came to be associated with the turkey or roasted beef on the table and not the other food that accompanies it.

In these contexts, *meat* as translated here in Genesis 1, using the Hebrew in which it was originally recorded, simply describes plant life as the source of food (sustenance), the primary energy supply for all life on Earth. This understanding is the one intended by the original Hebrew author. That the sacred text includes an understanding that the energy to support life of all forms ultimately comes from the Sun and is converted through photosynthesis in plants, a process entirely unknown to ancient man, is evidence of the divine source for this revelation.

> *"And I, God, saw everything that I had made, and, behold, all things which I had made were very good; and the evening and the morning were the sixth day." (Moses 2:31)*

Again we see this creative process at the end of Day 6 capped by the proclamation of God that the things He had made were good. By now we are well aware that these words follow a creative process that required a great length of time to complete. That is true of the entire process recorded in Day 6; the history of advanced land based animal life on Earth encompasses some 250 million plus years. This proclamation of completion, however, is different from the ones preceding it, for it includes the word, very. It was indeed, *very* good.

After the insertion of man into the Temporal sphere, the purpose of this unique creation was finally realized. The Plan of Salvation was finally empowered to provide for the benefit of the children of God in a way that could never have been fully realized before. Because of this, for the first time in this account, God proclaims His activity as *very* good. After proclaiming his works good five times over the last five days, including already once in this selfsame last, sixth day of creation, God now uses the word *very*; placing a stamp of final approval on the Temporal Creation.

When Moses sought to understand God, he beseeched God to "*inform me of Your way* (literally, path) *that I may know you*" (Exod. 33:13). God responded by saying, "*I will pass all My goodness before you.*" Here we observe Moses asking God to show him His purpose, His plan, even the *Plan of Salvation*. God answers by saying that He will pass a vision of *all His creation* (this Temporal sphere, His "*goodness*") before him. This is where the vision of the creation recorded in the Genesis 1 account comes from; it was provided in response to this request and exchange recorded in Exodus 33.[50] In Moses 2:31, the word *very* refers not only to the quality of God's creation: the Universe, life and mankind, but *very good* refers to the balance, order and connected relationships of everything there is.[51]

## LATE ADAM

Now, it does not require the most observant of readers to notice the fact that in detailing this account we have described a world, indeed an entire universe that is evolutionary in nature[52] and very ancient in origin. We have, as does the sacred record, left Man until last; it will be noticed, *very* last. That begs a question: When do Adam and Eve appear on the Earth?

The 24th verse of Genesis 1 does much more than just suggest that animals are on the Earth before Adam, it clearly establishes that as fact. The following Verse, 25, states "*all things were good*" meaning again a very long time had passed. It is in fact a *very* long time before man is considered (verse 26) and subsequently created (verse 27). We have shown that the Temporal is a mortal creation from the outset, subject to all of the direct and the nuanced laws of Physics, including the second law of thermodynamics and entropy — *death is attendant*.

---

[50] I believe the exchange of Exodus 33:13 led to the Genesis 1 accounts recorded by Moses. A careful reading of Moses, Chapter 1 reveals that the exchange recorded there closes with its last verse. Chapter 2 is not contiguous with Chapter 1 and itself is not what was received by Moses on Sinai. The exchange of knowledge shared in Exodus 33 occurs within the Holy-of-Holies of the Tabernacle, a setting sanctified and set apart for that reason. Today we are told the story of the creation in sanctified places within the Temple as well. God is ever consistent.

[51] Onkelos (c.35-120 AD) translates this phrase in Genesis 1:31 as follows: "*and God saw everything that he had made and behold it was a unified order.*" (Targum Onkelos, c.110 AD)

[52] The Apostle James Talmage, a well-respected scientist and scriptural scholar, was described by his geologist son, Sterling, as having expressed in 1920 a concept of pre-Adamites which "*went beyond anything that I had dared to think.*" After reviewing the initial treatment Sterling offered on the subject, explaining his father's ideas, Apostle John Widtsoe encouraged him in an endeavor to detail a doctrinal stance supporting evolution. Widtsoe wrote to him: "*it is very likely that the time is ripe for someone to begin right now to prepare a wise, temperate, scientific statement of the doctrine of evolution ... Evolution, as a law, seems to me to have been demonstrated*" (Widtsoe to Sterling Talmage, 20 Apr. 1934, S. Talmage Papers). It is not clear whether this series was not published because it could not receive the approval of the brethren or if Talmage voluntarily withdrew the manuscript. I, however, have here endeavored to fulfill Elder Widtsoe's fervent request.

Our detailing here of the scriptural and scientific record describes the presence of life on Earth and within the Universe, some 3.5 billion years before the arrival of Adam. On Earth we understand that life evolved and progressed from the simplest of forms, to the diverse spectrum and complexity we see all about us today. We affirm the ancient fossil record; throughout all of the billions of years recorded, living things have changed. Organisms of all forms sprang to life and subsequently died. There is no avoiding this fact. Because we have placed the Fall before and beyond the realm of the Temporal sphere, we have no conflict with doctrines that ascribe death to the fateful actions of Adam. That event happened before the creation of this Universe and its affect has been at play from the very moment of the Big Bang. Everything and anything is deteriorating, disintegrating, and dying. Within the Temporal confines we find ourselves in today, death always has been and always will be.

***...the scriptural and scientific record describes the presence of life on Earth and within the Universe, some 3.5 billion years before the arrival of Adam.***

The description of the Temporal Creation we have discussed here accepts entirely the fossil record of life on Earth. It embraces the evolution of such life and supports the true science that is done to understand it. But is that without limitation? What of the fossil record of the evolution of hominids? Who were Adam and Eve? When did they appear on Earth? Do they have any relation to early man in all of its evolutionary forms?

What I propose here is that the sacred record attests to what has been termed a "*Late Adam*." What that means and what its ramifications are, we will discuss here; though for consideration of the scope of this treatise, our discussion will of necessity be somewhat brief.

By "*Late Adam*" we mean that Adam and Eve were placed here most recently in the history of the Earth. We understand that the entire, almost 15 billion year, development of the Universe was completed to develop a perfect home for this privileged pair of human beings. The Universe, our Galaxy, this Solar System and Earth, along with all of the life forms on it, became, evolved, and progressed to a point in which conditions were just right for the introduction of our first parents to this planet by our Father in Heaven. Hominids of all forms and their evolution on the Earth preceded the introduction of Adam and Eve to this planet.

When that introduction *was*, is uncertain and debatable, and will remain so. We do not know if Adam's introduction to Earth was some odd 6000 years ago (in keeping with a somewhat uninformed understanding of what a "*literal*" reading of the Bible presents), or if it was a more palatable and probable 100 thousand plus years ago. What is clear, as we shall see from the Scriptures, is the introduction of our first parents was recent in the history of this planet and represented a break and a new direction taken from the natural processes that had already been at play for billions of years in this Universe. In this section, we will explore subtle hints to be found which support this scenario.

In answer to those who may belligerently question, "*If there is no life elsewhere in the Universe, why is it so vast?*" We can confidently respond in one or perhaps both of two ways:

First, it may be that Earth is not the only inhabited planet. Because the Fall of Adam occurred before this Universe was created, our theology remains intact, without need of modification or qualification, if God has placed, or even may place, His children on other planets. The Fall of Adam was not a recent, local event within the Universe. The actions of one Earthling, just a few millennia ago did not rewrite the laws of Physics throughout the cosmos, thereby, perhaps unjustly, entangling the lives of countless unidentified beings on other worlds with a consequence of death that was previously unknown to them. The Universe was created for the express purpose of Temporality and any and *all life within it* is either a direct development of, or is intentionally placed here to experience, the results of the laws of Physics.

Second, given the complexity of life itself and the incredible processes needed to provide the conditions and resources for its development, coupled with the almost random realities of a Universe empowered with the uncertainty needed to provide each of us with free agency, it is entirely possible that God needed all of what there is just to ensure He could have one planet (or just a few) like our own for the habitation of His sons and daughters.

Controversially a "*Late Adam*" scenario requires the existence of what has been termed Pre-Adamites (those that lived here on Earth, both animal and near human, before the arrival of Adam and Eve). Do the Scriptures provide any indication of this population? What can be said about them?

Previously, in Chapter 2, we discussed briefly a debate that raged over a century ago between the brethren regarding the existence of fossil evidence of ancient man and the ramifications those finds posed to the story of Adam and the effects of the Fall.

We have avoided obvious speculation in this work, and we must continue to do so. Let us, however, consider a few more buried scriptural truths that will shed some much needed light on the questions posed here before we venture into the fast current of conjecture that surrounds a scriptural understanding that speaks to humans living on Earth before Adam and Eve. We will reserve our discussion of this until the following chapter and until we have exhausted our examination of all that the Genesis story has to share.

## THE SONS OF GOD AND THE DAUGHTERS OF MEN

The story of Adam and Eve comes to a close with the last verse of Genesis, Chapter 4. This verse, however, contains an enigmatic phrase that represents an interesting transition: *"then began men to call upon the name of the Lord"* (Gen 4:26). What does this phrase mean and why does it appear at this position in the story; surely Adam and Able were calling on the LORD, even Cain was in direct communication with God? So why say, "*then began men*" to call on the LORD now? I would like to suggest a possible explanation for this. This explanation will require that we identify and define some terms often overlooked by even the most diligent searchers of the Scriptures.

The Hebrew translated in the KJV as *began* in Genesis 4:26 is *hu-hal* (הוּחַל); it is a verb meaning *began* or *begin* and only appears this one time in the Bible. Because it appears without a subject, one is provided. Tyndale used *men* in its broadest context as the subject. In the same manner, and just as accurately, *people* can also be used here as well. There are no bracketed words here in this verse and this is proper; the insertion of *men* or *people* is not due to an omission, but is because *hu-hal* is a compound word and the subject of the verb is intended. What is important to us remains intact. Who are these people or men that have been set apart in the text as different from those who had been calling upon the LORD before? The answer can be found by looking both forward and backward in the sacred text.

Before the mention of these enigmatic people who now know of the LORD and call upon him, the account of Adam and Eve had turned to their two famous sons, Cain and Able. We have already acknowledged that these two had been communicating with the LORD, but the story told continues by introducing the *Land of Nod* and the subsequent genealogy of Cain and his offspring there. It is this group of people who are referred to in Genesis 4:26.

Looking ahead in the text, we are presented with the generations following Adam through Seth, excluding any of Cain's lineages. As Chapter 6 begins, we are told that *daughters were born unto* the men of the earth; then in the text we find a very curious phrase: "*...the sons of God saw the daughters of men...*" (Gen 6:2). At this point, the text separates those on the Earth into two groups appointed by God, *men* and *the sons of God*. What is significant about this? Throughout Genesis Chapter 6, a careful reading of the text will reveal this separation has a profound meaning.

*Be-ne- ha-e-lo-him* (בְּנֵי־הָֽאֱלֹהִים), *the sons of God* stand as distinct and separate from the *people* of Genesis 4:26, who are the descendants of Cain and the population of Nod.[53] Here in Chapter 6, the daughters spoken of are *ha-a-dam* (אֲדָמָה), meaning, *of men*: literally the daughters of those who are of *the clay*, or of the earth.

---

[53] For a somewhat detailed examination of the land of Nod, Cain's entry there, and how that came to be, see Adam in the New World, in the Appendix, pg. 551.

Of these *men* of whom the daughters are, God proclaims: His Spirit "*will not always strive with*" them (*v3*), "*the wickedness of man was great*" (*v5*) and, "*the earth is filled with violence through them*" (*v13*). Finally He proclaims: "*I will destroy man ...from the face of the earth...*" (*v7*). Remarkably, in Verses 6 and 7 we see that the lamentations of God caused Him so much sorrow, He *repented* that He had made man in the earth.

How could this be so? Wasn't the entire Temporal Universe created just for that purpose? We see, however, that it is not *the sons of God* that are the source of His sorrow, but the *people of the clay*, those who came from the Land of Nod, those individuals and the interbreeding of those individuals with the sons (the children) of God.

In Genesis, Chapter 6, we observe two distinct populations discussed. The sons of God are clearly the descendants of Adam and Eve, distinct from the others at the outset as we would expect to find with a *Late Adam* introduction; and those of the clay, the people of Genesis 4:26, the inhabitants of Nod, the subsequent children of Cain who sprang from those (Pre-Adamites) that lived on Earth before Adam and Eve.

There is no doubt that the evil associated with the *men*, the *ha-a-dam* (הָאָדָם), was immense. It was of such scale that God decided to remove them from the Earth entirely. That He did so, that He always does as He says He will do, is telling; for we, *mankind*, remain here today. That God would do as He set out to do is without question, so what of their utter destruction if we still yet inhabit this world?

According to the Bible, all of mankind today is kin of Noah through his three sons and their wives. Ham's wife bore the Pre-Adamite genetic material of Cain within her, but through the flood event, or more likely the flood event in concert with multiple other destructive mechanisms, all traces of Pre-Adamite peoples and culture perished in short order. As evidence of this fact, Paleontology records many ancient branches to the hominid tree, but declares, inexplicably, that despite an evolutionary process that most often leads to diversity, today, atop a tree of human development Homo sapiens stands alone. In accord with the Biblical account, the archeological and fossil records show the time frame involved in the complete supplanting of all other hominids by modern humans to be short and dramatic.

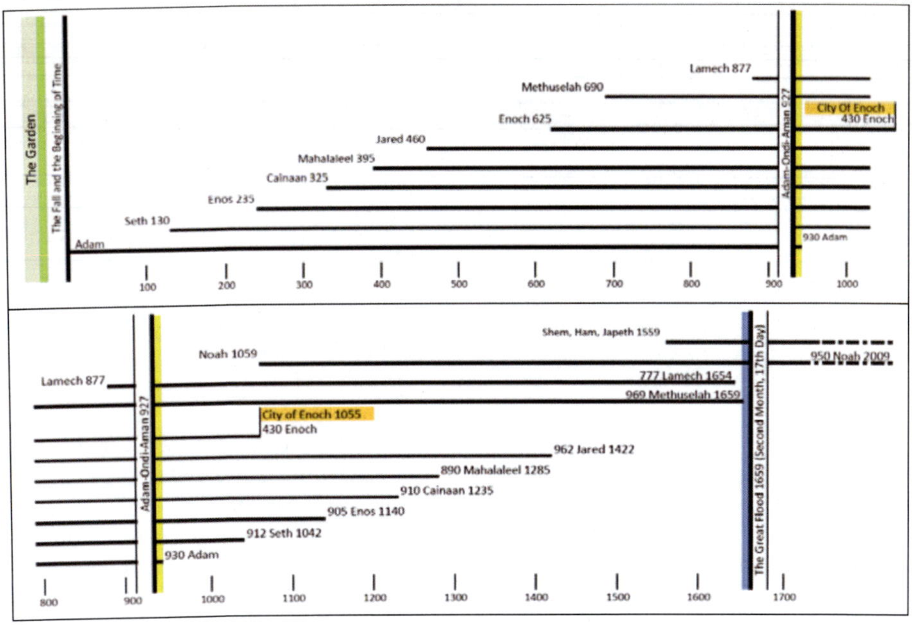


Figure 3-10 *The First Generation of Man. The lifespans of the antediluvian patriarchs and four key events of the first generation of the sons and daughters of God are charted here. Note when Seth is born, marking a turning point in the history of man, the hu-hal (הוּחַל) of Genesis 4:26 and the year of the Flood (1659) when the ha-a-dam (הָאָדָם), the people of the clay are destroyed from the face of the earth.*

## THE GENERATIONS OF ADAM BEGIN

Let us return to a question posed some time ago: When did Adam and Eve arrive on the Earth? We have no firm understanding. Many have reviewed the pages of the Bible with a determination to find an answer to this question and a few are thought to have found one, some in spectacular detail.

Some famous attempts have calculated the breathing of life into the nostrils of Adam to the precise year, day of the month, and even to the hour of that very day,[54] but in fact these attempts are silly and spurious, and not to be relied on, for they are

[54] In 1630, James Ussher, Archbishop of Armagh and Primate of All Ireland, working with a group of theologians calculated this exact time to be October 23, 4004 BC at 9:00am. He also calculated that creation itself began at "*the entrance of the night preceding the 23rd day of October ... The year before Christ 4004.*" It is unclear if he was using the Hebrew convention of a day starting at sundown or the, by his time, standard western European day beginning at sunrise, but it is clear that despite the literal interpretations of the Bible that must be used to make such a calculation of the timing of creation, by placing the initiation of creation at, 6pm October 22, 4004 BC, and the creation of Adam at 9am October 23, 4004 BC, Ussher allowed at most 39 hours for the first five days and most of the sixth day of creation to transpire. And he accomplished this feat without any understanding of the concept of the relativity of time as measured by different local observers.

flawed in their initial premise that the Bible includes information that is comprehensive and purposed for detailing history. In truth, the Bible does not presume to be a complete historical record, certainly not of the type we have become accustomed to today, of the type that would allow such calculations. The Bible never did provide that sort of detail nor did it ever intend to do so, especially in the many chapters of Genesis.

There are only 26 generations from Adam to Moses accounted for in the pages of the Bible, yet in the Psalms we read: "*He hath remembered his covenant for ever, the word which he commanded to a thousand generations*" (Psalms105:8). Where are these missing 974 generations? [55] How many years do their lifetimes encompass? It is impossible to tell, but it is certain that the Bible does not limit the history of the generations of Adam to a mere 6000 years.

Latter-day revelation gives some clear direction in calculating the ages and the time frame of the antediluvian Patriarchs, and with this information we can place the Great Flood with precise detail counting from Adam towards our time today. It occurred in the 1659th year following Adam's expulsion from the Garden and the beginning of the counting of the passage of time (*Figure 3-10*). We are, however, yet again left lacking a firm definition of how time was reckoned before the Flood and what a year may have meant to those who lived then. Taking what is recorded in the Psalms about 1000 generations, we are left with a woefully incomplete genealogy after the deluge, and along with an inconsistent understanding of the length in years of a generation, we are left with no dating of the flood event backward in time.

[55] For a more detailed look at the concept of missing generations in the scriptural account refer to the Appendix, Missing Generations in the Genealogy of Christ, pg. 547.

Taking an *Educated* Guess

Archeological evidence dates the most ancient urban civilizations back to circa 3400 BC[56] in Mesopotamia and Egypt, not long after that in the Indus Valley, and about 900 years later in Central America and China. So assuming that the Flood event of Genesis occurred before these civilizations existed, the creation of Adam would have been millennia before that, not just 400 – 500 years. Good archeological evidence places the initial forays into agrarian society with the planting of crops and domestication of animals to circa 8000 BC.

Everywhere you look in the regions below and surrounding the Indian Ocean and Southern Asia, as far to the West as the Arabian Sea and peninsula and all the way East to Japan and the regions around the South China Sea, you will find a layer of ash in the geological strata 2 ½ inches thick. It is the calling card of a catastrophic event that may have nearly wiped out the human race about 70,000 years ago. It was a huge bang; Toba was a super volcano eruption of massive proportions. Thirty five times larger than Tambora, the eruption that caused the 1816 year without a summer; the eruption of Toba at 2800 cubic kilometers, spewed 2800 times the amount of ash into the air than the Mount St. Helens eruption did in 1980. The eruption of Toba lifted 670 cubic miles of vaporized rock high up into the atmosphere where it hung and dimmed the Sun for perhaps up to six years. The explosion and its attendant ash disrupted seasonal rains. Lakes and rivers were choked as whole cubic miles of ash rained down upon the land, burying everything in sight.

---

56 In this section and indeed this entire chapter we have spoken of various events as occurring in years past. This is not a precise science and I do not mean to infer that it is. At best dating past events, prehistoric and early historic events,.with precision is difficult and should be understood as such. The confidence level should markedly decrease with the precision of the dating, especially as the length of time in the past increases; this, however, is entirely reasonable. All of us are familiar with, and forgiving of, a weather forecast that misses a precise mark. We understand that a prediction of 2.5 inches of rain in the next 12 hours when a storm is present and has been in progress is of more significance and more probably correct than the same prediction made of a twelve hour period a week or two in the future. We may have considerable confidence in a forecast of the expectation of rain given two weeks in advance when we consider the season, for we know to expect rain and or snow in the winter months and not in the summer. This type of reasonable assessment is made by us all of the time. Dating of past events should be offered the same consideration.

Dating, for example, something like the Bosphorus breach at 5600 years ago could be reasonable, though in error in precise detail. It could easily actually have been 5200 or even 7500 years ago; the margin of error is proportional in scale and is impacted by the amount and type of evidence and in its difficulty of interpretation.

The Bosphorus Breach date, though not precise, is also not at all likely to actually have occurred 500 years ago or 100,000 years ago, that margin of error is disproportional and untenable.

As we project farther into the past, the margin of error increases proportionately thus, 750 million could easily be 800 million. What is clear is that millions do not become billions, and certainly the thirteen plus billion year history of the Universe cannot be dismissed as six 24 hour days and the age of the Earth is measured in billions of years and not mere thousands.

Past dates provided here are best estimates of detailed, thoughtful scientific studies. I am fully aware of the margin of error in some of these numbers. We have not considered the margin in the text, but I am aware some error exists.

Tame in many respects when compared to some of the geological events known to us, Toba was devastating to man. The food chain was decimated, game was killed and became scarce, trees were destroyed and the fruits and berries they provided were gone. The Earth was in an *ice age* 70,000 years ago, and Toba made an already cool planet get much colder. The only tolerable climates on Earth were the equatorial regions; the Toba blast now dropped the average temperatures there by another 20 degrees, the chilling temperatures remained in place for many years. In Africa and in Asia, the small grassy plains that were surviving the ice age shrunk way back.

The best reasonable estimates suggest that the human population nearly went extinct. There were only about a thousand reproducing adults left alive. One study places the number of couples of reproductive age as low as 40. The remaining humans struggled and gathered together in little clumps hunting and gathering for thousands of years until the late Stone Age.[57] At 70,000 years ago, this event occurred about 10-20 thousand years before modern humans appeared on the scene, but these beleaguered individuals could easily be associated with the early ancestors of the people of Nod.

Other indications are provided suggesting that the human population had another choke point twice as far back as the Toba Event. Recent genetic studies,[58] based on very good science around Mitochondrial DNA for women and Y chromosome development for men, date the appearance of our first pair of parents between 148,000-209,000 years ago. This same research limits (surprising and problematic to evolutionists) the original population of these individuals to a group no more than a dozen or so in number, no more than 6 women for certain.

Though the timing of Toba and the genetic evidence of a constricted gene pool are much more ancient than the time frames we most commonly associate with dates and events in the genealogy of the sons of God, these dramatically diminutive numbers are strikingly consistent with the story of Noah and his small family engendering the entire extant human race. Perhaps there is some connection with these events in our distant past and the deluge of Noah told in the pages of Genesis.

---

[57] Recent studies suggest that the Toba event may not have had such a catastrophic effect on the planet as previously thought. However, that the Toba blast occurred and when it did is not debatable, what was the extent of the damage from its massive plume can only be inferred. The dramatic reductions in the Earth's animal and human population from this period are known. Linking these traumatic declines to the effect of the Toba blast is a very good assumption, as good as any made in science. For an opposing view see; *Ash from the Toba supereruption in Lake Malawi shows no volcanic winter in East Africa 75 ka* (Published May 14, 2013, Proceedings of the National Academy of Sciences, Christine S. Lane, Ben T. Chorn, and Thomas C. Johnson.)

[58] See *A Comprarative Study and a Phyogenetic Exploration of the Compositional Architectures of Mammalian Nuclear Genomes* (Published November 6, 2014, Dr. Eran Elhaik from the University of Sheffield and Professor Dan Grauer from the University of Houston and the Tel Aviv University). *Mitochondrial DNA and human evolution*, (1987, Nature 325 (6069) 31-36 Cann RL, Stoneking M, Wilson AC.)

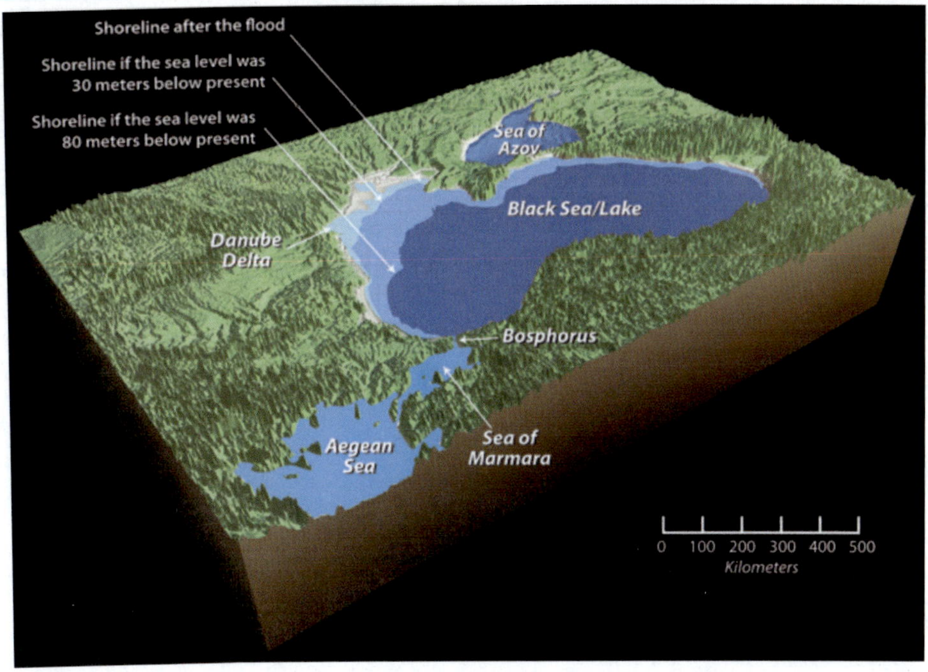


Figure 3-11 *Ancient Black Sea Shorelines. This illustration represents the Black Sea shoreline at 30 meters (98 Feet) and 80 meters (263 feet) an ancient coastline some 5000 - 7000 years ago has been discovered submerged at a depth of 400 feet.* (Woods Hole National Oceanographic Institute)

Intriguing and informative as these items are in detailing scientific findings which indicate a very recent restart in human populations, they only provide documented affirmation that the family of man can, and has, begun again almost from scratch. Now let us turn to an event that will sound much more familiar to readers of the Bible.

Today the foothills to the North of Mount Ararat settle gently into the waters of the Black Sea, but this was not always so. During the last (most recent) ice age, glacial meltwater made a vast freshwater lake where the Black Sea is today. Much smaller in size than today's Black Sea, the banks of this lake teemed with life as the glaciers pulled back at the close of the ice age. With the retreat of the glaciers, some of the rivers emptying into this lake rerouted to the North Sea. The rerouting combined with evaporation due to the more temperate weather, reduced the size of the lake even further.

9,000 to 10,000 years ago, the valleys and the slopes surrounding this lake were paradisiacal and made a perfect home for man (*Figure 3-11*). Large communities dotted the coastline, villages were everywhere.[59] However, to the West, silently, unnoticeably, there loomed pending destruction on an immense scale. The melting of tremendous (near continental size) pieces of ice all over the planet was causing sea levels to rise dramatically, including those of the Mediterranean. As the waters of the Mediterranean slowly crept to the East, up the Bosphorus, they eventually reached a rocky sill about 5600 BC and poured over it in catastrophic fashion breaking and shattering the precipice. It was like an unfathomable dam had broken. It is estimated that 10 cubic miles of water poured over the break *each* day, fully two hundred times the volume of Niagara Falls at their peak. The raging water tore a huge canyon as it crashed to the East, filling the once freshwater lake with salty sea water. The thundering flume surged at a full spate for at least 300 days until it raised the level of the once idyllic lake to the level of the oceans worldwide. Seawater covered the fertile, vibrant shorelines and surrounds, to depths that exceeded 400 feet.

The event occurred quickly and without warning. In what would have seemed almost overnight, it flooded 60,000 square miles (an area almost three times the size of Maryland) of land, which then would have been home to perhaps much of the world's human population. This scene was clearly an event of Biblical proportions, one that would be remembered and retold for generations. Add a little rainfall to this story along with prophetic words spoken by God to the Prophet Noah, and you have described the Great Flood of Genesis, Chapter 6, not a global flood, but one of immense proportions.[60]

---

[59] See Ballard Finds Traces of Ancient Habitation Beneath Black Sea, National Geographic, September 13, 2000, Lisa Krause. Though the archeological evidence obtained by Ballard of human populations destroyed by the rapid flood has subsequently been challenged, what was found continues to attest to swift destruction on tremendous scales caused by the torrential breach.

[60] There is much debate about the veracity of a global flood, even among Latter-day Saints. Though many affirm the reality of a global flood (some given the backing of *formal* acceptance, see *The Flood and the Tower of Babel*, by Donald W. Parry, Ensign, Jan. 1998), the Scriptural record does not in fact attest to a global event. There are many faithful, informed, Christians (including Latter-day Saints), who believe the flood of Noah was a large, catastrophic, local event (for a direct example, see *Noah's Flood: Modern Scholarship and Mormon Traditions*, by Duane E. Jeffery, Sunstone Magazine, October, 2004). Let me make this clear, I consider the Great Flood to be an actual historical event. However, I interpret the wording to entail an event of limited scale. Moses, despite his having received and recorded the sweeping revelation of the creation account, did not fully comprehend what he saw; he did not envision the full scope of the Earth nor did he imagine it to be spherical. In Genesis, Moses says, "*the waters were on the face of the whole earth.*" But it must be remembered his understanding of the *Earth* is a limited one. Consider also the following from the same prophet: He describes the LORD sending a swarm of locusts that, "*covered the face of the whole earth*" (Ex. 10:15). In the days of Joseph, Moses details a famine that "*was over all the face of the earth*" (Gen. 41:56). The Hebrew word translated in all of these verses is *eretz* (ארץ), it is the same word used in Genesis 12:1 where Moses records God's command to Abraham, "*get thee out of thy country*"; surely *eretz* is not always used to describe the *whole* Earth, God did not intend that Abraham leave the planet. Obviously these events were understood to be local, perhaps even *widespread*, as in the case of the famine, but all were *far from global* in scope, so too is the flood. Such a narrow view of the "*whole earth*" can be found elsewhere in other languages within the Scriptures as well. In Greek: "*A decree from Caesar Augustus, that all the world should be taxed*" (Luke 2:1). Devout Jews came to Jerusalem "*out of every nation under heaven*" (Acts 2:5). Within the book of Mormon, we are told that the book of Mormon peoples (constrained to *portions* of the Americas only) "*did multiply and spread, and did go forth from the land southward to the land northward, and did spread so insomuch that they began to cover the face of the whole earth*" (Helaman 3:8).

So, what are we left with to help us determine the date of entry of Adam into the Temporal sphere? Modern day revelation tells us that after his expulsion from the Garden, Adam built an alter at Adam-Ondi-Ahman in what is now central North America. We know, from what we have learned in this treatise thus far, that expulsion to be Adam and Eve's introductory point, or very near it, into the Temporal Universe we inhabit today. Modern day revelation also clearly defines the time of the flood in years following a point when the age of Adam begins to be counted. If we can associate the Bosphorus flume with the flood of Genesis Chapter 6, that would place the introduction of our first parents at about 7,300 years BC a little over 9,000 years ago.

How did the focus of the story of mankind in the Bible shift to what is now known as the old world? I propose this happened just after the translation of the City of Enoch, circa 6,300 BC. At that point, there had been over 1,000 years for the Sons of God to have spread through the Earth, intermingle and interbreed in the expansive, unbounded *Land of Nod*, with the daughters of the people of *the clay*. Much, if not all of the sons of God, were inhabitants of the City of Enoch and were taken up in that great translation event (*perhaps leaving no trace of their existence on the Temporal Earth on what may have been what North America is today, and we in fact do not have any archeological sites of such a civilization here)*. The remaining sons of God, including the children of Cain, infused with and dissolved into the outlaying wayward masses so criticized by and so troubling to God. Many of those masses were no doubt living at that time in the fertile valleys and plains of the ancient freshwater lake that is covered by the Black Sea today, completely unaware of their impending doom.[61]

It has been difficult to resist wading into speculation in the last few pages of this section detailing Day 6. I have forayed there only to provide some food for thought. I introduced the words of the Psalmist who spoke of 1,000 generations, and we never discussed further those unaccounted for 970 turns. We are left wondering, trying to piece together whole cloth from the threads of times and images of the past found written by God in the pages of His books, both those of paper, parchment and ink, and those of the strata of rock. It is all for us to consider, to be edified and strengthened by, and to learn.

## SUMMARIZING WHAT WE HAVE LEARNED, *Day 6*

Day 6 covers a lot of ground, much more than is apparent at first glance. In fact, though the Scriptures speak of a day of rest for God following his activities in Day 6, our examination here, the timing established by the CMBR,[62] and the surrounding Universe, tell us that the creative events of Day 6 continue even now in Temporality; the day of God's rest is still yet future. The Universe He created and watches over

---

[61] See <u>Adam in the New World</u>, in the Appendix, pg. 551.

[62] Applying the math we discussed previously, we see that our primordial clock places us today within Day 6 of the creation somewhere between the hours of 2:30am and 8:00pm. See Figure 3-3 on pg. 50.

remains hard at work fulfilling the measure of its creation. It evolves and changes still today in spectacular fashion, in both the tiniest and the grandest of scales.

Building upon Godly processes that began in Day 5, the creation (*bara*) of animal life continued evolving land animals which eventually led to the appearance of the animal, Modern Man. We discussed the shared heritage of all life on Earth. Despite enormous opportunities for incalculable diversity in structures and building blocks, life, *all of it*, is strikingly similar, impossibly so. This similarity is an artifact of the initial input of intelligence and law placed in the most basic building blocks (*bohu*) of the Universe by God as He *brooded* over them back in Day 1.

We learned that the last two verses of Genesis, Chapter 1 include a buried truth testifying of the divine source of the account, for in them we find a telling of the remarkable source of food and energy that sustains all life on Earth, the Sun. The details of photosynthesis are there in the words, though they escaped the notice of the writers and readers of the Scripture for millennia and were subsequently misunderstood for so long leading to false assumptions as to their meaning and the development of false doctrines misinforming our understandings of the Garden and the Fall.

The process of the development of life was long, as evidenced by the multiple appearances of the turn of phrase: "*And I, God saw that all these things were good.*" It appears not once, but twice in the verses covering the creative events of Day 6. The phrasing turns out not only to suggest a quantitative measurement of time but one that is qualitative as well, for as we know Day 6 is no longer than any of the other days. Through the first use of the phrase and its placement, we are shown that the arrival of animals on the Earth predate the introduction of man and Adam by a considerable length of time. In the second use of this phrase, God adds to good the designation *very*, signifying the completion of the Temporal sphere and its attainment of all the attributes it requires to perform its divine purpose.

The Temporal Creation, though complete, is not static and unchanging. It continues to evolve, change and self-create in the precise manner in which it was commanded. Because animals precede man on Earth by much time, the record allows for evolution (of the correct type) and the existence of long extinct creatures like the dinosaurs of the Mesozoic.

Man, (men) those who are *of clay*, are an extension of the evolutionary process though these creatures remain separate in distinction from Adam and Eve and their lineage, the sons of God. We looked forward in this book and back in time to discover the truths about human and animal Spirit, the *Nefesh* in animals and *Neshama* in man, *the sons of God* and the lineage of Adam.

Though we never discussed it in detail, it may be inferred that Pre-Adamite man on Earth lack the *Neshama* and are only bestowed with the *Nefesh* of the animals. In examining the Hebrew texts, we discussed the use of a double *yud* in the text to speak of the forming of man and how this unusual use is meant to set the creation of the sons and daughters of God apart from that of the animals and by extension that of Pre-Adamite man.

We discussed at length the introduction of Adam and Eve into this Temporal Universe and discovered that much is shrouded in ambiguity and even mystery. It was made clear that Adam is a late comer to the Temporal Universe, most recent in introduction, as if just a moment ago in the history of this almost 15 billion year old Universe. Though in that context, it seems like so much time had passed since the Fall that initiated this Universe, we have to remind ourselves that this is just Day 6, as measured by God in the frame of reference He chose to use in the telling of this story. We must also remember our previous discussion in Chapter 2 regarding the passage of time and God's existence outside of it.

We avoided with guardedness the ever changing landscape of *recent* scientific discovery often espoused by so many as fact beyond compare and *instead* developed our understandings here on science vetted by time, decades of experiment, and extensive peer acceptance. In a few cases, we endeavored to associate, if ever loosely, some things that are less certain to science with some of the storied even "*so-called mythological*" events in the Bible. Where it counted, in the precise detail of Scripture and the firm foundations of scientific fact, we found harmony and accord as well as mutually supportive evidence of events long past. Everywhere we looked, with each page we turned, we found that the best of modern scientific study correlated precisely with what was revealed to man by God so long ago and recorded in the early chapters of Genesis.

Day 6 began 650 million years ago by our accounting, and the six verses describing it methodically detail truths about how our biosphere sustains life, the interrelationship of animals and man, and man's relationship with our Father in Heaven. Directly correlating with the theories of science, the account of Day 6 properly sequences the appearance of animals and the most recent appearance of man. We are told of the process of photosynthesis and how through it plant life sustains all animal life of Earth. Our Temporal existence is not a curse; we are told that the Temporal Creation blesses Adam and Eve and all of their offspring because it allows for procreation and the magnifying of our abilities to learn and grow in wisdom and power.

*This ends the account of Day 6 and the story of the creation of the Temporal Universe.*

# CHAPTER 4

# THE IMPLICATIONS OF A TEMPORAL SPHERE

## Summary of the Temporal Creation

*"I felt it in my bones that this universe does not explain itself."*
*C. S. Lewis*

C.S. Lewis' bones had it right, the Universe does not explain itself, but rather God explains it, and has done so through it. In detailing all that has been revealed pertaining to the creation of the Temporal sphere and linking what has been discovered by science, we developed a structure that is remarkable to behold; a structure that discloses so much for us to consider.

### *Traditional Creation*

The traditional Judeo/Christian view of the creation story includes the Garden account as well. As it is traditionally told, God created an immortal universe and all that it would contain in just six days, culminating in the creation of our first parents, Adam and Eve. The Universe of tradition was transformed in response, and only in response, to the sin of Adam, becoming, at once, mortal. This transformation was somehow both unplanned by a God (who only intended man to live in the paradise of the Garden) yet had to have been foreseen by his omniscience. This is just the first of many problems the traditional view of creation holds.

The story of tradition does not explain why men were created in the first place; in fact, nowhere in the Bible can we find an answer for that question. We are left to wonder, if God is so good, then why did He create sin, or at least the conditions that allow its existence? Was the innocence of man our *intended* state? If so, then why are our minds capable of such soaring thought? What of the Universe around us, how does it fit into the simple story told by tradition? If it was created just for us, why is the Universe so big, so full of complexity, so beyond the grasp of our God given senses? Its machinations, in their vastness and sheer number expend so much energy, yet for most of mankind they provide absolutely no informative, educative, or moral value. Why does the Earth and life itself appear to be so old?

The traditional view has as its strength, *its simplicity*. The simplicity shown is not because, in its whole, its telling of the story of creation is beautifully concise, but it is because the story tradition tells is *woefully incomplete*. The traditional account is fraught with problems as it leads to many unanswered questions, leads to implications that are contradictory, and a review of the Scriptures reveals the story as interpreted by tradition is more a fairy tale, drawn from the imagination of man, than it is an accurate description of the Word of God in all of its beautiful detail.

This obviously flawed traditional story of the creation is not the exclusive property of sectarian Christianity, Jews and Muslims; it is also held as the definitive account by most Latter-day Saints, though with slight modification. Latter-day Saints resolve issues of the origin of sin, the purpose of life, why we are mortal and others, but leave intact dogmatic principles borrowed from centuries of error piled on by other theosophies.

Among the Latter-day faithful, you can still find many who believe the Earth to be just a few thousand years old. There are those that believe the Garden of Eden was located in what is now Missouri, that Adam was the first *man* on the Earth, and that there was no death on the Earth prior to his Fall. There are Latter-day Saints who insist that the theory of the evolution of life and the Universe is wrong, that the fossil record of life on Earth, stretching back into the millions of years is a false ruse, planted here by the deceiver. There are many who believe that the Earth, Sun, and Moon were shaped literally in the hands of God, formed as though they were fashioned of clay and hung in their orbits, without any process of physical law. Truly the view of Biblical "*literalists*" is not unknown to those who accept modern-day Scripture and revelation.

The errant doctrines found in the traditional view are subtended by a lack of understanding. This is explicable for the sectarian world, so much of what they accept as Scripture is in fact dogmatic interpretation repeated so often, for so long, it supplants the Word of God in their theological models. But to hold such views, the Latter-day Saint must discount entirely Latter-day Scriptures that reveal truths like: many Earths, all inhabited, that preceded and passed away before this Earth was created. When one considers the many Latter-day Scriptures that speak to such things, one must challenge the concept established by tradition that Adam was the first man on Earth. If Adam was the first flesh on Earth, then to which of the many, multiple Earths would he have been sent? Why here? What of the history of the other inhabited spheres?

Many incorrectly conclude that the Fall was an event that happened in *this* Universe in a Garden located on *this* Earth; we have seen this is also not so and will explore this concept fully later on in this work. The Scriptural account does not support this belief and there is no understandable means of transitioning the Universe as a whole, or any small part of it, from an immortal state to a mortal one. The Universe we call home was initially composed as mortal and is dying that way in whole. Many speak of the return of the Earth we stand on today to an immortal and paradisiacal state from which it changed as a result of the Fall. This too is in error, for the paradisiacal Earth is a construct of the Physical Creation; it never has been, nor ever will be, a component of the Temporal sphere.

## *Reading the Signposts of the Temporal Creation*

The creative processes for the Temporal sphere we detailed in Chapter 3, are in accord with Scripture and the correct observations of modern science. This is not surprising, for the Universe developed by a correct reading of the Genesis 1 accounts *is* the Universe we live in today. It is, as it must be, entirely mortal and ancient in form, not recent in origin. The Temporal Creation must *follow* the Fall and the planting of the Garden in Eden, not precede them. It is not the same creation that resulted in the Garden or any of the things within the Garden; that creation must precede the Temporal Creation we have described, and it does. The Temporal Creation follows the creation of Adam and Eve.

Because it was the intent of the Creator from the outset to produce an *uncertain* temporality which allows for free agency, the Temporal Creation eliminates objections to the existence of God, which are based on the implications of *perceived imperfection* in the intelligent design of the Universe. Because of the way it was created, this Universe is subject to the laws of physics. Indeed we have shown that the information necessary to instruct the matter (*bohu*) of the Universe was infused into it at the outset by God. Those laws provide the rules and structure that binds the Universe together, animates it, and allows it to evolve, operate, and respond to the actions of sentient beings. These laws allow for discoveries made by science. Our view of the Universe encourages such discovery and examination.

> *"...you can say 'I'm cool with science,' provided you allow science the berth to discover whatever it can about the Universe no matter what you say it should be discovering based on Scripture, ...If that's the case, then there's no conflict." Neil deGrasse-Tyson*

What we have done with our considerations in detailing the events of the Temporal Creation is a modification of Mr. deGrasse-Tyson's requested proviso. We have allowed science to discover, determine and explain. We have in turn shown that Scripture describes these very same truths; indeed we have exposed that it, in fact, had done so millennia ago. The Genesis 1 creation accounts we have seen are extraordinary just for this reason, for they remarkably describe details of the Universe only most recently known to science. The extraordinary nature of the Biblical record is even more marked when we understand that the things detailed were entirely counterintuitive and unknown to man of 3000 - 4000 years ago.

The Temporal Creation we described leads to many implications that we will address within this chapter:

- The Temporal Universe as we have seen was intelligently designed by its Creator.

- Our Universe was created with purpose; it has a *First Cause*, the Fall of Adam. This *<u>First</u>* cause is consistent with a preceding and *universal, <u>Eternal</u>* cause. We will examine the intricacies of these concepts and discover why our unique understanding both eliminates the pitfalls of the *first cause principle* while embracing the compelling logic it advances which insists there must be a creation at a fixed point in the past *and* a creator.

- Our understanding of the Temporal sphere: why it was made the way it is, that it is of itself only a component, a link in a chain, of events and constructs, all designed to facilitate the eternal progression of the sons and daughters of God, provides an intelligible answer to the existence of evil and unshackles God from incoherent logic that insists He is its creator.

- Through the decoupling of the eternal Physical sphere from the mortal Temporal sphere we developed an understanding that the Fall could, and indeed must, precede the creation of this Universe entirely. This decoupling allows the effects of the Fall to be justly shared with all of the sons and daughters of God who occupy this sphere; for their entry into it was accompanied with an awareness of the consequences, and an acceptance of the risks before they came. We (all of the sons and daughters of God) came here to be taught, gain experience, and be challenged, so that we might grow and progress in a way that could not happen otherwise. Because the effects of the Fall are universal, they cannot be avoided. To compensate for the destructive effects of the Fall, the Plan of Salvation was established and its central sanctifying principle, the Atonement, was instituted *before* the Temporal sphere was created.

- The Temporal sphere implies that there may be inhabited alien worlds. Scripture tells us there are many and that they are inhabited by sons and daughters of God. We will discuss these worlds and their separate histories, some whose pages opened and closed before ours began and perhaps some with histories yet to be written, but all within the physical confines of this Temporal Creation. We will discuss the many Scriptures that testify of these worlds and will explore in depth the ramifications of this understanding.

- We will discuss the implications of the Temporal sphere as they pertain to the Atonement. We will discover that the required events of the Atonement, the incarnation, death, and resurrection of Christ need only occur once, in one place within the Temporal sphere. We will show that place was here, on this Earth, in the environs of Jerusalem, just over 2000 years ago.

This implication, when coupled with yet another implication of the Temporal sphere, the probability of sons and daughters of God on other, distant and alien planets, leads to even more questions in search of answers. What does the Meridian of time mean? 7000 years have been appointed for man to strive on the Earth; how does that relate to the age of the entire Universe or to the independent histories of life on other planets? What about worlds that have passed away before? What of those that are yet to be? How is the Gospel message shared on other planets? What of the ministry of the Savior on these alien worlds?

- The Temporal Creation we detailed in the previous chapter describes the appearance of life on Earth billions of years before the arrival of man. In this chapter, we will discuss some of the implications that are a result of that discovery.

- Lastly our Temporal Universe is a construct that exists entirely within the broader eternal Physical sphere. We will discuss this briefly as an introduction to the next part of this work, *Part 2 The Physical Creation*.

## Intelligent Design?

One of the implications of what we have discussed about the Temporal Creation is that we unequivocally assert that an intelligent creator developed the Universe. In the previous chapter, we noted that six times in as many creative periods the records show the direct involvement of God (*bara*) in the physical processes of the forming of the Universe. It must be made absolutely clear that intelligent design (as we understand it here, and the infused intelligence that attends it) does not *replace*, but it informs, empowers, directs, initiates and provides purpose to *all* properly formulated scientific explanations for the development and subsequent evolution of the Universe and all life within it.

From the very moment of creation and throughout Day 1, we observed the *brooding* of God as He imbedded within the forming elements of matter the knowledge (information) they would need to build themselves from the bottom up into larger and more complex forms in accordance to the wishes and will of God. Our belief in this process is religious and not scientific, but it comes to us from the reading of Scriptures, ancient in origin, written by someone who had no understanding of the precise scientific details they hold, details which emulate those found within our most advanced understandings today. We assert that these ancient writers could only have known of these things from an external, informed source, indeed the very creator who performed them.

> ***...we unequivocally assert that an intelligent creator developed the Universe.***

The intelligence found in the Temporal creative process detailed here is not the same as, nor does it draw upon any of the elements of Intelligent Design espoused by many who oppose evolution for ideological reasons today. Though some elements of the Intelligent Design argument offer thoughtful challenges to standard evolutionary theory, we have not relied upon them here. However, I would like to consider the concept of *Irreducible Complexity* as a problem posed to the Theory of Evolution as one half of an illustration of conflicting views which can be brought into harmony and accord by the thorough reading of the Genesis 1 account we have conducted here.

In 1996, biochemist, Michael Behe, proposed that many biological systems are "*irreducibly complex*" at the molecular level. Behe gives the following definition of irreducible complexity:

> *"By irreducibly complex I mean a single system composed of several well-matched, interacting parts that contribute to the basic function, wherein the removal of any one of the parts causes the system to effectively cease functioning. An irreducibly complex system cannot be produced directly (that*

*is, by continuously improving the initial function, which continues to work by the same mechanism) by slight, successive modifications of a precursor system, because any precursor to an irreducibly complex system that is missing a part is by definition nonfunctional. An irreducibly complex biological system, if there is such a thing, would be a powerful challenge to Darwinian evolution."*[1]

Behe uses this definition to show that many biological structures and chemical processes are irreducibly complex and simply would not have any useful function unless and until all of the component parts or processes were in place. Behe contends that the gradual evolution taught popularly today, a theory that espouses and relies on a series of small improvements both in structure and effectual process adding up to advancement in the end result, cannot explain the appearance of these irreducible complexities. He argues that before the process or structure is complete, it has no purpose; the evolutionary process of selection then, would not have a reason to favor their repeatability over any other configuration. Why continue to reproduce, at cost, individual parts of a detailed mechanism if they have no purpose until they all were available and assembled?

Irreducible complexity became a buzz phrase of *creationists* and a bane to their detractors. Many attempts to refute it as a viable concept have been made, some of them laudable[2] and some quite cute, but for the most part all have failed to dispute irreducible complexity as an *a priori* principle[3] just as valid as the *a priori* principle of natural selection that empowers the theory of evolution. However, in the end, whether you accept the conundrum imposed by irreducible complexity or not is much more dependent on your philosophical bias and mindset going into the argument, than it is anything else. If you are an ideological believer in a-teleological evolutionary processes (many that were not discovered by Darwin and are not supported by the fossil record), then you will reject the premise. If, however, you believe in a creator and an intelligently designed, purposeful Universe, you may find irreducible complexity presents a chasm so deep and wide that it is impossible for naturalism to cross, let alone skirt around.[4]

---

[1] Darwin's Black Box: The Biochemical Challenge to Evolution (Michael Behe, Published by Free Press, 1996, pg. 39).

[2] See, From the Origin of Species to the origin of Bacterial Flagella (Mark J.Pallen and Nocholas J. Matzke, *Nat Rev Micro 4, No. 1*, 2006, pgs.784-90).

[3] It is often hard to adequately define what is irreducibly complex and what isn't, but it is something we *know* when we see it. Irreducible complexity is not a law, but it is even more fundamental. Even a child understands this concept when they smash their favorite toy beyond repair and recognize that the whole was something more than the sum of all of its parts. Once the function is lost its intrinsic value is gone; evolution assigns value only to the intrinsic advantage a random change produces.

[4] Our contention within this book places the intelligent designer/creator not so much *within* the processes of evolution/change over time as it places intelligence and purpose at the beginning, imbuing the very "*stuff*" of the Universe with knowledge and internal sense of form and function (commanding it to fulfill the measure of its creation). This is design at the most basic level. Protons know how to be protons and matter knows which way to fall in a gravitational field, even though in free-fall it cannot instantaneously feel any force acting on it. At the outset, the *bohu* was instilled with the laws of physics and empowered to develop into the complex structures we see today, with an instinct to survive and reproduce, and the ability to adapt and change to do so in an uncertain Universe.

Our understanding of the Temporal Universe leaves no such divide in ideology. We have fully accepted the scientifically verifiable components of biological and physical evolution of the Universe and the life within. In fact, we have embraced them, as we have shown that sacred Scripture describes these processes and did describe them millennia before they were conceived by science; we have taken nothing away. What we have added to the picture drawn by science is the driving mechanism for what the Universe does. We have filled in the gaps where theory had no answer, providing purpose and sense of direction to a process that had none, explaining why one variation might be *selected* over others when no valuable external input may have been available.

The stumbling point of evolutionary theory posed by Irreducible Complexity fades away when we add in the initial loading of the primary matter (*bohu*) of the Universe with knowledge, and examine the wording of Genesis 1, which details the intervention of God to direct and compose nature at critical times, the (*bara*) moments of the first six days. Where Darwinian evolution, as it has come to be understood today, fails in explaining punctuated equilibrium and providing the initial impetus of life, and indeed the force and reason behind the Big Bang itself, we have provided the answer. Through reading and rendering understandings of a text over 4,000 years old, we have now delivered a complete picture.

> ***Physical laws themselves cannot create anything; they are simply a description of what happens when previously existing things interact.***

The laws of physics are called upon by naturalists as the replacement for God. According to some, the laws of physics, not the will of God, provide the real explanation as to how the Universe and the life within it came to be; the Big Bang to them was an inevitable consequence of these laws. In his recent work, Stephen Hawking said, "*because there is a law such as gravity, the universe can and will create itself from nothing.*"[5]

This is not a new thought or groundbreaking discovery of Hawking, though he may claim authorship, he, in fact, has merely rephrased what has been said by others for decades; yet it is still wrong. Physical laws themselves cannot *create* anything; they

---

[5] The Grand Design (Stephen Hawking and Leonard Mlodinow, Bantam, 2010). The passage quoted continues: "*...Spontaneous creation is the reason there is something rather than nothing, why the universe exists, why we exist. It is not necessary to invoke God to light the blue touch paper and set the universe going.*" Of this book the famed and outspoken atheist Richard Dawkins wrote, "*Darwinism kicked God out of biology but physics remained more uncertain, Hawking is now administering the coup de grace.*" From, Another Ungodly Squabble, The Economist (September 2010).

are simply a description of what happens when previously existing things interact. Where do the things come from; how do they *know* how to act? To Hawking and others, the laws are somehow self-generating and all governing at the same time. This is misguided and it posits to some that God and the laws of physics are in mutual conflict, when they, in fact, are not. We do not have to choose between God and these laws. We know that He created law and placed those laws within the *bohu* back in the first day of the Temporal Creation.

How do some conclude that God and physical laws and processes are in conflict? Consider the following quotes of Neil deGrasse-Tyson:

> *"Any time someone describes their understanding of God, typically it involves some statement of benevolence or some kind of kindness... I look out to the universe and yes, it is filled with mysteries, but it's also filled with all manner of things that would just as soon have you dead. Like asteroid strikes, and hurricanes, and tornadoes, and tsunamis, and volcanoes and disease, and pestilence,"* he continues. *"There are things that exist in the natural world that do not have your health or longevity as a priority. And so I cannot look at the universe and say that yes, there's a God and this God cares about my life – at all. The evidence does not support this... I think of, like, the human body, and I look at what's going on between our legs," said Tyson. 'There's like a sewage system and entertainment complex intermingling. No engineer of any intelligence would have designed it that way."*[6]

Dr. deGrasse-Tyson is completely wrong, of course he is, but not where you would expect him to be, not about his definition of God. God *is* benevolent and kind and ultimately cares for and loves us, His children; deGrasse-Tyson got this right. He expresses, surprisingly, a proper understanding of the theological; where he is wrong, however, is in his understanding of the Temporal sphere and its purpose, the very foundation of the laws of physics, but, in this, he is not alone.

Though deGrasse-Tyson is an avowed atheist, the previous quote suggests he envisions the attributes of God to be the same as those held by Christians. His problem, the same one that manifests itself as an enigma to many Christians, is in the harshness and *apparent* imperfection of the Universe around us.

To most Christians this conundrum is simply chalked up to the mysteries of God, or as a manifestation of some form of divine justice for sin initiated by Adam, a consequence of the Fallen nature of the Universe.

For an unbeliever, a Universe that *is* dynamic, dangerous and even a hostile place is a deal killer for any belief in the existence of God.[7] For them, and specifically for

---

[6] The "*Nightly Show*" (November 4, 2015)

[7] This reasoning leads to a counter claim that runs like this: Despite the dangers we know within the Universe, we exist because our home, the Earth, has been shielded and meets a large number of extremely unlikely conditions prerequisite for life, the Rare Earth hypothesis. That these conditions exist, when they shouldn't, prove a designer; even if the design is simply to produce so many possible environments that statistically one would meet the requirements.

deGrasse-Tyson, their logic follows this line of reasoning: *God is perfect and creates only perfectly. The Universe and we are not perfect. Therefore God does not exist.* The argument has two premises and draws a direct conclusion. Though the argument is logical, the conclusion is incorrect because the premises are false assumptions.

The first premise is without a valid source; who says God only creates perfectly? This question is valid for as we have seen in this work, *it was God's perfect intention to create an imperfect Temporal sphere*, that imperfection allows for the free agency of man through quantum uncertainty. Those who draft the false premise that God only creates perfectly, do so because they do not understand the purpose and intentional design of the Temporal sphere. In effect, it is not so much that God creates imperfectly as it is that our understanding of what perfection is as it applies to the Universe we inhabit is skewed in the direction of error. This Universe is perfectly designed to allow for the free agency of man while affording life the adaptability it needs to survive in a world where uncertainty MUST be allowed.

Misunderstanding applies to the formation of the second premise as well. Contrary to the premise that the Universe and we are not perfect, we proclaim that the Universe and the life within it *perfectly* fulfill its purpose and declare it is incorrect to conclude otherwise.

An implication of the Temporal sphere is that it was designed to allow us to experience the effects of wrong choices and disobedience at every level. It is true the Universe is a dangerous and immensely hazardous environment, and to the under-informed or arrogantly biased mind, life may seem poorly designed to meet the challenges. But these things do NOT provide evidence of natural evolution through random events or some sort of argument against an intelligent, purposeful Creator, God. They are a result of God's desire to allow us to become like him; they are a necessity of our free agency and only appear to be *imperfections*. The scientific and sacred record both show that God spent much time getting the conditions just right so we can inhabit the Temporal sphere just a short while. We attest that our Fallen nature and the Temporal Universe are perfectly suited for His divine purpose. We were originally created in temporality, just as we are now, perfectly suited to meet the measure of our Temporal Creation. There is no imperfection of design here.

With both premises falsified, or more properly, understood through redefinition, we can see that the implications of the Temporal sphere, as it applies to the intelligence of its design are far reaching and provide us with a complete and coherent understanding of our Temporal physicality, our Creator and the purpose of life.

## The First Cause

A strong implication of the Temporal sphere is that there is a First Cause. First Cause is a logical argument that insists that the Universe must have a cause. Known as the argument from first cause, or the cosmological argument, it has been used for millennia to provide proof that there is a creator, God. Here we accept the First Cause argument and proclaim that this cause is the Fall of Adam and the plan of our Father in Heaven for the eternal progression of his sons and daughters.

What is First Cause and why is it needed? To answer this, we will look at the concept and rearward into history to flesh out our understanding. The argument is really very simple and intuitive. It is based on the instinctual sense we all share that everything needs an explanation, nothing can just be. Everything needs a reasoned, adequate, and sufficient explanation for its existence. This is what is known as the philosophical *Principle of Sufficient Reason*.

If we saw an object suddenly appear in a previously empty space, none of us would assume it came out of nowhere. We would all expect its appearance to have an explanation, a *cause*. We naturally, instinctively look first for a physical cause. Did someone put it there when we weren't looking? Did it fall from the sky? If no physical explanation can be found, we examine further, looking for psychological alternatives, perhaps we were mistaken and did not see it there before when in fact it was, perhaps we just *imagined* its sudden appearance. Lastly, we may turn to a supernatural cause, a miracle. Try as we may to explain, we may never find the reason, but we never deny the Principle of Sufficient Reason; we are absolutely certain there must be a cause. No one ever accepts that the object just came into existence for no reason at all. Our commitment to a necessary or *efficient cause*[8] increases proportionately to the complexity of the object requiring it.

The Universe exists, we exist within it, and these are two undeniable facts. In the middle of the 13th century, the theologian St. Thomas Aquinas used these facts to propose an argument for a creator from efficient cause, it went like this:

- *There is an efficient cause for everything; nothing can be the efficient cause of itself.*
- *It is not possible to regress to infinity in efficient causes.*
- *To take away the cause is to take away the effect.*
- *If there be no cause then there will be no others.*
- *Therefore a First Cause exists (and this is God).*

[8] An "*efficient cause*" is any action or thing which causes change, simply the thing that brings something about. For example, in the case of a painting, it is the person with the brush which transforms a blank canvas into a scene of beauty. Efficient cause need not be mechanical; the action of painting itself was *efficiently caused* by the painter's *desire to create* the work and his imagination, perhaps even the inspiration provided by a beautiful scene before him. An efficient cause is one link in a chain of efficient causes, all of which can be discovered and defined.

Later in this section we will discuss this argument in further detail, but for our purposes here, it is sufficient to note that Aquinas envisioned an eternal God creating an eternal universe. His argument uses our instinct of *Sufficient Reason* and is based on the first premise, that existence demands a cause that is not of itself; the eternal Universe exists therefore it needs a cause. Aquinas' next premise arbitrarily limits the number of causes there can be, seeking to get to a primal cause. The subsequent premises work towards pointing to a creation and then to a Creator.

In essence, there are two parts to Aquinas' argument: The first I have already alluded to; *we exist and so does the Universe; existent things must have an efficient cause; therefore, there must be a cause for the Universe*. The second part of the argument, however, was framed to insist that there must be a First Cause which precedes all other cause and is of itself not preceded. Together the two parts of the argument form a cohesive, strong case for the existence of God. The centuries following Aquinas, however, saw the weakening of the second premise as the eternal nature of the Universe was conceptually extended backwards in time.

Through some process of thought and understanding that is unclear, the creation story became localized. Seemingly overlooking the fact that the Genesis 1 account describes both the creation of the Heavens and the Earth in just six days, our scientific and theological traditions were allowed to envision and accept the immense depths of space itself as being eternal. For religionists, it came to be the eternal abode of God and things spiritual. The creation story for them involved only this Earth and its environs *within* an eternal universe. To science, the concept of things eternal was distilled down even more to simply leave a universe that was thought to be unchanging, in an eternally static and stable state.

As science progressed, cosmology was perfectly comfortable with the concept of an eternal, static and unchanging Universe. Einstein was so convicted that this was fact that when the brilliance of his initial relativistic calculations predicted a universe that could not be static, he repelled and reeled. To "*correct*" what had to have been an "*obvious flaw*" in his equations, he introduced a factor to compensate for it.[9] An eternal static universe had an implication; it did not have a beginning and could, therefore, skirt the second premise of Aquinas' argument from Efficient Cause, avoiding a First Cause altogether.

Einstein and science were not alone in their belief in an eternal static universe; there were others voices. At about the same time Einstein was formulating his original equations, Elder B.H Roberts was laying down a theologically focused argument that insisted there was no First Cause; he wrote:

---

[9] The correction was a cosmological constant, the insertion of which was later considered by Einstein himself to have been his greatest mistake. When Hubble and Lemaitre proved that the Universe was expanding, Einstein's original equations were vindicated as they precisely predicted what observation had proven. An interesting side note to this cosmological constant, fudged into the equations by Einstein, is that in most recent decades the Universe has been found to have an accelerating expansion. The acceleration has a value that precisely matches the constant introduced by Einstein, making his "*greatest mistake*" yet another triumph instead.

*"... we hold the universe to be self-existent and eternal. Duration, space, matter, force, mind – each infinite after its kind – are its prime and included factors. The universe is comprised of these.*

*Such a universe can have no first cause, since that would imply a time when there was no cause, and there has been and can be no such time. Causation is eternal, and in the eternal and infinite universe there has been operation always Eternal Cause."*

Don't lose track of what has happened here; to Elder Roberts, the Universe has become eternal and self-existent (not very much unlike the multi-verse described by Stephen Hawking today). Elder Roberts envisioned a Universe that is one continuous chain of causes without one that was *first.* He has supplanted a First Cause with an eternal one. In doing so he is detailing the concept of cause in an ever so slightly different way.

Elder Roberts correctly identifies that cause is not just linked in a simple way as in a chain, but it is also linked like the branches of a tree or more precisely like an engulfing vine. For Elder Roberts, *Eternal Cause* is understood as something that occurs continually, not just as something that happens in sequence. A chain like sequence can be described as follows: I am alive *because* my parents begat me and they *because* their parents begat them and their parents *because*...ad infinitum. It can be unending because, to the first (last) link, one can always add yet another.

For Elder Roberts, however, Eternal Cause is much more than the simple sequence we just described. It is as follows: I am also alive because: nuclear fusion in the Sun produces sun light, photosynthesis produces oxygen, I breathe and oxygen fills my lungs, my heart beats and the blood is oxygenated, my heart beats again and oxygen moves to my cells, oxygen is converted to...etc. I live because *all of this is going on all of the time*. Elder Roberts' Eternal Cause is a sustaining thing that is continuous; creation to him is an ongoing work of God. This concept described precisely the dynamic, evolving, yet static and unchanging (in the broad view) eternal Universe known to science of his day.

*Don't get too comfortable, things are about to change...*

ENTER HUBBLE

With just one day left in 1924, Edwin Hubble changed the Universe when he discovered other galaxies beyond our own.[10] By 1927, his discoveries had led

---

[10] I warmly remember time spent with Lois Keener Thome (1897-1994) as she interviewed me before nominating me to the American Astronomical Society. Mrs. Thome spoke fondly of her time working in the computing division of the Mt. Wilson Observatory from 1919 to 1924, during which she computed *by hand* the numbers used by Edwin Hubble to prove galaxies were beyond our own Milky Way. Those same calculations proved that the Universe was expanding. She spoke of being approached by George Hale and Edwin Hubble together when they announced their nomination of her for membership in the AAS almost a century ago; an amazing woman and wonderful story, I will never forget.

Georges Lemaître to prove that the Universe was expanding and the concept of a static, unchanging Universe was gone. It would only take another two years for the same Lemaître to propose the rudimentary form of a theory that would later become the *Standard Model* of the Universe, the *Big Bang*. In January 1933, Lemaître and Albert Einstein headed to California to present a series of seminars. At one, after Lemaître had detailed his theory that was to become the Big Bang, Einstein stood and applauded saying, "*this is the most beautiful and satisfactory explanation of the creation to which I have ever listened.*" *Creation?* Yes, that is what Einstein had said. Lemaître's theories left no doubt; the once static, eternal Universe was replaced by one that had a beginning; a beginning at a specific place in time somewhere in the distant past; a Universe that would now need a First Cause.

Almost overnight science had come full circle, back to a beginning and everything had to start anew. With a bang, a very Big Bang, science had evidence of a starting point for the Universe, and we had to begin to deal (again for some, and for some for the very first time) with a time *before* and what *caused* it, a *First Cause*. The discussion again turned then to Aquinas' arguments, and we will as well, here within this book.

Aquinas' second premise: "*It is not possible to regress to infinity in efficient causes.*" proved to be true in a very prophetic way, so much so that it is often lost in translation. The Big Bang theory begins with a singularity, by definition that is a point where space, time, matter and all the laws of Physics cease to exist. Without any place to act, no time to act in, nothing to act on, and no rules to act by, there is no action, and there can be no cause. Our internal instinct told each of us that cause could not regress to infinity. Aquinas codified that instinct into words and an argument, and 800 years later, science had finally confirmed it. At the Big Bang, with the universal singularity, the chain of efficient causes into the past ended.

Aquinas' third and fourth premises: "*To take away the cause is to take away the effect,*" and, "*If there be no cause then there will be no others.*" follow precisely the commonsense and instinctive logic of the *Principle of Sufficient Reason*. By them Aquinas argues, because we have the Universe where once there was none, it must have had a cause, and since we have causes within it, there must have been an *initial* cause. His conclusion is the same one we have made here; there must be a First Cause. Aquinas declares: "*Therefore a First Cause exists.*"

Why must there be a First Cause? Because the Big Bang insists the Universe had a beginning. A beginning without a cause (a first cause) violates the Principle of Sufficient Reason leaving the Universe and everything within it unexplained. Without a First Cause, very specific individual things and events in the Universe only have a *proximate*, short term, explanation (an efficient cause) for how they respond to things or forces in contact with them. Without a First Cause, nothing has an *ultimate*, long term explanation. The Universe as a whole then remains inexplicable. Every object, action or force, can reply in unison, "*I am just doing what I am told, what I have to, I am not the final explanation, I am just a mechanism. Something else caused me.*" Without a First Cause, we are left with an endless chain of passing the buck; a buck that ultimately stops nowhere.

So how did Elder Roberts come to the conclusion that there was no First Cause? Clearly he built his theology around his belief that thc Universe was eternal. He was a man of science as were his contemporaries, Elders Widtsoe and Talmage. Their understanding of a static eternal Universe stemmed as much from science as it did anything else, perhaps more so, *certainly Scripture did not reveal it*. They shared this view along with all the great men of science of their era: Einstein, Planck, and Eddington among them. For them all, there was no doubt the Universe was eternal, without either a beginning or an end. For Elder Roberts, his "*eternal*" Universe did not need a First Cause because it had no beginning. For him it was into an eternal Universe that temporality *came*, for within it the Fall *happened*. Elder Roberts projected the account of Genesis 1, as a local event, onto this sphere he assumed to be eternal. We have seen here that his was an incorrect assumption, but one that was to be expected of any learned man of his time.

The Temporal Universe we have described here was created with a purpose and a cause. An implication of our understanding of the Temporal sphere, is that it exists entirely *within* an eternal continuum that is the dwelling place of our Eternal Father. We affirm here that Elder Roberts' "*Eternal Cause,*" that which has been in "*operation always*," is the work of God, His purpose as revealed by Himself to Moses:

> *"For behold, this is my work and my glory—to bring to pass the immortality and eternal life of man."(Moses 1:39)*

## OF ATHIESTS AND LITTLE CHILDREN

The Principle of Sufficient Reason and the argument Aquinas built around it, is sound and intuitive. To dispute it one must *attempt* to be very clever, or incredibly stupid; take your choice. There is a difference between ignorance and the lack of intelligence. All too often those who seem most intelligent display complete and utter ignorance.

The most common objection to Aquinas' argument from efficient cause stems from his first premise and a circular argument that can be drawn from it. Nothing can be the efficient cause of itself. Opponents point at what seems to be a contradiction in this first premise. The question arises: Is God something or nothing? If God is something, then it can be asked: "*What caused God?*" If God is nothing, then God does not exist. If special privilege is extended to God (if God is something that has no cause), then, they say, the argument becomes flawed.

The question, "*What caused God?*" is a legitimate one when posed out of pure curiosity, as children sometimes do. But, it is often expressed as a statement of objection to theists, as a supposed logical wall put up by atheists and their like-minded. But on close examination their objection is not with the concept of cause or causation, but their argument is with the *causer* Himself. They seek to eliminate entirely the God that First Cause implies, preferring to conceive a Universe in which everything *just is*. This, however, runs counter to *every* observation that has *ever* been made and to *everything* we know to be true.

Those that challenge Aquinas merely posit that the Universe and the Laws of Physics are self-causing and self-existent. This they do by choice as they are in effect just unseating one thing, a creator, for another, an eternal nature.

This is not a superior logical conclusion; *removing one step from the stairs does not make the height of the climb any less*. By ascribing to nature the same attributes ascribed to God, nothing has been accomplished, and opponents of God have in fact produced yet another god, just this time one without a name. Choosing nature over God becomes a matter of ideological choice, not one that is based on a superior logic. If you are stuck on the fence, I have a little wager for you. Actually, Blaise Pascal, a 17th century French Christian philosopher, mathematician and physicist, has one for you to consider:

- *If I believe and God exists: I'll be happy for eternity = infinite gain.*
- *If I believe and God does not exist: I've lost nothing and gained nothing.*
- *If I do not believe and God exists: I'll be punished for eternity = infinite loss.*
- *If I do not believe and God does not exist: I've lost nothing and gained nothing.*

The reasoning here goes like this: by believing, we cannot lose anything, but if we do, we can win a lot. If we choose not to believe, we can lose a lot but cannot win a thing. So, the best wager is to believe.

***Choosing nature over God becomes a matter of ideological choice, not one that is based on a superior logic.***

Clever and true as the wager is, belief based on this type of logic (we know) is genuinely unfruitful. Because it is not sincere, it does not develop a relationship with God and does not grow our faith. Belief is the price of the wager, and it is impossible to believe something which deep down a person knows to be false. So true *belief* cannot be established by the logic of Pascal, no matter how much it may seem to make gambling sense.

The Temporal Creation, as we have discussed in this work, requires a First Cause because it began as a singular, locatable point in the past before which something was not and after which something came to be. Science and the Genesis 1 account of the Scriptures identify this very point as what we today call the Big Bang and its Singularity. We have seen this could not have happened without a First Cause. Our faith and the circumstantial evidence for the divine source of what is obviously advanced information in such an ancient record, demand that we accept the record as accurate in its detail of a Creator who is God. The Universe is not one of happenstance, occurring among a long chain of other universes. That there now is shown a need for a First Cause, there must be one. *That* First Cause must be an intelligent Creator, the one spoken of in the Genesis 1 accounts.

## The Root of All Evil

The workings of the Temporal sphere we have detailed here, with its intrinsic uncertainty, explains the mechanisms for evil in the world, and by implication makes it clear that God did not create evil. From the outset, let us understand that evil did not originate here in Temporality, but finds its beginning long before this Universe existed, at the time of the Falling of the *son of the morning* and the War in Heaven.

*Why does evil exist? Why does God allow it to flourish? What is the cause of evil?* In the contexts of where we are now and the Universe around us, these questions have been problematic to many for millennia. Many have drawn wrong conclusions because without a firm understanding of the purpose of this Temporal sphere and the ability to completely separate it from the Physical/Spiritual sphere (as we have done here), it is impossible to completely understand what evil is and its root cause.

To answer the preceding questions, we must first define evil. Evil is disharmony with the will of God; moreover, it is *any action* that is taken *independent* of the will of God. It does not have to be open, antagonistic rebellion or conflict, in fact most of the time it is not. In its simplest form, evil is merely doing, believing, or even thinking something that is in dissonance with the plans, thoughts, or will of God.

Evil is a consequence of, and is intrinsic to, the exercising of free will itself, regardless of the course of the decision, whether it is in discord or in harmony with the plan of God. Robert Frost said, "*Way leads on to way*;" this is true. Among the creatures that can exercise their will, there is a cascade effect that escalates evil from the simple point of free choice to the level of the horrors and atrocities we most willingly associate it with. The same is true of nature. A breeze blows, and through a chain of events, a hurricane forms; all of this occurs per the laws of God and without his direct intervention. Within the natural world, the possibility of *Free Will* requires uncertainty, and this uncertainty results in the elements themselves existing and functioning outside of the direct will of God.

Understanding evil as defined here, we can appreciate the context of King Benjamin's proclamation:

> *"for the natural man is an enemy to God and has been from the fall of Adam,"*
> *(Mosiah 3:19)*

It is a primary characteristic of nature for it to be independent and uncertain. Uncertainty is intrinsic to the structure of the Universe itself; the Creator designed it that way; it is a deliberate specification of the Temporal sphere. This natural characteristic is required to unshackle the effects of determination and allow *Free Will*. Uncertainty, by lowest common denominator, is the definition of evil itself. That God created the Universe with this characteristic means that He created the conditions for the presence of evil; but it must be understood, *He did not create the evil itself.* We are here to learn and experience, and by that, grow and progress. Those invaluable blessings, however, do not come without great cost: "*For we know that the whole creation groaneth and travaileth*" (Romans 8:22). Our benevolent

Father in Heaven recognized the price to be paid: He relinquished His control, He provided His Spirit and our Faith to give us direction; He ordained the Saving Grace of the Atonement to pay the wage of sin, unshackling evil from necessary eternal consequence that would burden His children forever.

The many manifestations of evil have been associated into groupings described loosely by type. In a sermon[11] he gave, John McArthur identified four of these types or forms of evil: natural, moral, supernatural, and the evil of Hell. He went on to say in that same sermon that God created evil itself. This conclusion is outlandish and not based on Scripture. Let us examine just two of McArthur's four evils as their characteristics directly relate to each of us and to the Temporal sphere we inhabit and have defined in this book.

- Natural Evil

Our modern perception of nature makes it difficult to describe any natural process as a form of evil. However, historically and scripturally, events we know today to have strictly *natural* causes were most often considered to be the results of evil. For our discussion here, we will consider that understanding as proper, allowing us to define evil in the broadest terms possible. Evil then, is something that causes pain, suffering, something that kills; the broader the sweep of these things, the greater the evil.

The Temporal sphere is full of hazards, both on the very small scale and in the tremendously large. The lethality of this Universe is everywhere, from viruses to supernovae and colliding galaxies. Death can be found on every corner: tornados, tsunamis, earthquakes and hurricanes. Death lies within, cancer; congenital disease and genetic mutation kill all too effectively. Deadly plagues and viral epidemics have killed millions and continue to threaten to recur today. Death waits without; wild animals hunt and kill each other and sometimes man. Malaria and typhoid runs rampant, even the more domestic and trivial ailments of cold and flu can kill.

The skies are full of things that can bring death and destruction. The stars bathe our atmosphere with radiation and particles that are lethal to life. There are comets and meteors and asteroids of all sizes and types that sweep along paths, which at times share the same point in space with the Earth. Convergence of their paths with ours can cause great destruction and loss of life. Asteroids have nearly brought about the complete extinction of life on Earth more than a few times in the past. Tectonic activity produces earthquakes and volcanism which have had, and can have, deadly results on massive scales. In these regards, nature can be seen as evil.

The natural world is without a doubt destructive, dangerous and deadly, but as we have discovered in this treatise, these characteristics are inherent to a Universe in

---

[11] John McArthur: Why Does God Allow So Much Suffering And Evil?, Published May 12, 2015, originally presented at the 2008 West Coast Conference, Tough Questions Christians Face. (Available for viewing on YouTube)

which free agency exists. The same Universe that so easily destroys also creates. It is wonderful and inspiring; it is full of bcauty and potential. God created this Universe with all that it entails, indeed He has created all there is. But He is not the creator of evil; He is not death or the destroyer of worlds; that is the attestation of a mythical Hindu god, not the Creator of this Universe.

The Temporal sphere was created for the exploration of the effects of discord with God; it is a schoolroom where there is much to be learned. To affect this learning, an uncertain Universe must exist. Allowing for consequence and aware of its negative effects, a loving compassionate God instituted the Atonement to correct whatever hurt and damage the necessary uncertainty allows.

- Moral Evil

Moral evil may at first glance seem completely unrelated to anything tangible and as such something that is independent of the Temporal sphere. However, the expression of morality itself requires a freedom of choice that can only be fully realized in the uncertain and tangible Temporal Universe we have described within this work; let me explain.

***In a determined universe we would not be capable of thought; the will of our Spirits would be shackled and imprisoned.***

This Temporal Universe, with its uncertainty, provides a framework for a future with choices. In contrast, within a universe *without* uncertainty, all actions are strictly known and defined forward and backward in time. A determined universe would be ordered, and any one action, anywhere, would control and rigidly fix all actions everywhere. None of us would be free to act on our own. Events that occurred millions of years in the past would determine in a precise and rigid manner everything that would happen at every moment today. In the same way events occurring today, would in turn predetermine and fix every succeeding event in the future. In such a universe, one might *imagine* thinking, desiring, will, and choices, but imagination would be all that it is. Within a physical reality without uncertainty, no independent action can take place; each of us would be bound in a body that we could not control.

A universe without uncertainty is even more restrictive than this, for within such an ordered universe even our thoughts would not be our own, imagination, thinking, will, desires, and choice could not exist.[12] The chemical processes that occur in our

---

[12] For an interesting take on the subject of Free Will and an argument that it does not exist even within the Universe we live in today, see *Free Will* (Sam Harris, 2012, Free Press, A division of Simon and

brain at the molecular level are the essence of our thoughts before they become the mechanisms of our actions. In a determined universe we would not be capable of original thought; the will of our Spirits would be shackled and imprisoned.

The Savior said that sin is in our thoughts and not in our actions, but we have seen that without the uncertainty that is attendant to the Temporal Universe we have described here, we would not be capable of independent thought or action. In a determined universe, one without uncertainty, moral evil would have no expression. We could not be held accountable for what we could not control nor were capable of avoiding.

After proclaiming that God created evil, which *can be* an interpretation of the facts declared in the Bible, John McArthur goes on to proclaim that, "*God created evil so he can show his glory by overcoming it.*" This reason for the existence of evil is completely untenable and is just the type of false doctrine that stems from an off-centered understanding of our Father in Heaven, His plan for our eternal progression, and any firm knowledge of this creation and its implications.

Such an understanding is devoid of any connection with our Father through association of His Spirit or the reason and Intelligence within us, which He gives. Such an understanding is as far removed from our Father in Heaven as the atheist understandings of the Universe we have described here; it cannot be advocated by any who profess to be Christian. Any theology that espouses a belief that God created evil, let alone that He did so for his personal gratification and glorification by overcoming it, is lame and bankrupt, wrong and immoral, and should be utterly condemned.

God did not create evil. He created a universe in which, due to the necessary attributes of uncertainty, evil may exist. Our Father in Heaven is no more responsible for the existence of evil than an expert carpenter who skillfully builds a beautiful, well fashioned staircase in the temple, is responsible for the broken leg of someone who clumsily falls while using the staircase. *In direct opposition of those who seek to accuse God of evil in any way, we state that God, the supreme architect, foresaw the challenges of the climb. To counter the effects of evil, which would develop without actions or designs of His own, our Father put in place a safety net before He built our staircase to Heaven, that safety net is the Atonement.*

> *"let no man say when he is tempted, I am tempted of God: for God cannot be tempted with evil, neither tempteth he any man." (James 1:13)*

*Why does evil exist? Why does God allow it to flourish? What is the cause of evil?* The answers to all three of these questions are to be found in the truths we have laid out here in our discussion of the Temporal Creation. Evil exists because of the

---

Schuster, New York.) Mr. Harris' work runs counter to the belief in the Free Agency of man we so firmly establish in this work, for he discounts the facts of uncertainty we affirm here. However, his book does present a fascinating exposition of what a universe without uncertainty would mean to free choice and the mind.

necessary disharmony attendant with freedom of choice that brings man and nature in an uncertain universe into dissonance with the direct will of God. Evil flourishes because of the great promise of potential that God has bestowed on each of us. The rewards are immense and the attendant costs great. Free agency itself, our application of it, and a Temporal Universe, planned, engineered, and constructed to allow for those choices, is its cause. That Evil *is*, is an implication of the Temporal Sphere.

## The Fall is Universal

The effects of the Fall are to be found everywhere in the Universe. Entropy and the arrow of time, the scientific terms equivalent to death and mortality are incumbent in this sphere. Everywhere we look, in any direction, the Universe and everything within it is dying. Entropy has been attendant from the very instant of creation. An implication of the Temporal sphere as we understand it is, that this death was brought upon it from the outside before it was created.

The Fall was universal and its effects were not (and are not) isolated to one corner of the Universe we inhabit. Immortality did not *happen* to one part of the Universe, nor did it begin here on Earth, affecting the entire Universe as it spread out from Adam and an Earthly Garden.

Because we understand the Fall to have occurred in the Physical Creation (the creation that preceded this Temporal sphere), the term *world* in the creation accounts (the "*world*" to which death is brought into) has a broader context than just the Earth we stand on. Death came by consequence of the Fall to the entire Universe.

Everything that has ever been, anywhere within this Temporal sphere, was conceived, existed, and will die immortal. Death is the admission price into the Temporal Universe; everyone and everything must pay the fee. A fact most strongly evidenced by this price being also paid by the Creator of this sphere Himself, Jesus Christ.

We understand the Fall to be universal with inescapable universal consequences that followed it. Because this Fall preceded the Temporal sphere, it can be implied that the Fall set up a situation in which all life entered this sphere under the same conditions, fully aware of the consequences and under the same demands of justice. By this we can accept, and even expect, that there may be life elsewhere within the Universe and that life on Earth is not alone in the Cosmos.

This implication does not strain our Christian theology, because all life in this sphere would have been placed here by the same Creator, with the same understanding we all shared in the pre-existence. Like us, all life, wherever it may be, came here with the same purpose: to receive instruction, gain experience, progress, and obtain the ultimate prize of exaltation.

## Is There Anyone Out There?

The structured view of the Temporal sphere as detailed in this thesis denotes that Adam's fall, and the consequences that followed it, are pre-existent to the creation of the entire Temporal Universe. This insists the traditionally held view, that Adam's transgression occurred in a Garden on *this* planet within *this* Universe, is demonstrably incorrect. Our understanding implies a set of circumstances that allow for the existence of other planets in this sphere inhabited by our brothers and sisters, sons and daughters of God.

The traditional reading of the Garden story and Adam's Fall dictates that death did not exist in this Universe until Adam transgressed. We are told that the effects of this consequence applied to all of creation as a result of Adam's actions alone, "*For since by man came death… as in Adam all die*" (1 Cor. 15:21, 22). Because this death (entropy) is universally observed (*all die*), we know that any life, anywhere in the Universe, suffer this consequence imposed by the Fall. The traditional view also holds that Adam was *entirely* native to this planet, the Earth we all inhabit today.

These traditional understandings combine to present a logical impasse. They suggest that life elsewhere in the Universe would be unjustly affected by the Fall of Adam as his localized actions brought consequences of death that overcame them as well. A common resolution of this impasse derives a conclusion that Temporal life does not and cannot exist beyond this planet and its immediate confines.[13] It determines that death is confined to a restricted locality influenced by the actions of Adam alone. This common resolution is wrong and has driven many incorrect theological constructs.

The proper reading of the Genesis 1 account reveals that this Universe was created to be Temporal; that from its inception all things within have been dying. This understanding removes the events of the Garden and the Fall from this sphere entirely. It places the origination of the Fall's consequential effects before the starting point of the Temporal Creation and connects all things in this Universe to a same shared fate from the outset.

The Genesis 1 accounts present no theological bulwark to the existence of life on other planets. Rather we understand that such life would be kindred to ours. All life, whatever it is, whatever form it may be, was created for the same purpose and through the same processes described in the passages of Genesis 1. Our story is their story; we share a common ancestry, a common Father; *there is one story of Life*. This implication of the Temporal Creation presents some important concepts that must be considered.

---

[13] See footnote 14 on pg. 35 for a very brief example of just one apostle's understanding that man is confined to this celestial sphere alone, because of his understanding of the facts of the Fall of Adam and its relationship to the Universe we live in today.

- Male and Female

What does it mean to be male and female? Our sexual identity is integral to who and what we each are. A male Adam and a female Eve were created as such in the Physical sphere. Gender then is an immortal, eternal attribute that has some divine purpose.[14] We were men and women long before this Universe and all that is within it came to be. We will explore this detail in subsequent parts of this thesis; however, for now, it is sufficient to state that male and female is a quality held by each of us as eternal, immortal, sons and daughters of God.

On Earth, the attribute of gender pertains to all advanced forms of life. That gender is pre-existent to this sphere is understood through recognition of both Adam and Eve existing in the Garden, itself a part of the Physical Creation. We are not certain that life elsewhere in the Universe would attain the qualities of advanced life or if the sons and daughters of God would be part of any habitation there; but if advanced life exists elsewhere, we would expect to find it both Male and Female, and any children of God there, to be our brothers and sisters, both men and women.

- In the Image of the Father

In the previous chapter, we discussed in detail what it meant to be "*in the image and likeness*" of God. We learned that the compound Hebrew word translated as image has a root meaning of *shadow*. From that meaning, it was shown that we derive our existence from God as an extension of Him. We also discussed the Spirit of God, relating it to the *Ruach*. We explained the *Ruach* to be the life-giving force that animates all living things, what we in the Latter-days associate most closely with the "*Light of Christ*." We fully expect these elements that define and empower us to define and empower life elsewhere in the Universe.

- The Coming of the Son

Perhaps the most interesting consideration of the implication that life may well exist elsewhere in the Universe is how it relates to the appearance of the Savior on Earth at the meridian of time, particularly in the contexts laid out by the Book of Mormon and His appearance to the peoples of ancient America.

While yet in the flesh in the land of Jerusalem, the Savior spoke of other folds of his sheep saying, *"them also I must bring."* He says, they *"shall hear my voice; and there shall be one fold and one shepherd"* (John 10:16). Added to this, Jesus Christ later proclaimed His partial fulfillment of this charge by his appearance to the righteous remnant on the American continent:

> *"And verily, verily, I say unto you that I have other sheep, which are not of this land, neither of the land of Jerusalem, neither in any parts of that land round about whither I have been to minister. For they of whom I speak are they who*

[14] See our discussion on this subject in Chapter 8, The Intelligences were Organized, pg. 298.

> *have not as yet heard my voice; neither have I at any time manifested myself unto them. But I have received a commandment of the Father that I shall go unto them, and that they shall hear my voice, and shall be numbered among my sheep, that there may be one fold and one shepherd; therefore I go to show myself unto them." (3 Nephi 16:1-3)*

We have come to associate the statements made by the Savior to those in the precincts of Jerusalem with His, then-still-future, ministry to the Nephites in America. But of whom does He speak when He makes His similar proclamations to the Nephites? Who are *they* that had, as of then, still not heard His voice, to whom He had not yet made Himself manifest?

Where are we to find those other sheep? Where can we locate this enigmatic location? The Savior said He had been commanded of the Father to "*go unto them*" that they may "*hear*" Him. He said He will go to show himself unto them. Can we, perhaps, extrapolate from these statements that the Christ was referring to life elsewhere in the Universe?

The LORD previously revealed, "*Many worlds without number have I created" (Moses 1:33).* Reading Verses 33 through 38 of the same book and chapter, we are left with the impression of an ongoing creation after the Fall. If all of these Temporal worlds will be, are, or (even then may already) had been "*inhabited*" (as the Scripture implies), would each world require a visitation and separate ministry of the Son?

At times in this thesis, we have expanded the parameters of concept to broaden our view in an attempt to comprehend the incomprehensible. The revelations of God are complete as they are presented to us, though in many respects they are impossible for us to completely understand. In this light, let us consider the following:

Returning again to the Savior's teaching in 3 Nephi 16, Verse 1, we are told, "*I say unto you that I have other sheep, which are not of this* ***land****, neither of the* ***land*** *of Jerusalem, neither in any parts of that* ***land*** *round about whither I have been to minister."*

Now, let us look at an interesting phrase used to describe what Moses saw millennia before Christ spoke these words to the Nephites: Moses records, "*And he beheld many* ***lands****; and each* ***land*** *was called earth, and there were inhabitants on the face thereof."*

Notice what I have bolded in both of the passages of Scripture; the word, **land**. What are we to understand by this considered and thoughtful choice of words?

What does *land* mean? Remember the source of both of these passages of Scripture is the same, the Word of God, the Savior Jesus Christ Himself. He was the one who spoke to the Nephites, and a careful reading of Moses, Chapter 1, reveals that it is the same revealing Word of God, not Moses, who used the word *land* to describe the other inhabited worlds.[15] The same direct source for the two phrases implies an intimate understanding of their meaning in their contexts. It seems entirely possible that Christ has revealed here a truth about this Universe that goes beyond what has been discovered before.

Multiple inhabited planets to which the Savior must go?[16] This now seems to be much more than just probable; but what of each of these worlds? What would be the mission of the Savior to them? We can imagine that it would be defined by the type of life extant on their localities; its attainment of the *Neshama* and not just that of the *Nefesh*. Our reading of the Genesis 1 account can easily provide for the development of life on many worlds through the processes we noted at play here on our Planet. The initial loading of information within the matter (*bohu*) of the Universe would lead those *informed elements* to produce life anywhere it could thrive; but what of advanced life, or more specifically, the sons and daughters of God?

If these other worlds, are the *lands* described by the Savior, then the fact that the Father has commanded Christ to go to them, that the inhabitants thereof might be brought into the same fold, strongly suggests that they are inhabited by the sons and daughters of God as well. That these children of God are in this Temporal sphere means they have the same purpose as we and the same need of a Savior and His redeeming Atonement.

Would that require of the Savior the same ministry and sacrifice on each of these foreign worlds? Would He (by such) be living out a near endless chain of incarnations, deaths and resurrections? How can this be so, given the understanding of eternal progression and our Spirits housed in Physical, Temporal bodies as taught by the Plan of Salvation?

For certain we are not yet blessed with the knowledge to provide definitive answers to these important questions, but we may find some guidance in forming them through our understanding of a truly *universal* Atonement.

- A Universal Atonement

With the possibility of life elsewhere addressed and understood, and with the precept established that it must share the same Creator and have the same purpose

---

[15] For a more complete understanding that the *many lands each called earth* revealed to Moses were in fact individual planets, and not just specific geographies on this our home planet, read Verses 29 through 35 of Moses, Chapter 1.

[16] This is one of many issues about which the Church has no official position. It may be helpful here to reiterate what President J. Reuben Clark taught while on assignment from the First Presidency: "*Here we must have in mind -must know- that only the President of the Church, the Presiding High Priest, is sustained as Prophet, Seer, and Revelator for the Church, and he alone has the right to receive revelations for the Church, either new or amendatory, or to give authoritative interpretations of scriptures that shall be binding on the Church....*"

as we: instruction, experience, progression, and ultimate exaltation, there too must also be a need for Atonement.

It has reportedly been said, "*This world was the only one corrupt enough to have crucified the Son of God.*" That may indeed be the case, but regardless of this statement being true or not, it may not be required of Christ to make the same type of sacrifice on other worlds, though our study here does not rule that out.[17]

We do know that Scripture tells us that Christ paid *once* the price of sin for all the sons and daughters of God, *"For Christ also hath once suffered for sins, the just for the unjust, that he might bring us to God"* (1 Peter 3:18). That His Atonement on the cross at Calvary was infinite is without doubt, "*For since by man came death, by man came also the resurrection of the dead. For as in Adam all die, even so in Christ shall all be made alive*" (1 Cor. 15:21, 22). And we fully understand that Man, *all of us*, were created both male and female in the immortal Physical sphere (spiritually) long before anyone or any living thing inhabited any land called Earth here in the Temporal sphere, "*...For I, the Lord God, created all things, of which I have spoken, spiritually, before they were naturally upon the face of the earth. ...And I, the Lord God, had created all the children of men; and not yet a man to till the ground; for in heaven created I them; and there was not yet flesh upon the earth...*" (Moses 3:5). But, can we find some more specific scriptural references that directly address the Atonement being universally based and dependent upon a onetime incarnation of the Son of God, on just this one planet, the one we call home?

Today scientific research suggests that many, if not most, of the 100 billion plus stars in our galaxy are encircled by at least one planet. This is a recent finding, because fewer than two decades ago the best guess among scientists would have been that from our vantage point on Earth, it would never have been possible to detect extra solar planets. Their observation, however new to science, was seen in vision by Moses over 3,000 years ago. Through inspiration, this vision was

---

[17] Elder Bruce R. McConkie concisely describes the universal Lordship of The Savior: "*Now our Lord's jurisdiction and power extend far beyond the limits of this one small earth on which we dwell. He is, under the Father, the Creator of worlds without number. (Moses 1:33). And...the atonement of Christ, being literally and truly infinite, applies to an infinite number of earths.*" Again, "*Just as the creative and redemptive powers of Christ extend to the earth and all things thereon, as also to the infinite expanse of worlds in immensity, so the power of the resurrection is universal in scope. Man, the earth, and all life thereon will come forth in the resurrection. And the resurrection applies to and is going on in other worlds and other galaxies.*" (Mormon Doctrine, Bookcraft, 1966, pg. 65, 642.)

President Marion G. Romney summarized the concept of the universal ministry of Jesus as follows: "*Jesus Christ, in the sense of being its Creator and Redeemer, is the Lord of the whole universe. Except for his mortal ministry accomplished on this earth, his service and relationship to other worlds and their inhabitants are the same as his service and relationship to this earth and its inhabitants... Implicit in the scriptures is the fact that the surest, if not the only, way to understand Jesus the Lord of the universe is to obtain an understanding of his relationship to this world and its inhabitants... I bear my own witness that these great testimonies to the fact that Jesus Christ is the Lord of the universe are true. That he is also our Savior, and that the gospel of Jesus Christ is universal—the only plan by which men ever have been or ever can be exalted.*" (Improvement Era, November 1968, pg. 46-49.)

translated and published in the first half of the 19th century, almost 200 years before science had advanced to a point where it could verify the truth revealed and recorded in Scripture so long ago.

> *"And worlds without number have I created: and I also created them for mine own purpose; and by the Son I created them, which is mine Only Begotten." (Moses 1:33)*

> "*...There are many worlds that have passed away by the word of my power. And there are many that now stand, and innumerable are they unto man..." (Moses 1:35)*

Jesus Christ, the Son, the Only Begotten of the Father then is the Creator of "*worlds without number.*" This truth revealed to Moses was also known by the Apostle Paul. Paul knew that God appointed Christ to be the Creator of worlds and he expressed that knowledge as follows:

> *"God, who at sundry times and in divers manners spake in time past unto the fathers by the prophets, Hath in these last days spoken unto us by his Son, whom he hath appointed heir of all things, by whom also he made the worlds." (Heb. 1:1-2)*

We don't yet have any scientific indication that any of the foreign worlds most recently discovered support life and certainly no indication that any support advanced life. But as we have seen, the 29th verse of Moses, Chapter 1, tells us that many of them do: "*And he beheld many lands; and each land was called earth, and there were inhabitants on the face thereof.*" Modern day revelation informs us that these inhabitants are, like we, the sons and daughters of God.

In 1832, while at the John Johnson home in Hiram, Ohio, the Prophet Joseph and Sidney Rigdon beheld what became to be referred to as the "*vision of glories.*" Of this revelation, they bore the following witness codified into Scripture:

> *"And now, after the many testimonies which have been given of him, this is the testimony, last of all, which we give of him: That he lives! For we saw him, even on the right hand of God; and we heard the voice bearing record that he is the Only Begotten of the Father— That by him, and through him, and of him, the worlds are and were created, and the inhabitants thereof are begotten sons and daughters unto God." (D&C 76:22-24)*

Alien to our world, these "*begotten sons and daughters of God,*" have different histories for sure, their home orbs, each called *earth* as declared in vision to Moses,[18] would be no doubt much different than ours. But, their presence here, in the Temporal sphere, within this Universe has the same purpose as ours, they are our brothers and sisters, and as such their eternal life depends on the Atonement through our Savior Jesus Christ.

---

[18] Moses 1:29.

The Prophet Joseph Smith plainly revealed his belief that such alien worlds, populated with our brothers and sisters, the sons and daughters of God, are included in the saving effects of the Atonement by the very same Savior as we. During his time in Nauvoo, the Prophet gave a poetic interpretation of his earlier "*vision of heav'n*" recorded as (D&C 76) and quoted above. The Prophet's poem declares:

*And I heard a great voice, bearing record from heav'n*
*He's the Saviour, and only begotten of God—*
*By him, of him, and through him, the worlds were all made,*
*Even all that career in the heavens so broad,*

*Whose inhabitants, too, from the first to the last,*
*Are sav'd by the very same Saviour of ours;*
*And, of course, are begotten God's daughters and sons,*
*By the very same truths, and the very same pow'rs.*
*(Times and Seasons, 4:82-83, stanzas 19-20)*

Theologically and in the contexts we have developed in this treatise, the sons and daughters of God are those independent Intelligences who existed in the eons past with our Heavenly Father and His First Born Son. They are those who received their immortal physical bodies at the same time as Adam, and are those who through participation in the effects of the Fall received Temporal, mortal bodies.

We understand that it was necessary for God to participate in mortality in order to bind deity to the mortal sphere with a bond so strong that it could overcome death for all and restore us to the perfection required in our eternal home. It must be understood that the, *we, those, they* and *us* in the preceding sentences, are the sons and daughters of God, all of us, wherever we may find ourselves within the Temporal Universe. We all became incarnate to come here, and so did the Son. We all only have one trip here to this sphere and so did the Son, *once and for all* He was the Sacrifice for sin, restoring all to the perfection of the Physical sphere:

> *"For Christ is not entered into the holy places made with hands, which are the figures of the true; but into heaven itself, now to appear in the presence of God for us: Nor yet that he should offer himself often, as the high priest entereth into the holy place every year with blood of others; For then must he often have suffered since the foundation of the world: but now once in the end of the world hath he appeared to put away sin by the sacrifice of himself. And as it is appointed unto men once to die, but after this the judgment: So Christ was once offered to bear the sins of many; and unto them that look for him shall he appear the second time without sin unto salvation." (Hebrews 9:24-28)*

- Upon This Rock

Near Caesarea Philippi in Northern Palestine, stand the very gates of Hell (*Figure 4-1*), or so it was told. By the time of the Son's mortal ministry on Earth, pagans had worshiped at the site for nearly a thousand years. Baal first, Pan later, and a plethora of other gods were thought to enter into Hell through a large water filled crevasse in a massive rock outcropping. For all intents and purposes, to any who looked down into it, the pool in the crevasse appeared to have no bottom, extending all the way to Hell itself.

Figure 4-1 *The Gates of Hell at Caesarea Philippi. This massive crevasse in the dominate outcroppings of rock at the base of Mt. Herman were thought to be the entrance of Hell by pagan worshippers who were actively practicing their religion and sacrifices here at the time of Christ.*

The rock outcroppings of the region, of which there are many, are enormous and dominate the landscape all around; they are the exposed bedrock of the area, the actual spine of all of Palestine. They are the uncovered base of imposing Mt. Herman; at over 9000 feet in elevation, its summit is the highest point in Biblical Israel. The crevasse itself splits a massive red and amber hued cliff within just hundreds of feet of springs that, today, rush out of the ground forming the headwaters of the River Jordan. At the time of Christ, however, the spring water streamed directly from the cave. The Jordan River flows from this very spot into the Galilee and beyond, past the meanders where Jesus was baptized by John and finally into the Dead Sea, itself marking the lowest point anywhere on the land surfaces of the Earth.

In first century Israel, this site was a bustling hub of decadent activity. Next to an open-air shrine that featured a statue of Pan with a large erect phallus, stood pagan temples that had been erected near and around the opening in the Earth. Surrounding the statue of Pan that was located in a large niche in the cliff wall were smaller niches for the gods Caesar and Nemesis and some that housed statues of Pan's attending nymphs.

On a shrine in front of this array, worshippers of Pan, a fertility god, would gather to join in sexual rites that even included public copulation with goats. Within the temples, alters were raised to accept offerings, for it was believed that in the winter Baal and Pan retreated through the massive cave into Hades. If the winter were to end and the crop bearing season of spring was to return, the pagans believed these gods needed to be coaxed back out of Hell through the spilling of the blood of innocents offered there. Infants and other living sacrifices were cast into the water within the pit of the crevasse. If blood appeared below, flowing from the springs, it was supposed that the sacrifice had been accepted and that the gods would return, bringing spring with them. If the spilt blood did not appear in the waters of the springs, more sacrifice was required.

It was with purpose for sure that the Savior brought His twelve disciples, themselves just in their teens and early twenties, to such a maleficent place, for teaching and for questioning.

We can easily envision the Savior standing on the rocks near this spot looking upon the horrid scene at the temples that had become such a part of life for the ancients. Here overlooking the commotion in the distance, the scene of pagan worship to many pagan gods, the Savior turned to His disciples and asked, "*whom do men say that I the Son of man am*?[19]

When the disciples answered Him telling that He was thought by many to be the reincarnation or reappearance of one of a variety of past prophets, Christ asked again, "*But whom say ye that I am?*" To this latest question of the Savior, Simon the son of Jonah (who as of yet had not been called Peter[20]), responded: "*Thou art the Christ, the Son of the living God.*"

What was to follow takes on deep meaning when we place the event in context with where the LORD and his disciples were, the depravity that was going on around them, and what the world believed that wickedness was supposedly accomplishing.

Christ, ever the master teacher, used the entire world around him to illustrate the truth with a depth of understanding that fills the senses and the Soul with wonder. Christ turned to Peter and said:

> *"Blessed art thou, Simon Bar-jona: for flesh and blood hath not revealed it unto thee, but my Father which is in heaven." (Matthew 16:17)*

Then keenly and intricately, Christ wove language and all of creation around him, to make a declaration of importance about his mission and the purpose of each of us here in the Temporal sphere.

The Savior stooped down to pick up a fist sized stone, declaring as he held it in front of the captivated disciple, "*this is a stone* (Greek, *petros*, πέτρος) *and I say unto you that you are Petros*." [21] Then, while standing on the rock they all were

---

[19] Matthew 16:13-19, see also the synoptic accounts of Mark 8:27-30 and Luke 9:18-21.

[20] Peter's original name was "*Shimon*" or "*Simeon*," in Hellenized Aramaic: *Šim'ōn or Sëm'ān* ("*Simon*" in modern English).

[21] There are no original documents in Aramaic, the tongue Jesus would have used here at Caesarea Philippi as he spoke to Simon the son of Jonah. The earliest manuscripts are in Greek, so we use Greek

standing upon, the rock into which that crevasse cleaved into, the rock of the whole region round about, no doubt even gesturing to the landscape, the autumn hued cliffs, and maybe producing a little stomp with his foot on the rock below Him to impress upon the disciples that He was indeed now speaking of the solid mass of the Earth beneath and around them, our LORD Jesus Christ declared: "*this is bedrock* (Greek, *Petra*, πέτρᾳ), *and upon* ***this*** *[bed]rock will I build my church.*" With a gesture and a glance towards the dreadful activity at the crevasse, the Savior then declared of His church, with both power and infinite sadness, "*and the gates of hell shall not prevail against it.*"

Christ, at that place, in that moment, with His words still ringing in the air, taught a message that was filled with a depth of meaning that continues to be absorbed and realized today. The double entendre employed was rich and intentional in *every way*.

Christianity took many of those meanings and used them to build theologies and the very church(s) Christ spoke of. Catholicism read into the statement that Peter was to be the foundation of the priesthood and the church it attended to. It made him the first Pope and today Catholicism traces its authority back to this very moment on the bedrock at Caesarea Philippi.

Protestantism saw in Christ's words that it was the revelatory power of the Spirit through the faith of that disciple that was to be the cornerstone of the true church of God. Realizing that Christ taught that man was small and so too would be anything he built of himself; for anything that drew its strength from an individual man (*a mere stone*), no matter how strong and rock solid that man's faith might be, was incapable of fulfilling God's need for bedrock upon which to build his Church.

It turns out that both of these understandings are true, but there was another message told *by* that place, *at* that time, and *through* those carefully selected words. That other message speaks directly to our discussion here: Christ's Church is universal; its members are indeed *all* of the sons and daughters of God. His church is not bounded by time, the age of the Earth, or the Solar System and Galaxy we call

---

here. **Petros** is the masculine form of the word used for rock. **Petra** is the feminine form. It is a universally accepted Greek grammatical convention to use the feminine form to suggest a source of something (i.e. literally big). Of course, Simon was a man and so we expect that he would be given a masculine form as a name. But to ascribe a gender explanation for the fact that two different words are used in the same breath here is misguided. That all of the original Greek manuscripts use two separate Greek words in the passage [*Matthew 16:18*] indicates that any underlying Aramaic/Syriac original also must have used two separate words. The Syriac or Aramaic word for "rock" is *Kepha* (Hebrew: כיפא, ; Aramaic ܟܺܐܦܳܐ), *Cephas* (Greek: Κηφᾶς). Catholic theologian Rudolf Pesch argues that the Aramaic *cepha* means "stone, ball, clump, clew" and that "rock" is only a connotation; that in the Attic Greek *petra* denotes "grown rock, rocky range, cliff, grotto;" and that *petros* means "small stone, firestone, sling stone, moving boulder." A properly reconstructed Aramaic/Syriac rendition of this passage would be: *"You are KE'PHA' (a movable stone) and upon this SHU`A' (a large massive rock) I will build my church."*

home. It includes all of God's children, wherever they may be in the Temporal sphere.

Our LORD was aware of all of this as He stood there on the bedrock of Northern Palestine so long ago. With all of that in mind, He looked at the unspeakable sin of the pagans in the distance and made a declarative statement that He had come here to ***this*** rock, our rock, the Earth, the third rock from our Sun, to build His Church.

He declared that it was here to which He had come to take on a mortal body. Here He had come to live a perfect life and die unblemished for the sins of all of those who stood alongside Him in the Physical sphere as Adam partook of the fruit of free agency in the Garden. Here He would overcome a Universe of death by rising from the dead one single, glorious, Sunday morning.

There on the bedrock beneath His feet, overlooking what the pagans thought were the very gates of hell, Christ declared that He would overcome hell, shutting its gates once-and-for-all. This miraculous feat was to be accomplished for all of the sons and daughters of God through what He had done and was about to do, just once in the history of the Universe, on a tiny planet in the vastness of space, the planet we call Earth.

The Scriptures leave no room for doubt regarding the existence of life on other planets. An implication of the Temporal sphere is that such inhabited planets may be embedded within the Universe around us and perhaps may someday be discoverable.[22] Latter-day Scripture and Latter-day theology is the only Judeo-Christian theology that would remain intact and endorse the finding of extraterrestrial life. Such life of any kind is completely compatible with our theology and in no way represents any challenge to our concepts of God and our relationship to him. Discovery of life on extraterrestrial worlds would not represent a crises or trial to the world view of Latter-day theologians using the understandings provided in this work. This remains so despite it being clear that *every* other Western theology would have to be rewritten to adjust for such a find.

---

[22] The Universe is expanding and that expansion is now known to be accelerating. Because of this, there is a distance from us beyond which light emitted will never reach us. This distance is called the Cosmic Event Horizon. At the Event Horizon, the speed of expansion exceeds the speed of light and anything moving from any object beyond that point will never be able to reach us. Even though the *absolute* distance of the Event Horizon from us (5Gpc, some 16.3 billion light years) continues to increase ever more rapidly, objects (galaxies and stars) at a similar ever increasing rate, move beyond that horizon, *or more correctly, at the horizon, from our vantage point, light from such objects redshifts to infinity, slowing down their time until it comes to a halt, effectively freezing any information available to us from them at a point in the objects history after which we will never know any more about it. Light from the object will diminish as it continues to redshift until the frozen in time image becomes undetectable to us.*

Objects escape from our view over the horizon at an ever increasing rate. Today, and as I am writing and you are reading, entire galaxies are disappearing from us never to be seen again. This process will continue forever, at an ever increasing rate, until we are, as it seems, alone in an infinitely large, empty Universe. This is the perspective of the Universe as seen by us. The effect is increased because our point of view is from one so long after the Universe began. This view would appear the same today for an observer in any location, and God could, in the planning and placing of his children on habitable planets beyond each other's Event Horizons, preclude mutual observation and contact from the outset.

That life probably does exist on other planets has been established, and *that* life may even have predated life on Earth and even the Earth itself for we are told: "*there are many worlds that have passed away...*" (Moses 1:35). Unequivocally the Temporal Creation as discussed here implies that there was both life and death on this Earth well before the appearance of Adam, perhaps, and most probably, for billions of years. Because of our understanding of a separate and unique Temporal Creation and the Scriptures behind that understanding, we can without compromise or apology discard arguments to the contrary, no matter their source.

We have seen that the entire spectrum of life evolved in whole, thrived, and in parts failed, in accordance with the character of an uncertain Temporality that had been infused with knowledge. That process, though long by our reckoning, in short order by the viewpoint of God, fashioned a planet, perhaps many planets, teeming with life, which were perfect for the introduction of Man, the sons and daughters of God. It is at this point in the creation story that the scriptural accounts take a new direction, where the story of the "*generations of Adam*" begins, "*in the day that God created man, in the likeness of God*" (Gen. 5:1).

We have described a "*Late Adam*" theology and we must now discuss some of its implications. We will also describe some clarification those implications provide for many enigmatic scriptural and scientific contexts.

- And it Repented The LORD

Other than the Gospels and the accounts of the Creation and the Garden, perhaps the most crucial and informative part of the Bible is the sixth Chapter of Genesis. For in it we are told of the flood and what led up to the deluge. Among many revealing passages, the most overwhelming follows:

> *"And it repented the LORD that he had made man in the Earth, and it grieved him at his heart. And the LORD said, I will destroy man from the face of the earth..." (Gen. 6:6, 7)*

A Repentant God? This thought is so disturbing to some that for millennia it has been discounted as incorrect even by those who think the Bible is inerrant. It is foreign to the nature and character of God as many view Him, that He would ever second guess His own actions. He is perfect and omniscient, He knows all and can foresee all; how can it be possible for Him to *repent* (Hebrew, *way·yin·nā·ḥem*, וַיִּנָּ֫חֶם)? Joseph Smith, faced with a direct understanding of the word proffered through inspiration, deflected the subject of the recorded emotion away from God and placed the feeling of remorse into the mind of Noah instead. Joseph's inspired translation of the Bible aside,[23] there is no need to question the fact that God was

[23] The Joseph Smith Translation in its entirety has not been canonized, nor has it even been submitted for approval of the brethren. It does not replace the inspired KJV except in the places where its passages

pained and remorseful of the situation man had developed. To deny that He was, would be to imply that it was the *intent* of God all along to develop this world, allow it to bring forth millions, no doubt billions, of specialized species of flora and fauna, including advanced life of all forms, introduce Adam and Eve, let their offspring reproduce until there was perhaps by some estimates, almost 5 billion Souls, only to destroy all of it, save 8 persons and a boat-load of land based life.

For God to have foreseen such an event and not have been sorrowful or feel regret for its necessity, strains our understanding of a compassionate, loving Father in Heaven, and our relationship to Him as his children.

No, in that context (in the true and full context) the original Hebrew must stand as it was written, without need of adaptation or revision, no matter how disturbing it may be to someone's preconceived notions of a perfect all-knowing God. A God who can repent when He sees the painful results of free agency is the God of whom we are the offspring, the God we worship, and the God described by Scripture.[24]

***For God to have foreseen such an event and not have been sorrowful or feel regret for its necessity, strains our understanding of a compassionate, loving Father in Heaven…***

Why then the Flood? Why the drastic narrowing of the gene pool? Where did "*Man*" go? We know full well the effects of uncertainty in the Temporal sphere, the Toba Blast and the Bosphorus breach were natural calamities (though geologically recent and historic, as defined as happening after the first presence of man on Earth as opposed to pre-historic events that occurred before man arrived); they did not need God as their direct cause. The sixth chapter of Genesis, however, describes something different than God just letting nature take its course. God is actively involved with purpose and indeed should be because repentance does require a change of direction and action.

It will be remembered that we identified the source of God's sorrow. The reason for His reflection was the *ha-a-dam* (הָאָדָם), the *people of the clay*, those who were

---

have been added to the cannon in the Doctrine and Covenants and within the Pearl of Great Price. In those places it does not replace the KJV Biblical account but stands side-by-side with it, in equal authority. Those non-canonized passages provided in the footnotes of our standard works are to be considered only for clarification and an aid in understanding. They do not stand with any higher authority than any other utterance of a Church leader from any era.

[24] See our previous discussion of the nature and attributes of God in Chapter 2 of this work.

from the Land of Nod, who through Cain began interbreeding with the *Be-ne- ha-e-lo-him* (בְנֵי־הָאֱלֹהִים), *the sons* (the children) *of God*.

Let us start with a point of clarification. It is our understanding here that there is an important difference between the sons and daughters of God and *Early Man*, the most advanced and recent of which (Modern Man)[25] we have very liberally associated with the *ha-a-dam* (הָאָדָם). We understand here these, *people of the clay*, to be nothing more than extremely intelligent animals, the very pinnacle of evolution of life on the Earth. The *ha-a-dam* (Modern Man) received the *Nefesh* and were not blessed with the *Neshama* as only the sons and daughters of God are given that. Of course, all precursory iterations of man before these creatures are also only just animals, just like modern day Gorillas and Chimpanzees are.

The baseness of sin and corruption, the depravity that had become continuously attendant with man, was exacerbated by a gross intermingling of these animals with the *Neshama* held by the Children of God. The *people of the clay* did not, could not, fully know right from wrong, not in a spiritual sense; they were not inherently moral creatures. They acted on base instinct but were empowered by association with the sons and daughters of God with knowledge beyond what they should have known. Almost criminal from birth, there is no doubt that the evil associated with the *ha-a-dam* was immense. This then is the state of affairs that brought the future of the sons and daughters of God to a precipice, the reason for the repentance of the LORD. The corruption was of such scale that God proclaimed, "*I will destroy man from the face of the earth...*" (Genesis 6:7).

The events that led up to the flood are mostly unseen and largely unknown. It is generally accepted that gross sin was the reason for God's actions, but the details are obscured as if left shadowy to the sight, somehow shrouded in the muddy, murky waters of the flood itself. With the passage of time, the *ha-a-dam* became a distant memory and that was fitting for God had eliminated them. Like any sin committed by man that is forgiven, they are to be remembered no more.[26]

---

[25] The population of beings we name in this work both *Modern Man and Modern Human,* are what is defined by science as *Homo sapiens sapiens*. They are the latest grouping in the chain of evolutionary development leading from the *great apes* through *Hominina* and *Hominini*, Modern Man being the culmination of the progression of Hominins. Though science places the first appearance of Modern Man on Earth at about 200K years ago, it is the exhibition of behavioral modernity which is placed much more recently (est. 50K years ago or more recent) that most interests us in this work. We accept here the evolutionary processes leading up to *Homo sapiens sapiens.* We understand that this physiological evolution of the animal that Hominin was, is prerequisite to the introduction of Adam and Eve on Earth, an event which corresponds to the marked, rapid advancement in behavior evidenced in the archeological record. For further details of our understanding here, refer to the Appendix, The Introduction of the Sons and Daughters of God, pg. 543.

[26] It is not my intent to suggest that God sinned, it was man that did so. I intend here to convey that repentance expunges the wrong entirely. We understand that God's actions, a result of His repentant feelings, resulted in the elimination of early man (the *ha-a-dam* and the *hu-hal*) entirely from the Earth. They, their interaction with the sons and daughters of God, are to be remembered (and repeated) no more. Repentance is most commonly understood to be a turning away from sin. But that is a "*specific*" application of the word. Repentance, from the Greek, *metanoia* (μετάνοια) (Hebrew, *way·yin·nā·ḥem,*

As the Scriptures have been handed down to us, *man*, the sons and daughters of God and the daughters of men become hopelessly entangled through shared definitions that have become intertwined over thousands of years of misconception and neglect. Today without the careful reading we have done here, the true meanings of the phrases and the distinction of the groupings are lost. But the effects of these different populations and their intermingling led to consequences that affect us all in very real ways today.

- The Days of our Lives

The lifespan of man in the generations following Adam and before the flood exceeded 900 years, but by the time of David, "*the days of our years are threescore and ten* (70) *and if by reason of strength they be fourscore* (80) *years" (Ps. 90:10).* The lifespan of man dramatically changed and shortened over time, why? An answer to such a question can begin to be formed by reading in the sixth chapter of Genesis, in the passages that appear just before the account of the flood:

> *"... My spirit shall not always strive with man, for that he also is flesh: yet his days shall be and hundred and twenty years." (Gen. 6:3)*

What does it mean that the Spirit of God will not always strive with man? We know that the Spirit is something we experience intimately in our lives when we are righteous, and pure, and clean, but is that state sustainable for long? Is it possible, here in this verse, taken in the context of the lamentation of God that we can see, as He did, the wearing effects of our mortal probation?

Perhaps God felt the need to repent for making *man* (the sons and daughters of God) in the Earth because the trials of living on the Earth, being cut off from the presence of God, were overwhelmingly difficult for man, straining the bonding effects of the *Neshama* to the point of breaking. The challenges to man's Spiritual focus are abused by exposure to the world and its carnality, ever increasing over time until, for the natural man, it becomes nearly impossible for the Spirit to find an unblemished part within to abide? The thought is transformative and its realization by God transformed the world and us with it.

Through the purging of the Earth and its subsequent rebirth, we observe the actions of God that are responsive to the needs of His children in a world that was created with uncertainty fundamental in its charter. Perhaps in recognition of the struggles the anti-diluvian patriarchs experienced living nearly 1000 years in Temporality, God rethought the risks of exposure and reduced the time of man on Earth by tenfold. This understanding presents a *spiritual* reason for a *temporal* effect that is in harmony with the purposes of God; an understanding that provides an

---

וַיִּנָּחֶם), means "*a transformative change of one's mind*" (or heart). In this case, God, truly "*transforms*" the Earth and man's history, as He changes the conditions on the Earth in response to the actions of Fallen men. He completely eliminates early man from the Earth and reduces the age of man, minimizing our exposure to the effects of sin on our individual Spirits.

explanation to the question we posed at the outset: What does it mean that the Spirit of God will not always strive with man? There is, however, another possibility for this change in the lifespan of man.

As we saw in the previous section, Genesis, Chapter 6, describes an intermingling and interbreeding of the Sons of God, the offspring of Adam and Eve, with the daughters of men. This association led to a disintegration of spirituality and the eventual destruction of man in the Earth and, as we saw, this interbreeding and its negative effects on the children of God may have been the primary reason God brought the flood upon the Earth. Interbreeding may also have had another destructive effect on mankind through detrimental changes in the DNA within each of us.

It can be assumed that Adam and Eve were formed here in this Temporal sphere with perfect Temporal bodies, free of any defect.[27] Since they were late comers to the Earth and were not the offspring of any Temporal creature, there would be no deficiency, none of the negatives that would have necessarily come along with 3.5 billion years of evolution on the planet. It is easy to imagine such a body and the offspring of such bodies could live to extremely old age. It could take hundreds of years of the effects of uncertainty in the Temporal sphere to damage perfectly fashioned genes, and the cells formed by them, to a level where which disease and old age could take hold.

Interbreeding with existent man (*ha-a-dam,* הָֽאָדָם, the *people of the clay*) (Modern Man and perhaps even Neanderthal[28]), the *Be-ne- ha-e-lo-him* (בְנֵי־הָֽאֱלֹהִים), *the sons* (the children) *of God*, would have very quickly adulterated a flawless gene pool, having drastic effects. It can easily be imagined that those effects would lead to rapid deterioration of genes from a perfect and pristine state to one that is more natural and more practically suited to life in Temporality. Life perfectly suited to thrive here in this Universe is what 3.5 billion years of adaptation developed, but the adaptability and responsiveness to uncertain threats have a cost; length of life span is one of those costs it seems.

This natural explanation may be the true cause of our diminished life expectancy, or it may have been through the unilateral actions of God to counter the effects of sin on His children, or both; but today it is clear, as the Psalmist says, we live to be 80, "*by reason of strength*."

---

[27] Of Adam's intellectual gifts, President Joseph Fielding Smith observed: "*The first man placed on the earth was an intelligent being, created in the image of God, possessed of wisdom and knowledge, with power to communicate his thoughts and language, both oral and written, which was superior to anything found on the earth today.*" We declare here that Adam and Eve, when sent to this earth from the Garden, were in fact the first of the Be-ne- ha-e-lo-him (בְנֵי־הָֽאֱלֹהִים), *the sons* (the children) *of God.*

[28] See The Introduction of the Sons and Daughters of God in the Appendix, pg. 543.

- A Brief History of the Daughters of Men

The processes of creation described in the Genesis 1 accounts lead to life that evolves, and as such, we fully expect to find the evidences of early hominids that we do. To deny their existence on scriptural grounds is indefensible and just plain wrong, we have shown that. Creationists, who do, display an ignorance of both fact and the Scriptures. By doing so, they are not just denying the existence of evolutionary processes, they are more importantly denying the gifts of reason and intelligence that God has given them. A correct understanding of the Genesis 1 accounts imply the existence of what has been termed, *Early Man*, and allows us to consider how they, as creatures of our Heavenly Father, relate to us, the Children of God.

Let us reiterate again a very important point we have already made. The *ha-a-dam*, the *people of the clay* (Modern Man, Neanderthal), are nothing more than very intelligent animals. They inherited the *Nefesh* only, the Godly force that gives them life. They are instinctual having nothing more than the ability of a created thing to fulfill the measure of its creation. Evolutionary man, in all of its iterations, through all of its advancements, was never given the *Neshama*, only the sons and daughters of God are blessed with that.

The population known to science as *Modern Man* most closely mirrors the civilization described in the Bible that was coexistent with Adam; those who do not follow the commandments of God, the *ha-a-dam*, the *people of the clay*, the fathers of the daughters of men spoken of in Genesis 6. In addition to Modern Man, Neanderthal man is an interesting population to consider as it closely relates to modern humans but yet can be clearly shown to be a species set apart – *other-than-they*.

It is clear Neanderthal existed before Modern Man appeared and by most accounts was pushed out of Europe by Modern Man, more correctly far to the East and South into what is modern Portugal (the most recent habitation of these creatures having been found there). Much of what has been determined about Neanderthal is still hard to substantiate. Truthfully, all information known about them, anatomy, culture, genetic make-up and history is based upon very limited datum. Scientific conclusions have been drawn about this whole species based upon numbers of individuals and data collection of such limited quantity as to be laughed at if submitted to the FDA for approval of a drug or new food source. Still, in fairness, the work that has been done, with so little, is impressive, and I consider the conclusions sound, even if they may not be unshakable by yet future discoveries.

Of interest here are a couple of findings: The unearthing of Neanderthal burials have shown that they had advanced so far as to make them capable of conceptualizing a world view that included a religious component as early as 50,000 years ago,[29]

---

[29] Remember here the previous footnote discussing the uncertainty of dating past events precisely (*see footnote 56, page 99*). We are not certain of the date of entry of Adam to the Temporal sphere. Perhaps

perhaps they learned it from Modern Man who in-turn learned it from the first sons and daughters of God. Are these the men who *began* to call upon the name of the LORD, spoken of in Genesis 4:26, the *hu-hal* (הוּחַל)? Perhaps they are.

Modern Man and Neanderthal are shown to be intermingling culturally and interbreeding 55,000 years ago. *It is not surprising to me that the earliest sites of this cohabitation, indeed really the only such sites of intermingling known, are in Palestine.*

In 2008, a bulldozer clearing land for a development northwest of the Sea of Galilee in Israel opened a marvelous find, a prehistoric cave that had been occupied for a very long time. Since a collapse of the cave's entrance some 30,000 years ago, its contents had been sealed within, untouched and frozen in time.

The cave called Manot is more than 60 feet deep, 150 feet wide and over 300 feet long. Since its discovery, it has yielded an impressive haul of human remains, charcoal pieces, stone tools, and bones from over seven species of animal that indicate the cave had been occupied during a period spanning thousands of years.

However, the most important find in the dusty confines and darkness of the cave is that of an almost complete human skull found perched atop a stone shelf, deep in the opening of the Earth. Dated at just short of 55,000 years ago, the skull is the oldest known modern human fossil found outside of Africa and connects them (Modern Man) to Neanderthal, who occupied the same cave system[30] during the same several thousand year period. That modern humans and Neanderthal shared the same gentle rolling hill landscape of what is now Galilee, suggests that here is where the trace DNA of Neanderthal may have found its way into the DNA we have today.[31]

---

Cain may have been the first or among the first to introduce religious concepts to Modern Man and Neanderthal Man through his entry into the land of Nod.

[30] Neanderthal remains have been recovered from Amud cave just 24 miles East and Kebara cave 30 miles South of Manot. These sites are in the same formations of rock we discussed in the section titled, -Upon This Rock, pg. 131.

[31] See Appendix, The Introduction of the Sons and Daughters of God, pg. 543.

## This Is Not All There Is

It is implied that the Temporal Creation is a subset of a larger, dynamic reality. The Temporal Creation exists as a subtraction (TzimTzum) from a complete whole.

This Universe, what we see of it, was in fact created *ex-nihlio*, literally, from a singularity, but was done so within the structure and in the image of the Physical (spiritual) sphere that pre-existed it and which continues to subtend it today. It formed in a void created in the Physical sphere when the presence of God was removed.

The Universe in its Temporality is all we can see and touch physically; it sustains our lives and presence. But it is only an apparition of something much more; like Plato's shadows on the wall, our Temporal world is an inadequate representation of a more marvelous reality that can only be imagined from the feeble image we see in front of us.

Most of us don't have a clue that what we see and feel every day is not really there in any sense that we imagine it to be. Solid matter is really extraordinarily empty space. The "*stuff*" we attribute to be all that there is, makes up only 5 percent (by some estimates only 2-3%) of the mass and energy in the Universe, with preternatural dark matter and energy making up the remaining 95-98 percent. The heavens we see above us are delayed signals presenting a beautiful performance that came to a close billions of years before any of us took our first breath. Our eyes are only capable of sensing a tiny fraction of the electromagnetic activity that surrounds and bathes us. Truly for reasons far more than the limitations of thought, knowledge and mind, the Universe is incomprehensible to us.

The Universe we have described here is a subset of something *much larger*. That "*larger*" overlays what and where we are today, we have connection to it, it is the Physical sphere, the product of the second creation we will discuss in the second part of this work. Theologically this is a truth that is familiar and is represented as spiritual activity, the *metaphysical worlds* of science and philosophy.

Because of the Scriptures, we know that the Physical sphere has a corporeality and substance. All things were first created in this Physical sphere (which predates the Temporal) and are destined to exist there forever in immortality. Our Savior previewed what a resurrected immortal physical body looks like when (after His crucifixion) He appeared to His disciples and said, "*Behold my hands and feet... handle me*" (Luke 24:39). That we are destined to share the same immortal corporeality, a glorious eternal body of *Flesh and Bone*, is promised to us by the Apostle John, who proclaimed, "... *we know that when he shall appear, we shall be like him; for we shall see him as he is*" (1 John 3:2). Because of these passages and others like them, belief in a physical resurrection is an *essential* tenet of the Christian Faith. Without a doubt, a corporeal immortality is real.

The resurrected Savior's appearance to His disciples in that room in Jerusalem long ago indicates another truth: the Physical sphere and the things within it are tangibly sensible to us here when the conditions are right. In the next chapter, we will explore the possibility that some specific structures and effects of the Physical sphere may have already been sensed by modern science.

Left to its own and given time, the *apparition* we perceive as the real, tangible Universe will evaporate away as the echo of its existence fades into nothingness. Soon after that, it will be just a distant memory of a long lost past in the eternal story of each of us as we continue to progress and experience the blessings of being the sons and daughters of God. We are, and always have been, since our creation long ago, eternal, embodied spiritual beings, citizens of the Spiritual and Physical spheres, the realms we truly call home.

# PART 2

# THE PHYSICAL CREATION

*...why do thoughts arise in your hearts? ...It is I myself: handle me, and see; for a spirit hath not flesh and bones, as ye see me have.*

*LUKE 24:38, 39*

## CHAPTER 5

# A WHOLE NEW WORLD

### Physical Creation

In the previous chapter, we discussed in some detail the processes and resultant effects of the Temporal Creation described in the synoptic accounts of the POGP and Genesis Chapter 1; the Creation we inhabit today. In contrast, Genesis 2:4-5, Moses 3:4-5, and Abraham 5:4-5 provide a glimpse of a Creation that is much different than the one we previously discussed. These brief passages provide a preamble of sorts to a story that remains mostly untold in the pages of recorded Scripture. As we will see, each provides an account of the Physical Creation. Though these scriptural accounts do not detail how this creation was accomplished, they each speak of when it occurred, stating: _all_ things were created spiritually _before_ they were *in* the Earth.

In the previous chapter, we learned that the term "*earth*," as used in the creation accounts, is not limited to the planet we inhabit today, but encompasses a much broader understanding, meaning matter itself and the Temporal Universe as a whole. So in this proper context, things created spiritually before they were in the earth, are things created before this Universe and the matter that composes it was formed, before the event known to science as the Big Bang.

***…things created spiritually before they were in the earth, are things created before this Universe and the matter that composes it was formed, before the event known to science as the Big Bang.***

The recorded Scriptures, however, offer little information about the processes of this creation, and this is understandable, because the single revelation upon which the synoptic accounts are largely based begin with a qualifying statement that the Lord will only provide *a detailed account of the Universe we inhabit today.*[1] Because of this limiting parameter given to Moses, and quite probably also to Abraham, the specific creative processes recounted in the Scriptures are limited to the events, details, and chronology of events for the Temporal sphere exclusively.

So what of the Creation alluded to in Genesis 2:4-5, Moses 3:4-5, and Abraham 5:4-5? What can be known of the event spoken of with these few words? Do these passages refer to a separate creation than the one detailed in Genesis, Chapter 1, and the POGP synoptic accounts? Answers to these questions can be found in a thorough examination of a much different creation account well known to faithful, Temple attending, Latter-day-Saints. Though familiar to many, the specific details of the Temple Endowment account may not be fully appreciated for the insight and evidence they provide of the Physical (spiritual) Creation.

In the next chapter, we will closely examine the Endowment account, gleaning from its details evidences of the reality of the separate and distinct Physical Creation, learn of its differences from the Temporal Creation, and come to an understanding of the processes it involved. We will see that the account provided to us in the Endowment narrative is the one spoken of in the passages of Genesis 2:4-5, Moses 3:4-5, and Abraham 5:4-5. However, before we begin to discover how the Physical Creation was accomplished, we need to step back and discuss some things that will ground us on just exactly what the Physical Creation or *Physical sphere* is.

## A PHYSICAL CREATION?

Why call this a *Physical* Creation? That is a very good question, one that requires an explanation. For millennia the sphere we are to discuss in this chapter has been considered the *spiritual realm*, and so it is.[2] However, that description falls well short of what the Gospels, the Plan of Salvation, Latter-day revelation and perhaps most surprising of all; modern science teaches us about the very real physicality of this enigmatic sphere.

---

[1] "*And now, Moses, my son, I will speak unto thee concerning this earth upon which thou standest; and thou shalt write the things which I shall speak.*" (Moses 1:40)

[2] I struggled with identifying this important sphere as the *Physical Sphere* for quite some time being fully aware of the confusion it would initially cause to the readers of this book. In the end, however, it is ultimately important for us here to firmly establish the true tangibility and physicality of things in this sphere. When we (later in this book) develop a complete understanding of the *being* that is termed a "*Spirit*", it will be seen that it is in fact proper to say that "*spiritual*" things were created when this sphere was formed; see our discussion on this subject, The Spirit of Man, pg. 500.

WHY ARE YE TROUBLED?

One of the earliest aberrations to the Christian faith involved a rejection of a physical resurrection, indeed many believed that Jesus Christ himself, because He was truly God, did not really have a physical body while here on Earth, nor could He.[3] To many, the physical was something that was completely incompatible with the Spirit. These false conceptions found their root in Greek Philosophy, which had pervaded religious thought for several centuries before the Christian era. The early Church Fathers had to remain ever vigilant in suppressing the heretical thoughts that crept into church doctrine in the early centuries of the church. Today many Low Church believers still teach of a resurrection into the spiritual world without an encumbrance of the "*filthy rag*" of a physical body. However, the Savior, the Apostles and the early Christian Fathers all taught clearly of a true resurrection with a true Physical body.

Perhaps the most direct teaching on the reality of a Physical resurrection (beyond of course, the empty tomb) was given by the Savior as He suddenly appeared in a closed room to His Apostles. Speaking to them who were sorely afraid because they were gazing upon a "*spirit*," the resurrected Jesus Christ demonstrated the reality of His immortal physical presence by having them touch and feel Him and subsequently he asked that they bring Him food to eat.

> *"¶And as they thus spake, Jesus himself stood in the midst of them, and saith unto them, Peace be unto you. But they were terrified and affrighted, and supposed that they had seen a spirit. And he said unto them, Why are ye troubled? and why do thoughts arise in your hearts? Behold my hands and my feet, that it is I myself: handle me, and see; for a spirit hath not flesh and bones, as ye see me have. And when he had thus spoken, he shewed them his hands and his feet. And while they yet believed not for joy, and wondered, he said unto them, Have ye here any meat? And they gave him a piece of a broiled fish, and of an honeycomb. And he took it, and did eat before them." (Luke 24:36-43)*

In addition to the events in the closed room, after His resurrection, to emphasize that He was resurrected in the same physical body in which He was crucified, the Savior displayed His hands and feet with the nail prints in them (Luke 24:39, John 20:27). The Gospels record that the two Marys, who came to the sepulcher, held the risen Savior by His feet and worshipped Him (Matt. 28:9). As the reports of Jesus' resurrection spread, Thomas, who was doubtful of the resurrection as it was told to him by the other Apostles said, *"Unless I shall see in His hands the imprint of the nails, and put my finger into the print of the nails, and thrust my hand into his side, I will not believe."* (John 20:25). Later, Jesus appeared to Thomas and said to him, *"Reach hither your finger, and behold my hands; and reach hither thy hand, and thrust it into my side; and be not faithless, but believing,"* (John 20:27).

---

[3] For a more detailed account of many of these early heresies (Adoptionism, Docetism, Substitutionalism, Patripassianism), refer to our discussion later in this work, The Firstborn of Every Creature beginning on pg. 317.

Through these actions, the Savior declared that in His immortal state He has a body of *Flesh and Bones*; His form is eternal corporeality. This body, though *physical*, exhibits attributes that clearly do not belong to the matter of the Temporal sphere.[4] The Savior appeared to the troubled Apostles in a completely closed room as a spiritual being capable of passing through solid stone walls and closed doors, but also with a body made of solid, tangible, matter though no doubt more pure than what we see commonly around us every day.

Despite several heresies that infiltrated Christian thoughts as early as the opening decades of the second century, Christian orthodoxy began to define the absolute certainty of an immortal spiritual world where all Christian believers, like the Savior who led the way, would be resurrected into a corporeal body that was tangible and physical in nature. As the following quotes from many of the Church Fathers show, this has been the historic and continuous teaching of the faithful on this matter since the very beginning of the Christian era.

**Second Clement**

"Let none of you say that this flesh is not judged and does not rise again. Just think: In what state were you saved, and in what state did you recover your [spiritual] sight, if not in the flesh? In the same manner, as you were called in the flesh, so you shall come in the flesh. If Christ, the Lord who saved us, though he was originally spirit, became flesh and in this state called us, so also shall we receive our reward in the flesh. Let us, therefore, love one another, so that we may all come into the kingdom of God" (*Second Clement* 9:1–6 [AD 150]).

**Justin Martyr**

"Indeed, God calls even the body to resurrection and promises it everlasting life. When he promises to save the man, he thereby makes his promise to the flesh. What is man but a rational living being composed of soul and body? Is the soul by itself a man? No, it is but the soul of a man. Can the body be called a man? No, it can but be called the body of a man. If, then, neither of these is by itself a man, but that which is composed of the two together is called a man, and if God has called man to life and resurrection, he has called not a part, but the whole, which is the soul and the body" (*The Resurrection* 8 [AD 153]).

---

[4] Additionally, our LORD's body retained the wounds of the crucifixion, wearing those wounds as a testimony of His sacrifice for and the love He has for each one of us. Along the road to Emmaus, the Savior was able to avoid immediate recognition through some means that may also have been a distinctive attribute of a resurrected body.

**Tatian the Syrian**

"We believe that there will be a resurrection of bodies after the consummation of all things" (*Address to the Greeks* 155 [AD 170]).

**Theophilus of Antioch**

"God will raise up your flesh immortal with your soul; and then, having become immortal, you shall see the immortal, if you will believe in him now; and then you will realize that you have spoken against him unjustly. But you do not believe that the dead will be raised. When it happens, then you will believe, whether you want to or not; but unless you believe now, your faith then will be reckoned as unbelief" (*To Autolycus* 1:7–8 [AD 181]).

**Irenaeus**

"For the Church, although dispersed throughout the whole world even to the ends of the earth, has received from the apostles and from their disciples the faith in . . . the raising up again of all flesh of all humanity, in order that to Jesus Christ our Lord and God and Savior and King, in accord with the approval of the invisible Father, every knee shall bend of those in heaven and on earth and under the earth, and that every tongue shall confess him, and that he may make just judgment of them all" (*Against Heresies* 1:10:1–4 [AD 189]).

**Tertullian**

"Therefore, the flesh shall rise again: certainly of every man, certainly the same flesh, and certainly in its entirety. Wherever it is, in the safekeeping with God through that most faithful agent between God and man, Jesus Christ, who shall reconcile both God to man and man to God, [and] the spirit to the flesh and the flesh to the spirit" (*The Resurrection of the Dead* 63:1 [AD 210]).

"In regard to that which is called the resurrection of the dead, it is necessary to defend the proper meaning of the terms 'of the dead' and 'resurrection.' The word 'dead' signifies merely that something has lost the soul, by the faculty of which it formerly lived. The term 'dead' then applies to a body. Moreover, if resurrection is of the dead, and 'dead' applies only to a body, the resurrection will be of a body. . . . 'To rise' may be said of that which never in any way fell, but which was always lying down. But 'to rise again' can only be said of that which has fallen; for by 'rising again' that which fell is said to 're-surrect.' The syllable 're-' always implies iteration [happening again]. We say, therefore, that a body falls to the ground in death . . . and that which falls, rises again" (*Against Marcion* 5:9:3–4 [AD 210]).

**Cyril of Jerusalem**

"This body shall be raised, not remaining weak as it is now, but this same body shall be raised. By putting on incorruption, it shall be altered, as iron blending with fire becomes fire–or rather, in a manner the Lord who raises us knows. However it will be, this body shall be raised, but it shall not remain such as it is. Rather, it shall abide as an eternal body. It shall no longer require for its life such nourishment as now, nor shall it require a ladder for its ascent; for it shall be made a spiritual body, a marvelous thing, such as we have not the ability to describe" (*Catechetical Lectures* 18:18 [AD 350]).

**Epiphanius of Salamis**

"As for those who profess to be Christians . . . and who confess the resurrection of the dead, of our body and of the body of the Lord . . . but who at the same time say that the same flesh does not rise, but other flesh is given in its place by God, are we not to say that this opinion exceeds all others in impiety" (*The Man Well-Anchored* 87 [AD 374]).

**Augustine**

"Perish the thought that the omnipotence of the Creator is unable, for the raising of our bodies and for the restoring of them to life, to recall all [their] parts, which were consumed by beasts or by fire, or which disintegrated into dust or ashes, or were melted away into a fluid, or were evaporated away in vapors" (*The City of God* 22:20:1 [AD 419]).

"God, the wonderful and inexpressible Artisan, will, with a wonderful and inexpressible speed, restore our flesh from the whole of the material of which it was constituted, and it will make no difference to its reconstruction whether hairs go back to hairs and nails go back to nails, or whatever of these had perished be changed to flesh and be assigned to other parts of the body, while the providence of the Artisan will take care that nothing unseemly result" (*Handbook of Faith, Hope, and Charity* 23:89 [AD 421]).

It should be clear that a true physical resurrection of the *Flesh and Bone* physical body in a perpetual, immortal and tangible state is fundamental to Christian doctrine and has been from the meridian of time. Latter-day Scripture and teaching re-establish what was taught in times long past and adds a clarity attendant to the time frame, language art, and understandings that are associated with the modern era.

Latter-day theology declares that our coming to earth (remember earth, in this larger context, is entry into the Temporal sphere) and receiving a physical body is a blessing of our having chosen Heavenly Father's plan long before the creation of the Universe. Before we were born on Earth, all Spirits who would ever live on

Earth chose to follow the plan of our Father rather than join in Lucifer's rebellion against it (Abraham 3:23–28). As a result, we received a mortal body and, through the gift of Christ's resurrection, will be resurrected with an immortal physical body.

Latter-day revelation teaches that both our Spirit and physical makeup need to be bonded together for our eternal happiness, "*spirit and element [a physical body], inseparably connected, receive a fulness of joy*" (D&C 93:33). We are told of those who had died and were waiting in the Spirit World for Christ's Resurrection, "*had looked upon the long absence of their spirits from their bodies as a bondage*" (D&C 138:50).

The Book of Mormon clearly teaches that it is the power of the resurrection that allows us to enter God's presence to be judged according to our works.[5] We have been told that a resurrected immortal, physical body is required for salvation. Joseph Smith taught, "*No person can have... salvation except through a tabernacle* [a physical body]." A resurrected, immortal, physical body is one of the things that allows us to become like our Father in Heaven and His Son Jesus Christ, for, *"The Father has a body of flesh and bones, as tangible as man's; the Son also"* (D&C 130:22).[6]

The concept of a tangible physical immortality is without a doubt firmly planted in Christian thought, of that there can be no doubt. Even before the time of Christ on Earth the Biblical account records the appearance of Heavenly beings occupying tangible physical forms; for example, the visitation of the LORD and two other messengers as they ate with Abraham on the plains of Mamre.[7]

What immortal physicality means, the understanding of what comprises the eternal "*stuff*" that can be seen, handled, and ultimately hoped for through the resurrection of the body, had been left enigmatically undefined for millennia. In May of 1843, the LORD, through the prophet Joseph Smith, provided a key to understanding the very real properties of the eternal sphere in terms that would ultimately fit into the beautiful lock that would be formed by modern science and an advanced understanding of the cosmos.

---

[5] See Alma, Chapters 40 and 41 (specifically, 40:21).

[6] *God is a Spirit: and they that worship him must worship him in spirit and in truth. (John 4:24)* This passage, some would say, is in direct opposition to an understanding of physically resurrected beings eternally worshiping their Father in Heaven and would indeed be so without the correct understanding that God is a Spirit with a tangible eternal body. We, all of mankind, clearly have corporeal bodies. We have seen in this chapter that we will have physical, tangible bodies in the resurrection, yet within each of us is our Spirit with which we are told we worship our Heavenly Father. The resurrected LORD appeared as an embodied Spirit in the closed room to his disciples, so likewise is the Father embodied.

[7] Genesis, Chapter 18.

Latter day Scripture reveals, "... *All spirit is matter, but it is more fine or pure, and can only be discerned by purer eyes; We cannot seė it; but when our bodies are purified we shall see that it is all matter*" (D&C 131:7, 8).

What is meant by this declaration spoken through inspiration by Joseph Smith?

Joseph clearly taught here that the Spirit World has a physicality that was testified to by the Savior and the Apostle John in the Gospels. He defines in a clear manner the quality identified as Spirit by the Western religions for over four thousand years is in fact a tangible. So real in its physical attribute, the things of Spirit are to be equivocated with matter itself. Through Joseph this revelation restates the truths taught by the Savior to his frightened Apostles in that closed room so long ago. This should be no surprise to readers of the Gospel accounts and to any Christian who believes in the resurrection.

Joseph Smith considered the "*matter*" that is Spirit to be eternal in nature and indestructible. Joseph spoke early of the eternal nature of element but in context clearly referring to the "*matter*" of the revelation recorded as D&C 131; the Scientist and Apostle John A. Widtsoe said of Joseph's understanding:

> *"As early as May, 1833, the Prophet [Joseph Smith] declared that 'the elements are eternal,' and in a sermon delivered in April, 1844, he said, 'Element had an existence from the time he [God] had. The pure principles of element are principles which can never be destroyed; they may be organized and re-organized but not destroyed. They had no beginning and can have no end.'"(From Joseph Smith As Scientist by John A. Widtsoe).*

In its most rudimentary form, and taken in the broadest of contexts, the Prophet Joseph Smith's teaching about the nature of matter that is spiritual and eternal is a statement of the Physical Laws of Conservation of both Mass and the Energy matter is composed of. But he extends these scientific principles beyond the matter and energy we readily detect and measure in this era of the Temporal Universe. Joseph's understanding of this *indestructible* matter was one that most closely described a manifestation of something spiritual in nature, like that of the Savior's body when He appeared after His resurrection in the enclosed room.

With an understanding that there is a tangible reality associated with the Physical (spiritual) sphere and as such understood that *matter*, however pure, occupied a form of *space*, Brigham Young taught about where the *Spirit World* was located:

> *"Is the spirit world here? It is not beyond the sun, but is on this earth..." (Journal of Discourses, 3:372); "Where is the spirit world? It is right here."*[8]

---

[8] "Many of us have heard the statement made—and ascribed to either Joseph Smith or Brigham Young—to the effect that if a person could see the glory of the Telestial Kingdom, he would commit suicide to get there. If only we could get the fundamental doctrines across to Church members as rapidly as we get

*(<u>Journal of Discourses</u>, 3:369); "When our spirits leave our bodies where do they go to? They do not pass out of the organization of this earth on which we live... But where is the spirit world? It is incorporated within this celestial system... Suppose the Lord should touch your eyes that you might see, could you then see the spirits? Yes, as plainly as you now see bodies, as did the servant of Elijah."* (<u>*Journal of Discourses*</u>, *3:368*)

If the *Spirit World* is all around us and Spirit is matter, is it possible we can sense it? Have we already developed the purer eyes the Prophet Joseph spoke of? If we in fact could see that world from here, what would it look like? Is it possible modern science has already caught a glimpse of it, no matter how fleeting the glimpse may be? Let us explore this exciting possibility.

- Ta Horata Kai Ta Aroata

*" for by him were all things created that are in heaven, and that are in earth, visible and invisible..." (Col. 1:16)*

This singular phrase spoken by Paul so long ago reveals truth that continues to be increasingly realized today. In Greek the descriptive words translated in the KJV as visible and invisible are, *ta horata kai ta aroata* (τὰ ὁρατὰ καὶ τὰ ἀόρατα) literally, *the seen and the unseen*. Today the most recent advances in science, with our most modern of instruments, reveal that most of the Universe is unknowable to us with our given, natural senses; it has been invisible to us and as such is unseen.

Things discovered, however, often include the otherwise detectable, if they were not so distant and removed from us in space and time. An example of such on the very large scale is the (now estimated to be) nearly two trillion galaxies that have formed over the age of the Universe, only about one trillion of those still in existence (*Figure 5-1*), and only a handful of those that are visible in the night sky to our God given eyes. We are all aware of the microscopic world teeming with life of all forms, much of it keenly important to us for health and much extremely hazardous to our lives. Even more distantly removed from our natural senses are the molecular, atomic, and subatomic realms.

The detectable Universe includes more than the tangible and observable if however large, yet distant, or tiny and small they may be. The things we can detect (what we

---

across rumors, everyone would be saved. Am I saying that's a rumor? Well, I am saying this, that over a period of many years, I have combed everything Joseph Smith said and wrote, and I can't find it. Hugh Nibley has done the same with Brigham Young's words, and he can't find it. It is hard to prove a negative, of course. What I can say is that we have found a statement from Joseph via Wilford Woodruff that says something else that is close, and I suspect it is the origin of the alleged statement (see Diary of Charles C. Walker, August 1837, in Church Historical Department). Elder Woodruff said the Prophet taught this, roughly: that if we could see what is beyond the veil we couldn't stand to stay here in mortality for five minutes. And I suggest from the context that he was not talking about the telestial kingdom. He was talking about what it was like to be in the presence of God and the family." Truman G. Madsen (The Radiant Life, pg. 91)

Figure 5-1 *The Distant Universe Through the Eyes of Hubble. This impressive image of thousands of galaxies was taken by the Hubble space telescope of a single patch of space the size of the red square. This patch of space was selected because no object (no light) had ever been detected in it before. Every object seen in this picture was previously unseen, and all but seven of the light sources here are far distant galaxies and not a star within our own Milky Way.*

know of) include many intangibles. Among them, the forces of nature, gravity and electromagnetism, and things expressed in mathematical terms such as $E=P_E+K_E$.

We can *feel* untouchable heat and *see* light that cannot be held. We can even *hea*r and *see* the transformed results of radio and other frequency transmissions in our cars, on our radios and on our television screens. Though these things cannot be touched, they can be known; they are *visible* to us by their actions.

There is much about our Universe that is perceptible, that which can be seen. But Paul spoke of *ta aroata (*τὰ ἀόρατα), the invisible and unseen. No doubt he understood these things to be beyond our view, no matter what instrumentation we may eventually develop. He spoke of the *spiritual* as he defined it, what we have understood here to be things of both the Spiritual and Physical spheres.

What is the Universe made of? (*Figure 5-2*) Only 4% of the Universe is the "*stuff*" we call matter (most of that 3.6% of the "*stuff*" of the Universe is unorganized matter in the form of interstellar gas and free baryons in a plasma state). Only .4% of the Universe is the matter that makes the stars and the material comprising planets. The large atoms, capable of forming the molecules, the rudiments of life, comprise in total just .008% of all we can see in the Universe. 23% of the Universe is Dark Matter, mysterious and unknown; indeed by definition it is *unknowable*.

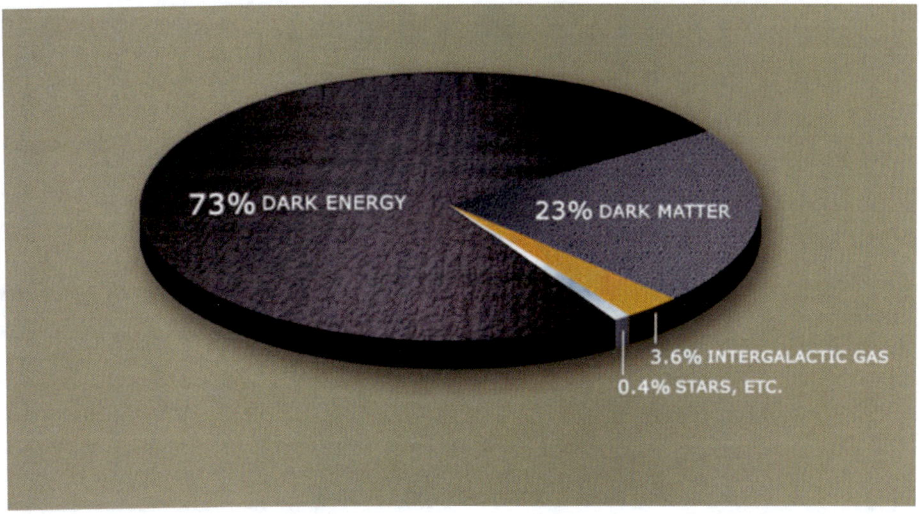


Figure 5-2 *Makeup of the Universe.*

Dark Matter is over 5 times more abundant than the matter we can see, yet it will forever remain outside of our grasp and understanding because it does not produce or interact Electromagnetically, meaning it does not give off light of any wavelength nor absorb it. We only know of the existence of Dark Matter through the gravitational effects it has on "*regular*" matter. However, because it is so abundant, much more so than regular matter, the combined gravitational effect of Dark Matter is the primary driver of gravitation in the Universe. Dark Matter becomes apparent only through the way it binds Galaxies together both internally and externally, one to another. Dark Matter is the source of massive gravitational lenses warping and twisting the light from sources far beyond in the distance, resulting in majestic images now familiar because of the Hubble space telescope.

*Light thinks it is faster than anything, but it is wrong. No matter how fast light travels, it finds darkness is there ahead of it, and is waiting for it*

*-Sir Terry Pratchett*

So far in our discussion here, we have accounted for only 27% of what makes up the Universe, what of the rest? A full 73% of the Universe is in the form of raw energy, Dark Energy, as it has come to be known. Dark Energy does not produce anything measurable in the electromagnetic spectrum but is apparent to us for the effect it has on the size, shape and destiny of the Universe. Dark Energy is the source of entropy and time,[9] in that it is driving the Universe outward from its original singularity and is doing so at a pace that is actually increasing with time and not diminishing.

***Because of Dark Energy, we all have a future and a past that is behind us. Dark Energy ensures that the last moment is no longer where it was when we left it…***

Because of Dark Energy, we all have a future and a past that is behind us. Dark Energy ensures that the last moment is no longer where it was when we left it, making it impossible for us to return to the where we once were.

Dark Energy ensures we have choices and free agency, both physically (*as it is known to science and physics)* and spiritually *(for its value to our eternal progression and the reason we are mortal today)*. Dark Energy, however ephemeral, intangible, and incomprehensible it may seem, is the most powerful and important element of the Temporal sphere and indeed along with Dark Matter is an overlay on to Temporality from the "*eternal, spiritual*" Physical sphere.

In Chapter 1, we touched upon Dark Energy in a brief discussion of the *Zero Point Field* as it relates to the voice of God in creation. I propose here that Dark Energy is the result of diminishing quantum energy fluctuations in this field. These fluctuations are on the tiniest of scales, one hundred million billion times smaller than the electron. Though extremely small in dimensional scale, the field comprises the very fabric of space/time, and it occupies literally all of everything there is.

Ordinary matter fills an infinitesimally small fraction of the volume of the Universe. Even where matter exists, the space occupied by that matter is almost entirely empty. When we examine the space occupied by any singular atom made up of a massive nucleus and dispersed electron shell, we find that the particles themselves comprise a tiny fraction of it, leaving the atom itself 99.9999999999999% empty. It

---

[9] See footnote 2 on page 8 wherein I propose and detail the concept of a new physical law for the equivalency of energy and space. Though the *Zero Point Field* spoken of here is, as of yet, beyond the means of detection and experimentation, and in my view forever will be to those in the Temporal sphere, its effects can be seen in the realization of the arrow-of-time, and it is the reason for Entropy in the Universe.

must be remembered that the relatively massive electrons and quarks that make up everything we can sense within the atoms are themselves filled with the *Zero Point Field*; there literally is no place, *anywhere*, where it is not.

Total energy is defined as the total of Potential energy and Kinetic energy ($E=P_E+K_E$). Matter is a confined form of energy (a special, tangible, form of Potential energy), and we see this to be true. In the extreme fireball of a nuclear blast or the processes at play deep within our own Sun, a small amount of matter releases vast amounts of energy when it is converted. In this, despite the enormous amount of energy released with forceful effect, we must realize that the total $P_E$ of matter is not fully converted in this process but *much* is dispersed and spread out, remaining within the *Zero Point Field.*

Matter and energy are equivalent so, when we do the math, when we sum the energy that forms detectable matter (both bright and dark), and subtract out what can be attributed to energy in the *Zero Point Field* within the space that matter occupies, we can account for only 24% of the energy/mass in the Universe. By that, it is revealed that fully 76% of the energy/mass that comprises the Universe is tied up in the miniscule fluctuations of the *Zero Point Field.*

The matter that exists in the Temporal Universe is but a manifestation of energy spikes on this field (the peaks between the valleys of the waves moving through the field).[10] The fluctuations draw up the field (the fabric of space/time itself) into a fourth dimension, constraining the field's size in the other three. As the field energy dissipates and the fluctuations diminish, the other three dimensions enlarge and expand, taking the Universe and all of its matter along with it.

At the moment of the Big Bang, the *Zero Point Field* was at infinite energy and zero entropy, space was non-existent, completely confined within a singularity. In the intervening eons, the *Zero Point Field* has dropped in energy and increased in entropy producing the arrow of time in the process. Since the publication of Einstein's theory of general relativity, the world has become accustomed to visualizing time as a fourth dimension. Through the *Zero Point Field,* time in fact takes its form and is a direct result of the diminishing value of fluctuations in the fourth dimension.

In the discussion we previously had of the second verse of Genesis in Chapter 1, we spoke of the Hebrew words *tohu* (תֹּהוּ) and *bohu* (בֹּהוּ). The verse reads: "*And*

---

[10] Here, I expose my *personal affinity* towards the mechanics of an alternate theory of the observations attributed to Quantum Mechanics. The *De Broglie Bohm Pilot Wave Theory* most closely resembles the concepts described within this work of vibrations in the Zero Point Field giving rise to matter, directing its actions and propagating information. Most describe this theory as deterministic, rather than uncertain, however, in fact, the same limitations of determining an outcome apply. In *Bohmian Mechanics* uncertainty remains within its *Contextuality*. Bohmian Mechanics describes a universe that looks as uncertain as we observe ours to be (it wholly adopts the Schrodinger equation), but is in fact determinable if all of the facts (unknowable to us) are known, as they would be by an omniscient deity (perhaps a description of the *middle knowledge* of Molinism). I envision our Universe to be modeled per Bohmian constructs but accept that it is uncertain allowing the agency requisite to the Temporal sphere.

*the earth was tohu and bohu.*" We remember that the word *bohu* has been most commonly translated in English as "*void*" and is done so in the KJV in Genesis 1:2. *Void* is a good description of the emptiness within the Universe that is filled with the *Zero Point Field*, making the rendering *void* appropriate, but remember in Hebrew the word *bohu* means something a bit more. *Bohu* is a composite word, as are many Hebrew words of the Bible, meaning: *bo* – in it, and *hu* – there is. Both the Talmud and the Jewish scholar Nahmanides (Rambam) state that *bohu* (in it, there is) can be understood as meaning "*filled with the building blocks of matter.*" So here in this singular verse, we understand the void (*tohu*) is filled with the most fundamental "*stuff*" that the tangible "*stuff*" of the Universe is made up of. This is *exactly* in line with what the concept of this field as described here should be.

Now we must turn again to the Hebrew word *tohu* (תֹּהוּ), for its meaning takes on greater detail that is important to us here. In addition to its meaning of "*formless or chaotic,*" the Kabalah teaches that *tohu* is the solitary primordial substance (substrate) from which all that is material (tangible) was to be formed. So adding *tohu* to *bohu* (בֹּהוּ), as was done in Genesis 1:2, we find that, "In it (in the *tohu*) there is *potential.*

> *And the earth (spiritual matter) was the single primordial substrate within which was the building blocks (potential) of all that there is.*

From the placement of these Hebrew words together, at the very beginning, we see that the single substance (substrate) that fills the void, the *Zero Point Field*, is the "*stuff*" from which everything came to be, and it contained the potential to form and infuse all that there is with the structure and information that has become the material of the Temporal Universe.

This is a treatise on things pertaining to the Creations of God and not a primer on Cosmology or Quantum Physics, so we will not delve into these subjects further, but in closing, consider the following about the *Zero Point Field*:[11]

- It exists outside of time and actually provides the engine for time itself.
- Because it exists outside of time, it is eternal.
- All that exists or ever will be is embedded within it.
- Communication within it is instantaneous at any distance.
- It is intimately a part of everything, yet is entirely undetectable by physical means.

---

[11] The *Zero Point Field* is an entirely theoretical construct; it is not possible to prove its existence through experimentation for its confines and actions are both smaller than Plank length and time. However, the *Zero Point Field* provides the mechanisms needed to explain physical phenomena detectable through indirect means. It also accounts for things we have come to call spiritual. The Word of God is established through "*the mouth of two or three witnesses*" (Matt. 18:16). Through the recent discoveries of science and the revealed Word of God to prophets, both ancient and modern, we receive strong testimony of evidences attained through broadly dissimilar disciplines of the existence of the *Zero Point Field.*

So why be concerned here about all of this? What does this have to do with the Physical (spiritual) Creation? It was said previously that Dark Energy is the most powerful and important element of the Temporal sphere and indeed along with Dark Matter is an overlay on to Temporality from the "eternal spiritual" Physical sphere. This is the beautiful lock that is to be opened by the key given to us by the LORD through Joseph Smith when he said, "*All spirit is matter, but it is more fine or pure, and can only be discerned by purer eyes.*" It is entirely probable that Dark Matter, the undetectable matter that holds the Universe together, is the Temporal manifestation of that undiscernible *spiritual matter* spoken of in the revelation. Dark Matter is five times more abundant than the matter we can see, giving us an idea of relationship between the Physical (spiritual) and the Temporal. The *Zero Point Field* can be ultimately identified as something definable as Spirit. In this context, let us turn to the words of the Prophet Zechariah:

> *"The burden of the word of the LORD for Israel, saith the LORD, which stretcheth forth the heavens and layeth the foundation of the earth, and formeth the spirit of man within him." (Zechariah 12:1)*

Here, in a manner we would expect to find in a physics textbook, the LORD reveals to his prophet key characteristics of the *Zero Point Field* we have discussed here. Through the diminishing amplitude of the energy within it, the field is stretching, taking the heavens along with it. The field gives rise to the very matter (the earth) of both the Temporal and Physical spheres, the matter that is the foundation of everything there is both mortal and immortal. The *Zero Point Field* is that which is within us and through us; by our understanding, it is what we can envision as the "*stuff*" of Spirit from which all of our Spirits are made. In contrast to the tangible things in our Temporal Universe, the *Zero Point Field* is something that is *in* the Universe, but not *of* it. Yet at the same time, the *Zero Point Field* is the very essence of everything that is, while forever remaining just beyond the veil.

## SUMMARIZING WHAT WE HAVE LEARNED

The Endowment tells of unorganized matter being formed into eternal worlds by Jesus Christ and Michael. Through our understanding of the ancient Hebrew (*tohu* and *bohu*), new discoveries of the accelerating expansion of the Universe and the teaching of Joseph Smith, we can equate the (spoken of) unorganized matter with Dark Energy and the *Zero Point Field*.

Since the Temporal Universe is a subset of the greater Physical (spiritual) realm, we see the imprint of the Physical realm through the manifestation of Dark Energy and the *Zero Point Field*. The Physical building blocks that were organized into tangible, eternal, Physical (spiritual) worlds, materials, plants, animals and the children of men (Moses 3:5) are manifest to us today by the effects of the eternal Physical Matter that comprises them on the Temporal Universe around us. These effects are those we attribute to Dark Matter in the Universe today. We are told by Latter-day prophets that the Spirit World is all around us.[12] If it can be sensed at all Temporally, the manifestations of Dark Energy and Dark Matter are what we would expect the sensibility to be.

I must note here that the term Dark as it applies to Dark Energy and Matter does not in any way derive its name from any association with the Dark forces of the evil one, Lucifer, the father of all lies. Remember the term Dark is applied scientifically and only refers to the attribute of non-detectability through any direct means by the Temporal sphere. Those of us who are familiar with the spiritual are fully versed in the realization that the spiritual can be addressed by, and influences, the Temporal every day.

Finally, it is interesting to note a comment made by Job over 3000 years ago. Reflect back to what was said about the fluctuations of the *Zero Point Field* in the fourth dimension constraining the other three dimensions and read these inspired words:

> *... For I am full of matter, the spirit within me constraineth me.*
> *(Job 32:18)*

Surely, the word of God is a wonder to behold...

We have learned that the resurrection of wholly physical, tangible, bodies belonging to the faithful, eternally in the presence of God the Father, is an essential and core doctrine of the Christian faith. We have seen that this concept was taught by church leaders both ancient and modern and demonstrated by the Savior Himself. We have discussed modern scientific discoveries that illuminate the existence of physical matter and fields that remain beyond the reaches of what we know to be the mortal Universe around us. We know that Dark Matter and Dark Energy influence us every moment of every day. It makes our Temporal existence possible.

Through the evidences of disparate disciplines, scientific and religious, we are assured that there is a very real, immortal, Physical sphere. This sphere and the eternal elements within it is the Physical Creation, the second of the Three Creations of God, the one spoken of and detailed to us in the Temple Endowment.

---

12 Joseph Smith taught that the Spirits of righteous people "*are not far from us, and know and understand our thoughts, feelings, and motions, and are often pained therewith*" (Teachings of the Prophet Joseph Smith, pg. 326). Brigham Young said: "*Where is the spirit world? It is right here*" (Discourses of Brigham Young, pg. 376).

# CHAPTER 6

# THE SECOND CREATION

## The Generations of the Heavens

Tucked neatly into the gap between the telling of the Temporal Creation and the passages revealing the story of Adam and Eve in the Garden of Eden can be found in just two verses (in each of the three recorded scriptural accounts) that ground the events to follow *deep* within a context that is fundamentally different than the creation story that preceded. As we discussed in Chapter 1, this abrupt disconnect and restart in the creation account has been the subject of discussion and confusion for much of three millennia. Despite its position in the text, following the "*first*" account, the "*second*" account is broadly believed to be the older of the two, predating the telling of the "*first*" account by perhaps a thousand years. That this is true is largely held without dispute.

That the *second* is considered to be the more ancient of the two accounts[1] is fortuitous and coincidental, in that it actually tells of a time and events that precede those in the passages it follows by perhaps eons. Attempts to explain away the apparent discrepancies of these two accounts stem from a misunderstanding of the creative processes and failure to recognize that there are, in fact, two separate stories being told. The *second* account is far less problematic to Latter-day Saints because we have an understanding that includes a Spiritual (physical) Creation before the creative events described in Genesis 1 and its parallel accounts in the POGP.

---

[1] Extensive analysis of its style and content have led scholars to conclude that the second story beginning with Genesis 2:4 was written in about the Tenth Century B.C, while the first account (Genesis 1) was probably written sometime around 530 BC by the priestly caste after the Hebrews had returned from Babylonian captivity. These two accounts of creation include: The first, a cosmocentric account of how God created the Heavens and Earth out of the void, repeated from Babylonian myths learned during the Hebrew Captivity, and the second, an anthropocentric account of how humanity came to populate the Earth based upon the Hebrew mythology. Here in this work, we affirm a divine source which revealed these two dissimilar accounts to man, though we accept the order of these accounts as assessed through scholarly study.

| Genesis 2 |
|---|
| 4 ¶These are the generations of the heavens and of the earth when they were created, in the day that the Lord God made the earth and the heavens,<br>5 And every plant of the field before it was in the earth, and every herb of the field before it grew: for the Lord God had not caused it to rain upon the earth, and there was not a man to till the ground. |
| **Moses 3** |
| 4 And now, behold, I say unto you, that these are the generations of the heaven and of the earth, when they were created, in the day that I, the Lord God, made the heaven and the earth,<br>5 And every plant of the field before it was in the earth, and every herb of the field before it grew. For I, the Lord God, created all things, of which I have spoken, spiritually, before they were naturally upon the face of the earth. For I, the Lord God, had not caused it to rain upon the face of the earth. And I, the Lord God, had created all the children of men; and not yet a man to till the ground; for in heaven created I them; and there was not yet flesh upon the earth, neither in the water, neither in the air; |
| **Abraham 5** |
| 4 And the Gods came down and formed these the generations of the heavens and of the earth, when they were formed in the day that the Gods formed the earth and the heavens,<br>5 According to all that which they had said concerning every plant of the field before it was in the earth, and every herb of the field before it grew; for the Gods had not caused it to rain upon the earth when they counseled to do them, and had not formed a man to till the ground. |

Table 6-1 *The Second Creation. I have intentionally highlighted the first part of Genesis, Chapter 2, Verse 4 setting those words apart from the rest of that verse in that properly, Verse 4 starts at "in the day." The underlined words from the beginning of Verse 4 belong to Verse 3. How the verse got to be assembled the way it has been is a lengthy story involving many factors all well documented.*[2]

[2] For a very good treatment of this issue, refer to commentary in (*The Jewish Study Bible*, Featuring the Jewish Publication Society, Tanakh Translation, Berlin. Adele and Brettler, Marc Zvi, editors, Oxford University Press 2004). That an iteration of this misplaced phrase appears in both the Moses and Abraham accounts muddles up the scholarly conclusion. Perhaps that the phrase appears in the inspired

It is important for us to realize that these verses are not only in reference to the processes that result in the creation, development, and filling of the Garden with plants, animals and man, but they speak of events from a time predating that. These verses are the only vestiges to be found in the written pages of Scripture of the creation story told in the Temple Endowment. Following them, the scriptural story continues with a fully formed Adam in a completed Garden at a time before Eve joins him there.

As stated earlier, these verses record a preamble of sorts to the Garden story, though in truth they are so much more. Much more than an introduction to the story of the Garden; the preamble verses describe, in the broadest of strokes, an entire creation that is much different than the account given for the Temporal world. We are told of generations, both of Heaven and of Earth. Here we see a creation of plants before they were in the Earth and herbs before they grew, of a time before there was rain upon the Earth, or man to till the ground. The Genesis account ends with these mystical phrases, but the Moses account adds something of significance that embeds the words of Genesis in a firmer understanding:

> *"For I, the Lord God, created all things, of which I have spoken, spiritually, before they were naturally upon the face of the earth. For I, the Lord God, had not caused it to rain upon the face of the earth. And I, the Lord God, had created all the children of men; and not yet a man to till the ground; for in heaven created I them; and there was not yet flesh upon the earth, neither in the water, neither in the air;" (Moses 3:5)*

In this passage, we are given an explanation of a different creative process than that we have discussed for the Temporal sphere. Because all things were created *spiritually*, in the Physical sphere, they existed in whole before they were *temporally*. Plants and herbs were fashioned entirely (perhaps in their mature state without the developmental process of growing), capable at once of fulfilling the "*full measure*" of their creation. That this may very well be so can be understood by considering the phrases: "*before it grew...*" and "*had not caused it to rain...*"

Also of extreme significance here is that we are told the LORD God had created all the children of men, before there was a single man to till the ground. How could this be so given the traditional understanding of a single man and woman placed in a Garden, located somewhere on the very Earth we all now inhabit, being the progenitors of all of mankind? The answer is to be found in an understanding of a separate and distinct creation of eternal Physical (spiritual) beings, children of men, plants and animals at a time *before* the Garden story and *before* the creation of the Temporal Universe we all inhabit. The past tense "*had*" is used by the LORD in Moses 3:5 to describe the creation of *all* of the children of men, firmly placing that event or process at a time *before* the events of the Genesis 1 accounts we discussed

---

renditions of verse 4 indicates that it is properly included there and that it was removed anciently from verse 3 for the reasons detailed by the Hebrew Biblical scholars. Those who did the removal found justification in doing so, because the phrase was repeated in the very next verse. If there was repetition in the original texts as inferred by me here, this would add further strength to our belief that there are indeed separate creation accounts recorded in the Bible.

in the previous chapter. *Before* the Universe we live in today was created, *before* any living thing was on the Earth or in the water, and *before* the time when a singular Adam occupied a singular Garden planted somewhere Eastward in Eden; The LORD God says, He created *all* of the children of men[3] in Heaven, *all of them*.

## SPIRITUALLY AND NATURALLY

The preamble verses delineate between the spiritual and natural. In Verse 5 of Moses 3, we read that all things were created spiritually "*Before they were naturally upon the face of the earth.*" This key phrase underscores a very vital difference between the Temporal and Physical spheres. In the Bible the term natural describes the Temporal state. None of the things listed as being created in these verses (plants and herbs and the all the children of men) existed in a natural state at this time; they were spiritual, having been created in the Physical sphere. The *natural* analogue, the impermanent twin of all things first fashioned spiritually, had not yet been created in the parallel Temporal sphere. Living (natural) things did not exist *until* their Temporal counterparts were formed on the Earth through the processes of the Temporal Creation we detailed in Chapter 3.[4]

The Scriptures describe the *natural state* as one that is in direct conflict with the state of God. A state that is separated from the presence of God until it is put away and the natural is *restored* by sanctification through the Atonement of the Son of God.

> *"For the natural man is an enemy to God, and has been from the fall of Adam, and will be, forever and ever, unless he yields to the enticings of the Holy Spirit, and putteth off the natural man and becometh a saint through the atonement of Christ the Lord..." (Mosiah 3:19)*

The insertion of "*before they were naturally...*" in Moses 3, identifies the quality of the things created spiritually as being different from the state they were to obtain after the Fall in their re-creation within the Temporal sphere.

## THE DIVERSITY OF LIFE

The variety of life we observe in the world around us appears to be a result of the needs of survival, protection, nutrition and reproduction, the sustaining of life itself in a harsh natural and Fallen world. Perhaps the Physical (spiritual) Creation is not so diverse, not needing the range and assortment we see about us today.[5] We are

---

[3] This is one of the places where the entanglement of terms, a result of millennia of transmission of the sacred record, has confused the common term *men* with the sons and daughters of God. That *men* here infers the sons and daughters of God and not the people of the clay, the *ha-a-dam* (הָאָדָם), of Genesis 6 is clear from the context.

[4] There is one marked exception to this rule that we will address at length within this work – that of the special circumstances associated with the Garden Temple.

[5] After the Fall, the "*ground*" (the earth, the substance of matter in the Temporal sphere) was to "*bring forth*" "*thorns and thistles*" in a manner that was foreign to the precincts of the Garden in the Physical sphere. (Gen. 3:17, 18)

told, however, that the range of life in the Physical sphere is varied and with beauty and that it was created that it might be pleasing to man, "*...all things... are made for the benefit and the use of man, both to please the eye and to gladden the heart*" (D&C 59:18).

The very variety of life of the Physical (spiritual) sphere as spoken of in the Endowment narrative and the Garden account, testifies of the certainty of physicality in what has traditionally been considered a strictly *spiritual* (nonphysical) province.

It is perhaps reasonable to envision a spiritual analog of rational sentient human beings, the existence of which, though distinctly different from their mortal twins, would have intrinsic value to a traditional God who requires worship; but what of the spiritual analog of all the rest of this creation? Certainly spiders and moths, flora and bacterium, asteroids and pulsars can offer little in the way of worship and companionship to the Creator, but we are expressly told of the creation of *all things* spiritually. The spiritual creation of things like these must include something more than mere apparitional manifestation, some ethereal intangible substance, usually associated with things spiritual. Our understanding here is that these things were created spiritually to be part of the eternal experience of our eternal selves in an idyllic habitation that mirrors the troubled and dangerous Temporal sphere we inhabit today.

After the Fall, the diversity of life expanded within the Temporal sphere; through the evolutionary processes of creation we discussed in Chapter 3, variety ensued. A small and unassuming reference that speaks to this fact rests hidden in the Scriptures: "*Thorns also and Thistles shall it bring forth...*" (Gen. 3:18). Here we see the expansion of the variety of life to come forth with purpose. Once that door was opened, powered by the information imbedded by God within the *bohu* of the Universe, diversity has continued to multiply the many forms and functions of life.

## THE PRE-EXISTENCE OF MAN

The things spoken of in the preamble verses are immortal, capable of being in the presence of the Father without restriction and wholly unnatural. Yet they include *all* the children of men. Without a doubt this is one of the most direct statements of the pre-existence of man found anywhere in the Scriptures. The implications of this statement are far reaching when taken in the context of its chronological positioning in the narrative.

The Scriptures state that all of the children of men were created spiritually before they were naturally on the Earth *and* before there was yet flesh upon the Earth. Here we have all the children of God created with, in what is now clear, tangible, physical, Spirit bodies before the Temporal sphere exists and before Adam is placed in the Garden. In the next section of this chapter, we will see that this happens on the sixth day of the creation story told in the Endowment.

That pre-existent man did have a physical body is without question. For it is fundamental to our understanding of the Physical sphere and consistent to our notion of returning again, through the resurrection, to a state and existence we once held. Christ, in an antemortal appearance to the brother of Jared, spoke of the reality of the spiritual body:

> *"Behold, I am he who was prepared from the foundation of the world to redeem my people. Behold, I am Jesus Christ. I am the Father and the Son. In me shall all mankind have life, and that eternally, even they who shall believe on my name; and they shall become my sons and my daughters. And never have I showed myself unto man whom I have created, for never has man believed in me as thou hast. Seest thou that ye are created after mine own image? Yea, even all men were created in the beginning after mine own image. Behold, this body, which ye now behold, is the body of my spirit; and man have I created after the body of my spirit; and even as I appear unto thee to be in the spirit will I appear unto my people in the flesh." (Ether 3:14-16)*

Here in these verses stand Christ, a "*begotten*" created being, for we are told by the apostle Paul, He "*...is in the image of the invisible God, the firstborn of every creature*" (Col. 1:15). Christ appeared as an *embodied Spirit*, the "*firstborn among many brethren*" (Rom. 8:29), the very, "*beginning of the creation[s] of God*" (Rev. 3:14) as revealed to John.

> ***The bodies we will attain in the resurrection are not simply made up of the fallen materials of Temporality, but rather are those spoken of by Christ to the brother of Jared.***

The Savior's spiritual embodiment has a human form, for He declares, "*even as I appear unto thee to be in spirit, will I appear unto my people in the flesh.*" This is not just an apparition, a mere *appearance* of the LORD in bodily form for the convenience of relating to man, as in the manner believed by the Docetists, but a true representation of Himself in immortality and a true representation of all men in the form in which we all were created in the Physical sphere; "*all men were created in the beginning after mine own image...man have I created after the body of my spirit.*"[6]

The "*body of my spirit*" revealed to the brother of Jared has a true physicality. It is a commonly expressed belief among Latter-day Saints, that among other reasons, we come to Earth to receive our physical bodies. The belief is often conveyed in stories frequently relating to stillborn babies or the premature death of infants. It is commonly articulated as a component in the detailing of the Plan of Salvation. But this belief, as it is simply understood and spoken, is incomplete and incorrect in whole. We were fully

---

[6] Ether 3:16.

embodied physical entities before we came to the Temporal sphere. The Temporal bodies that enclose our Spirits here on Earth are only a small part of our eternal inheritance. It is true we have argued for the reality of a physical, tangible resurrection of our Earthly bodies previously in this work, and we continue to stand behind that argument here. However, the bodies we will attain in the resurrection are not simply made up of the Fallen materials of Temporality, but rather are those spoken of by Christ to the brother of Jared. They are the bodies we were given in the Physical sphere *plus a little bit more*, the addition of perfected (immortal) Temporal matter.

Our Temporal bodies are a lesser part of the whole we are to be and a fleeting image of what we once were. They are only a Temporal representation of what we really are and will again be.

The bodies we attain in the resurrection are a joining of the Temporal elements with the Physical elements in a way, and through a process, still unknown to us. We can conclude that this joining empowers in a way that was not attained before. For this reason we find demons seeking to indwell even the bodies of swine for an apparent advantage. From our previous discussion regarding evil and the freedom of choice attendant to the indeterminacy of Temporal matter,[7] we can surmise that possession of matter and control of events in an uncertain reality promotes the attainment of abilities that were not otherwise available to our spiritual entities.

The uncertainty of Temporal things provides for a future with choices. Without it, our thoughts and will cannot be expressed. In a universe without uncertainty, all events are strictly known and defined forward and backward in time; no independent action or thought can take place. The blending of our Spirits (our embodied eternal Intelligences) with the Temporal matter that makes up our Temporal bodies constitutes our living Soul, "*...the spirit and the body...living souls*" (2 Nephi 9:13).[8] This three part synergy-of-self touches on Deity, providing the divine gifts of eternal Intelligence and immortal physicality and the means of independent expression through the control of Temporal matter. This is the everlasting glory spoken of by the LORD, that which was "*added*" to Adam and the sons and daughters of God, after the Fall, *"...they who keep their first estate shall be added upon; and they who keep not their first estate shall not have glory in the same kingdom..."* (Abr. 3:26). This is what is denied of Satan and those who chose to follow after him. Later in this work, we will discuss the reality and potential of an eternal Soul.[9]

In our discussion of the Temporal Creation, we understood *Heaven* (as used in Genesis 1:1) referred to the macro structures of the Universe, galaxies and clusters of galaxies, and more specifically the expanse of the heavens seen in our night sky,

---

[7] See Moral Evil, pg. 124.

[8] That our living Soul is a Temporal entity as well as a manifestation of immortality after the resurrection can be seen in the contexts of Alma 16:9, 3 Nephi 5:1 and of course Gen. 2:7.

[9] See Chapter 11, What Is Man, pg. 469.

our home galaxy, the Milky Way. Heaven is used again in the first half of Genesis 2:4 "*These are the generations of the heavens and of the earth when they were created.*" In the context provided here by the preamble passages, its meaning (though similar) includes something more. "*The generations of the heavens and the earth...*" is a specific reference to the creation of all of the sons and daughters of God, those to whom Earth was to become home and those who would live on other worlds, alien to ours, worlds out amongst the heavens stretched above us.

WHAT IS SPIRIT?

Perhaps the most common proof text used by Christians today to defend their belief in the Greek philosophical conception of God[10] is the New Testament passage, "*God is a spirit...*" (John 4:24). But this is no proof text at all for the concept of an incorporeal God; in fact, it is just the opposite. Let me explain.

The true meaning of this passage, as dictated by the Apostle John to his scribe Bishop Papias, has been lost both in translation and in doctrinal expectation. Because of Platonic influence on our notion of deity and the actual changes that influence has made to language itself, the original intent of these four words have been not just altered, but altered *entirely*.

In his seminal work, De Principiis, the Greek speaking and renowned, prolific, early Christian theologian, Origen (a Platonist himself), believed in an incorporeal God and strongly defended his belief against those who thought otherwise. He wrote:

> *"I know that some will attempt to say that, even according to the declaration of our own Scriptures,*[11] *God is a body, because...They find it said...in the Gospel according to John, that God is a Spirit, and they who worship Him must worship Him in spirit and in truth. ...Spirit, according to them, [is] to be regarded as nothing else than body."*[12]

---

[10] The concept of a disembodied God without body and parts is of modern origin. For the ancients and the early Christian church, God was always *corporeal*. Christ they understood to be God, though in body He tangibly walked among men in Palestine. In His resurrected, immortal form, He remained embodied; He begged them to touch and feel Him, they understood His physical nature well. Even at Nicaea, in the fourth century, the debate centered on the "*substance*" of God, *not* his insubstantiality. For those employed there, the Son was corporeal and so too was the Father; the only thing left to debate was how exact the match was for their divine *substance*. In fact, it was the understanding of both an embodied Father and Son that necessitated the defining of the third member of the Godhead, the Holy Ghost. Those that espoused non-corporeal views of the Father and the Son were termed "*Heretical.*" Corporeality was never the question, until Greek teaching replaced divine truth.

[11] Origen uses the term "*our own Scriptures*" to distinguish the Apostolic, Christian writings of the New Testament from the Hebrew Old Testament writings. It has significance in his argument here because he has made a concession that the Old Testament writings were replete with references to a corporeal God. He had dismissed those references as interpolations made through millennia of cultural influence. Here he must address the fact that the then modern apostolic writings were also used to teach the corporeality of our Father in Heaven and begins his defense of the incorporeality of Deity by admitting the grammar used by John before he attempts to explain it away.

[12] Origen, De Principiis, eds. Alexander Roberts and James Donaldson, *The Anti-Nicene Fathers*, 10 vols. (Grand Rapids, Mich.: Eerdmans, 1951) 4:242.

To our modern mind this statement makes no sense at all. We justly reason; how could any argue that Spirit is, "*nothing else <u>than</u> body*?" That this is so confusing to us today is precisely the point to be made here; our concepts of Deity and even the definitions of our words, as we know them today, have been immutably altered by Greek philosophy.

Origen wrote what he did because *Early Christians almost exclusively believed God to be embodied*; the concept of a corporeal God was the common view. Origen, a Platonist, was defending his incorporeal position against the masses that held his view to be wrong. Origen begins by shrinking the opposition when he says "*<u>some</u> will attempt to say*," but this is misleading in context. The corporeal view was indeed the typical view held by the body of the church at this time. Though Origen's view of incorporeality would become orthodoxy in the following few centuries, at the time he penned (early in the third century), <u>De Principiis</u> it was written to defend a novel and *minority* view held by very few (the intellectual elite) at the time.

It very well may have been entirely accurate to state that *some* would use this passage as a proof text, for few could offer such a challenge to their commonly held belief with such a level of intellectual sophistication, especially against a mind so blessed as Origen's. *But the strength of the text in this passage of John, as proof of an embodied, corporeal, God, is clear and powerful*, though that reasoning is entirely lost to us today. How can this be so?

The *basis* for the original reasoning addressed by Origen rests in the Greek word *pneuma* (πνεῦμα), translated in the KJV and virtually everywhere else today as *spirit*. To those who wrote the Gospel, those who read the Gospel then, and to Origen, all Greek speakers by birth, and educated with a world view much different than the one held by us today, *pneuma* literally meant, air or breath. Because of this, what to our ear appears to be an extremely confusing statement made by Origen is easy to explain.

Air, and by extension breath, was one of the four basic *elements* making up the <u>*material*</u> *world*. Pneuma then is material, tangible, and measurable; to the Greek world, it was fundamental and was *substance*, and in its original context, it still is today.

<u>De Principiis</u> was directed to early Christians who, with few exceptions, believed in a corporeal God. To them this passage merely confirmed what they already knew to be true. God being *pneuma* was embodied; John revealed that his form was made of the purest and most ethereal of all the elements. That fact was expected and would be surprising to none, for God is perfect in form. For Latter-day Saints, this concept should strike a very familiar chord, for Joseph Smith said, "*All spirit is matter, but it is more fine and pure, and can only be discerned by purer eyes; We cannot see it*" (D&C 131:7, 8).

With all of this in mind, the passage recording the Savior's words spoken to the Samaritan woman at the well[13] can most properly be translated as follows: *God is pure element, and those that worship Him must do so in their pure element and in truth.*

## EASTWARD (IN?) EDEN

Eden is a place or region, but where? We are told in Scripture that the Garden was placed *in* Eden – not East *of* it, but *in* it, somehow eastward from some internal reference point. It has been traditionally assumed that by *eastward*, the Biblical scribes intended to convey that Eden was a place to the East of the Holy Land. In that perspective, they took not only the Garden to the East of Jerusalem, but the whole of Eden away to the East as well. This precept has led to many fanciful claims locating Eden in oriental locales all over central South Asia and beyond.

Some Latter-day Saints take statements made by the Prophet Joseph Smith to infer that he said Eden was located in what is Missouri today. A careful search will turn up no such claim made by Brother Joseph.

It is our understanding that the Garden is entirely an immortal physical construct within the Physical sphere and should not be thought to be located anywhere within the Temporal Universe. This conclusion is strongly implied and indeed *must* be prerequisite to the existence of the Temporal Creation. That this is true can easily be deduced from the Scriptures in the context of the Three Creations we have detailed in this work.

The Garden itself will be discussed thoroughly in the next chapter, but of interest to us here in our discussion of the Physical sphere as a whole, is the possibility that by placing the Garden *in* the Physical sphere, specifically somewhere eastward within it, we are left with an entire geography of a place called Eden to consider. Perhaps Eden refers to the entirety of the Physical Sphere and as such is the name of the place traditionally called Heaven.[14]

The proper context of the word *heaven* is in association with the Temporal Universe and not with the abode of things spiritual. It refers to the expanse and contents of the Universe itself, literally all that there is, beyond the local confines of the planet Earth itself.

The word heaven is not used in the Endowment narrative, but in that narrative we are told of the creation of stars in an expanse which function to provide light and

---

[13] John 4:24.

[14] Some credence can be given to this suggestion by the Targum of Rabbi Jonathan Ben Uzziel (Palestine): "*...and the garden from the Eden of the just was planted by the Word of the Lord God before the creation of the world, and He made there to dwell the man when He had created him.*" Here we read that the Garden was *from the Eden of the just, planted before the creation of the world* (the Temporal sphere) so Eden is a place name of a locale in the Physical sphere.

govern the orbits of the several worlds around them. The Physical Creation resulted in Celestial and Terrestial objects without number. The Physical Creation detailed in the Endowment narrative developed an immortal Earth with equally immortal Sun, Moon and stars within its skies. The Physical Creation not only produced one such Earth but many, each with their own appointed skies filled with suns, moons and stars.

Though powered in some way much different than the process of fusing Temporal particles to one another, which has a deteriorating effect that ultimately leads to the death of each star, the stars of the Physical sphere do exist in an immortal way, for they were revealed to Abraham:

> *"And I saw the stars, that they were very great, and that one of them was nearest unto the throne of God; and there were many great ones which were near unto it; And the Lord said unto me: These are the governing ones; and the name of the great one is Kolob, because it is near unto me, for I am the Lord thy God: I have set this one to govern all those which belong to the same order as that upon which thou standest. And the Lord said unto me, by the Urim and Thummim, that Kolob was after the manner of the Lord, according to its times and seasons in the revolutions thereof; that one revolution was a day unto the Lord, after his manner of reckoning, it being one thousand years according to the time appointed unto that whereon thou standest. This is the reckoning of the Lord's time, according to the reckoning of Kolob." (Abraham 3:2-4)*

This vision was seen through the use of the Urim and Thummin, and in it Abraham witnessed the significance of a great governing one, named Kolob, because it is "*near unto the Lord God.*" The LORD said that Kolob itself was "*after the manner of the Lord.*" Without doubt our God and His Son Jesus Christ do not permanently inhabit the Temporal sphere or any part of it, so Kolob along with the "*many great ones*" are constructs of the Physical sphere, eternal orbs hung in an expanse above eternal worlds, all within the Celestial realm of the Physical sphere and not part of the Temporal, Fallen Universe we inhabit today. Perhaps, as suggested here, we truly will call our eternal home Eden and come to know that Eden was its name all along.

## THE INTRODUCTION OF JEHOVAH

The preamble verses introduce for the first time the name of Jehovah, as Creator. The original splitting of what today is Genesis 2:4 into two parts is significant and was done with purpose. The first half of Genesis 2:4 (highlighted in purple lettering and underlined in Table 6-1) properly belong affixed to the end of the Temporal account and should end the verse we identify as Gen. 2:3. (The same realignment of verses should apply to the Moses account as well.) When properly positioned (standing on its own), the first phrase of Genesis 2:4 results in starting the discussion of the Physical Creation with an introduction of Jehovah as its creator. "*...in the day that the Lord God made the earth and the heavens.*" The Temporal account declared (in Hebrew) the Creator to be God, "*Elohiym*" (אֱלֹהִים). The preamble verses of Genesis and Moses refer to the Creator as the "LORD God" or

"*YHWH Elohiym*" (יְהוָה אֱלֹהִים), "*YHWH*" (יְהוָה) being Jehovah the Son of God. This is of significance and must be understood.

Jehovah has been co-creator with the Father in all things since the very beginning, for as John correctly revealed: "*All things were made by him; and without him was not any thing made that was made*" (John 1:3). Through the sequence of events revealed in the Temple, great care is given to sequencing the direction for and accomplishment of the creative acts which result in the Physical sphere. In every case, on each of the first 5 days, our Father Elohiym, directs Jehovah and Michael to complete a precise task which they complete as told and report back that they had done so.

Why "*Elohiym*" (אֱלֹהִים) in the Genesis 1 accounts? Why Jehovah, "*YHWH*" now in the Physical Creation? Is there a difference? If it is an error, how did the error creep in? What do we know about the Father in creation? A review of all the creation accounts reveals that the Father always takes the lead through personal participation in any creative process that involves man in any iteration.

Genesis 1 through Genesis 2:3, uses "*Elohiym*" alone, then beginning with the preamble verses, changes to "*YHWH Elohiym.*"[15] It is important for us to understand here that throughout the scriptural accounts (Genesis and Moses) of the Garden story, "*YHWH Elohiym*" is always used. The serpent, however, never recognizes "*YWHW*", expressly speaking to Eve using the name "*Elohiym*" alone.

The Endowment narrative differs from these accounts in that Elohiym and Jehovah appear in the Garden together to Adam and Eve, but it is Elohiym alone that questions our first parents, curses Lucifer and proclaims to Adam and Eve the consequences of the fall, not Jehovah. The Temple account and the Scriptures only realign as the events of Verse 20 of Genesis, Chapter 3 begin, at which point the narration returns again to "*YHWH Elohiym*" when the coats of skins are made.

Throughout the entire Endowment narrative, Jehovah takes direction in all things from the Father (Elohiym); Jehovah never acts independently. The pattern continues even after Adam and Eve enter the *Terrestrial world* until finally both the Father and Son disappear altogether from the narrative itself.

Within the pages of Scripture (both ancient and modern), once the story of the Garden is complete (with the last verse of Genesis, Chapter 3 and Moses, Chapter 4) the name "*Elohiym*" never again appears. This is not without significance in that God's dealings with Man in the Temporal sphere are accomplished, with few exceptions, through the interaction of the Son, "*YHWH*" alone. When the Father does choose to reveal Himself, He is always accompanied by the Son, and His purpose at these times is to proclaim the divine calling of His Son. Why is this so?

---

[15] The Endowment account tells of the Father (Elohiym) and the Son (*YHWH*) in the Garden together with Adam and Eve.

Over the millennia, many answers have been put forward in response to this single question. The first attempts were founded upon Greek philosophical concepts that dictated God was immovable. As such, He could not leave His place in Heaven to take up personal contact with Man. He could not feel, show mercy or suffer as Christ did on the cross. These beliefs, led to many heretical theologies but also explained nicely the existence and requisite for both the Father and the Son. In addition to the many false doctrines Greek philosophy gave rise to, it also helped form the ideas that led to Christian and Latter-day conceptualizations of the relationship of the Father and the Son. In this treatise, we have described a God who weeps and is personable; the Father is not immovable nor is He detached by necessity from our Fallen world.

Our theology teaches that the Son is God, thus He would have the same needs of separation from the Fallen Temporal sphere (per Greek traditions) that the Father would; so that explanation for division in the Godhead does not stand. We also understand that a combining of the immortal Spirit with Temporal elements is what ultimately makes the Divine whole. These two facts bring God and Man together in one family and that relationship with Christ as our older brother (the one who stood before us willing to sacrifice all for our restoration) is what provides a partial answer for why the Son *represents* the Father in all things to us here.

While in the flesh, Christ made several declarations of the witness of the Father. We are told of the requirement that all things be established through the voice of multiple witnesses "*...that in the mouth of two or three witnesses every word may be established*" (Matt 18:16). It was with full recognition of this divine principal that He proclaimed, "*It is also written in your law, that the testimony of two men is true. I am one that bear witness of myself, and the Father that sent me beareth witness of me*" (John 8:17, 18) and again as recorded through inspiration by John the beloved:

> *"I can of mine own self do nothing: as I hear, I judge: and my judgment is just; because I seek not mine own will, but the will of the Father which hath sent me. If I bear witness of myself, my witness is not true. ¶There is another that beareth witness of me; and I know that the witness which he witnesseth of me is true. Ye sent unto John, and he bare witness unto the truth. But I receive not testimony from man: but these things I say, that ye might be saved. He was a burning and a shining light: and ye were willing for a season to rejoice in his light. ¶But I have greater witness than* that *of John: for the works which the Father hath given me to finish, the same works that I do, bear witness of me, that the Father hath sent me. And the Father himself, which hath sent me, hath borne witness of me. Ye have neither heard his voice at any time, nor seen his shape."(John 5:30-37)*

Clearly another part of the answer to why it is the person of Jehovah, the Christ, the Son of God, who interacts almost exclusively with man on Earth is the ability it affords the Father to stand as a witness of the Son because He is separate and distinct from Him.

In the Tanakh[16] (Hebrew Bible), the personal *Name of God* (Theonym) is "*YHWH*" (יְהוָה). The Theonym appears more than 5600 times and can be found in every book of the Tanakh except Esther and Song of Solomon. This is the Ineffable Name or Unutterable Name of the God of Israel. Because it is composed of just four Hebrew letters, the Theonym is also called the Tetragrammaton (from the Greek τετραγράμματον, meaning "*[consisting of] four letters*.")

When God called Moses to be His voice in the liberation of Israel from the bondage of Egypt, Moses asked, "*...What is his [thy] name?*" (Ex. 3:13). To this request God simply answered "*Ehyeh Asher Ehyeh*" (אֶהְיֶה אֲשֶׁר אֶהְיֶה) (I am who I am). The word "*Ehyeh*" (I am) is a derivative of the Hebrew verb (היה) "*to be*," "*to exist*," "*to cause to become*" or "*to come to pass*." This indicates an intimate connection between the name "*YHWH*" and being itself. "*YHWH*" is the source of all being and has being inherent in Himself. He is *necessary being*, everything else is *contingent being*, that is, something that derives its existence from Him. The name "*YHWH*" also speaks to the complete transcendence of God. In Himself, God is beyond all "*predications*" or attributes of language. "*YHWH*" is the source and foundation of all possibility of utterance and thus is beyond all definite descriptions. To express it yet another way, John correctly called Him the Word, for He is the Source of all that is communicated, and can be, communicated to man. As the very *Source of Words*, no words can adequately describe Him.

The Theonym is combined with others to describe God's relationship with man. "*Ehyeh Asher Eyeh*" (I am who I am) is to stand as a "*memorial unto all generations*" of His love for us, the sons and daughters of God:

> *"And God said moreover unto Moses, Thus shalt thou say unto the children of Israel, The LORD God of your fathers, the God of Abraham, the God of Isaac, and the God of Jacob, hath sent me unto you: this is my name for ever, and this is my memorial unto all generations." (Ex. 3:15)*

Exodus 20:7 commands: "*Thou shalt not take the name of the LORD thy God in vain.*" Because of this prohibition, the Masoretes[17] ensured that the authentic Name of the LORD could not be taken in vain by actually changing what was written in the text, literally forming a new word (essentially a new written image) for the Theonym and in effect obscuring the *Name of God* entirely from our view. To affect

---

[16] The Tanakh is assembled from three parts and includes the five books credited to Moses (the Torah), the Prophets (the Nevi'im) and the Writings (the Kethuvim).

[17] The **Masoretes** (Hebrew: בעלי המסורה Ba'alei ha-Masora) were Jewish scribe-scholars of the 6th and 10th centuries AD. The Masoretes are responsible for the form of the Hebrew Scriptures we have today in that they devised the vowel notation system for Hebrew, as well as the trope symbols used for cantillation. This system of diacritical notes on the external form of the Biblical text standardized the pronunciation, paragraphing, verse divisions and cantillation of the Jewish Bible, the Tanakh, for the worldwide Jewish community.

their purpose, the Masoretes placed vowel marks under the letters for "*YHWH*" (יהוה) with those associated with *Adonai* (אֲדֹנָי) in the running text, thereby creating the image for the Theonym we see today (יְהֹוָה).[18] This image is the one that appears in all Hebrew renditions of the sacred text and is the image we have been using throughout this work as well. The Masoretes added, in a sense, graffiti to the text, with the intent of obscuring from the reader the correct pronunciation of the word, while at the same time ensuring any copying of what was written would not be a transcription of the sacred name.

> ***The Masoretes ensured that the authentic Name of the LORD could not be taken in vain by actually changing what was written in the text, … in effect obscuring the Name of God entirely from our view.***

Everywhere the Theonym appears in the Hebrew text we read today, the vowel marks of Adonai from the *Qere* (what is to be read) are in the place of those that may have originally been in the *Ketiv* (what is to be written). There are no copies anywhere of what an original Ketiv of "*YHWH*" may have looked like, if ever one existed at all. As it was intended, this using of vowel markings from one word with the letters of another has led to incorrect pronunciation of the Theonym ensuring it cannot be spoken, even unintentionally.

Additionally, there are good reasons to speculate that the Masoretes further obscured the Theonym by reversing the order of the vowels in the text for Adonai as well when they applied them to the letters (יהוה) in the running text. Based on this assumption, it has become popular in some circles to "*correct*" the pronunciation of the Theonym from the traditional Jehovah to "*Yahveh*." This form is incorrect as well though the construct of the initial syllable *Yah* might in fact be part of the original pronunciation.[19]

Since the original ancient Hebrew did not use vowel markings (they were not instituted until the 5$^{th}$ Century AD), the actual pronunciation of the *Name of God* simply cannot be known. There is no source for audible memory throughout the

---

[18] Adding the vowels (Hataf Patah, Holem and Kametz) a, o, a, from *Adonai* in order, between the consonants *YHWH* is what gives us the familiar form Yahovah. It must also be remembered here that in the "*old*" English (from which we get Jehovah) "*J*" was pronounced like a "*Y*." Hebrew vowel markings, in fact, did not exist in the text before the Masoretes put them in place, intentionally changing the "*sound*" of the Holy name.

[19] E.g., see Psalm 68:4 and Isaiah 26:4.

ages either; it has long since been lost from memory of anyone who may have heard it spoken. The Theonym was only vocalized 10 times, on one given day, once per year, by the High Priest on Yom Kippur in the Temple at Jerusalem. After the Temple was destroyed by the Romans in 70 AD, now almost 2000 years ago, the *Name* of the LORD has quite literally never been uttered.

Today, Orthodox Jews and those of religiously observant conservative Jewish traditions do not pronounce "יְהֹוָה," out loud or even silently to themselves, nor do they read aloud transliterated forms such as *Yahweh* or *Yahuveh*; instead, when addressing or referring to the God of Israel, substitutions for the Theonym or its forms are used. The most frequently used Hebrew replacements are *hakadosh baruch hu* (The Holy One, Blessed Be He), *Adonai* (The LORD), *HaShem* (The Name) or *Shem HaMeforash* (The Ineffable Name).

In the contexts shown here, the labored attempts to revere the name of God, perhaps we can imagine a probability that a similar replacement of terms occurred anciently and like Adonai is used today in place of "*YHWH*," similarly "*YHWH*" (יְהֹוָה) may have been an ancient replacement for a more original name of God "*Elohiym*" (אֱלֹהִים) with "*LORD God*," or "*YHWH Elohiym*" (יְהֹוָה אֱלֹהִים) representing a transitional phase in the record. This is more than just plausible, but may be likely.

For some Latter-day Saints it is all too easy to point at the Biblical record and place emphasis on flawed processes and apparent errors that may have crept in over the millennia. But in this context, one can take no solace in any such attempt. Latter-day revelation is rife with references to Elohiym and Jehovah in their several and separate roles as Father and Son, God and Savior, respectively. Despite the expected clarity one would anticipate from the use of modern language and most recent revelation, there exists no such flawless vision; Latter-day Scriptures are not at all clear on this issue either, often themselves irretrievably confusing terms and compounding several definitions:

> *"And I, the Lord God, said unto mine Only Begotten: Behold, the man is become as one of us to know good and evil;..." (Moses 4:28)*
>
> *"Thus saith the Lord; for I am God, and have sent mine Only Begotten Son into the world for the redemption of the world, and have decreed that he that receiveth him shall be saved, and he that receiveth him not shall be damned—"*
>
> *"Behold, I am Jesus Christ, and I come quickly. Even so. Amen."*
> *(D&C 49:5, 28)*

Returning to the words we used earlier in this discussion, in consideration of all that has been said here, after all that we can do, it truly does seem that God is beyond all "*predications*" or attributes of language. He, even His very *Name*, is beyond all definite descriptions. As He is the ultimate *Source of Words*, language itself, we find ourselves unable to call Him by name and wanting of words to adequately describe Him.

## The Endowment Creation Narrative

Our Temple Endowment provides a narrative that details a creation that is much different from the Temporal Creation described in Part 1 of this work. The Endowment narrative lays out a creative process that was completed by Jehovah at the direction of Elohiym, adding in also Michael as co-creator, assisting Jehovah throughout the creative periods only to disappear from the record as the sixth day of creation begins.

In this chapter, we will closely examine the Temple Endowment account, gleaning from its details evidences of the reality of the separate and distinct Physical Creation. As we examine this account, we will find that, at times, the most significant things we can mention are merely the differences between this account and that of the record of the Temporal Creation. We will learn from those differences and develop an understanding of the processes involved. We will see that the account provided to us in the Endowment narrative is the one spoken of in the preamble passages of Genesis 2:4-5, Moses 3:4-5, and Abraham 5:4-5. Throughout, it will become very clear that the details provided here are substantial in establishing the certainty of a separate Physical Creation. As we go along we will note the significant evidences and return to them in the end.

THE DAY OF CREATION

The Endowment account specifically details six creative "*days*," yet the preamble verses all speak of "*the day*" of creation.[20] This distinction, along with the introduction of Jehovah as the Creator, is the primary identifier that the preamble verses encapsulate an event that is separate and distinct from the one described in detail within the passages of the Genesis 1 accounts. But how can we harmonize "*a day*" used in the preamble verses with the six days spoken of in the Endowment narrative? Because we have no insight into the translation processes used by Joseph Smith in forming the choice of words used in Moses and Abraham, We cannot draw any conclusions there. However, the original Hebrew of the Biblical record does provide us with a great deal of guidance.

Throughout Genesis Chapter 1, the Hebrew *yôm* (יוֹם) is used six times to signify a time period we showed to be truly representative of a *single* day. We discovered that this in fact was true when calculated in a reference frame for time established at the point when pure energy began to coalesce into the matter of the Universe. In Genesis 2:4, the Hebrew scribes, however, used something quite different to describe a day, and this is of upmost significance here.

---

[20] Some early Christian commentators understood this passage to assert an instantaneous creation of the whole. Augustine, for example, deduced Genesis 2:4 to mean a simultaneous creation of all things, which was established by his understanding of Sirach 18:1. Augustine, however, did not know the Hebrew language and was unaware of the meaning of *beyôm* used here. Sirach 18:1 is a passage from the apocryphal book, *The Wisdom of Joshua Ben Sirach*, also known as *Ecclesiasticus*.

In Genesis 2:4, we are given a different construction of the word *yôm*. The actual construction is *beyôm* (בְּיוֹם), meaning "*in (the) day (of)*," or simply "*when*" (as it is correctly translated in the New International Version). Because of this construction, the word "*day*," as used here, does not have its normal connotation; rather, it is part of an idiomatic phrase that does not signify specifically a day with twelve hours of sunlight, one with twenty-four hours as measured on the clock, or any other constrained length of time.

Each time *beyôm* is used in the Hebrew Bible, it has the same idiomatic meaning, "*at the time when*." Compare, for instance Numbers 3:1, "*IN THE DAY that the LORD spake with Moses in Mount Sinai*" (we know that Moses was spoken to by the LORD throughout a period of forty days and forty nights while he sojourned on Sinai, and that the generations spoken of in this verse are not attendant to a single period of just 24 hours). It is with the same understanding that *beyôm* is used in Numbers 7:84; "*This was the dedication of the altar, IN THE DAY when it was anointed, by the princes of Israel...*" (The offering of the sacrifices of the princes actually took twelve days). *Beyôm* again describes a period of time other than a single day. 2 Samuel 22:1 reads, "*IN THE DAY that the LORD delivered him out of the hand of all his enemies...*" (Obviously it was not in one day that the LORD delivered David from all his enemies). In its use of *beyôm* in the preamble, the Hebrew text clearly intends to convey that the creative period spoken of includes something more than just a 24 hour period.

In keeping with the intended brevity of the preamble verses, *for their purpose was not to describe the Physical Creation, but merely to provide a summary context upon which the story of the Garden could be affixed*, the idiomatic phrase of *in the day* was used by the Hebrew scribes to capture and enclose the entire 6 part creation detailed in the Endowment narrative.

MICHAEL

The first readily discernable difference between the Temporal Creation accounts and the account given in the Temple Endowment is the addition of Michael as co-creator. This is extremely significant and should not be dismissed or accounted to an inspired addition to the previously existing narrative. I propose that Michael is not one of the *Gods* mentioned as Creators in the Abraham fragment, because of the following: The Abraham account details a different chronology than the Endowment account that includes Michael. In the Endowment narrative, Michael is no longer present at the time Man is created, whereas in the Abraham account, we are still told of *Gods* (plural) creating Man.

In the Endowment, Michael and the Son create, at the direction of the Father, an Earth from unorganized matter. Through a five period creative process, the world is organized and formed, plant life is introduced and animals are created. Throughout all of this, the Son (Jehovah) and Michael create together. Then in the sixth period, Michael no longer responds, as the Father and Jehovah alone create Man.

We are left with an impression that unorganized matter predates the initial period in some way and for some time,[21] but this may not be necessarily so. That it may is not at all incongruent with our understanding of the spiritual realm, for Joseph Smith taught, "*... All spirit is matter, but it is more fine or pure, and can only be discerned by purer eyes; We cannot see it; but when our bodies are purified we shall see that it is all matter.* (D&C 131:7, 8) Whether the matter the Prophet spoke of is discernable before it is organized, or if it can be comprehended only after it is organized into forms and structures is unclear, but immaterial to our discussion here. What is clear, however, is that Michael himself predates this creative period and significantly so.

At the outset of period one, Michael is established as an individual entity that has an established relationship with both the Father and the Son, though it is meticulously illustrated that the Son is an intermediary between Michael and the Father. This indirect contact with the Father so painstakingly identified in periods one thru five, however, breaks down and disappears altogether by the time Michael re-emerges as Adam in the Garden; the relevance of this detail is unknown.

***Michael clearly existed as a distinct entity in the, perhaps, eons of time leading up to the Physical Creation.***

The pre-existence of Michael at the outset of period one has to be of significant length for Michael to have established a relationship and history that sets him apart and warrants his inclusion in the creative process itself. We must consider not only his inclusion in the process but the fact that he and Mankind are of such import to the Father that the creative process was initiated in the first place, with the specific intent of creating a habitation for Michael (Adam) and those that were to follow.

Michael clearly existed as a distinct entity in the, perhaps, eons of time leading up to the Physical Creation. That Michael *is* included in the Endowment account and *is not* found in the synoptic POGP accounts or the Genesis 1 account is **Evidence 1** that the creation told of in the Endowment is not the same as the Temporal Creation account.

---

[21] Of significance is the instruction from God that Jehovah and Michael were to "*organize... a world like unto the worlds we have hereunto formed.*" Throughout this work, we will construct a chronology that will lead us to understand that the Endowment narrative not only details the creation of the entire Physical sphere but also provides information that reveals many worlds were created therein, some at least which predated the assembly of the one detailed in the narrative we receive. We will establish a chronology that will support an understanding that the physical sphere is the location of the rebellion that is called the War in Heaven, conducted there by contestants perhaps not yet embodied as Spirits.

The preamble verses of Genesis and Moses, identify only the LORD God, or "*YHWH Elohiym*" (יְהוָה אֱלֹהִים) as the maker of "*the earth and the heavens,*" the creator of all things "*spiritually, before they were naturally upon the face of the earth*" including "*all the children of men.*" However, the Abraham account declares that this act of creation was completed by the Gods. What can this mean? Is there a discrepancy here?

The preamble verses are in association with the Garden account to follow, and they themselves follow the account of the Temporal Creation. Both accounts, as recorded in Scripture (both ancient and modern) speak of a *plurality* of Gods, which aligns precisely with the Abrahamic account. The Endowment narrative reinforces this plurality providing additional detail as it describes the creation of the Physical sphere being completed by both Jehovah and Michael at, and through, the direction of Elohiym. Let us consider this plurality and what we can learn from it.

First, the name/title "*Elohiym*" (אֱלֹהִים) itself speaks of a plurality.

Secondly, the "*Let us*" and "*as one of us*" phrases of Genesis 1:26 and 3:22 are much more than grammatical capitulations to the plural contexts of "*Elohiym*" we have detailed here; they are declarations with theological intent.

It would surprise most to find that Jewish scholars of ancient Hebrew language and beliefs determine these phrases to express a resolute pronouncement of the existence of a plurality of divine beings. A commentary for Genesis 1:26 can be found in The Jewish Study Bible,[22] which reads: "*The plural construction (Let us...) most likely reflects a setting in the Grand Council (cf. 1Kings 22:19-22; Isa. Ch. 6, Job Chs. 1-2) God the King announces the proposed course of action to his cabinet of subordinate deities, though He alone retains the power of decision...*" Here, one of the most respected sources of Jewish exposition speaks of "*subordinate deities.*" How could this be so, considering the source is Hebrew Scripture, the foundation documents of the world's great monotheistic religions?

This commentary does not stand in isolation as a work of a single group of scholars either; there is much more to be found in Jewish teachings over the centuries that express this surprising reading of the original Hebrew.[23] This cannot be more fully

---

[22] The Jewish Study Bible, Featuring the Jewish Publication Society, Tanakh Translation, Berlin. Adele and Brettler, Marc Zvi, editors, Oxford University Press 2004. This published version of the Tanakh is highly regarded as one of the finest resources of Jewish thought and exposition available to both the Jewish and non-Jewish communities.

[23] Commentary on this verse in the Jewish Study Bible *ibid* continues: "*The Midrash manifests considerable uneasiness with God's proposal to create something so capable of evil as human beings are. Playing on Ps.1:6, one midrash reports that God told his ministering angels only of 'the way of the righteous' and hid from them 'the way of the wicked' (Gen. Rab. 8.4). Another one reports that while the angels were debating the proposal among themselves, God took the matter in hand. 'Why are you debating?' he asked them, 'Man has already been created!' (Gen. Rab. 8.5).*"

understood than through a scholarly examination of the Shema itself, Deut. 6:4; it reads:

> *Hear, O Israel: the LORD our God is one LORD*

The Shema, named for the first Hebrew word of the declaration, "*Sh'ma Yisrael Adonai Eloeinu Adonai echad*" is most properly rendered: *Hear, O Israel! The LORD is our God, the LORD alone.* (NJPS)[24] Here, "*The LORD...alone*" correctly departs from the more familiar translation of the King James shown above. The original Hebrew allows either "*YWHW, our God, YWHW is one*" or "*YWHW is our God, YWHW alone.*"

This verse, through its choice of "*echad,*" actually makes not a quantitative argument (about the number of deities) but a qualitative one about the relationship of God with Israel.[25] Almost certainly, the original dynamism of the verse, as the medieval Jewish exegetes recognized it, was to show exclusive loyalty to "*our God, YHWH*" – but not thereby to deny the existence of other gods which ancient and middle Israel clearly did not do. Commentary about this found in The Jewish Study Bible, reads: "*In this way it assumes the same perspective as the first commandment which, by prohibiting the worship of other gods, presupposes their existence.*"[26]

The first, more familiar translation of this verse as used in the King James, makes a statement about the unity and indivisibility of God, but does not do full justice to the original meaning of the Hebrew text.

How this single rendering found its way to preeminence and became almost exclusive is interesting. It is particularly so when consideration is given to the fact that Christianity has adopted its reading so freely.

The eventual monotheistic interpretation of the Shema did not begin to take hold until late. Though the concept may be evident as early as the third century BC where the Septuagint translation of the verse appears in the now familiar form: "*the LORD is one,*" monotheism did not take a foothold in Jewish thought until the Christian period began, when the *oneness* of the God of Israel became *a necessary polemic against Christianity* and its insistence that Jesus of Nazareth, though separate and distinct from the Father, is the Divine Son of God.

---

[24] New Jewish Publication Society of America Tanakh.

[25] John Calvin in his book, Commentary on the Gospel According to John expressed a similar view of yet another famous monotheistic declaration in the Bible. Of John 10:30, Calvin explained, "*the ancients made a wrong use of this passage to prove that Christ is...of the same essence with the Father. For Christ does not argue about the unity of substance, but about the agreement which he has with the Father.*" John Calvin's analysis of the Greek language in this verse is of the same structure of argument made here for the Hebrew of Exodus 3:4.

[26] The Jewish Study Bible, Featuring the Jewish Publication Society (Tanakh Translation, Berlin. Adele and Brettler, Marc Zvi, editors, Oxford University Press 2004, pg. 380).

The initial use of the interpretation of the verse as a polemic against Christians notwithstanding, once Christianity became established, and the Trinity of the creeds fully ingrained, medieval translations of the Bible, in both the Christian and Jewish worlds, relied heavily on the Septuagint in rendering the verse, resulting in the common understanding and now "*traditional*" reading found today.

The Abraham account simply reads, "*And the Gods came down and formed these the generations of the heavens and of the earth*" (Abr. 5:4). As such, the Abrahamic account conforms to what we have learned about the Genesis 1 reading.

The Endowment narrative describes a concerted effort of Elohiym, Jehovah and Michael. In the Endowment, it is the Son and Michael who carry out the actions of creation themselves until Michael disappears from the account before reappearing as Adam.

Just before the withdrawal of Michael from the account, there is a transition that takes place as the Father (Elohiym) declares that He now, for the first time in the narrative, will go down with Jehovah and together they form man. As we have discussed before, whenever the sons and daughters of God are created, in any of their forms, the Father is directly involved; this is clear in both Day 6 of the Endowment narrative describing the Physical Creation and of Day 6 of the Temporal Creation described in the Genesis 1 accounts. This is not without significance, and we will return to this subject again later in this work.

THE CHRONOLOGY OF CREATION OF THE PHYSICAL SPHERE

The Endowment narrative presents a creation account that re-sequences the order of the creative events from those given in Genesis 1 and the POGP synoptic accounts. In the Endowment, this change is almost imperceptible and is given without fanfare or announcement. Though it is repeated many times in well over one hundred and fifty temples each day, most, if not very near all, who hear it don't even notice the alternate order presented.

Why is the order of events different from the Genesis 1 account? What can we learn from the differences and what do they intend for us to know? Let us start by looking at the differences themselves (*Table 6-2*), and then we will examine their significance.

That the order of the creative events is different in the Endowment narrative from that presented in the Genesis 1 accounts is **Evidence 2** that the creation account of the Endowment is not the same as the Temporal Creation account.

| DAY | Temple Endowment (Physical) | Genesis/ Moses/Abraham (Temporal) |
|---|---|---|
| ONE | World Organized | Earth Organized, Light Created, Light and Dark Separated |
| TWO | Water and Land Separated, Earth Formed | Firmament Created |
| THREE | Light and Darkness Divided -Lights in the Firmament-Night and Day Separated, Greater and Lesser Lights | Land and Water Separated – Plant Life Created |
| FOUR | Plant Life Created | Lights in the Firmament-Night and Day Separated |
| FIVE | Creation of Animal Life - Land, Air and Sea | Sea Life and Birds Created |
| SIX | Creation of Man | Land Animals and Man Created |

Table 6-2 *Differing Sequence of Creation Events.*

> The Endowment first day includes organizing the world from previously existing unorganized matter. It does not include a separation of the light from the darkness. There is no brooding over the waters and there is no waiting for the creation to obey.

Day 1 of the Endowment account is primarily distinguished from the Temporal account for what *it does not include*. Though both the Temporal account of the Scriptures and Physical account of the Endowment detail the organization of the *world*, the Endowment narrative does not include the creation of elements (*bohu*) which we saw were "*brooded*" over in Day 1 of the Temporal Creation.

Earth is not mentioned at all in Day 1 of the Endowment narrative; it is, however, spoken of in the Temporal account. This difference is not insignificant and should be understood. In the Temporal account, the *earth* spoken of is to be understood as Temporal matter, literally the elements our planet and all the Universe is composed of today. In the Temporal account of Day 1, the planet we live on, what we call Earth, our home, is not yet existent. *"And the earth was without form, and void"* (Genesis 1:2). The context of "*earth*" as it is used in Day 1 of the Temporal account is to be understood as it would be when we may say: *at death our mortal remains return to the earth from which it came*. Earth, here, is element or more correctly, elements. The same is true of the Physical account. The *planet* Earth (Spirit Earth) itself, is not created independently but is formed crudely (at first) from unorganized spiritual matter. The "*world*", we are told, is organized in a manner "*like unto*" other "*worlds*" which were previously "*formed*" as well.

The Physical account clearly records that the building blocks of the Physical (spiritual matter) are pre-existent and are not created at this time but merely organized. In the Physical account, there is no separation of the light from the darkness (for as we discussed in the previous chapter) that referred to a specific phase transition in the infant Temporal Universe when the particles composing Temporal matter formed. This transition has no correlation in the Physical Creation.

In the Physical account, there is no brooding over the face of the waters; this is of double significance. First, there is no primordial Hydrogen (plasma) sea like the one which existed in the Big Bang and birthed the Temporal Universe. And secondly, God does not (in the Physical sphere) need to carefully and intently "*load*" foreign, imperfect Temporal matter with the physical laws (knowledge, information) that are required in the Temporal Universe.

What is not included in the Endowment day one account speaks volumes of the differences in the Temporal and Physical Creation and in detail clearly identifies the accounts to truly be of two separate creations, producing fundamentally different environs.

Michael is part of this Physical creative process, and after it is complete, it is he who occupies the Garden of Eden as Adam.

## *Day 2*

> The Endowment second day account details the separation of land masses and seas on the face of the Earth, while the Temporal account focuses instead on the creation of the firmament.

In Day 2 of the Endowment account, the first major chronological shift occurs from that of the Temporal account. Why is the order of the creative events changed? Not by chance, carelessness, ignorance, or misunderstanding, but by design and with purpose.

The Genesis 1 scriptural accounts detail the creation of the Temporal sphere. *Physical laws dictated the order there, for they control the development and evolution of the Universe from the Big Bang to the living Earth we have today.* We must remember the Physical Creation is immortal and has no such dependence on these laws of Physics. There are no constraints placed on the Physical (spiritual) elements by entropy and the passage of time that accompanies it.

The Endowment account of Day 2 details the terra forming of the Earth itself far in advance of the time it is done in the Temporal account. This is because the formation of the Temporal Earth and its subsequent development of seas and continents followed many billions of years of star formation and galaxy building required by the very laws of Physics that were loaded into the Temporal Universe by God at its outset. In the Physical sphere, those time-consuming evolutionary processes were not needed.

There is no mention of the firmament in the Endowment account at all. This is so because the matter that comprises the Physical (*spiritual*) Earth (though initially unorganized) was fully formed and did not (like the Temporal elements) need to be forged in the furnaces of a thousand, billion, trillion stars. The skies above the Spirit Earth were already teeming with immortal, Heavenly objects and Celestial orbs of every variety before the planet came to be.

In these marked differences of the two accounts, we can glean subtle truths about the two separate creations: the Temporal and the Physical. Truths that punctuate marked differences between the immortal Physical materials and their mortal Temporal doubles.

> Again the Endowment account differs from the Temporal Creation chronology. At this point, the Physical Creation turns to the day and night. Here the light is separated from the darkness in a way that is associated with the formation of a Sun, Moon and stars. The focus is on association with the hierarchy of the celestial objects as they will later apply to the degrees of Celestial glory. The Temporal account finally gets to the point where the Earth forms and includes creation of the earliest life on the planet. On Day 3 of the Temporal account, the appearance of the Sun, Moon and stars in the Temporal Earth's sky is still to come.

The Endowment account continues to follow a creative chronology that represents a process much different from that of the Temporal Creation detailed in Genesis 1 and the POGP synoptic accounts. This is because the Physical Creation has no need to follow an order defined by the laws of physics and the development of the needed elements which will result in Temporal life on this planet. The Physical account begins with unorganized matter being formed into a locale suitable for the habitation of immortal Physical beings. At the outset the locale formed is not called Earth, but it is called a world (this same statement applies to multiple formed locales as well).

Note this significance. *World,* as used in Scripture, has a much broader definition than one that would mean a single planet: as in Adam brought death to the world, "*Wherefore, as by one man sin entered into the world, and death by sin; and so death passed upon all men, for that all have sinned*" (Romans 5:12). It is important to remember that death is scientifically defined as entropy, and entropy is a characteristic of the *entire* Universe, a Universe containing perhaps billions of *worlds* (planets), all of which are dying. So given this understanding, the organization of Physical (spiritual) matter into a *world* (perhaps more succinctly a "*sphere*") could connote many, perhaps nearly an infinite multitude of individual locations.

By Day 2, the Son and Michael are directed to call the dry land earth as they focus on forming a habitation for Mankind specifically. Their focus is on creating a habitation through day three when celestial objects are created, completing the needed elements necessary to sustain life. Though immortal Physical Spirit beings do not need these elements to sustain their life, to fully *enjoy* life, they are nonetheless created, for the Endowment tells us several times these things were to be *pleasing* to man.

The Physical Creation then, begins with the organization of matter, planet forming, followed by the creation of celestial objects and stars. This chronology is much different than that of the Temporal Creation which describes, in proper sequence, the processes necessary in forming the Universe we inhabit today: primordial fire ball, phase transition, creation of elements and free photons, formation of stars and galaxies, then the formation of the Earth and the almost immediate occurrence of life upon it.

## Day 4

> Here the Endowment account details the creation of plant life on the Earth in all its variety. The Temporal account remains on a differing chronological course as it records the lights in the sky and the beginning of night and day.

The creation of the Physical sphere turns to living organisms for the first time in Day 4. We are told that plant life in all its variety is created now and placed in the Earth. The Temporal account introduces plant life back in Day 3, before the cycles of night and day begin, and the Sun, Moon and subsequent stars appear in the sky in its Day 4.

Reviewing ground we covered in Chapter 3, the chronological sequence of the Temporal account is coherent to the scientifically informed of this century. Today we now know that elementary plant life appeared on the Earth almost immediately after it was formed, well before the vast clouds of dust and debris of the solar accretion disk was cleared and the nascent atmosphere became oxic and transparent. Those changes finally exposed the surface of the Earth to unfiltered light from the Sun, clear views of the Moon, and eventually views of the stars of the Galaxy in the night. As this dramatic transformation occurred to the Earth and its environs, plant life evolved from simple, single cell, forms to more complex varieties.

The sequence of events needed to accomplish the Temporal Creation was not required in the Physical Creation and, therefore, is not detailed in its account. Plants in all their varietal forms were created and placed on the Physical (spiritual) Earth during one creative period and they were all in place before animal life and Man were created.

## Day 5

> Animal life on the land, in the air, and sea are created in Day 5 of the Endowment account; while in the Temporal accounts, sea life and birds are created exclusively.

Animal life, of all forms, is created in one single period in the Physical world. The Temporal account details the creation of sea life and birds first, followed later by more advanced forms of life. The Temporal sequence is in harmony with the evolutionary processes known to science. That this sequence is properly identified in the Genesis 1 account, as well as the Synoptic POGP accounts of the Temporal Creation and is absent in the Endowment account of the Physical Creation, is evidence of the true existence of the two separate creations and signals another fundamental difference between the Physical world and the Temporal world.

The Physical Creation was not dependent on physical laws building the genome of life from the bottom up, lesser complexity followed by the greater. Here in the eternal sphere, the requirements of the laws of Physics, Chemistry and Biology need not exist. There were no restrictions placed on the creative processes used by Jehovah and Michael as they worked together to create the diversity of life.

> As this day opens in the Endowment account, God the Father and the Son create Man on the Earth. In the Temporal accounts, advanced land based animal life and Man appear as the age of the Universe approaches that of the present era.

At this point, Michael disappears from the creative process only to soon reappear as Adam on Earth just beyond the confines of the Garden. Lacking from the Endowment account is the breathing of life (the *Ruach*) into Adam, for the *Neshama* was already attendant to him as an eternal spiritual being and did not need to be added to him when he was placed on the Physical (spiritual) Earth and within the Garden. In contrast, the Temporal Adam, at the point of his creation from the clay (the elements of the Temporal Universe), did not have the *Neshama*. He lacked that necessary connection with God, the joining that set Adam and his descendants, the *Be-ne- ha-e-lo-him* (בְנֵי־הָאֱלֹהִים), *the sons* (the children) *of God* apart from the *ha-a-dam* (הָאָדָם), the *people* of the clay, those who were from the Land of Nod of the Earth from a time long before Adam's arrival. The lack of any mention of the *Ruach* in the Endowment account and its pointed inclusion in the Genesis 1 accounts is an important difference between the creation stories and represents significant evidence that they describe two separate Creations. This is **Evidence 3.**

The Endowment narrative does not address the dietary needs for man and beast as was the case in the Temporal accounts. We previously discussed photosynthesis, as the energy source for life on Earth; recognizing it as the singular means of capturing the energy of the Sun providing for all life on this planet. We marveled at the way this incredible process, only recently understood by modern science, was included in an ancient text some 4000 years old. As significant as the inclusion of this detail is within the Temporal Genesis 1 accounts, the fact that any like passage is missing from the Endowment account is equally revealing and important. An immortal, eternal, physical body does not depend on the same chemical reaction for its sustenance and the omission of the details of diet in the Physical Creation marks that obvious fact.

- A Proper Introduction

The Endowment narrative introduces a pause in the story told at this point that must be noted. We will return to the events encompassed in this pause in depth throughout the remainder of this work, but let us address them here in overview for it is important for us to be aware of them before we move forward.

The Endowment narrative endeavors to tell the story of each of us through the story of just a single eternal pairing of a man and a woman. And in keeping with that, we are introduced now in the story to both Adam and Eve. Because the story is presented in the form of a play, and as such we need to know the characters involved, there is a necessary liberty taken in the sequencing of events to present the images and the lesson with continuity; that liberty appears here with the naming of Eve. Though her creation and appearance is properly placed at this point in the story, for it is significant that her appearance in the story is done now, Eve is not actually named by Adam at this time, though we are shown it being done.[27]

What the Endowment intends to reveal here is that Adam and Eve are both created at this time. Indeed through the Moses account, we know that all of the children of men were "*created*" at this time as well: "*...And I, the Lord God, had created all the children of men; and not yet a man to till the ground; for in heaven created I them; and there was not yet flesh upon the earth...*" (Moses 3:5).

Before the Garden, after the creation of the Physical sphere by both Jehovah and Michael, all of the sons and daughters of God were created. The Endowment narrative describes here, through the symbolism of "*Adam's rib*," the intimacy of the pairing of both male and female in embodied Spirit. This same symbolism would again be used later in the Garden to describe that same intimacy of pairing for the *Flesh and Bone* Adam and Eve. We are created in pairs, intimately bonded and eternally inseparable.

In their, now, Physically Embodied state, Adam and Eve stand on a Spiritual World (an embodied tangible Planet). Though we are told Adam had already forgotten everything, we are also told that both he and Eve are taught and informed.

While not yet in the Garden, they are told both of its presence and their pending sojourn there. Adam and Eve are informed of the purpose of the Garden, what challenges they will face there and of the rewards that will come by addressing those challenges. In effect, the story here represents the detailing of the Plan of Salvation to all the Children of God and their acceptance of its challenges in whole before entry into the Temporal sphere.

We will return to these several themes later in this work, providing much detail, but for now just be aware of this significant detail in the Endowment narrative that is so often overlooked.

---

[27] When Adam and Eve are created here, God, in recognition of their eternal pairing calls them both by a single name, Adam. "*Male and Female created he them; and blessed them, and called their name Adam, in the day when they were created.*" (Gen. 5:2). Though the timing of Adam naming Eve is misaligned in the Endowment narrative, it of course correctly represents that it is Adam who first calls his eternal companion Eve. God merely confirms it. The Scriptural records detail that it is Adam who names Eve as well, though they place that naming in its actual sequence after the Fall at the time our first Parents receive their Temporal Nature (represented by the *coats of skins* they receive from God) and in fact Eve first becomes capable of being the *mother* her name represents. (Gen. 3:20; Moses 4:26)

## CREATION THROUGH ADDITION

Unlike the Temporal Creation, the Physical is a creation of addition, one of making things that did not exist before. In the Temporal sphere, all that there is, is the result of creation through subtraction (TzimTzum). What we sense around us today is a lesser image, *a simulacrum*, of something eternal that it mirrors, only a part of a greater whole. The six days of the Physical Creation produced that eternal physicality and represents the true beginning of the story of the sons and daughters of God.

***...what we sense around us today is a lesser image of something eternal that it mirrors, only a part of a greater whole.***

Here in the Endowment narrative and in the preamble verses, we are to find true Genesis.[28] With the creation of the Physical sphere, our individual intelligences were added to and housed within immortal Physical bodies each now an entity we call a *Spirit*. This process of adding things together (eternal intelligence and physical bodies) gave us *immortal life*. Immortal life is something that stands distinct from *eternal life*, which is an attainment of potential received through the resurrection after the judgement. Eternal life is earned through the attainment of knowledge and understanding we gain by experience in the Temporal sphere and is bestowed by the grace of God through the atoning sacrifice of His Son Jesus Christ.

---

[28] Ancient Hebrew writers would have been puzzled by the way modern readers refer to their work as "*Genesis*" for several reasons. First, our modern title comes from a Greek word meaning "*creation*," but Greek is a language that would have been unknown to the original Hebrew writers. Secondly, the convention of giving written works an actual title did not appear until centuries after the time the text was composed. If ancient Hebrew readers had to talk about a text, they would refer to it by its opening lines, which in the case of Genesis would be "*Bereshith*," the Hebrew words for "*In the beginning*."

## The War in Heaven

Our discovery of a three part creation has significant impact on our view of the War in Heaven. By dividing the creations into three, we have uncovered a sequential framework that has obviously always existed but had not been previously discussed nor understood. Our study here now means we have some things to consider regarding the event that is called the War in Heaven:

> *"And there was war in heaven: Michael and his angels fought against the dragon; and the dragon fought and his angels, And prevailed not; neither was their place found any more in heaven. And the great dragon was cast out, that old serpent the Devil, and Satan, which deceiveth the whole world: he was cast out into the earth, and his angels were cast out with him." (Rev. 12 7-9)*

> *"for behold, the devil... rebelled against me, saying give me thine honor, which is my power; and also the third part of the hosts of heaven turned he away from me because of their agency;" (D&C 29:36)*

When did it take place? Until now it was enough to say, *in the pre-existence*. This, however, is insufficient now. Was it before or after the creation of the Physical sphere? Answering that question impacts our definitions of the first and second estate of man and perhaps even our understanding of the *bodies* Satan and his followers were denied.

Why was the War contested? A careful consideration of this question will reveal much. Surprisingly, much that we expose may challenge the simple views we have held on to for so long. By constructing an eternal chronology, one that has a *middle period*, external to the frame of reference we view now, we can no longer speak of things as being before the beginning and of after. We can no longer take solace in glossing over the transition, the *Genesis Point*, allowing it to obscure important details of who we are, why we are here, the purpose of the Atonement, our Universe, and why all of it was fought for, at such great cost, so long ago. And speaking of the cost, it will be discovered that what agency is, or more correctly how it has commonly been defined, must be reassessed as well.

*"It ain't what you don't know that gets you into trouble.*
*It's what you know for sure that just ain't so."*

*-Mark Twain*

So let's look at what we know, or at least what we *think* we know.

First, we must frankly realize that we have not been told much. We are familiar with the small plates of Nephi spoken of in the Book of Mormon.[29] They, we are told, contained a *constrained* record, focused primarily on the ministry of the people and not providing the "*full account*" which was recorded elsewhere. In a like manner, what remains of the account of the War in Heaven in our sacred record today, is much constrained and by no means a full report.

What we have been left with is an ephemeral record that primarily speaks to the spiritual ramifications of the struggle. However brief the record is in its telling of spiritual things, it is even more fleeting in historical detail. Revelatory passages regarding the War in Heaven are to be found scattered almost everywhere in both ancient and modern Scripture, never really addressing the account full on but providing just glancing blows that infer there is much more to be known that has not been said.

Let us first attempt to examine the timing of the conflict. The order of the events recorded for the Physical sphere is not made clear. It is certain that the 6 days of creation are sequential and the planting of the Garden follows them. Adam and Eve are embodied and stand on the spiritual Earth, and then they are placed within the Garden just after that. We understand the War in Heaven must precede the placement of Adam in the Garden and the childlike state he assumes when given that name, but by how much? Does it fit in after the creative events of Day 6, those that occurred before Adam was introduced to the Garden? Or, was the struggle over long before the six creative periods began?

The Scripture, "*And there was war in heaven: Michael and his angels fought against the dragon;*" (Revelation 12:7) places Michael in that war, and of course he would be, as we all were. Michael, the use of the name, and the inference that he was a leader of significance for the just cause in the struggle, confirms that Michael fought in that confrontation, but does not speak at all to what his form, his mode of existence, may have been at the time, and indeed nor do the Scriptures speak of ours.

Was the great struggle conducted before the embodiment, the creation of our Spirits? Did we all stand together as Spirits in that great Grand Council? Or must we place the War at a time before we all received our immortal bodies? We will explore the details of this in the next chapter. What we can of certainty say here is, the Grand Council and the following confrontation occurred before Michael disappears from the Endowment narrative and reappears in a state of innocence as Adam in the Garden, midway through the events of Day 6.

---

[29] 1 Nephi 9:2-4.

When God questioned Lucifer's[30] appearance in the Garden, he defended his presence there by saying he was merely doing what had been "*done in other worlds.*"[31] Satan's declaration is enigmatic for sure, but it speaks to the existence of conflict between Lucifer and the Father from a former time.

Another indication that the origination of dispute was even then in the past can be inferred by Michael's participation in creation alongside Jehovah. Perhaps the valiant effort and leadership on the part of Michael in the past war may have been instrumental in securing him the privilege of co-creating the Physical sphere. Though not definitive, these observations perhaps suggest that the Great War in Heaven was conducted before the six periods of creation, which formed the immortal physicality of the Physical sphere.

The Scriptures provide little detail about what Satan suggested or how he presented his ideas in the Grand Council; however, they are clear that at its core the plan of Satan "*sought to destroy the Agency of man*" (Moses 4:3). It is telling that a careful reading of all Scripture pertaining to the adversary in this premortal time reveal only two things: references to his desire and need for power and honor because of his pride, and that he sought to destroy our agency. It is significant and should be understood that one would not come without the other. It is essential then that we learn just what agency is, what Satan could have done to destroy it and why he desired to do so.

## FREE AGENCY "A MATTER OF RISK"

We have been told, "*There is a law, irrevocably decreed in heaven before the foundations of this world, upon which all blessings are predicated—*" (D&C 130:20). Though we are left unknowing as to the particulars of this law, we are given some clue, for when asked what the greatest commandment was, the Savior replied:

> *"Jesus said unto him, Thou shalt love the Lord thy God with all thy heart, and with all thy soul, and with all thy mind. This is the first and great commandment. And the second is like unto it, Thou shalt love thy neighbour as thyself. On these two commandments hang all the law and the prophets." (Matthew 22:37-40)*

---

[30] There should be no inferences made due to the name of Lucifer and Satan. These names are not associated with a transition like that of Michael to Adam or the new name as spoken of in Revelation 2:17, *the one so familiar to Latter-Day Saints*. The names of Satan, Lucifer and the devil are constructs of language and culture. Lucifer is Latin for the Greek rendering of his name "*bringer of dawn*" (referring to the morning star), found in the Septuagint. In Roman astronomy Lucifer was the name given to the morning star, which we now know by another Roman name Venus. The name Lucifer in Latin derives from lucem ferre, bringer, or bearer, of light. Satan is derived phonetically from the Hebrew word for *adversary*. Devil has evolved from the Greek Diablos used in the New Testament. See our subsequent discussion of this topic in, What Is In A Name, pg. 240.

[31] This mysterious statement may possibly support a historical framework that suggests that the physical sphere was created long before Lucifer's appearance in the Garden and one that could support many fully inhabited spheres within that construct. See our detailed discussion of this topic in, That Which Has Been Done (Then Again Once More), pg. 286.

Despite these words spoken by the Savior, our sense of agency is so strongly imbedded that any survey of Latter-day Saints as to what the unknown decreed law pertained to would reveal (in an unqualified way) some answer centered on our actions, their effects and accountability.

As freedom loving Americans, our liberty, our right to live according to our personal desires and to exercise our agency, even if our choices are detrimental to us, is of utmost value. We reverence our moral agency so much that any effort to circumvent, undermine, disregard, influence, regulate, or eradicate it most certainly leads to confrontation. This passionate sense of liberty is not just borne by Americans either; it is a fundamental attribute of being human. Battles over agency, both big and small, have been fought for ages here on earth, and the right of agency was likewise contested in the Grand Council.

As we said earlier, scriptural references to the War in Heaven are heavily influenced by the spiritual aspects of the great confrontation. Any discussion of the agency of man, in the context of that struggle, as it took place anciently and continues to this day, follows along with the theme of divine law and its consequences, presenting definitions that reflect those decrees and costs. In this regard, consider the following from Elder Bruce R. McConkie:

> *"Agency is the ability and freedom to choose good or evil. . . . Four great principles must be in force if there is to be agency: 1. Laws must exist, laws ordained by an Omnipotent power, laws which can be obeyed or disobeyed; 2. Opposites must exist—good and evil, virtue and vice, right and wrong—that is, there must be opposition, one force pulling one way and another pulling the other; 3. A knowledge of good and evil must be had by those who are to enjoy the agency, that is, they must know the differences between the opposites; and 4. An unfettered power of choice must prevail." (Mormon Doctrine, Bruce R. McConkie, Salt Lake City: Bookcraft, 1966, pg. 26)*

Agency is much more than an option to do wrong, to do the forbidden. Agency is the ability, the power to act for oneself. As we have described here in this work, to make Agency even possible involves the very structure of matter in the Universe itself. The first three principles in the quote of Elder McConkie above are baffling, for they miss the mark entirely. They have nothing to do with the power of agency.

In elucidating his essential principles, it is apparent Elder McConkie has conflated the *requirements of agency* with the *justification of consequences* for its exercise. He has described the validation of punishment through the notions of law, disobedience, foreknowledge, and justice. He first presents the *concept* of free agency then details the foundation for the progression it allows through the exercise of proper choice weighed against the application of reasoned judgement. In effect, he provides *principles* which *control* and *delimit* agency. Though the quote purports to be a definition of *Agency*, it, in fact, does little in the way of explaining what Agency *is*.

I propose here that it is a *common mistake* in the Church to confuse Agency with the judgement of God that is necessarily attendant to its application. To clarify, let us look at a helpful analogy:

I have the ability to travel from point A to point B at any time I choose. This ability to move is analogous to my freedom of choice; it is equivalent to free agency. What gives me the ability to move is my body, my legs, maybe a vehicle if the distance between the points A and B is too distant. There is no corresponding equivalent of law or opposition or knowledge of right and wrong at this point in this analogy, nor should there be; *they are not a component of my ability to move freely*.

Now, if I want to exercise my ability, my agency and move from point A to point B in heavy downtown traffic, in a city full of people with like abilities and the desire to use them, I would be helped along by laws, sidewalks, stop signs and roads. That these laws and the elements that come from and support them exist, assists me in the exercise of my ability to move by clearing and controlling traffic, but my ability to move is not contingent upon, nor does it draw any of its actuality from them. *I can move from point A to point B with or without Law*.

Should I find the laws and traffic controls too confining, I may choose to ignore them when I move from point A to point B, and that may put me in opposition to my own safety and to that of others. Depending upon how hazardous my actions are, they could be described as being evil. Again my choosing good or evil in the way I move about town has nothing at all to do with my original ability to move (*my agency*). My agency, my ability to move, was mine and fully available to me *prior* to my selecting the manner of my movement. Had I chosen to drive or walk safely, looking out for my own self as well as others as I went along, I would have gotten to my destination without any problem, without any opposition. *My ability to move does not require opposition*.[32]

Additionally, as I zip along exercising my ability to move (*my agency*), I may not even be aware of black ice on the road or the suddenness of a blind turn ahead. I may be entirely oblivious of how hazardous my driving may be, or how unsafe it is, still, I have put myself and others in mortal danger. By missing the signs along the road warning of the dangers ahead, I have been in opposition to law (laws of Physics) without my having any knowledge of it. Still my unsafe, ignorant actions can and probably will kill someone, perhaps even me. Unfortunately, my complete lack of knowledge of the hazards did not prevent me in any way at all from moving along at perilous speed. *No matter how unaware of the opposite affects my actions would produce, I still have the ability to travel. I do not need to have knowledge of good and evil to freely move from point A to point B.*

So, now to address Elder McConkie's fourth principle; unlike the others, this one does hold up in the analogy, but it is of itself sometimes dependent on the other principles.

---

[32] At this point, I must say at the risk of stealing thunder from an idea I will submit very soon after this, that I am completely aware that the laws of motion as described by Newton do in fact require opposing forces to move objects, but this detail, though one I will make later, is not what Elder McConkie had in mind when he framed his second principle of Agency.

In our analogy, I may choose to move through a city where laws exist. If I break any of those laws, I may suffer consequences. However, the consequences affect me independently of my being aware of the violated laws that bring them about.

In this analogy, I could be ignorant of *natural* laws; driving at excessive speed may take my car off of a cliff, and the horrible accident may take my ability to move away from me through severe injury or the loss of my life.

My irresponsibility may not have been severe enough to have caused a mishap, but I may have been in violation of *imposed* rules or law, as in driving recklessly or speeding too fast. In this case, governing authorities (themselves, necessary extensions of the imposition of law) may choose to imprison me after making a determination of my intentions and my cognizance of the law. Such imprisonment will deny me of the fourth principle, the unfettered power of choice.

Through imposed law, I can be denied my ability to move even though my careless actions may not have been serious enough for me to have denied that ability to myself through the effects of an incapacitating accident.

Through analogy[33] then it seems, that the fourth principle is foundational, in that without it (unfettered choice), I truly do not have agency; in this analogy, without unfettered choice, I would not be able to move freely from point A to point B. Through this analogy, however, we see that the first three principles, contrary to Elder McConkie's insistence, in no way empower or provide me with any ability to move freely, they are only restrictive and limiting. Because principles 1, 2 and 3 exist, and *only* because they exist, I can be denied my ability to move freely (*my agency*). *The law, opposition, and knowledge of good and evil, right and wrong, are*

[33] Some may argue that agency, *the unfettered power of choice*, is not a stand-alone attribute, but it is only existent through a combination of all four principles. In such a case, the fourth principle in this analogy would be like a driver's license, not in affect the "*ability to move*" but merely a "*permit to move*," revocable at any time by failing to comply or complete the terms of the other three principles This understanding is incorrect.

There are many arguments that can be made against the alternate form of the analogy just described. First, the fourth principle is a *principle* and as such should stand independent of the others, a fourth leg of a chair not itself something supported by the other three. Second, there are uncountable instances in the scriptures where individuals have violated the laws of God, knowingly, in dramatic ways, yet there is no instance where agency (in any form) has been rescinded, i.e. despite his being cast down and his receiving the most rigorous kind of condemnation from God, Satan still exercises his agency today by standing in opposition to the Plan of Happiness. Third, for the fourth principle to be "*unfettered,*" it truly needs to be free of any constraints and qualifications. As we discuss in this work, the freedom of choice, true freedom of choice, is what allows God to encounter an infinite array of experiences.

In this work, we understand the ability to choose (agency) is inherent to Eternal Intelligence; it is a principle attribute of Deity. Because our Intelligence is sourced from God Himself, agency is inherent to each of us as well. Within this Universe, the Temporal elements respond to our eternal consciousness. The Temporal sphere was designed from the outset to unfetter the elements themselves from constraints that bind actions to eternal consequence, allowing for us, the children of God, a truly "*free*" agency. Here, with the restoration and the eventual end of time, all things that are "*done*" can be completely "*forgotten*" through repentance (a sincere change of heart in the conscious) and the redeeming effects of the atonement.

*hindrances to my agency, not qualifying principals providing for its existence.* We see that only one of the four principles is truly necessary for agency and is in fact assaulted and put in peril by the existence of the other three.

The purpose of the preceding exercise of analogy was not to criticize, but to point out the errant views that have all too commonly become ingrained among us. Agency is a power of God, and through Him, it is a power held by the sons and daughters of God. Agency was held by us and exercised by us long before our sojourn on Earth, before the Garden, and before the War in Heaven. Satan sought to curtail or restrict the Free Agency that God *HAD* already given us; he did not seek to prevent us from getting it, nor could he.

> ***Agency is a fundamental attribute of the sons and daughters of God. We held and exercised it before mortality, before the Ten Commandments were given to Moses and before Adam and Eve were given the law in the Garden.***

Agency is a fundamental attribute of the sons and daughters of God. We held and exercised it before mortality, before the Ten Commandments were given to Moses and before Adam and Eve were given the law in the Garden. Each of us were autonomous, active, free agents long before the initial physical laws of the Universe were infused into the very substance of matter (the *bohu* itself), through the brooding of God in the opening day of the Temporal Creation.

So what of law and temptation? What do scriptural passages like these, "*And it must needs be that the devil should tempt the children of men, or they could not be agents unto themselves*" (D&C 29:39) mean? What sense can we make of interpretations of the word of God which say, "*...if we did not have opposition, we could not exercise our free agency by making choices?*"[34] Can we really properly infer that our free agency is *dependent* upon our being tempted to break the laws of God? How then would a righteous recipient of a place in the eternal precincts of the Celestial Kingdom hold his/her agency there? Do those in the Celestial Kingdom lose the agency they once held? Does God have agency? How would God be tempted? How then does He exercise His agency; what are His temptations?

---

[34] Dallin H. Oaks, "Free Agency and Freedom," in The Book of Mormon: Second Nephi, the Doctrinal Structure, ed. Monte S. Nyman and Charles D. Tate Jr. (Salt Lake City: Bookcraft, 1989), pg. 8.

Lehi explained to his son Jacob, "*it must needs be, that there is an opposition in all things*" (2 Nephi 2:11). This *opposition*, as the Prophet Lehi detailed, provides for righteousness, wickedness, holiness, misery, and it allows for both the good and the bad. Lehi further states that all these things must be "*compound in one.*" He says without this dynamism, what we scientifically describe as uncertainty or, more correctly, the freedom of action that stems from it, that *one* (unified/inert) body "*must needs remain as dead.*"

The Nephite Prophet speaks of the governing laws (the knowledge within the *bohu*) when he explains: If there is no law, there is no opposition, and if there is no opposition, "*man could not act for himself*" (2 Nephi 2:16). To repeat that in terms we understand today in the contexts of this work: *If there were no laws of physics to govern the Universe, there would be no structure for uncertainty. Where there is no uncertainty (opposition), there can be no independence of thought and action; free agency cannot exist.*

After the Fall, the boundaries of our agency changed in a fundamental way. Temporality unshackled our choices from the restrained selection available in the Physical sphere. Here, a near unlimited variety of experience produced an environment that is rich in potential for growth and learning. In the Physical sphere, our proximity and intimacy with our Father of necessity *restricted* our choices to right or wrong, in a black or white, yes or no, manner.

Agency, restricted in such a way, though fully allowing us to choose between right and wrong could not teach us what right and wrong was. In a world where only good existed, we lived in harmony with it or were cast out. There we were never part of defining what good and evil was, and as a result, we were never fully capable of comprehending either of them. In this world (the Temporal Universe), the Fall would make things different; here we would know misery, so we could recognize joy; we would know the consequences of sin, so we would want to do good.[35] In this Universe, we experience evil and grow when *we* act to create law which forbids it.

To affect such learning, our choices are to be complicated and rarely, if ever, clear. The story of the Garden itself vividly demonstrates how Agency in this sphere is to work. Adam is told not to partake of the tree of Knowledge of Good and Evil, yet the tree is prominently placed directly before him. He is also commanded to cleave unto Eve. In due course, these two disparate commandments are brought into conflict through the actions of Eve. Now the choice is not right or wrong, black or white, but a determination of intent and a weighing of principles, determining what is the greater good over a lesser evil.

Adam had to develop for himself the value of the good and the costs of the evil, and through that process, he learned for the first time what they truly were. It is of extreme significance that this story (the story of weighted/competing choice) is

---

[35] 2 Nephi 2:23.

directly associated with the Fall itself. For it is this weighing of choice and the variety of decisions needed to make them, that is the purpose of the Temporal world.

What are the risks of free agency in the Temporal sphere? The Temporal state estranges us from God, and puts in play the possibility of complete, eternal, separation from God our Father in Heaven. That this is so, is evident in the mortal life and mission of His Son Jesus Christ. The incarnation of the Son speaks as much about how great the divide between our Temporal state and our Eternal state is, as it does of the infinite love our Father and His Son have for us. The Atonement, the necessity of it, and the pain and effort associated with it, demonstrate how great our risk is.

The moment we entered into the Fallen state of Temporality, we crossed a great divide that is impossible to pass over again through our efforts alone. The chasm is so deep and wide that it required the crossing of God himself. Through the incarnation, Christ bridged the expanse, extending a crossing, *literally as it turned out*, for it is through His death on the cross that we may return, and only through that death on the cross. We receive the blessing of the Atonement only through action on our part, failure to do so will leave us short of the attainment of our goals. Such is the risk we each face.

Any reading of the need for opposition in the Scriptures can be understood in the direct context of uncertainty within the Temporal Universe that we have detailed in this work. This, however, does not diminish the reality of the workings of Satan, his desire to deceive men and draw them away from God. There is no doubt that we, and indeed all involved in the Grand Council, understood the workings and risks of agency better than it is possible for us to do today, so what was the case and proposal made by Satan? What did he have to gain? The most unanswerable of all unanswerable questions is why and how so many of the "*stars in Heaven*" chose to follow Satan in rejecting the will and plan of our Father in Heaven. What were they thinking? What did they understand they had to gain or keep? More telling perhaps is what did they fear?

Satan proposed that he could "*redeem all mankind, that one soul shall not be lost*" (Moses 4:1). We can assume this was believed to be a reasonable promise; that the plan he had presented included a way for him to do this, and that at least those who chose to follow him thought he was capable of delivering on it. So how could this promise be accomplished and what was the plan?

The usual answer follows along these lines: *Satan based his plan on restricting the agency of man, denying us all, of our freedom to choose, thereby guaranteeing none would fall and ensuring that all could safely return to our place in Heaven*. But, as we have previously discussed, our desire for personal liberty is of such measure that it is unfathomable to accept that any would embrace a plan that openly proposed to take it away. History has been witness to many episodes where tyrants have come to power on a wave of populism, promoting vacant promises of prosperity and great gain for all, while in actuality seeking to control, imprison and bring under subjection. So it is not unimaginable to envision Satan doing the same thing, duping in some way even up to a third of the host of heaven into believing him. However, though the Scriptures provide us with Satan's intent, they are silent about what his public promises were, what he *said* he was planning to do.

The Scriptures are replete with the workings and deceptions of Satan. We can learn much from them in our attempt to reconstruct Satan's first promotion. We are told that Satan was "*an angel of God who was in authority . . . who rebelled against the Only Begotten Son*" (D&C 76:25). He is the "*father of all lies*" (Ether 8:25). The devil desired the "*honor*" and "*power*" of God (D&C 29:36). He wanted not only to "*exalt [his] throne above the stars of God*" (Isa. 14:13), but he "*exalteth himself above all that is called God... shewing himself that he is God.*" (2 Thes. 2:4). These passages describe what was and is still his *intent*. But they likely are not describing what he said he was attempting to do. His presentation would have been cunning, and (no doubt) in it he appeared to be selfless, or at least protective of his fellow beings.

The same influences and schemes Satan uses so effectively today, served him well in the premortal existence. The gospel according to the deceiver has consistently been "*Eat, drink, and be merry, for tomorrow we die; and it shall be well with us*" (2 Nephi 28:7). "*Do this, or do that, and it mattereth not, for the Lord will uphold such at the last day*" (Mormon 8:31). Certainly chords of this strain were played as Satan presented a plan that laid out a life in mortality without the dangers of accountability. We could each individually, and through each of us we could all collectively, experience good and evil without the restrictions and consequences of law. Law is a necessary confine of the Temporal sphere, and Satan's plan must, therefore, have precluded the Fall, Temporality, the Resurrection and the Atonement. He would replace all of that with what had to be a trade-off of responsibility and accountability for a lack of progression. He sold a *safe* status-quo, no doubt in his mind ensuring his already established place of esteem above the rest of the hosts of heaven.[36] By presenting a plan that was without risk, he ensured it would also be without progression. By convincing us that progression was not worth the inherent risks, he hoped to prevent our advancement toward becoming like our Father in Heaven and by that he sought to ensure his place of glory above us all; he would ensure he was without challenge or challengers to him.

President Brigham Young said, "*This is what Lucifer, the Son of the Morning, wanted. He wanted to save the world without law, to redeem the world without order.*"[37] Indeed Satan hoped to eliminate law and the progression it affords us. This scenario explains the continued activities of Satan amongst us today, seeking to stumble and destroy us, now attempting to preclude our success and subsequent return to the precincts of our Father where those of us who are successful will attain a position of exaltation above what Satan held then as the "*son of the morning*."[38]

So this was *and is* the campaign; what is at stake? What was Satan trying to prevent us from achieving? Why was it; why is it, so valuable? The '*it*" in these questions is something each of us has today, attributes we hold in the Temporal sphere, ones we attained after the Grand Council. They are not anything we held before the War in Heaven began, for to be certain the struggle was not over things that had been past but of things that would be future.

Working backward from what we have now to the conditions of each of us before, we can peel away layers of additions and the effects of those additions. When we eliminate any of the additions that are not truly new, for they were *in effect* a reissuing of previous qualities and capabilities, we are left with two significant attributes unique to our mortal beings and of enough value to be worth fighting for. They are the seemingly endless array of choices we have and our ability to procreate. It is fitting because next to the love and oneness associated with the bond between a man and woman, it is these, and only these two attributes, that are mentioned in the story of the Garden.

In the Garden story, our first parents are commanded to multiply and replenish the Earth; they are also spoken of as having become like God, in that they can know good from evil and exercise agency in choosing between them. That these two attributes among the many we hold are the only two that can be directly attributed to our mortal existence is beyond coincidence.

---

[36] The apocryphal work ***Vita Adae et Evae*** (Life of Adam and Eve) records the details of an angelic rebellion. The narrative tells of God calling the angels together following the creation of Adam to admire and "*worship the image of God.*" The ***Vita*** says that Satan refused, proclaiming, "*I do not worship Adam... I will not worship one inferior and subsequent to me. I am prior to him in creation, before he was made, I was already made; he ought to worship me.*" From "*The Life of Adam and Eve" [Vita], the Old Testament Pseudepigrapha*, edited by James H Charlesworth, translated by M.D. Johnson (New York; Doubleday, 1985).

[37] Brigham Young, in Journal of Discourses, 14:280–281.

[38] Isaiah 14:12.

Satan did not frame his offers around agency in the way that has been accepted by most teachers; he could not have. Agency is a gift from God bestowed before the confrontation began, and as such, it could not have been prevented nor rescinded. The hosts of heaven knew full well what agency was and that Satan could not curtail it or supplant it in any way. Rather Satan sought to convince that there was another less perilous way for our eternal progression, literally a less *uncertain* way.[39]

Satan sought to prevent the creation of the Temporal sphere (with its law and order [*boker*]) and the expansion of free agency it afforded us. He wanted to prevent our attaining the Temporal bodies that would be so much a part of our immortal Souls.

He attempted to convince each of us that these things were accompanied by enormous dangers, risks far too costly than the potential reward. Satan's argument would have been that others (the Father and perhaps the Son) were taking unnecessary risks at our expense for the attainment of their own glory. He taught that his way would be better and was no doubt convinced, when everyone recognized what he had proposed, that his plan would be accepted and he would be vindicated and honored for it.[40] However, the debate quickly became a struggle and the choice made clear; it was to be Satan's way or the way of our Father.

Our Father experiences through us everything we experience; He shares in every pain and joy. When we became like Him at the *instant* when He declared that we had *"become as one of us*" (Gen. 3:22), there was a metamorphosis that connected us to Him in a way not realized before.

Each of us, already a very real extension of Him (for our intelligences are literally *of* Him), could now experience life *jointly* with Him. Each instant of every day of our lives is factually a moment we share wholly with our Father in Heaven. Each of us is an actual extension of Him, living as much for Him as we do for ourselves. Our experiences and the learning we achieve through them *are* His as well, for each of us provides an unimaginable array of choices and circumstances that adds to His omniscience.

---

[39] Satan sought to prevent the development of the Temporal Sphere itself. The law in its full context then, includes not only moral and spiritual law, for those have always been in place preceding the War in Heaven, the Grand Council, Satan and our intelligences themselves, but the very laws of physics which define our Universe, govern the resurrected components of our Immortal Souls, and provide the quantum uncertainty necessary for the expansion of our free agency through an infinite array of choices not possible in the Spiritual sphere.

[40] When we chose to accept the dangers of Temporality, Satan was keenly aware of the constraining effects of the law on free agency, and began, from the outset of the decision in the Grand Council, to use the law to prevent our progression. Satan has been actively pursuing a course that ensures we Fall under constraint through condemnation of the law from its inception. In the Garden, with the sons and daughters of God before the age of the patriarchs, with the Savior through His 40 days of temptation, and to this day with each of us, Satan works to entice each of us to transgress, bringing ourselves under condemnation of the law, and he seeks to separate us from the saving grace of the Atonement. The struggle is real; with each success, the deceiver adds another subject to his tragic kingdom.

This ability, to continue to progress, and create newness is divine,[41] as it is, *and only can be*, accomplished through an intimate association with God. The experiences God shares with us, however, are truly ours, a result of our free agency; for without our freedom, the variety of action and consequence through which progression is achieved, would be diminished, limited only by the immensity of the mind of God admittedly, but nonetheless limited.

The world teaches us, and indeed to a certain extent, so does our *now common Latter-day* understanding of agency, that we each are agents unto ourselves; that what we do, as long as it doesn't hurt anybody else, affects us alone. We alone suffer the consequences; this is of itself, a very cornerstone of pride. Common sense tells us this independence of action is not real. None of us can act in any way that does not affect another. We are all interrelated in very real, tangible, physical ways by the laws of this Universe. We are interconnected in civil, political and social ways. We are interconnected through family and by love.

***Each instant of every day of our lives is factually a moment we share wholly with our Father in Heaven. Each of us is an actual extension of Him, living as much for Him as we do for ourselves.***

These facts are undeniable, yet because of pride, self-image and inflated self-worth, each of us, in varying degree, choose to ignore these obvious facts each and every day. Here we see, that to these so oft ignored facts, we can add another; we share our lives with a personal God in a most intimate way. In this aspect as well, we are not alone, agents unto ourselves. Our agency is His agency; and one day, we in like manner, will share our agency with our sons and daughters as well.

*THAT is what eternal progression is. This is the purpose of life, the reason for Creation, all three of them, and the reason we have come to the Temporal sphere.*

---

[41] Recognizing this creative spark for what it is, and for its infinite value, A.D. Lindsay wrote: "In this morality of grace, the situation says, here is a mess, a crying evil or a need, what can you do about it? We are asked not to say 'Yes' or 'No,' 'I Will' or 'I Won't,' but to be inventive, to create, to discover something new. The difference between ordinary people and saints is not that saints fulfill the plain duties which ordinary men neglect: the things saints do have not usually occurred to ordinary people at all. 'Gracious' conduct is somehow like the work of an artist, it needs imagination and spontaneity. It is not a choice between presented alternatives, but the creation of something new." ("The Two Moralities: Our Duty to God and to Society", A.D. Lindsay, 1940, Eyre & Spottiswoode, London).

## The Evidences and Implications of the Physical Sphere

We have covered a lot of ground in our discussion of the creation of the Physical sphere and have much more to cover in our following discussion of the Garden. We have discovered many elements that shed significant light on a realm beyond the Veil. In our current state of existence, the Physical sphere seems distant to us, yet still somehow welcomingly familiar. That we sense this welcome should not be surprising because the Physical sphere is our home. It is here, in this Temporal sphere, that we are wanderers, strangers in the land, finding our way through unaccustomed places. The sense of belonging to the Physical sphere we feel, testifies to our Spirits of the reality of this home, though the telling of its separate creation may have been a revelation to many.

Through our study here, we identified three evidences of this separate creation and that of its telling through the Endowment narrative. As promised, we will review those evidences before we close this chapter.

The first, and most apparent, evidence of a separate telling of a distinct Physical Creation is the inclusion of Michael in the Endowment account and his complete absence in any of the multiple tellings of the Genesis 1 account. Michael's absence as co-creator in the Genesis 1 accounts was not an oversight, it was not meant to obscure a sacred association, and he is not one of the unnamed Gods spoken of in the passages of Abraham. The Endowment narrative, by including Michael, is not filling in information that had been left out of the written record. Michael is not included in the records of the Temporal Creation because he simply was not involved there. That he is integral to the account told in the Endowment testifies of the individuality of the two creation events.

The second evidence we discovered of the certainty of a separate Physical Creation was that the order of the creative events is different in the Endowment narrative from that presented in the Genesis 1 accounts. This difference of arrangement is not accidental for we have discovered here that there is intended significance to the change of sequence.

The Endowment narrative is one that represents a top down sequence for creation, starting at the point one might expect a builders list of priorities to suggest. The creative chronology is anthropocentric, focused on developing a home for man, a man that is eternal and immortal. The Genesis 1 accounts are Cosmo centric in order and detail, the focus was developing an entirely new Temporal sphere, literally from nothing, as it turns out.

In the Physical Creation described in this chapter, there was no waiting for the creation to obey. Because it is not *uncertain*, Physical (*spiritual*) matter does not struggle with perfection and obedience in the way the materials of this Temporal Universe do. An example of the differences between the Physical and Temporal creations is provided through the variation of the accounts for day three and four. The Temporal account details the creation of plant life before the Sun, Moon and stars *appear* in the sky. In Part 1 of this work, we explained that this counter-intuitive detail was indeed the way it is scientifically understood and can be traced to the conditions present on the surface of the early Earth and within the young Solar System. The Physical narrative identifies the creation of the celestial bodies in the Heavens before the creation of plant life on the surface of the Earth.[42] The sequence of the Endowment narrative throughout indicates the creation of a domain that does not have to allow for the Physical laws to act and take effect as they must do in the Temporal Universe.

The third evidence we discussed in this chapter not only speaks to the individual reality of the Physical Creation, but it also focuses our attention on the uniqueness of a truly Physical (*spiritual*) sphere. There is no mention of the breathing (*Ruach*) of life into Adam nor to any of the animals created. This is because in the Physical sphere the *spiritual* creations were not disconnected from the presence of the Father in the same manner as they are here in Temporality. There is no need for the *Nefesh* and *Neshama* in *spiritual* beings.

The *Nefesh and Neshama*, so necessary for life here in this Universe, are the cement that binds the Spirit to Temporal bodies. For animals, including the *ha-a-dam* (הָאָדָם), the people of the clay (early man), the *Nefesh* only is given in Temporality providing the attributes of instinct. In addition to the Nefesh, the *Be-ne- ha-e-lo-him* (בְנֵי־הָאֱלֹהִים), *the sons* (the children) *of God*, however, receive also the gift of the *Neshama*, creating a Soul, as the *Neshama* binds us to God. We found that through the sharing of the *Neshama* in the Temporal sphere, we are connected, sharing all that we are and experience, with our Father in a most intimate way.

The Endowment narrative continues through six creative periods until the Father himself joins Jehovah and Michael to form man, but though Michael is there at the outset of the process, something happens that changes the pattern of creation repeated through each of the preceding five days. It is at this point where the creation narrative of the Physical sphere ends and the story of Man, yet to be placed in the Garden, begins.

The Endowment narrative takes us to the point we are told of in the Preamble verses, that of the creation of "*all the children of men,*" but it provides no detail of how that is carried out. What we are made aware of is, God the Father, *Elohiym*, says He will join with Jehovah and Michael and through His utterance, "*let us make man in our own image,*" we are made aware that the Father will be personally involved with the next creative event, in striking contrast with His pattern of distant directorial involvement during the previous five periods.[43] Note carefully that Michael receives instruction at the outset of Day 6 but never responds, as we had become accustomed to hearing him do throughout days one through five.

---

[42] Let us note here that in fact the order of creation of the Sun, Moon and stars before life on Earth as described in the Endowment narrative of the Temple, is the actual order of creation in the Temporal sphere as well. The order told in the Genesis 1 accounts, however, represent a chronology as *viewed* from Earth. The Endowment narrative represents an order that is told from the perspective of the Creator.

[43] Later in this work, I will suggest a scenario that proposes the involvement of Heavenly Mother at this point in the creative process. See our discussion, Hearing the Softest of Voices, Our Heavenly Mother, beginning on pg. 449.

The Endowment narrative of the creation and the associated Preamble verses do not mention gender until that unique relationship is discussed when Eve is created just prior to the entry of our first parents to the Garden. We will explore this in detail in the following chapter. However, it is clear, through the separation of the Physical sphere from the Temporal sphere, and God creating both Adam and Eve in their distinct male and female bodies within the eternal Physical construct, that gender is specific to the Spirit and is a divinely appointed attribute of everlasting significance. Gender is not biologically allotted, nor culturally influenced. It stems from something far more fundamental and important.

Though the Garden belongs to the Physical sphere and never was in the Universe we call home today, we have yet to discuss this special place within that eternal realm in any detail. This is intentional. In the next chapter, we will describe that it was planted by the LORD after the Physical Creation was completed within that same eternal sphere. We will discover that God imbued the Garden placed in Eden with elements specific for the task of providing the immortal with the ability to Fall.

CHAPTER 7

# THE GARDEN AND THE FALL

## Day Six and the Garden (A Special Place in Paradise)

Without doubt the most cherished and celebrated story in all of human history is that of our first parents in the Garden called Eden. This story of the Garden, known as Gan Eden (גַּן עֵדֶן), in original Hebrew, is foundational for its establishment of principles so important to theology. For the Latter-day Saint this fact should be obvious, in that *of all the things we could have been taught in the Endowment*, it is this story that was chosen to be repeated sacredly within Temple walls.

Figure 7-1 *"I Drove Out the Man, and I placed at the east of the Garden of Eden, cherubim and a flaming sword...*

As we will see, the importance placed on the telling of this story and its many retellings is not an effort wasted. The Garden, its form, placement, and the events that occur there, are the nexus of the history of the sons and daughters of God, His creations and eternity itself. Through our proper placement here of the Garden, *outside* of the Temporal Universe and *within* the eternal Physical sphere, it occupies a unique position mirrored today by Temples on Earth. A powerful message of continuity is taught through the telling of this story within Temple walls (*Gan Eden*, literally translates as *walled/enclosed paradise*) for as the Temples of the LORD provide a sacred link between the Temporal and the Heavenly, Physical realm, so to was the Garden a link, a special place, where the veil was lifted between the Heavenly realm of God and the

Temporal world that was then yet still to come.[1] The Garden was, is, and forever will be, an inverse of Temples here on Earth. The Garden is a Temple dedicated to this Temporal Creation of God. Here in this Universe, we build Temples that are dedicated to the spiritual things of the Physical. It is fitting that we are endowed with heavenly knowledge on Earth today within the Temple through a rendition of the Garden story, for it brings to memory a time long ago when, from the precincts of the Garden, our first parents were provided knowledge of the Temporal that was yet to be. The Garden provided Adam and Eve with the means of achieving the blessings gained by this Temporal sphere. Our Temples on Earth today provide us with the means of attaining the blessings of the Physical sphere. In this chapter, we will examine fully the things that have been revealed to us about this special place in paradise.

The Endowment narrative, Genesis 2, Moses 3 and Abraham 5 are synoptic in that they are in complete accord in the details of the account of Adam and Eve in the Garden. As we have done before, we will defer to the account of Moses as our guide, pointing out the minor differences found in other accounts when they occur.

Though we are told the Lehi party carried with them to the new world brass plates containing *"...the five books of Moses, which gave an account of the creation of the world, and also of Adam and Eve"* (1 Nephi 5:11), and (separately) an account of the creation and the Garden down to the great tower was included on the 24 plates of the Book of Ether (Ether 1:3), the translated Nephite record does not cover the creation story in significant detail. The Book of Mormon does, however, mention our first parents in the Garden and provides some significant details thereof in two places of note; we will refer to these resources when appropriate. It must be pointed out here that The Book of Abraham does not present any information of the actions of our first parents within the Garden and surprisingly does not detail the circumstances of their Fall at all.

## *Adam Is The First Man*

Before we begin our discussion of the Garden in earnest, we must first close out the story of the Physical Creation by covering the events of Day 6 of the Endowment account and by looking at a verse from the written record that we have not yet discussed in comprehensive detail. An examination of these teachings reveals that Adam was not *created* in the Garden. That process was completed before the Garden was planted and its grounds were set apart and committed to the conditions required for the Fall.

---

[1] Temple/Temporal, share the same root word meaning *template*, or "*image*" of the final intended or original object. The Garden is exactly that, a true Temple, a template of the Temporal sphere that was yet to be created. All things within it, vegetation, animals, Adam and Eve themselves were images of what was yet to be. In this one case, the images were more perfect than the "*real*" we know today. Through the process of creation by subtraction, God, in His withdrawal, left all that is today, less than what it once was.

As the account of Day 6 begins in the Endowment, man is not yet to be "*found on the earth.*" The Father details what must yet transpire and declares to both Jehovah and Michael, "*Come, let us go down.*" Here in Day 6, for the first time, the Father goes down to perform the acts of creation. But that is not the only change, for we hear only the response of Jehovah as He says, "*we will go down*;" the, by now familiar, reply, "*we will go down*" is not heard from Michael. In fact, Michael is never heard from again, because he is completing an important transition; he is becoming Adam in the transpiring process.

The Endowment narrative next details both the Father (Elohiym) and the Son (Jehovah) forming man in their own likeness, a phrase that is familiar from the scriptural accounts, but this process is followed by Elohiym declaring to Jehovah that they had just "*organized*" man, and this forming and organizing of the creature in their own "*likeness*" and in their own "*image*" was completed before the Spirit of the man was within him. This detail is so often overlooked, yet it is extremely significant; in that it means, unquestionably, that *there MUST be a correlation between the physical form of man, to the image and likeness of the Father and the Son*. The likeness and image Adam shares with God is more than just that of Spirit as the traditional Christian view holds; the similarity between man and God is much more than that, very much more. Here God declares Adam to be in His likeness and His image, before the man's Spirit is within him. By this we understand that Adam shares a *form* and *physical organization* that is in the likeness and image of God. More importantly here, the Endowment account reveals that God has a likeness and image that is equivalent with the physicality of Man.

After man is formed and organized, the Endowment tells us that both the Father and the Son "*put into him his spirit, the breath of life, that he may become a living soul.*" This is the *Ruach* (the breathing) of the *Neshama* (the Spirit) into man of which we are by now (in this work) very familiar, but I want to point out here one additional hue to the meaning of *Neshama* because it sheds light on the truth we revealed in our previous paragraph. *Nefesh* and *Neshama* both carry with them a meaning of *being*, as in the sense of a corporeal body, something tangible capable of life. Understood in this sense the Spirit is embodied, a concept we will discuss in much greater depth and detail later in this work.

Briefly, we must only note a few important details here, because we will get back to them all in depth later. The Endowment account also includes the creation of Eve from Adam's body at this time. In agreement with the written accounts, the making of Eve does not involve the *Ruach* or *Neshama*. And Adam names Eve here (in the Temple Narrative) before he is recorded as doing so in the written accounts.

This all takes place in Day 6 of the Endowment account BEFORE Adam is placed within the Garden. It must be noted that there is no day of rest associated with the Endowment creation narrative (**Another evidence that the creation story told here is not the one told in the Genesis 1 account**), *that a day of rest is associated with the Temporal may have some relationship to the Father withdrawing and the rising preeminence of the Son as the mediator between man and the Father.* All of the details we have just addressed are included in just one short verse in the Biblical account, Genesis, Chapter 2, Verse 7. Comparing it to what we have just covered demonstrates how much was missing.

> *"And the LORD formed man of the dust of the ground, and breathed into his nostrils the breath of life; and man became a living soul." (Genesis 2:7)*

Here in the Genesis account, Adam is said to be "*formed*" because the text uses the Hebrew *way-yat-sar* (וַיִּיצֶר, meaning, *and formed*), again with the double *yud in yat-sar*[2]. Specifically, the Hebrew, *bara* or *barau* is not used, informing us that Adam is *not* created at this time. By using *yat-sar* the text intends to imply a molding or fashioning of pre-existent substance from something else, to borrow a phrase from the Endowment, an *organizing*. It will be remembered that *Yat-sar* is used to describe the molding of a vessel from clay; as such, Adam was fashioned from the *dust*, meaning the *substance* of spiritual matter. Here the Bible and the Endowment account describe a progressive, tiered creation, one that is organized from building blocks previously in existence.

Through the Moses account we know that all of the children of men were "*created*"[3] at this time as well:

> *"...For I, the Lord God, created all things, of which I have spoken, spiritually, before they were naturally upon the face of the earth. ...And I, the Lord God, had created all the children of men; and not yet a man to till the ground; for in heaven created I them; and there was not yet flesh upon the earth..." (Moses 3:5)*

*All of the children of men were created spirtually (Physically) as Spirits before the Garden was planted*; we will discuss what this means later in this work, but for now let it suffice us to understand that *we all* became embodied Intelligences (Spirits) at this same point in time. What is to be done with Adam, however, is something much more, for he will be (upon his entry into the Garden) given an immortal body made of Temporal "*stuff*." Adam is here identified (before the creation of the Temporal world) as being *ha-a-da-mah* (הָאֲדָמָה), *of the earth* (embodied in a way not at all unlike that which we are to be in the resurrection).[4] Adam becomes the first man, *here called*, because he becomes the first of the sons and daughters of God to become a *living soul*. Indeed it is because of this unusal set of circumstances, this preview of Temporality, that the Garden was planted in the first place. In the Garden, within its confines, Temporality was to exist in the eternal Physical sphere.

---

[2] See the brief discussion of *yud* beginning in the middle of page 88.

[3] The Moses account is an inspired translation of the Bible directly into 18th Century English; as such it may not have represented the entire nuance of the original Hebrew. Here (sadly) it seems to conflate, as our modern language often tends to do, the meanings of two distinct words (*Yat-sar and Bara*) into one that on the surface seems to mean the same thing, but in truth is entirely inadequate in detail.

[4] Adam, and indeed all things within the Garden, had not Fallen. Adam and Eve had not yet known of good and evil, they had not yet tasted death, though they were "*living Souls*" embodied now within the Garden with immortal "*Temporal*" materials like a resurrected being, like our Father; they were indeed not resurrected. For we know, "*But now is Christ risen from the dead. And become the firstfruits of them that slept*" (1 Cor. 15:20).

Accompanying this transition, taking a step towards full Temporality, Adam fell into a state of innocence. This fact cannot be overlooked or reasoned away in search of some foundation for the just application of consequence; it is of upmost significance for it ties our first father to us by way of shared experience. The first vestiges of what would later become a more complete veil was placed between Adam and his memory of what he had once been.

In the Endowment, we are told Adam arrived at the veiled Garden walls "*having forgotten all*," that he became as a "*little child.*" Yet in this state, he was still in direct association with God, this dichotomy could only exist in a special place that hung somewhere between two spheres. The Garden, through its setting apart, was such a place, an eternal world within which existed the seeds of Temporality, the formative substance of the natural.

In preparation for his entry to the Garden, the Endowment account says Adam became a *little child*; he did not become *like* a child or *childlike*, but he became a child, specifically a *little* child. What is the implication of this? Are we to take this literally? Perhaps?

Could it be that Adam was indeed allowed to mature from a child to a man within the Garden or in the time immediately before his placement there?[5] If so, this may have significance in regards to how he first appeared on the Temporal Earth within this Universe. We will discuss and make a case for this probablity later, but of significance, and to our point here, is this phrase, vocalized in the Endowment, at least speaks to the cognitive capabilities of Adam when he is placed in the Garden. Adam no doubt came to the Temporal world empowered with great knowldge and wisdom befitting his station among the other life on Earth, that would have been necessary for his survival. Certainly in the ages before the Garden, Adam as Michael was a great leader of stature. But all of that stands in stark contrast here with the clear, concise, statement that after awaking as Adam "*he will have become*

---

[5] We must remain somewhat vague in our discussion of the events attendant to Adam *in and around* his "*placement*" in the Garden, standing, as it were, before the Garden walls. We are certainly aware of Adam's Spiritual embodiment *before* his entering the Garden. Adam and Eve are informed of what challenges they will face in the Garden before they are placed there. Of the timing of these events we are adamant.

We will discuss later in this work Adam's receipt of a body of *Flesh and Bone*, its relationship to the "*new name*" he received and what that means to each of us. We also know that with the receipt of a body of Flesh and Bone, Adam received the *ruach* (the Breathing), the Breath of Life. We know Adam became as a "*little child*," that he had "*forgotten everything,*" and we have more than inferred here, that this is to be taken literally (allowing for an understanding that Adam was "*born*" an innocent child, like all of us, and grew into manhood and gained intelligence and knowledge). We know these events all occurred in **direct** association with his entry into the Garden Temple. We propose *these* events to have taken place *after* entry in the Garden, but admit (because continuity of the story narrative of the Endowment sequences some events out of order, e.g. the naming of Eve) the actual timing remains unclear. It may be possible that some of the events on this second list may have been completed preparatory to (immediately before) Adam's entry to the Garden. This may seem inconsequential, for all of these events are tied directly to Adam's placement as a *Temporal Being* in that perfect Temple of God, but in the interest of accuracy and complete clarity of fact, I point these things out here.

*a little child.*" We will communicate more about the significance of this in our discussion of the Fall, Transgression and Sin to follow.

Adam is a new name for Michael. Though it is not specifically said, it is implied. It has a connection to his Temporal identity, for it is given to him as he is formed of the clay and as he is prepared to enter the Garden. The new name is familiar to Latter-day Saints, and I suggest here, that by receiving a new name (Adam) with his transition from Physical (spiritual) to Temporal (of an immortal form, like our resurrected bodies), Michael experienced the *reciprocal* of what we experience here on Earth when we receive our new name attendant with our Temple Endowment. Michael entered the Garden Temple, a temple built by the very hands of God and received a new name just as we receive a new name when we enter the Temple here on Earth.[6] This too is another parallel the Garden shares with our Earth bound Temples

Adam is *placed* in the Garden, not *formed* there, but *placed* there *after* he was formed. Notice the tense used in Verse 8. There is a chapter break before Genesis, Verse 8, designating a pause or break in the story. It is a delineation, separating the *before* events of Verse 7, from the *after* events of Verse 8. The written and Endowment accounts clearly inform us that Adam was formed, organized and became a living Soul, *then* placed within the Garden.

> *"And the LORD formed man of the dust of the ground, and breathed into his nostrils the breath of life; and man became a living soul."¶And the LORD God planted a garden eastward in Eden; and there he put the man whom he had formed." (Genesis 2:7, 8)*

-A Second Witness

The Endowment account breaks with the chronology of the written word here, in that Eve is both created from the body of Adam and named before either of them are placed into the Garden. This account confirms that the time in which ... *all the children of men,* were created preceded the Garden and confirms that those children were embodied at this time. A careful examination of the Endowment teaching found in the statements immediately following the creation of both Adam and Eve details much that on first glance may appear confusing, but the fine points, *as always*, are revealing.

The Father says in *future tense* that "*We*" (this is a royal We, signifying both Father and Son) will place Adam in the Garden and once there, "*We*" will command you and Eve to multiply and replenish the "*earth*." Why not command Adam and Eve presently? They both are created at this point and Adam understands that Eve "*is*" the mother of all living, not that she *will be*. That Eve "*is*" the mother of all living is a fact confirmed by none other than the Father himself. Note here that the

[6] See our discussion, <u>The New Name</u>, beginning on pg. 376.

commandment, when it will be given, will be to replenish (fill) the Earth, *not* the Garden.

The Endowment narrative next presents the Father speaking to Adam and Eve within the Garden, saying, "*we have created this earth for you*," and "*placed upon it all kinds of vegetation and animal life.*" The Father says that Adam is to be "*LORD over the whole earth, and all things on the face thereof.*" Then comes the before-promised commandment (to multiply and replenish) as the term of locality changes from that of *earth* to that of the *Garden* when the Father declares, "*We have also planted for you this garden.*"

Why all of this detail that is so different from the inspired written accounts? It is instructive, very much so. Here we are being told of the time of instruction, of the recounting of the plan to our first parents before the veil of forgetfulness was fully placed around them, before their entry into the Garden. They are being told of intricate workings of the task before them, *before their innocence*, while they are yet capable of accepting the consequences with full knowledge of choice. The commandment to multiply and replenish and its countering partner, "*Thou shalt not eat of the tree of the knowledge of good and evil,*" were to be repeated and given only after the veil was firmly in place, for that would ensure the decision would be freely made with its subsequent consequences falling directly on to Adam and Eve.

Adam and Eve are told of the Earth, created with Michael spiritually, but still yet to be made and then inhabited Temporally by both of them and their children; note that plant and animal life predate them on Earth. Our first parents are told that procreation is to be part of the reason for the Temporal Creation and their entry into it. They are to fill[7] the Earth, not the Garden. Note that this commandment does not appear in the scriptural Garden accounts, it is given to man in the Genesis 1 accounts of the Temporal Creation which follow the Fall, in that context the commandment properly refers to the Temporal Earth, but is recorded out of sequence and as such (if actually spoken then) could not have had any part at all in Eve's decision to partake of the fruit so that she may fulfill her calling as a Mother.

The Endowment account, however, restores continuity as it tells of the initial knowledge of the commandment and of the knowledge of Adam and Eve that she was to be the mother of all the sons and daughters of God on the Earth. The Endowment account confirms that the commandment was given again, within the confines of the Garden, *for the second of the three times it was to be given*. Eve, while in a state of innocence and forgetfulness, no longer had a memory of having received the commandment before, by its retelling in the Garden; Eve had the knowledge to act upon her understanding when Satan *pressed* her, as we will see later, perhaps quite literally.

---

[7] The scriptural accounts and the Endowment use the Old English rendering of "*replenish*" for the Hebrew, *Umilu* (וּמִלְאוּ), which properly means to fill or populate for the first time; it is not to be a repopulation.

The narrative reveals that Adam and Eve were placed in the Garden in a state of innocence where they were commanded and instructed by the Father, who told them to multiply an Earth they were both (while *in* the Garden) not yet and no longer a part of.

Of significance is the fact that the commandment not to eat of the tree of Knowledge of Good and Evil was not discussed beforehand, consider the reason for this?

Our Father intended from the outset that we, his sons and daughters, grow in knowledge and wisdom. The Plan of Happiness is built on that premise; it is a fundamental of our faith and an imperative in any foundation of a belief in God. As this is true, we expect that Father would have never commanded that we not obtain knowledge; preventing us from gaining knowledge was Satan's plan. So there could be no discussion of a commandment curtailing our attainment of knowledge spoken of in the perfect precincts of the Physical and Spiritual spheres where there is to be found no context for the experience of Evil.

The fruit of the tree is a metaphor for choice. Truly, "*free*" Agency is an opportunity to explore something other than the Divine Will of God. A commandment given (more appropriately a warning) against *choosing the knowledge of good and evil*, even as a path towards progression towards Exaltation, had no intelligible understanding outside of the Garden or the Temporal sphere. The commandment to not eat (consume evil) had to wait until it could be actually realized and understood.

It must be understood that it is explicitly because the knowledge of evil and its effects could not be had in the Spiritual sphere, that first the Physical and then the Temporal spheres were created. The attainment of such knowledge was so valuable that it was worth the immense cost of perhaps losing Spirit Children to its effects and worth the extreme sacrifices of the Atonement. If evil could have been experienced or even known in the precincts of God, none of what we have been discussing in this book would have been necessary.

An understanding of this part of the Endowment narrative, its sequencing and its deviations from the written accounts, reveals a rich tapestry of truth. We find that justice is upheld, there is no consequence without informed choice. We see that our first parents were both fully informed and completely innocent. There is divine truth revealed of the qualities and attributes of God, man and the precincts of the Garden, and its relationship to the Physical and Temporal spheres.

The precision of the account, the intricacy of its detail, its filling in of gaps in logic and information when the gaps were not even perceptable,[8] speaks to the divinity of the message given within Temple walls. We find that there is no disharmony with the written accounts. Instead we find accord, and in fact confirmation of the informed choice required by justice in enacting consequences for the Fall. Our picture of this critical time is for certain not complete and exact detail is not known, but clearly we have an alternate chronology again to learn from. At this point, we will return to the written records and follow through their narrative returning time-to-time to the second witness we have explored here when it becomes appropriate.

| **The Garden — Moses 3** |
| --- |
| 8 And I, the Lord God, planted a garden eastward in Eden, and there I put the man whom I had formed.<br>9 And out of the ground made I, the Lord God, to grow every tree, *naturally*, that is pleasant to the sight of man; and man could behold it. And it became also a living soul. For it was spiritual in the day that I created it; for it remaineth in the sphere in which I, God, created it, yea, even all things which I prepared for the use of man; and man saw that it was good for food. And I, the Lord God, planted the tree of life also in the midst of the garden, and also the tree of knowledge of good and evil. |

## *The Garden is Planted*

The Scriptures tell us that the Garden was planted, and the Endowment uses the same word. Why was it planted, and not created? The process of making the Garden a place for Adam to dwell was something fundamentally different than the creation processes we have described before; here there is no speaking of things into existence, but a physical action is described, God "*planted*" a Garden. This is a process of organization, not of creation, for the things of the Physical sphere were already created. The process described is the delineation and a dedication of a place set apart from the rest of somewhere known as Eden.

Hebrew religion and culture is eastward oriented. Because the day breaks in the East with the rising of the Sun over the distant hills on the horizon, the Temple doors at Jerusalem opened toward the dawn. Our Temples today are oriented to the East as

---

[8] Many detractors criticize the Endowment from outside for perceived errors in the story it tells. They sharply criticize what they know of it, calling it unlearned and ignorant (see footnote 4 on page 25). Through this section, we have shown the Endowment to be sophisticated, detailed and a masterpiece of logic and storytelling, revelatory on multiple levels, beyond the understanding of its "*supposed*" originators, and certainly from God.

well. At the top of the Temple spires, the Angel Moroni looks to the morning Sun[9] heralding its arrival at the start of each new day. A planned exception is the rebuilt Temple in Nauvoo, which has a replica of the angel that sat atop the original Temple there. This Moroni now looks to the West over the plains in the direction of the Saints departure when they were driven from their Temple and homes to resettle in the mountains of Utah. In all other instances, Moroni looks back to the East in anticipation of the return of the Savior who will come from the East. The LORD's return from the East is revelatory for it bespeaks to the orientation of ancient Hebrew tradition placing the precincts of God to the East, a tradition which finds its tap root right here in the placement of the Garden.

***...the tree of knowledge of good and evil remained in the Garden though it was infused with death, decay and entropy.***

The Garden is specifically formed as a proving ground where, in the innocence of forgetfulness, Adam and Eve can "*Fall*" by exercising free will, "*I gave unto him that he should be an agent unto himself...*" (D&C 29:35). Free will requires a Universe with Quantum Uncertainty; without it, all of our paths are dictated from the beginning. Without an uncertain Universe, we would all be inseparably locked into a fixed future formed like stone from a determining past. The Universe as a whole, every single particle and structure, is designed for the *Free Agency* of Man. Free[10] agency is NOT possible in direct contact with the sphere of the Father. Because it allows us to Fall, free agency requires an Atonement to make us whole. Free agency is an attribute, which through its exercise, separates each of us from the Father.

## *A Natural Tree in Paradise*

Verse 9 contains a statement that appears enigmatic, "*natural*" trees, we are told, are to be found in this eternal paradise. What are they doing here in the Garden? What at first seems problematic, in the end is infinitely revealing.

---

[9] *"According to scripture at the second coming the Savior will come from the East (Matthew 24:27). The Church's guideline concerning the placement of the angel Moroni figures is that where possible, they should face Eastward. Sometimes, however, the angel Moroni figure may face another direction in order to align it with the orientation of the temple. Originally the figure on the Los Angeles Temple was placed facing Southwest, the same direction the temple faced, but the figure's orientation was adjusted to face due East upon instruction from President David O. McKay."* Looking Up to Moroni (*Wendy Kenney. November 2009 new-era on lds.org/new-era*)

[10] We have described "*Free*" Agency as something more than agency in this work. Agency expressed freely with an unlimited variety of possible actions and consequence is something that can only be experienced in a sphere separated from God.

That the Garden was wholly a construct of the eternal Physical sphere can, as of now, be beyond doubt. BUT, the Garden represents something special: it is a place where an immortal Adam and Eve walked freely in thc presence of the Father and Son, yet they were in a state of innocence and blanked remembrance. In the Garden there existed the potential for the Fall, a critical form of consequence for choices freely made. What set the Garden apart in this manner? Verse 9 provides an answer for this as well, "*I, the Lord God, planted the tree of life also in the midst of the garden, and also the tree of knowledge of good and evil.*" Further we are told, "*But of the tree of the knowledge of good and evil, thou shalt not eat of it, nevertheless, thou mayest choose for thyself, for it is given unto thee; but, remember that I forbid it, for in the day thou eatest thereof thou shalt surely die*" (Moses 3:17).

Speaking of the tree "*that is pleasant to the sight of man,*" the Scripture says it "*became also a living soul.*" The wording here remains somewhat enigmatic. We will discuss at length elsewhere the Natural and Fallen things of this the Temporal sphere, but here in the Garden, before the Fall, we are told of a tree, identified by the singular, *it*, that is already "*natural.*" The "*natural*" is a Fallen state having attributes which are fundamentally much different than what is normally found in the immortal Physical sphere. Within the Garden, where immortality reigned, with the exception of both Adam and Eve (after the Fall), "*natural*" things did exist.

This tree, a single one, but perhaps one specific kind among many, had within it the potential for death, the most fundamental characteristic of the *natural* world. This tree, more than any before or since, truly encapsulated the very essence of the *natural* (Temporal) world. It may have stood out, most probably did, from every other object in the Spiritual sphere. Verse 9 informs us that the tree of knowledge of good and evil is more than an allegorical object, but a very real one with serious effects. Also of note are the adjectives used to describe it. For those who have experienced the Endowment, it is telling: "*pleasant to the sight of man,*" a phrase which the narrative tells is used in a very destructive manner.

Of note also in Verse 9 is the curious statement, "*For it was spiritual in the day that I created it; for it remaineth in the sphere in which I, God, created it.*" Here God speaks of an individual tree, or a specific type of tree; "*it*" is used deliberately, three times, instead of the proper "*them*" which should have been used if He was speaking of trees in general. Also of serious consideration is His revelation that *it remaineth in the sphere in which I created it,* as if it possessed some polarity of existence. God said it "*remaineth*" as if He is emphasizing that it no longer belonged to the sphere in which it remained. That somehow it was now foreign and even perhaps no longer welcome. Perhaps what is expressed here, is the tree of knowledge of good and evil remained in the Garden though it (in itself) was infused with death, decay and entropy. This may lend a much more literal meaning to the eating of its fruit allowing in a very real sense the entry of degenerate (Fallen) Temporal matter into the perfect immortal bodies of Adam and Eve. Remember, our brief discussion of the need of God Himself to incarnate when entering fully into the Temporal sphere. We spoke of death being the price to pay for a passport to this Universe. Perhaps the incarnation and the Fall are attached to much more literal physical requirements than we have imagined.

**The Garden — Moses 3**

10 And I, the Lord God, caused a river to go out of Eden to water the garden; and from thence it was parted, and became into four heads.
11 And I, the Lord God, called the name of the first Pison, and it compasseth the whole land of Havilah, where I, the Lord God, created much gold;
12 And the gold of that land was good, and there was bdellium and the onyx stone.
13 And the name of the second river was called Gihon; the same that compasseth the whole land of Ethiopia.
14 And the name of the third river was Hiddekel; that which goeth toward the east of Assyria. And the fourth river was the Euphrates.

## *A River Flowed To It*

Moses 3:10-14, Genesis 2:10-14 and Abraham's abbreviated account tell us: "*There was a river running out of Eden, to water the garden. And from thence it was parted and became four heads*" (Abr. 5:10).

Today our Temporal world subsists through the life sustaining water flowing from a divine source. "*...whosoever drinketh of the water that I shall give him shall never thirst; but the water that I shall give him shall be in him a well of water springing up into everlasting life*" (John 4:14). The Garden, because it is a bit of Temporality within the Physical sphere, requires spiritual nourishment as well. This detail of the story symbolically illustrates that truth, in a sense doubling down on the imagery.

Many have sought to use the geography of the verses in this section of the Garden story as some sort of *treasure map*, charting a way to a secreted paradise, hidden somewhere either right in front of us or lost in our past. To them the Garden is or perhaps was somewhere we can identify on Earth today, in the cool woods or the verdant mountains of a no doubt familiar and wonderful land. Alas, I would that such was so, but we know from what we have discussed here in this work that the truth is in fact much different.

What we see here in these verses, is a map of a real place found on a world far away. It is important to us for two reasons because it describes a place we all know, and metaphorically it teaches truth. The river feeding the Garden, as we have seen, is analogous to the spiritual water coming from the precincts of Heaven we all depend on today. And the four heads are associated with the ancient concept of the four corners (or pillars) of the Earth. The Hebrew here alludes to geographic fullness or totality, expressing perfection of the Garden, and by full extension Eden, what we know to be the entire Physical Creation itself.

I would like to point out here something that is all too often misunderstood and misrepresented. Here the Scripture plainly says that a river flowed *from* Eden *to* water the Garden. It does not say that the river or any of its four heads flow out of the Garden. As worded here, we can confidently conclude that Eden is a geographic entity *other than* the Garden.

As we discussed previously, it is futile to attempt to identify any of the geographical features associated with the Garden account with any known feature on the Earth we inhabit today. The Garden of Eden was not, is not, located on this planet, nor is it to be understood to be located anywhere within the Temporal Universe.

The spiritual Earth, the one of the Physical sphere, though recognizable as home to us is an image of the one we inhabit today, it is not identical; there are no areas of desolation, of filth, and destruction to be found there. The Earth of the Physical sphere has not suffered the natural changes of plate tectonics, earthquakes, volcanism or glacial scarring. Vegetation and animal life have no need to evolve and change. That Earth remains in a perfect state. Where the Garden is, where its rivers are, have a different geologic history than any locale to be found here on our Temporal home planet.

Our body changes as we grow, it deteriorates with age, bones break, and teeth come and go. Our hair lengthens and grays and sometimes disappears. Wrinkles come and our stature wanes as we stoop and yield to gravity and the weakening of our muscles. We recognize these effects on our physical body as we look in the mirror with the hope and understanding that what we see does not represent the fullness of our eternal physical reality.

Why then would we expect our home world, just shy of 5 billion years old, to remain forever like the one eternal Physical sphere it is an image of? We must remember Adam and Eve did not exist in this Universe until they were cast out of the Garden and entered the lone and dreary world. This Earth was not the first home of our parents and neither is it ours.

The Garden is eternal, not mortal; it is NOT to be found within the Universe around us. The Garden is a bridge between the Physical and the Temporal, a great Temple built by the hand of God, walled off from the rest of the eternal Physical sphere by a veil of forgetfulness; it encloses and harbors something foreign to its home world, the seeds of Temporality.

By placing the Garden and the actions that occurred beyond the confines of this Temporal sphere, we understand and demand that the consequences, as they apply to the Temporal sphere, which was created as a result of them, are universal in scope and effect. This simple adjustment to our traditional thinking, placing the locale and events of the Garden in the Physical sphere, allows us to avoid errant understandings that prove problematic. Consider the following that starts out correct enough, but in the end misses the mark, in that its understanding of the Universe is itself limited and not broad enough:

> *"Some have argued that the paradisiacal glory of which we speak was confined to the Garden of Eden while the evolutionary processes were taking place through the rest of the earth. The great difficulty with this idea is that it confines the effects of the Atonement to forty acres (or whatever size the Garden of Eden was). The plain testimony of the scripture is that the entire earth and all created things were affected by the Fall and thus recipients of the blessings of the Atonement." (*Joseph Fielding McConkie, *Straightforward Answers to Tough Gospel Questions,* 1998, Salt Lake City, UT, Deseret Book, pg. 161.)

Though in this quote Brother McConkie does not provide the reasoning that the Atonement would be confined only to the contents of the Garden, we will agree here in this work that he is correct in its assumption that "*all created things were affected by the Fall.*" However, we must understand, in order for this assumption to be true, the Fall must encompass environs broader than that of just the Earth, our Solar System or even our Galaxy; it must address the entire contents of the Universe. Brother McConkie here, as well as millions of others, work from a commonly held premise that the Garden was on the planet we inhabit today; thus their logic is confined by that assumption.

What started out as a good and proper argument of McConkie's against those who hold a very narrow view, is in fact found to be made from a premise that is itself unnecessarily limiting. Through his view, a view held by most in fact, the effects of the Fall pertain to all life on Earth; justice is satisfied, perhaps, because all life on Earth shares a common heritage with its founding father Adam, death's dire consequences justified by his having dominion over all life on this orb. But the logic of this justice being extended beyond an Earthly Garden fails when we consider that everything in the Universe is dying, death being a consequence and *caused* by the Fall. The Universe was in a state of death and decay billions of years before this planet existed, and it would be billions of more years before a Garden could have been planted on it. What of that life found elsewhere in the Universe and that which preceded Adam here on Earth? How are the effects of a Fall, that occurred only *lately* on this planet, justified for that early life and that to be found on distant, disconnected locales?

The only way this can be reasoned is to maintain the integrity of the position of Adam as the grand head of all that there is in this sphere and placing his choice at a time before this Universe and the death that everywhere accompanies it was in existence. We must locate the confines of the Garden at a place and time before all that is Temporal ever was.[11]

---

[11] I would be remiss to continue to avoid the elephant in the room, Jackson County Missouri… Many Saints know (all too well) that early in the history of this church a number of brethren taught that Eden was located in this County of Missouri on the North American Continent. Some of them very confident in their belief: Heber C. Kimball proclaimed, "*...the spot chosen for the garden of Eden was Jackson County, in the State of Missouri, where Independence now stands; it was occupied in the morn of creation by Adam.*" Several of the claims are not without seeming *prophetic authority*, even if not directly revealed. Brigham Young said, "*...in the beginning, after this earth was prepared for man, the Lord commenced his work on what is now called the American continent, where the Garden of Eden was.*" Adding more to this, President Wilford Woodruff recorded in his diary, "*President Young said the*

| The Garden — Moses 3 |
|---|
| 15 And I, the Lord God, took the man, and put him into the Garden of Eden, to dress it, and to keep it.<br>16 And I, the Lord God, commanded the man, saying: Of every tree of the garden thou mayest freely eat,<br>17 But of the tree of the knowledge of good and evil, thou shalt not eat of it, nevertheless, thou mayest choose for thyself, for it is given unto thee; but, remember that I forbid it, for in the day thou eatest thereof thou shalt surely die. |

## *A Choice to be Made*

The choice was presented to Adam before Eve was present in the Garden;[12] perhaps this speaks to Adam's inability to choose, perhaps because of complacency? Perhaps

---

*Prophet told me that the Garden of Eden was in Jackson Co. Missouri, & when Adam was driven out of the garden of Eden he went about 40 miles to the place which he named Adam-Ondi-Ahman, & there built an Alter of Stone & offered Sacrifice. That Alter remains to this day. I saw it as Adam left it as did many others, & through all of the revolutions of the world that Alter had not been disturbed.*" We will simply say here, these statements are wrong. Perhaps they derive from misunderstandings of things the Prophet Joseph said.

These statements all position the Prophet Joseph Smith as the source of what has been revealed; no one claims direct revelation of this for themselves. A thorough search, however, will reveal that a statement by Joseph to this affect does not exist but for second hand and more commonly, third hand accounts – All recorded by men who never claimed to have heard the Prophet speak the words. This false teaching is propagated through repetition, by those that don't know. Now 180+ years after the Prophet's death, tertiary witnesses, which provide no quotation or evidence themselves but merely repeat something heard from somewhere else, are used as primary sources in books and articles purporting them as authoritative.

What *is* recorded is a single revelation naming a site: Spring Hill, of Davis County, Missouri as Adam-Ondi-Ahman (D&C 116). There is nothing contradictory to our treatise here in that revelation. We without exception believe that Adam's life and the events recorded of them after he was driven from the Garden were accomplished on the same Earth we inhabit today. It is the locating of the Garden here in Temporality that we object to. That Spring Hill and its surrounds will be the site of the, yet future, closing Grand Council on Earth we attest here, for it was revealed. We are resigned to understand that some associate the location of the future event with that of the one past; this may in fact be a correct assumption, but the location of Adam-Ondi-Ahman is <u>not</u> that of the Garden. In fact, Scripture and the Endowment liturgy insist it is not. We understand them to be separated by a vast gulf; one of unimaginable distance only alluded to by great differences in description: one, the "*Lone and Dreary World*" where death, toil, suffering and Satan reign; the other, a paradisiacal, immortal garden where the Father strolls in the cool of the day. Is it conceivable that the distance between the two is a mere 40 mile walk? Surely our God given reason says no.

That the prophet could locate a site for an alter built by Adam and position the location of the ancient Grand Council in Missouri is consistent to what we have shown here to be true, no matter how unlikely it may seem to some ears. That those locations would be found here is no less probable or unbelievable than those that have located the Garden in any number of old world locales. We submit here that there is no record of the Prophet placing the Garden in Missouri or anywhere else on this planet, nor should we expect one.

[12] In the Endowment account, the commandment regarding the tree is given to both Adam and Eve together. The addition of Eve's name to this part of the narrative has occurred only in the last decade and

it speaks to an expectation, anticipation, hope of the Father that Adam would, could Fall on his own. We have already discussed the possibility that the Scriptures and the Endowment actually speak of Adam being placed in the Garden, or its very near precincts, as a very young child, allowing him to grow and mature. The process of incarnation of the Savior included a birth and infancy, His growth in stature and wisdom. It is fair to at least accept that Adam went through the same process of development (perhaps excepting birth though not necessarily so). More than just accepting this possibility, perhaps drawing upon the example of the Savior and an absolute need for each of us who come to Temporality and have full exposure to all of its phases and experiences, it is fair to assume it to be a *requirement*. If so, when did the incarnate Adam experience infancy, childhood, and adolescence? Was it after he was cast out with Eve into the lone and dreary world? *Of course not.* We are then left with Adam growing and developing into adulthood within or near the Garden.

This scenario satisfies the requirements of multiple theological expectations stemming from diverse quarters. Contrasting with the Endowment account which states that Adam, when he entered the Garden, became a *little child* are the seemingly contradictory modern day musings of many Latter-day teachers that insist Adam entered the world much wiser and more knowledgeble than any creature since.[13] Understanding that it is possible that Adam grew and matured within the safe confines of the Garden and its environs, under the Father's care, through childhood to adulthood, harmonizes the dichotomies of innocence and informed choice. It provides the basis for the justice of consequences from the Fall and accounts for the singular choice made in response to deception and beguiling.

There is an additional benefit of the passage of time associated with allowing Adam to grow within or around the Garden. That time allows us to also harmonize the Scripture with the Endowment narrative where they diverge in regard to the timing of events, the Scriptures being more chronological and sequential, and the Endowment more summary.

The *choice* is laid out here in Moses and in the Endowment. In contrast, the parrallel Abraham and Genesis verses, do not decribed a choice but a prohibition. As we have discussed in this treatise, Agency (the "*free*" agency we came to the Temporal sphere to experience) is not rooted in simply choosing between two opposites, it is

---

was not originally there. It can be noted here as well, that the written accounts that include Eve's actions in the Garden (Genesis and Moses) do not call Eve by name until Adam gives her that name *after* the Fall. In the Endowment narrative, the naming of Eve is advanced to a much earlier point in the story, simply as a means of facilitating the dialog and as such is not revealing.

13 An example of this teaching: "*The first man placed upon this earth was an intelligent being, created in the image of God, possessed of wisdom and knowledge, with power to communicate his thought in a language, both oral and written, which was superior to anything to be found on the earth today.*" (Doctrines of Salvation, Joseph Fielding Smith, Bookcraft, Salt Lake City, UT, 1998, vol. 1, pg. 92).

fundamentally about developing the *sense* of right and wrong, creating our own choices and developing our understanding of forming those choices.[14]

Here in Temporality choice is a requirement, it cannot be avoided; we all must choose. Most often the options are not in the extreme, a clear choice between right and wrong is all too frequently not the option; of this truth we are all very much aware. The clear choices are the easy ones. What we most often find, in fact almost *always* find, are options that are neither good or evil, but are segregated by degree of importance. We are almost always left with an amount of doubt and uncertainty. Our task is to choose the one that concerns the good of others rather than always choose benefit for ourselves. We are to learn to live by the greater law and not the lesser.

This is the choice described to Adam here in Eden, soon he would have to choose; fortunately for us, he chose well. This type of choice, an interacting one: the exercise of our own will in forming right and wrong, living with the results, and learning from bad choices, is simply not possible within divine precincts and in the presence of the Father.

| The Garden — Abraham 5 |
| --- |
| 13 But of the tree of the knowledge of good and evil, thou shalt not eat of it; For in the time that thou eatest thereof, thou shalt surely die. ***Now I, Abraham, saw that it was after the Lord's time, which is after the time of Kolob: for as yet the Gods had not appointed unto Adam his reckoning.*** |

-Adam's Reckoning

The Abraham account adds a detail not discovered in any of the other accounts, and it proves to be very revealing. Abraham was something of an astronomer and cosmologist it seems; very much like me, he finds interest and pays attention to such things. So, after he describes the commandment forbidding eating from the tree of the knowledge of good and evil, Abraham provides a commentary about the timing of the commandment and of when it would be carried out. He says it *"...was after the Lord's time,"* and of even more significance, Abraham says, *"for as yet the Gods had not appointed unto Adam his reckoning."*

[14] In many respects, we are not here so much to experience the effects of choice but to create a composition, a book that records the development of law in response to unknowable circumstances, a collective diary of sorts detailing the reason for law and the forming of it at a most basic level.

The book of Abraham is replete with astronomical awareness.[15] Abraham was provided with a view of the cosmos unrivaled in the annals of human history, viewing the Universe in its fullness and with comprehension that is still not available through the advances of science today.

Long before Einstein did, Abraham had a working understanding of the relative nature of time and how the motion and relationship of celestial objects affects its perceived passage. To him it was clear that the reckoning of time was defined by physical forces, set independently of any other factor. The reckoning of time on Earth is a function of orbits and revolutions that are fixed by gravity, momentum and mass.

Given that understanding, how could Abraham say that Adam had not yet received his reckoning of time while in the Garden? Look at the creation sequence in all of the accounts. The celestial orbs and their "*times*" are created and in place before Adam was in the Garden or is on the Earth. If the Garden was on this planet (the one we inhabit today with its lights in the sky, and days and nights), Adam would have had his time reckoning, even if his death would never come because he was immortal.

When we place the Garden at a locale within the Physical sphere, however, the reckoning of time there would not have to be the same as it is here on Earth in the Temporal sphere, in fact it seems it wasn't. Wherever the Garden was, the reckoning of time for Adam within it was not yet what would be his on this planet.

Through this detail, that Adam had not yet been given his reckoning, that time as it was measured was not the time reference frame that would be Adam's, Abraham (perhaps unintentionally) reveals a verity that further testifies of the truths we have discussed here. This single statement, more than perhaps any other, firmly positions the Garden somewhere else, outside of Temporality, and deeply into the immortal, eternal Physical sphere.

---

[15] This verse is also famous for its mention of Kolob. Much has been said about this orb by those opposed to the church, prompting much to be said that is not known or understood in defense. Our understanding here of the placement of the Garden and the paradisiacal Earth beyond the confines of the Temporal Universe applies to Kolob as well. This Celestialized locale is not to be found in the Universe we inhabit, and as such, by its definition then, should draw no more ire in its concept than the images drawn in scripture of a heavenly throne for God suspended in clouds above.

| The Garden — Moses 3 |
| --- |
| 18 And I, the Lord God, said unto mine Only Begotten, that it was not good that the man should be alone; wherefore, I will make an help meet for him.<br>19 And out of the ground I, the Lord God, formed every beast of the field, and every fowl of the air; and commanded that they should come unto Adam, to see what he would call them; and they were also living souls; for I, God, breathed into them the breath of life, and commanded that whatsoever Adam called every living creature, that should be the name thereof.<br>20 And Adam gave names to all cattle, and to the fowl of the air, and to every beast of the field; but as for Adam, there was not found an help meet for him. |

## *An Help Meet for Adam*

In Verse 19 of Moses, the animals are commanded to present themselves to Adam. The verse represents them as being formed, not created here and the *Ruach* of God is used to apply the *Nefesh* making them living Souls.[16] As we have discussed, this is all done to create the unique environment of the Garden. We know from the Endowment that the Physical Creation of plant and animal life on the earth of the Physical sphere was accomplished before the Garden, before the creation of Adam, and before the Garden was planted. A living Soul is comprised of an embodied intelligence (called a Spirit) and a body made of Temporal substance; we will discuss this in detail later in this Chapter.

Animals are first introduced to Adam here as possible help meets. They are brought before him one at a time as if he is being asked: "*Adam, is this a good help meet for you? How about this one? Or maybe this one?*" By bringing the already formed animals to Adam in such a way, God taught Adam. One-by-one He illustrated that the animals were not a perfect fit. In doing this, He ensured Adam would know how perfect Eve would be for him.

The truthfulness of this understanding comes when we fully understand the meaning of the phrase "*help meet*." A proper reading of the Hebrew renders something akin to *a power or strength equal to*, *a helper that is opposite to but corresponding with*, *a perfect pairing*, or even *the second half of the same whole.* However helpful and strong and loyal the animals may have been, they were not any of these things for Adam; Eve, however, would be.

---

[16] See our previous discussion of the *Ruach* and *Nefesh* in, In The Image Of God, beginning on pg. 84.

By instructing Adam as He did, God assured Adam understood each of the animal's strengths and weaknesses in meeting his varied needs. When Eve finally arrived, Adam knew she was perfect and had been made just for him.[17]

Abraham, records a different sequence of the events. This is perhaps true because the Book of Abraham presents a somewhat abbreviated version of the Garden account. The Abraham account does not include the Satan's tempting or the Fall for instance. In Abraham, the Gods bring the animals to Adam; Moses records that the animals obeyed a command of God to present themselves to Adam. In the Abraham account, because Eve is created before the animals are brought to Adam, the account becomes internally confused when it declares, "*there was found an help meet for him,*" seemingly in direct connection with the presentation of the animals:

> *"And out of the ground the Gods formed every beast of the field, and every foul of the air, and brought them unto Adam to see what he would call them; and whatsoever Adam called every living creature, that should be the name thereof. And Adam gave names to all cattle, to the fowl of the air, to every beast of the field; and for Adam, there was found an help meet for him." (Abr. 5:20, 21)*

We may also infer from these verses (Moses 3:19, 20) that the animals did not express their natural instincts; they were neither predator nor prey. Through the presentation of the animals to Adam for his naming and in the declaration that the animals were "*an help meet for him,*" we are informed that the animals were made specifically for him.

> *"For, behold, the beasts of the field and the fowls of the air, and that which cometh of the earth, is ordained for the use of man for food and for raiment, and that he might have in abundance," (D&C 49:19).*

> *"Yea, all things which come of the earth in the season thereof, are made for the benefit and the use of man, both to please the eye and to gladden the heart" (D&C 59:18).*

---

[17] This wonderful instructive detail is only revealed in the written record. There is no trace of it in the Endowment rendition, yet this marvelous facet remains in scripture for us to absorb and learn from as Adam did. The Endowment presents its own story of the Garden providing inspiring and marvelous truths missing from the scriptures themselves. The differences between the two accounts are not contrary, but complementary filling in the gaps where the other story may have missed something. In this case, the Endowment speaks of an Earth already populated with animal life and an Adam already fully aware of Eve. We have seen, however, that his knowledge of, and relationship with, Eve predated their entry into the Garden and its attendant innocence and veil of forgetfulness. The Endowment tells of Adam helping in the creation of the animals spiritually, so he had intimate knowledge of them. The written records of the Garden, however, speak of another time, a bit later in the history of Adam, after he is placed within the confines of the Garden, after he has forgotten all that he once knew. There, as a little child, perhaps actually growing physically and mentally from childhood into young adulthood, he was instructed by his Father and introduced to the animals one-by-one, as recorded here, until finally, when the time was right, he was reintroduced to his perfect companion; this is the fullness of the story. As was true with Eve being a help meet for Adam, the Endowment and the Scripture are *a perfect pairing, equal halves of the same whole.*

The last verse here provides a key phrase that applies directly to the diversity of life that was specifically put in the Garden, and in general, the diversity of life in the Physical sphere; the verse speaks of cattle and fowl and beast. In immortality, the practical needs for diversity (as that attribute of life on Earth efficiently provides for life's ability to adapt, resist disease and survive, and the need of a broad food chain, with both predators and prey and even scavengers) is not necessary. In the Physical sphere, as is also true of the Garden, life in all of its variety and splendor, serves to provide for the joy of God and His sons and daughters; it is to "*please the eye and to gladden the heart.*"

In Hebrew, Adam is most often a common, collective noun, meaning alternately *man*, or *human*. When it is used in its common form, it is used with a Hebrew *hey* (ה), meaning "*the*." In Genesis 2:19, Adam appears for the first time in the Scriptures as a proper noun without a *hey*. It remains a proper noun from this point forward. The next time it reappears with a *hey*, as a collective noun, it will refer to an entirely different class of Earthly being.[18]

If we are to assume our assertion that Adam indeed began his stay in the Garden or its immediate precincts as a little child, the presentation and naming of animals would most probably have occurred through a recurring chain of individual events over the course of his young life. Adam no doubt used his native Adamic language when giving names to the animals, and the naming of the animals would have to have happened after he had learned to speak and at least had rudimentary language skills.

---

[18] See our previous discussion, THE SONS OF GOD AND THE DAUGHTERS OF MEN, on pg. 95.

**The Garden — Moses 3**

21 And I, the Lord God, caused a deep sleep to fall upon Adam;
and he slept, and I took one of his ribs and closed up the flesh
in the stead thereof;
22 And the rib which I, the Lord God, had taken from man,
made I a woman, and brought her unto the man.
23 And Adam said: This I know now is bone of my bones, and
flesh of my flesh; she shall be called Woman, because she was
taken out of man.
24 Therefore shall a man leave his father and his mother, and
shall cleave unto his wife; and they shall be one flesh.
25 And they were both naked, the man and his wife, and were
not ashamed.

## *Two Shall Be One*

It is clear we existed as individual Intelligences before the Physical Creation. Formed separately in the Physical sphere before the Garden, both the man and the woman are individual and distinct, created as independent beings there as well. This independence of being is again affirmed within the Garden through the details of the story of the partaking of the fruit of the knowledge of good and evil.

We understand that not only were the actions of Adam and Eve independent of one another, but so too would have been their fates if Adam had not finally partaken of the forbidden fruit as well. With this independence affirmed, let us put it aside for a moment and examine the incredible lengths these verses have gone to, to describe a singular closeness of this first pair and explore how this archetype was meant to apply to each of us.

-Under His Arm, at His Side, Near His Heart

In forming the immortal, yet Temporal, Eve, literally *from* (in the case of the rib) Adam in the Garden, God ensured a continuity that reveals something about the transition of Adam and Eve from the Physical sphere to the Temporal sphere.

All of us today on Earth have a Temporal body that is a product of two individuals and in fact many more. Each of us contains a genetic history that is traceable to our first parents on Earth, but with Adam and Eve that was not so. Adam was formed by God directly out of the basic blocks of spiritual matter within the Garden in the Physical sphere. Adam was formed from the ground up, literally of the dust; he was not a form that was a product of evolution, not while he was yet in the Physical sphere and not here after he came to the Temporal.[19]

As such, Adam had no mother and father, in the sense we understand to be true here in the Temporal Universe.[20] To answer an age old question – no, Adam did not have a navel. Within him there was no Temporal genetic history to be found. The same was not to be the case for Eve. She was to be intimately bound to Adam in a way that has not been repeated since. Eve was not comprised of a shared gene pool, half from her father and half from her mother. She was *all* from Adam. She was *of him* and he knew it; he knew she *was* him.

It is significant that this intimate physical binding, this genetic accounting was accomplished in the Garden, within the Physical sphere, instead of here in the Temporal. This speaks to an understanding that we have an eternal genetic code as well. Though the binding of Adam and Eve occurred in the Garden, a Temple to the Temporal sphere and as such mirrored the Temporal, it was accomplished to complete a pairing that was already formed, one that has eternal implications.

---

[19] It is our understanding in this work that both Adam and Eve upon their expulsion from the Garden Temple in the Physical sphere, entered the Temporal Universe as they were, fully functioning intelligent, informed, adults. They had knowledge and wisdom at this point beyond any mortal since. We understand their bodies of *Flesh and Bone*, from the Garden encapsulated genetic materials not much unlike our own and as such, not very dissimilar to the structure of other life on Earth. This is to be expected, for the same creator who fashioned Adam from the dust (the perfect materials of physicality) *informed* the *bohu* of this Universe (itself just a subtracted form of Physical matter) with laws that would develop the blueprint of life, DNA. Today our chromosomes have been corrupted and diluted through entanglement with other Hominid karyotypes; this we have discussed in this work was the cause of great concern with God and precipitated the Great Flood and other events which eradicated all forms of Hominid other than Modern Man (the sons and daughters of God). This intermingling is perhaps evidenced in the fact that Humans have 46 chromosomes and the apes have 48. Who knows how this came about or what process this represents? Perhaps Adam's DNA originally assembled to form 24 chromosomes in his perfect karyotype within the Garden (a single set of 12 pairs) and what we see inside most of us today is largely, intrusive material from Temporal additions. It is not the form or numbers of chromosomes that matter but the coded information they carry; for example, despite superficial similarities between ape and human chromosomes, they each contain significant differences at the molecular level. In that aspect, it is the information that makes a difference. It is clear that Man stands above all other life on Earth in a significant way. That significance is the imprint of the original *perfection* of Adam and the *Neshama* within us all.

So how did the tangible, perfect, genetic coding of Adam make it into the Temporal Sphere? First, it must be remembered it did not do so in whole; the *Tzimtzum* of the creative process removed some of its divine nature, adding uncertainty, disorder and decay. What was perfect did not remain so. The actual process of Adam and Eve's entry into the Temporal sphere was a kind of transfiguration, *only in reverse*. Where the Bible and Latter-day scripture describe Temporal things being elevated to a higher plane through transfiguration, we can imagine a process whereby the Holy Ghost "*strengthens*" and "*binds*" the Temporal, healing it, and closing the gaps made through subtraction, allowing the Fallen nature to withstand the full effects of the Physical sphere. Here I propose that Adam and Eve arrived on the Temporal Earth in effect as transfigured beings, and then the presence of God was withdrawn and their nature assumed a Fallen, uncertain state. What would have been witnessed then was the image Peter, James, and John saw on the Mount of Transfiguration when the Christ completed his mission there, on the summit, and returned again to His human nature to complete His ministry (Matt. 17:1-9).

[20] We will explore an alternate understanding of this scenario later in this work. See Hearing the Softest of Voices, Our Heavenly Mother, pg. 449.

*What is not told here in the story of the Garden, is that (at this same time) the rest of the sons and daughters of God had also been embodied as well.*[21] Adam and Eve were paired eternally before the Garden at the same time we all were; the binding that took place therein for them was a completion of what had been started before. When we were embodied, we were created in divinely appointed pairs as well, for we read, "*Nevertheless neither is the man without the woman, neither the woman without the man, in the Lord*" (1 Cor. 11:11). We have eternal families and are bound into the twelve tribes in the Physical sphere. This is of significance and bespeaks a truth about the chain of life and genetic heritage of all of mankind that goes beyond the veil into the Physical sphere. This genetic link remains unbroken today and includes the incarnate Jesus Christ.

***Adam and Eve were paired eternally before the Garden; the binding that took place therein was a completion of what had been started before. When we were embodied, we were created in divinely appointed pairs as well…***

It is also of significance that the first marriage was solemnized in the Garden, a Temple within the Physical sphere foreshadowing Temple marriages here on Earth. This delineates the differences between the sealing of us *one to another* into the family of God and the special binding of Temple marriage between a man and a woman. Though Adam and Eve (and all men and women who have since been joined in this sacred and eternal manner) were separate and independent beings, they are to be one in the eyes of God, one in heart and Soul. This is a truth that cannot be taken literally enough.

-She Shall be Called Woman

After Eve was brought to him Adam said, "*This I know now is bone of my bones and flesh of my flesh.*" The depth of his understanding of this statement is expressed in the carefully chosen words to follow. In them truth is revealed in careful detail.

In the third Chapter of Moses, Verse 24, we read: "*Therefore shall a man leave his father and his mother and shall cleave unto his wife and they shall be one flesh.*" In the parallel Genesis account, the Hebrew word translated as "*one*" here, is *Echad* (אחד). The LORD uses the Hebrew word *Echad* to describe the oneness of Adam

---

[21] *...And I, the Lord God, had created all the children of men; and not yet a man to till the ground; for in heaven created I them; and there was not yet flesh upon the earth…" (Moses 3:5)*

and Eve. *Echad* is descriptive and revelatory in its use here because it characterizes a closeness of two Souls that is indescribably powerful and permanent.[22] *Echad* is used one other place in the Hebrew passages of the Old Testament; 258 pages later in Deuteronomy, Chapter 6, Verse 4, God reveals a truth about His nature and that of His Son. There He proclaims: "*Hear O Israel The LORD our God is **one** LORD.*" Divinely inspired, the word carefully chosen to describe the eternal relationship of the Father and the Son is also *Echad*, the same one used here to describe the bonding between the man and a woman, Adam and Eve, within the Garden. *Echad* describes an inseparable bond that stresses loyalty and promise as in the eternal covenant *YHWH* made with Israel.

*Echad* is the word used in the ancient texts and its meaning remained in place and unquestioned for thousands of years after it was first penned. However, the meaning of the word was changed in the 12th century through association of this verse with the work of Rabbi Maimonides. Rabbi Maimonides developed 13 articles of the Jewish faith to define what Jews believed. In the article describing the Jewish concept of God, rather than use terms very familiar from the verse in Deuteronomy, he borrowed from the concept of the Christian trinity and used the Hebrew *Yachid* (יָחִד) instead. He chose to do so, because *Yachid* was quantitative rather than qualitative and developed a similar context in describing the oneness of God understood by the Christians.[23] Because Maimonides used *Yachid* to describe the oneness of God, the meaning of *Echad* was forever diluted and transformed in theological thought.

Verse 24 employs yet another Hebrew word that bespeaks of intimacy in the strongest of terms; Adam uses cleave (Hebrew *wedavak* וְדָבַק). Cleave comes from the root *davak* (Hebrew, *debek*, דָּבַק) which means to "*cling*" or to "*stick*" and indicates an inseparable closeness (tightness) like that of flesh on bone. Adam understood this relationship was to be closer than any that he had. Adam was to be closer to Eve than to any other, even closer than his relationship with God.

There remains a vestige of this unusual intimacy hidden today within the Hebrew Biblical text that goes unnoticed by most; it is often interpreted away or glossed over in many translations. God calls both male and female Adam: "*Male and female he created them; and blessed them, and called their name Adam, in the day when*

---

[22] The usual translation of Deuteronomy 6:4 "*...which makes a statement about the unity and the indivisibility of God, does not do full justice to this text (though it makes sense in a later Jewish context as a polemic against Christianity). The word Echad makes not a quantitative argument (about the number of deities) but a qualitative one, about the nature of the relationship between God and Israel. Almost certainly the original force of the verse, as the medieval Jewish exegetes (Rashbam and Ibn Ezra) recognized, was to demand that Israel show exclusive loyalty to our God, YHWH—but not to deny the existence of other gods!*" The Jewish Study Bible, Featuring the Jewish Publication Society, (Tanakh Translation, Berlin. Adele and Brettler, Marc Zvi, editors, Oxford University Press 2004, pg. 380)

[23] Maimonides perhaps had the concepts of Wisdom Theology in mind when he chose to describe the oneness of God in quantitative terms. See our subsequent discussion of this topic as a development of the first century Jewish philosopher, Philo of Alexandria; Wisdom is God Himself, pg. 274.

*they were created*" (Gen. 5:2).[24] The proper noun Adam is used here, the one signifying Adam the individual. This occurrence is not the common noun used for the collective, Man; for it is recorded here without a *hey* (ה). As such, we understand that God called both the man and the woman, *the two who stood in front of Him and walked with Him within the Garden*, both of them individually, Adam.[25] God addressed *two* as though they were *one*, for indeed in His eyes they were.

Religionists have debated the personage of God for centuries, Trinitarians, Unitarians, Subordinationists, all have attempted to describe what the use of *Echad* in these verses mean and what the word attempts to convey. *Echad* describes a relationship of oneness that is most visibly understood in the context of the Father and the Son. Some early Christians sought to prove that the Son was equivalent to the Father, identical in every way; they were *Homoousian* from the Greek *Homoousios* (ὁμοούσιος) which meant "*same substance*," or *Consubstantialis* in Latin.

Other, equally versed, early Christians were *Homoiosian* from the Greek *Homoiosios* (ὁμοιούσιος) which meant "*similar substance*." They insisted the Son was merely *like* the Father, Deity but separate and distinct from him. This debate, one that began millennia ago, still rages in some camps today as we wrest to understand how two (and even three) can be one.[26] The debate always overlooks the context demanded here that Adam and Eve were *Echad* as well. If it did not, the debate would have been over centuries ago.

Of extreme significance, among a basket full of significant details, is an omission. Even a cursory reading will reveal that there is no breathing *Ruach* of the *Neshama* into Eve. No need to check; this is also omitted in the Endowment. God does not breathe the *Neshama* into Eve. What is within her came from Adam; it was imputed.

Imputation is the assigning of an attribute to someone by virtue of a similar quality of another. Through the Atonement, Christ's righteousness is imputed to us. In relation to Adam and Eve, imputation is usually mentioned in the doctrine of original sin. That doctrine states that because of the Fall, the guilt of our first parents is imputed to us, thereby bringing each of us under condemnation as well. Though the doctrine of original sin is a human abstract, these verses, through their

---

[24] *See also Moses 6:9.*

[25] It must be remembered for its significance that it was first Adam who called his companion Eve.

[26] This long debate is over for me. In tragedy, I have been blessed. I have an understanding of just how two separate persons can be wholly and completely one at the same time. I completely understand in a way no one can understand until they have been through it. By the time I lost my wife of thirty years to cancer, I had learned many things that have made it into this work, but none of the things I learned were more impactful or more important than what I learned in her passing. When my wife passed away, I died with her. Without a doubt, Lynda died far too young, but her dreams and Spirit indwell me with a determination to accomplish all that she hoped for me, to be all that she wanted me to be. There is more of *her* left in me than there is *me* left in me. Mark, the person I was, what I lived for, my dreams, my hopes, my desires, all died when *she* drew her last breath. Who I truly was, died with my Soul mate. Because of Lynda, I know what it is like to be one with another, what it is like to be separate and distinct and independent, but to be less than whole without the other.

omission of any mention of the *Ruach* and *Neshama* as they relate to Eve, leave the imputation of the Spirit from Adam to Eve as our only remedy.

This imputation is not exclusive to Eve (she like each of us) had within her, her own unique Spirit. The *Ruach* is the sole imputed element, and the same is true with each of us. Through an unbroken chain, starting with our first mother Eve, the *Neshama* has been passed along generation to generation. Eve's creation *from* Adam eliminated the requirement of the *Ruach*, but that creation, from the materials of Adam, also ensured Eve was to be the epitome of the creations of God.

- Eve is the Crowning Glory of Creation

Latter-day teaching expands upon the role of the female in God's plan. Though she may be derived from the man, she is in no way inferior to him. Eve is the capstone of the creation of God in the Garden and elsewhere. This has been an unwritten, yet fully expressed, doctrine of the Church. Elder Jeffery R. Holland articulates this consideration as follows:

> *"We believe that the creation of a woman was the crowning and final and most glorified moment of human creation. That we start with light and dark and land and sea, we move through fish and fowl and beasts of the field and we get to Adam and it's still not good enough. And only when Eve was created, this is our theology ... that the crowning creation and the glory of the human experience came with the creation of Eve. ...Certainly doctrinally, theologically... I would put the position of a woman in the Church of Jesus Christ of Latter Day Saints up against the dignity and worth and merit and wonder of a woman anywhere on this planet." (Elder Jeffery R. Holland, address to Students and Faculty at Harvard University, 2012)*

We have broached the fact that Eve's creation out of Adam in the Garden is reflective of an eternal pairing from the outset, giving a truer, deeper, meaning to the all-too-casually used phrase *Soul-mate*. What we have not yet emphasized is that the binding of Intelligences, Spirits, and eventually Souls, is more complex than is commonly thought. It is a binding that goes much deeper than marriage and even sealing. This eternal pairing is something beyond the comprehension of the world today and even our theology.

These verses describing the making of Eve in the Garden disclose the intent of God to bind this first pair in an intimate way. They are a reflection of things done before the Garden was created and they provide a prototype of the relationships God intends all of us to build here in Temporality with our eternal companions.

As we noted here, theologians and scholars have debated the oneness of the Godhead for thousands of years, seeking to harmonize the concepts of *same*, and *similar*, *individual*, and *oneness*. We find that in these verses, the model in its fullness existed all along. This is truly *Homoousian*, for Eve is of the substance of Adam in every way, but she remains in possession of a Spirit of her own. She is independent and agent, yet still united with Adam in such a way that while being fully Eve, God can properly call her, and both of them together –Adam. This is reflective of the relationship of the Father and the Son, both separate and distinct, yet each truly and singularly God.

Through Adam and Eve we learn who we are and what we may become. They are the image of what we aspire to be and the example of how we should act in all things, when we are under assault from the adversary and when we are both seeking and in the presence of God. Through all they experienced, they acted with one heart and one mind. They would face the new world of the Garden and challenges that lay ahead – *together*.

"*And they were both naked, the man and his wife, and were not ashamed.*"

**The Garden — Moses 4**

1 And I, the Lord God, spake unto Moses, saying: That Satan, whom thou hast commanded in the name of mine Only Begotten, is the same which was from the beginning, and he came before me, saying—Behold, here am I, send me, I will be thy son, and I will redeem all mankind, that one soul shall not be lost, and surely I will do it; wherefore give me thine honor.
2 But, behold, my Beloved Son, which was my Beloved and Chosen from the beginning, said unto me—Father, thy will be done, and the glory be thine forever.
3 Wherefore, because that Satan rebelled against me, and sought to destroy the agency of man, which I, the Lord God, had given him, and also, that I should give unto him mine own power; by the power of mine Only Begotten, I caused that he should be cast down;
4 And he became Satan, yea, even the devil, the father of all lies, to deceive and to blind men, and to lead them captive at his will, even as many as would not hearken unto my voice

## *Satan, the Father of All Lies*

The story of the Fall is told in Moses, Chapter 4 and Genesis, Chapter 3. It is recounted in the Endowment beginning with Lucifer's approach to Adam immediately following our symbolic entry into the Garden.

The first four verses of Moses, Chapter 4 record an introduction of sorts of Satan to Moses. Moses is told who Satan is. In fact, the verses are more of a backstory than an introduction, for Moses had already had a personal encounter with him.[27] God informs Moses of the ancient rebellion and its cause. What is not included here, though we may assume it was covered, is an explanation of the Plan of Happiness.

It is telling that the LORD God uses this information to lead directly into the discussion of the serpent beguiling Eve. By it, we are led to believe that Satan and the serpent have at least a cursory relationship in the story. Indeed, in the very next verse, the Moses translation provides detail of this relationship not found in Genesis today. Through the Moses translation, we are left without doubt that though the temptation in the Garden comes from the *serpent*, it is one that is called Satan who is behind it.

---

[27] Moses 1:12-24. It is interesting to note though Moses had this previous experience with the Adversary, he still required God to inform him of who Satan is at this later time. Consider this in relation to the presence *felt* by Joseph Smith in the Grove prior to the First Vision and how that relates to the evolution of the story told by the Prophet of the first encounter. Joseph, like Moses here, learned through consecutive revelation, which only informed of the whole picture over time.

In the Endowment, it is one named Lucifer who comes to lead both Adam and Eve astray. We have mentioned this elsewhere,[28] but it is worth repeating here, because it is a detail that is at once so familiar that it may not be realized; it is only in the Endowment account that we find Lucifer approaching Adam before he approaches Eve.

I have switched to using Lucifer instead of Satan here, and that was intentional, for in the Endowment Lucifer is his name.

-What Is In A Name?

Our KJV renders the Hebrew *Heylel* or *Hevlel* (הילל) as *Lucifer*. This proper name appears only two times in the Standard Works of the church, once in the Bible; Isaiah 14:12 reads: "*How art thou fallen from heaven, O Lucifer, son of the morning! how art thou cut down to the ground, which didst weaken the nations!*" and once in the Doctrine and Covenants which reads similarly, equating the name *Lucifer* with yet another, *Perdition*, "*And was called Perdition, for the heavens wept over him—he was Lucifer, a son of the morning!*" (D&C 76:26). *Hevlel* most literally means, "*Shining One, or light-bearer,*" for in the Isaiah verse above, the complete Hebrew passage reads; *Heylel ben Shachar* (הֵילֵל בֶּן־שָׁחַר), "*Shining One, son of the morning*"

The name *Hevlel* (Shining One) became *Lucifer* through a curious two-step process. Interestingly, the Hebrew was first rendered into Greek, then the Greek into Latin as follows:

In the third century BC, the translators of the Greek Septuagint rendered *Hevlel* as ἑωσφόρος (*heōsphoros*), a proper name, compounded from two words, literally "*bringer of dawn.*" The seventy translators purposefully referenced the *morning star*, for it shines brightly in the sky in the morning, but of a lesser greatness than that of the Sun (symbolic of a son to a father) (*perhaps it was here in the Septuagint that the name Hevlel was for the first time associated with the bright "star" that appears on the horizon before sunrise*). In the Latin speaking world, Roman astronomers called that same morning star "*Lucifer,*" which comes through a contraction of the two word phrase *lucem ferre* (bringer, or bearer, of light).

When Jerome composed his Latin Vulgate[29] text late in the fourth century AD (almost 700 years after the Greek Septuagint translation was completed), he used

---

[28] Lucifer approached Adam first in the Garden before he approached Eve (see Neither Shall Ye Touch It, pg. 251 herein). Ancient Hebrew Midrash contain commentary that suggests that the serpent first attempts to deceive Adam, but beyond these obscure references, no other sources record this detail included in the Endowment narrative.

[29] Saint Jerome, Eusebius Heironymous Sophronius was born 345AD in northeast Italy. He studied for years as an aesthetic monk in a Syrian Desert monastery before becoming the secretary to Pope Damasus in 382. By that time, he was widely considered the greatest Christian Scholar in the world. He rendered the four gospels into the ascendant Christian language of Latin, and this Gospel set is the one that sits upon the Alter at Rome in St. Peters Basilica. The common language of the church in Jerome's day was

the logic of this process to render the name *Hevlel* as *Lucifer*[30,31] in the one single place in the entire Biblical text that it appears. Today we know this heavenly light the "*star*" that brings the dawn, by another Roman name, we call it the Planet Venus.

Unlike the complex development of the word *Lucifer*, in which the Hebrew word was replaced first by a Greek, then a Latin substitute, *satan* in use today, is direct from the Hebrew without any change, and it is as phonetically identical to the original as it can be. We say *satan* today exactly as it was pronounced in Hebrew three thousand years ago.

The Hebrew word *satan* (שָׂטָן), or *ha-satan* (הַשָּׂטָן) (a satan) occurs only nine times in the entire Hebrew Bible, our Old Testament. *Satan* is Hebrew for "*adversary*" or "*accuser*." Except for one location, where it appears as a proper noun without the *hey* (הַ) preceding it, *satan* is used as a common noun describing the subject by its function or task, in much the same way we use *Police Officer or Lawyer* today. Five of its nine appearances in the text describe a Temporal human individual or

---

Greek and the Bible, both the Old and New Testaments were widely available in Greek and anyone who knew Greek had access to the Scriptures, but the church was shifting demographically from the East to the Latin speaking West. The Western population knew no Greek and many early translations in various western languages were in use. Damasus proposed that Jerome produce an authoritative Latin translation of the entire Bible as he had previously done with the Gospels. In response, Jerome penned a letter explaining his frustration with finding a suitable base text from which to translate, he wrote: "*If we are to pin our faith to the Latin texts, it is for our opponents to tell us which; for there are almost as many forms of the text as there are copies.*" To illuminate the rampant inaccuracies Jerome proposed: "*...why not go back to the original language and correct the mistakes introduced by inaccurate translators, and the blundering alterations of confident but ignorant critics, and, further, all that has been inserted or changed by copyists more asleep than awake?*" Damasus agreed and Jerome set about to learn Hebrew so he could translate the Old Testament directly from that language into Latin, and he sought amongst the hundreds of versions of the Greek New Testament those he considered to be most original. The dialogue between Damasus and Jerome highlight a reality that is much different than what is believed and taught by those that preach "*sola scriptura*," espousing a doctrine that insists that the written word of God is His singular, divinely, appointed form of communication to man; that the scripture is inerrant because of some supernal process. Such would have us believe that the "*word*" of God, He who was described by John as being "*in the beginning*" is fully manifested today by ink on paper in the characters of Hebrew and Greek. To them just ask: How this manifestation is in anyway eternal, perfect or unchanging?

[30] As Christian tradition developed, it came to use the Latin "*Lucifer*" the name of the "*morning star*" as a proper name for the devil; more specifically, the name he had before he fell from grace, Hevlel (Shining One) as it was rendered in Hebrew.

[31] Most recent translations have abandoned the use of "*Lucifer*" for *Hevlel* (הֵילֵל) in Isaiah 14:12 instead rendering it: "*shining star*" (New Living Translation), "*Day Star*" (New Revised Standard Version) "*morning star*" (New American Standard Bible, New International Version, Good News Translation, Common English Bible, New Century Version, Contemporary English Version, Complete Jewish Bible) "*daystar*" (English Standard Version, New Jerusalem Bible), or "*Shining One*" (JPS Tanakh). Note here, for it bears significance to the use of the name Lucifer in the Book of Mormon, that *all* of these translations with exception of the Jewish Tanakh, draw on the astronomical association of the name with the planet Venus.

The name Lucifer appears in the Book of Mormon in 2 Nephi 24:12; this is a re-recording of the same Isaiah verse (Isaiah 14:12). That Lucifer appears here in the Book of Mormon can only be understood in the context that Joseph Smith rendered the name Lucifer in 1820, because it most properly represented in the familiar terms of the language of our day, what the original scribing of Mormon meant to convey.

organization, either a political, legal or military adversary. Of the eight times *ha-satan* (הַשָּׂטָן) is used in the text, only three of those occurrences describe a being from beyond the Temporal sphere. In at least one of those instances, the *satan* (adversary) is an angel of the LORD "*¶And God's anger was kindled because he went: and the angel of the LORD stood in the way for an adversary against him. ...*" (Num. 22:22). Here the Satan *ha-satan* is an angel who stands in the way of Balaam, turning him from his perverse way.

The word *satan* (שָׂטָן) appears only once without *hey* (הַ) ("*the*") in front of it in the entire Hebrew Bible. 1 Chronicles 21:1 reads, "*And Satan stood up against Israel and provoked David to number Israel.*" Here *satan* is in an adversarial role against Israel and imaginably this may indeed be Satan, the individual, whom Moses had commanded. It, however, cannot be known for certain, because the Hebrew language does not use indefinite articles, or uppercase letters. *Ha-satan* appears in the story of Job, and because of our Christian mind-set, we ascribe this common noun meaning *an adversary*, to the *Satan* of our modern Christian understanding. However, there is nothing directly in the text that would indicate that in fact this is the case.[32] Indeed it is markedly true that the noun *satan* (proper or common) does not appear in the book of Genesis at all, and neither does *Hevlel*, the *Shining One*.

Perhaps it was the turmoil of a rapidly changing world that caused the rise of beliefs in demons, angels, a final apocalyptic battle, and the hoped for arrival of the Messiah, but before long, these things began to change the theological landscape of the Jewish world. In the two centuries before the birth of the Savior, the concept of the devil began to appear in Jewish texts and thought for the first time. In the apocryphal book 1 Enoch, Eve was "*led astray,*" not by the serpent of Genesis 3 but by the "*angel*" Gadreel, the one who "*showed the weapons of death to the children of men.*" Gadreel was not directly adversarial to man, but he was to God. Even so, he is not described as *Ha-satan*. The earliest known mention of the *devil* as the serpent in the Garden is found in yet another apocryphal book from the same period known as the Wisdom of Solomon. This book tells us that, "*through the devil's envy death entered the world, and those who are on his side suffer it.*"

As the Christian era approached, the use of Satan as the proper name of the devil became fixed, and he is not only identified by name but he is envisioned as a

---

[32] For an alternative view of the accuser in Job, consider the following: "*In Job, 'the satan' is a member of God's heavenly council—one of the divine beings, whose role in Job's story is to be an 'accuser,' a status acquired by people in ancient Israel and Mesopotamia for the purposes of particular legal proceedings. In Job's case, what's on trial is God's assertion that Job is completely 'blameless and upright' vs. the satan's contention that Job only behaves himself because God has rewarded him. God argues that Job is rewarded because he is good, and not good because he is rewarded. The satan challenges God to a wager that if everything is taken away from poor Job, he won't be so good anymore, and God accepts. Though a perception of 'the satan' as Satan would make this portrait of God easier to swallow, the story demonstrates otherwise; like Yahweh's messenger in Numbers 22, this satan acts on YHWH's instructions (and as a result of God's braggadocio) and is not an independent force of evil.*" How the Serpent Became Satan, Shawna Dolansky, Bible History Daily, Biblical Archeology Society, April 8, 2016.

supernal force in direct, active, opposition to the will and purpose of God. This was an evolution of thought; it represented a significant change and it did not happen overnight. Among the scrolls found at Qumran, in the cliffs near the Dead Sea (many composed as early as 200 BC), are writings that tell of Satan threatening humanity by leading the forces of darkness. The people of Qumran hoped for a coming salvation that would conquer Satan and evil. As Christ began his ministry, one now known by name as Satan, was considered by most of whom the Savior would meet, to be the one who controlled a world of demons and the leader of the forces of evil.

Despite what Moses knew (what we know from the restoration of his writings by Joseph Smith), *Satan*, as the proper name of *he* who was known as the *Shining One*, did not remain in the Genesis account and was evidently lost very early on for it not to have influenced the remaining books of the Hebrew Bible.

It is not certain how Nephi, the Son of Lehi, had the very Christian understanding of Satan that he did as early as 600 BC, but it can be surmised that this understanding came along with the sweeping visions he had been given.[33] Beside the three references to Satan by Nephi, the word Satan is not used by any other Book of Mormon writer in the Nephite record until Alma refers to Satan by name again about 80 BC. Alma's use and understanding of the word *satan* is concurrent with its emerging use in the Old World as well. In the century before Christ's mortal ministry and those that would follow, the name Satan and who he is, begins to take on the same meaning to readers and hearers of the Word of God on both Earthly hemispheres.

As the first roots of the Christian church took hold in the late first century, the concept of *satan* fully formed into the one the believers in Christ commonly share today. The developing canon of the New Testament depicted an individual being by that name. *Satan* is the ruler of the kingdom of darkness, the opponent and tempter of Jesus (Mark 1:13). *Satan* is *Beelzebub* the chief of the devils, an opposing force to God (Luke 11:15–19; Matthew 12:24–27; Mark 3:22–26). *Satan* and his legion of demons have power to enter and possess people (Luke 22:3; Mark 5:2–13; Luke 8:2).

Why have we gone into such detail here? We have looked at it intentionally not only to highlight a century's long process, but to completely circumscribe it for a purpose.

---

[33] Refer to the visions incorporating, 1 Nephi 13:29 and 2 Nephi 30:18. See also 1 Nephi 22:26 for an additional reference to Satan by name made by Nephi. Also of consideration, it must be remembered the plates as received by Joseph Smith were actually the work of Mormon and Moroni, both of the 4th and 5th centuries AD. As such, their understandings of what they transcribed were highly influenced by Christ's visit to the new world. This is an understanding that explains the presence of very many Christian interpretations being present in the recorded teachings of others who themselves lived and died before the Christian era commenced. This process is akin to one repeated commonly today, when Christians "*see*" Christ *everywhere* spoken of in the writings of the Old Testament.

The name we use today, find in the Scriptures, and hear in the Temple, has little relationship at all to the truthfulness of the word of God. The one who fell from such heights was the "*Shining One*," a "*son of the morning*." From his name and his inclusion among those called a "*son of the morning*," real truth can be gleaned.

There is nothing at all to be understood from any perceived difference of timing or locality in the use of Satan over Lucifer besides the meanings of those words we have discussed here. This individual is an accuser, an adversary, and he *was* the Shining One of those that were sons of the morning. Lucifer was not his name in the pre-existence, within the Garden or anywhere else, that name is entirely of *human construct*; his name was not changed to Satan because of his Fall. Satan did not begin as a proper noun, though it came to be understood as one. *Satan* was meant to describe what this individual does, has done and will forever do, accuse and oppose our Father in Heaven. When the Savior addressed the adversary as Satan, and when the inspired Book of Mormon prophets used that name centuries before it was used in the Old World in the same way, they did so to identify the *intent* of the evil one; it was not to be understood as his name. Having said all of this, we will out of respect for long-standing and well-intentioned, if not uninformed, tradition; continue to use both of these names to refer to the one commanded in the name of the LORD by Moses. As long as we agree to understand what we have discussed here.

Isaiah identifies the "*Shining One*" (Lucifer) as he who sought to be "*like the most High*," but was "*brought down to hell.*"

> "*How art thou fallen from heaven, O Lucifer, son of the morning! how art thou cut down to the ground, which didst weaken the nations! For thou hast said in thine heart, I will ascend into heaven, I will exalt my throne above the stars of God: I will sit also upon the mount of the congregation, in the sides of the north: I will ascend above the heights of the clouds; I will be like the most High. Yet thou shalt be brought down to hell, to the sides of the pit. They that see thee shall narrowly look upon thee, and consider thee, saying, Is this the man that made the earth to tremble, that did shake kingdoms; That made the world as a wilderness, and destroyed the cities thereof; that opened not the house of his prisoners?" (Isaiah 14:12-17)*

The first four verses of Moses, Chapter 4, reveal that the one "*cut down to the ground,*" the "*Shining One*" (Lucifer), is also known to God as Satan and the devil. The Moses translation ties all three of these names into one:

> "*That Satan, ... is the same which was from the beginning, and he came before me, saying—Behold, here am I, send me, I will be thy son, and I will redeem all mankind, that one soul shall not be lost, and surely I will do it; wherefore give me thine honor.* [the Shining One , Lucifer]... *because that Satan rebelled against me, ... I caused that he should be cast down; ... And he became Satan, yea, even the devil, the father of all lies, to deceive and to blind men, and to lead them captive at his will..." (Moses 4:1-4)*

Note that Verse three uses Satan as the common noun; he is identified here as *an* adversary not yet *the* adversary. In Verse 4, Satan and the devil became his proper name as the eternal consequences of his initial rebellion and pride transformed him from a being who opposed God, into one that leads an active and eternal opposition.

We have noted here that the Hebrew meaning of *satan* includes *accuser*. We already addressed this important attribute of the name and how Satan earned the appellation. In our extended discussion of the War in Heaven, we said that in the Grand Council, Satan's argument included an accusation that the Father and Son were taking unnecessary risks at our expense for the attainment of their own glory.[34] Satan stood before us accusing God, trying to convince us that his plan was best for us, in that he perhaps first began to deceive. He took to his role as the deceiver well as he next appears in the Garden deceiving both Adam and his wife, Eve.

| **The Garden — Moses 4** |
|---|
| 5 And now the serpent was more subtle than any beast of the field which I, the Lord God, had made.<br>6 And Satan put it into the heart of the serpent (for he had drawn away many after him,) and he sought also to beguile Eve, for he knew not the mind of God, wherefore he sought to destroy the world. |

-The Serpent

Though Moses was fully aware of who Satan is and what his destructive mission entailed, the prophet's narrative in the Moses translation repeats what is recorded in Genesis, that it is the *Serpent* that beguiles Eve in that Garden, providing a direct and undeniable witness that the Genesis account of the serpent in the Garden is correct and unaltered from the original. The Moses translation, however, declares, "*Satan put it into the heart of the serpent.*" So, knowing that the Endowment account portrays Lucifer, the *Shining One,* as *the* Satan (the Accuser) who tempted and coerced, deceived, and beguiled Eve and Adam within the Garden, we must find a harmony with the written accounts that insist on portraying him as a serpent.

We can find some clues to this mystery of sorts through the realization that the LORD declared, "*he became Satan, yea, even the devil, the father of all lies, to deceive and to blind men….*" In his role of liar and deceiver, Satan entered the Garden with the intent to mislead. We have record of him doing that on another occasion as well; after Adam and Eve were in place in the Temporal sphere, Satan approached their young children displaying his familiar talent of misleading by twisting the truth. Of his activity then, we read:

---

[34] See Send Me beginning on pg. 204 of this work. Also footnotes 39 and 40 on pg. 206.

> "*And Satan came among them, saying: I am also a son of God; and he commanded them saying: Believe it not;* (The gospel of redemption and eternal life as taught to them by Adam and Eve) *and they believed it not, and they loved Satan more than God.*" *(Moses 5:13)*

The Endowment account tells of this same type of half-truth deception as Satan first approached Eve in the Garden. Eve asks, "*who are you?*" and he responds "*I am your brother.*" Though correct in fact, his statement meant to misrepresent that he was the *Son of God*, in much the same way he would later falsify his credentials to Eve's children in the Temporal world. What does this have to do with the serpent?

We have some record of a pattern in the Scriptures wherein a serpent is used as a symbol, a representation of Jehovah the Son of God. While the Israelites wandered in the Sinai desert on their generation long transit to the Promised Land, their faith waivered and the LORD cursed them with a plague of poisonous serpents. The prophet Moses interceded in their behalf. In reply to Moses' intercession and to save the unfaithful, God instructed Moses to: "*...Make thee a fiery serpent, and set it upon a pole: and it shall come to pass, that every one that is bitten, when he looketh upon it, shall live*"[35] (Numbers 21:8).

Figure 7-2 *The Brazen Serpent* by P. Ball. From "Royal Academy Prize Designs," *Illustrated London News* (January 1848)

Moses did what the LORD commanded and, "*Moses made a serpent of brass, and put it upon a pole, and it came to pass, that if the serpent had bitten any man, when he beheld the serpent of brass, he lived*" (Numbers 21:9) (*Figure 7-2*).

There can be no doubt that this "*savior*" image (that of a serpent raised high on a pole) is a "*type*" of the Christ. Those afflicted with certain death because they had been bitten by a *serpent*, would be saved by and through another *serpent* raised high on a wooden pole. Those repentant of their lack of faith and needing salvation could

---

[35] The *action* that was required by those who sought salvation from death that came with the serpents was merely to *look*. This has critical meaning when applied to an understanding of the attributes of Satan; "*And he became Satan, yea, even the devil, the father of all lies, to deceive and to blind men, and to lead them captive at his will...*" (Moses 4: 4). The remedy for the blindness of Satan is simply to *look* upon the Son; to prevent men from doing so, Satan seeks to blind men.

just look to the icon (The Christ) with faith and live. Indeed this typology is confirmed by none other than the Son of God himself; Christ proclaimed, "¶*And as Moses lifted up the serpent in the wilderness, even so must the Son of man be lifted up*" (John 3:14). The Savior taught His followers that the brazen serpent of Moses' day was a divinely authenticated type of His Crucifixion that was to follow.

The literary parallel here is overwhelming: In the first instance the serpents that had stricken the people represented Satan and the serpent in the Garden; in the later, it is shown that the saving brazen serpent was a representation of the raised Son of God who through the Atonement would undo the working of the serpent in the Garden.[36] The Endowment account details that Satan misrepresented while he was misleading. Though truthfully proclaiming, "*I am your brother*," there can be no doubt that Satan was attempting a, not so veiled, reference to Jehovah our oldest brother.

I propose here that the representation of the serpent in the Garden story is symbolic of Lucifer's attempt to represent himself as the Son of God. It is a metaphor that is reflective of a great chain of parallels stemming back from the Grand Council where Lucifer said to the Father, "*I will be thy Son*" and perhaps even before. The first lethal strike from that *false son of God, the pretender to the throne*, did not originally happen in the sands of Sinai. The fangs did not first puncture flesh in the Garden, but the initial fatal blows hit their mark when the seeds of the rebellion were sown, eons ago, before the sons and daughters of God were embodied when the Physical Sphere was created. From that instant, and maybe before, our all-knowing Father in Heaven had positioned our LORD Jesus Christ, the true Only-Begotten of the Father, to be the true Savior of man lifted high on the cross providing an infinite Atonement for all who would just "*looketh upon*" him.

The preceding presented the story through literary parallel and symbolism. However, there is more to be gleaned of the truth contained in the Garden story by studying the etymology of the Hebrew and looking at the complex history of the various recordings.

The Hebrew word for serpent is, *nachash* (נָחָשׁ). The KJV renders it as serpent or snake, but it has other meanings as well. *Nachash* is a root word that forms the basis for three other words, one a noun, one a verb, and one an adjective.

As a noun, in its common form, *nachash* means snake, for reasons we will discuss in a moment. The verb form of *nachash* displays its action, and means deceiver or

---

[36] It is necessary to note here an additional pairing of a serpent image with that of the Savior. Though it is certainly unknown how the Mayan mythology of Quetzalcoatl came to exist, it may have developed as a Lamanite adulteration of the Hebrew stories told here. The god Quetzalcoatl is depicted in his temple as, and is called, the feathered serpent, or the precious twin. (He has an evil brother.) He is the god of intelligence and self-reflection (the reason within us, equivalent to the Greek logos) he is the patron of the priesthood. Quetzalcoatl is a primordial god of creation, the giver of life. He is the god of the corn providing the sustenance (the bread) of life. He created the world with his *opposite (as in Adam 'Michael' all men die, so even in Christ shall all be made alive)*. He was born of a virgin and is associated with the color white. He created our current world by using his own blood. Quetzalcoatl is the "*Lord of the East*" and the "*Lord of the Star of the Dawn.*"

diviner, as in one who falsely claims to reveal divine knowledge. As an adjective, *nachash* is the attribute of shining, as in the bright shine of bronze or a brazen object. Sharing the same Semitic tongue with Israel, Chaldee uses the same word *nachash* to mean copper or brass. When used as a descriptor, something that is *nachash*, is in fact a "*Shining One*," or something that shines.[37]

Throughout the third chapter of Genesis, *nachash* is used as an adjective in all instances except one, where its usage is not clear (Genesis 3:4). In all but this one instance in Genesis, Chapter 3, *nachash* should be rendered as "*Shining One*."[38] This is revelatory considering what we learned of the name Lucifer above.

In etymology it can be difficult to discern what came first, the chicken or the egg; this is never truer than in this case before us. A snake came to be called *nachash* (נָחָשׁ) because its scales glisten and shine in the sun, its movements are deceptive, and it appears cunning in the way it stalks its prey, never in the open, always lying in wait to kill.

Did these attributes of the snake *inform* the other meanings of the word, or was the serpent given that name because its attributes matched a previously extant meaning. The latter seems to be the most common practice in use today; we name our children and locales based on their attributes, both apparent and hoped for. A daughter is named Hope or Charity because of how she makes us feel or because of what we dream she may become; it is rare when it works the other way around.

Consider the following chain of circumstance.

Moses is told that a great *satan* (an adversary, accuser) rebelled against God, seeking to destroy the agency of man. That *satan* was cast down and became the *devil* by name. Next the LORD God dictates the story of the Fall to Moses and the prophet records that it is a *Shining One* (son of the morning) who deceives Eve and that it was the *spirit of the adversary* (*satan*) that was at the heart of this *Shining One*.

Over a thousand years pass and the *Shining One* becomes a snake in the common mythology which focused mostly on the attributes of the deceiver and most specifically on the curse pronounced in the story. Isaiah, in vision sees the *Shining*

---

37 Another confirmation that *nachash* was considered to be a supernal *brilliant* being by Moses is buried in the language of the brazen serpent on the pole of Numbers, Chapter 29. When Moses plead for relief of the suffering Israelites in the desert, the LORD God instructed him to "*make thee a fiery serpent*," using the Hebrew word *Serap* (שָׂרָף). These are the same heavenly beings seen attending to the throne of God in Isaiah 6:2, the *Seraphim* (שְׂרָפִים) and kin to those spoken of as guarding the tree of life in the Garden, the *Cherubim* (הַכְּרֻבִים). In response to this command, and with the acceptance of the LORD God, Moses fully complied with the request to make a *serap* by constructing a brazen serpent, a *nachash* (נָחָשׁ).

38 This is how it is rendered in the International Standard Version of the Bible; e.g. "*Now the Shining One was more clever than any animal of the field.*" (Gen. 3:1); "*You certainly will not die!*" the Shining One told the woman." (Gen. 3:4).

*One* (son of the morning) fall from heaven and records that name directly again, for the first time in over 500 years. Isaiah's rendering remains there in his 14th chapter today, and a copy of that same chapter went with Nephi to the new world on the Brass Plates of Laban.

With the approach of the advent of the Savior, the word *satan* started to take on a religious meaning, that of the leader of the dark forces, for the first time. This is not surprising because it was never used as a proper noun at all; remember *satan* was the *reason and action* associated with the rebellion of the *Shining One*,[39] and the verses of Moses that mentioned the history of the *Shining One* did not remain in the Hebrew Bible. In both the Old and the New Worlds, *satan* now became the focus of attention, as this *Adversary* stepped up his efforts to foil the mission of Christ. The Son of God rightfully declared the Adversary (*Satan*) to be the bane of mankind. This is true because it is not only the work of the one we now know here as the *Shining One* that keeps us from returning to God, but it is his spirit (one of rebellion) within us and our own individual, adversarial, Spirits that accomplish that act.

Through a twist of fate, the last vestige of the name (הֵילֵל בֶּן־שָׁחַר) *Hevlel ben Shachar,* the *Shining One*, found in Isaiah, was translated first to Greek then to Latin where, without reason, its chain of translation serendipitously stops. *Shining One* first became day star, then bringer of the dawn, then light-bearer (*Lucem ferre*), then Lucifer.

Without any knowledge of this millennia long process, a young uneducated man in upstate New York, barely 20 years old, began to translate golden plates eventually developing not only a church, but a religion. Before him he saw foreign words that brought understanding to his mind. Without any training, a Hebrew word *Hevlel* (הֵילֵל), carried by Nephi to the New World and rewritten into yet another language by Mormon a thousand years later was rendered by Joseph (after yet another 1400 years) as Lucifer in just one spot (2 Nephi 24:12) the exact same verse recording the text of Isaiah 14:12. All of the other references to the deceiver and adversary in the Book of Mormon are rendered appropriately as *satan*.

A little more time would pass before Joseph, through inspiration, would restore the lost passages of Moses, bringing us full circle back to where this chain of circumstance began. That restoration would include the word *satan* used in its proper form as an adjective, though by now Joseph Smith was, *himself*, using *Satan* as a noun, like the rest of the world was. That restoration also restored the proper

[39] Ezra Taft Benson reminds us, "*In the primordial council, it was pride that felled Lucifer, 'a son of the morning.' ...Most of us think of pride as self-centeredness, conceit, boastfulness, arrogance, or haughtiness. All of these are elements of the sin, but the heart, or core, is still missing. The central feature of pride is enmity—enmity toward God and enmity toward our fellowmen ...Pride is essentially competitive in nature. We pit our will against God's.*" Beware of Pride (Address given at the April 1989 General Conference of the Church of Jesus Christ of Latter-Day Saints). President Benson here describes the adversary (the adversarial spirit) which drove the rebellious actions of Lucifer to be, pride and enmity.

noun *Devil* to the verse narrative, but *serpent* was left in place within the story, until the Endowment. In the Endowment, the *Shining One* as *Lucifer* finally returned to his place in the Garden. It must be noted, the allegory apparently is too deeply rooted to be entirely let go; for in the Endowment, Eve, when pleading her case before God (after having seen Lucifer for who he is, and calling him by his proper name) *still* says it was the *serpent* who beguiled her.

What a remarkable and epic turn of events. It testifies of the hand of God at work among men who have little knowledge of their impact in this incredible history.

-Returning to the Story

So we have the *Shining One* (*Lucifer*) in the Garden with Adam and Eve. With the understanding we have developed here, we see that this is confirmed in the written accounts and in the Endowment. Lucifer was "*cast down*" and is Fallen from his previous glory. It is in his Fallen state that he approaches Eve; Eve's observations and her responses to his questions are framed from this point of view. Lucifer presents himself to Eve as her brother and can do so because the glory that made him the *Shining One* is gone. It is only after Eve *joins* Lucifer in his Fallen state that she is able to recognize him for who he truly is.

Verse 6 tells us, *"he had drawn away many after him."* This phrase alludes to the fact that Moses was aware of the details of the War in Heaven and of the third of the hosts (a part of them actually, not truly 33%) that had followed Satan in the rebellion he was told of back in Verse 3.

The Scripture here in Moses records that "*Satan knew not the mind of God.*" The account given in Genesis as well as the Endowment narrative, are completely silent on this detail. How is it possible that Lucifer, *a son of the morning*, would not have known of the mind of God as it pertained to the Fall? Mustn't he have had full knowledge of the Great Plan when he stood in the Grand Council in opposition to it? Doesn't our understanding of agency and informed choice demand that? Of course it does; so what does the phrase "*he knew not the mind of God*" mean to us here?

What were Satan's intentions as he first sought to beguile Adam and then Eve? How did the *Shining One* (who must have been a most vocal advocate of his position in the rebellion and possessed with such intelligence that he certainly understood full well God's intent with the Fall) come to be an unwitting participant in bringing about the Plan of Salvation and the entry of Man into Temporal mortality?

What was Lucifer doing in the Garden? Conundrum seems to be a word specifically developed for this single question. For it truly is a mystery as to why Satan would act in any way to further along the Plan of God. We will address this on-point later in this chapter as we discuss the Fall and the deception of Adam and Eve.

| The Garden — Moses 4 |
|---|
| 7 And he said unto the woman: Yea, hath God said—Ye shall not eat of every tree of the garden? (And he spake by the mouth of the serpent.)<br>8 And the woman said unto the serpent: We may eat of the fruit of the trees of the garden;<br>9 But of the fruit of the tree which thou beholdest in the midst of the garden, God hath said—Ye shall not eat of it, neither shall ye touch it, lest ye die.<br>10 And the serpent said unto the woman: Ye shall not surely die;<br>11 For God doth know that in the day ye eat thereof, then your eyes shall be opened, and ye shall be as gods, knowing good and evil.<br>12 And when the woman saw that the tree was good for food, and that it became pleasant to the eyes, and a tree to be desired to make her wise, she took of the fruit thereof, and did eat, and also gave unto her husband with her, and he did eat.<br>13 And the eyes of them both were opened, and they knew that they had been naked. And they sewed fig leaves together and made themselves aprons. |

-Neither Shall Ye Touch It

God had commanded only that Adam and Eve not *eat* from the Tree of Knowledge: "*But of the tree of the knowledge of good and evil, thou shalt not eat of it: for in the day that thou eatest thereof thou shalt surely die*" (Gen. 2:17). However, in her response to Satan when he asks of the tree, Eve says "*of the fruit of the tree which thou beholdest in the midst of the garden, God hath said—Ye shall not eat of it, neither shall ye touch it, lest ye die*" (Moses 4:9). The addition, more correctly *extension* of the prohibition to not eat of the tree, to include not even *touching* it, appears in the record for the first time here, and it does so without explanation.

This detail has not been overlooked and has been given great consideration by searchers of Scripture for several millennia. Hebrew Midrash containing traditions dating to the Hasmonean period and before includes commentary of this particular phrase that developed an understanding of the text in various ways. Some textual traditions assign the addition to Eve herself, alternately attributing it to either her lack of understanding of the original command from God, or to her intentional amendment of it through the addition of this new restrictive component.

One tradition holds that Eve may have elevated the addition in her mind to a position of importance above and beyond the original commandment of God; in doing so, she opened a door which Satan entered to deceive. This tradition teaches that when the Serpent heard Eve had added to the commandment of God, he sought the advantage by pushing Eve against the tree, challenging her faith by showing she was wrong. As he held her against the tree he said tauntingly: "*Here, you have died!*" Eve, no doubt expressed confusion as her fear of death subsided. Then the Serpent pressed on, saying: "*Just as you did not die by touching it, so, too, you shall not die by eating of it*" (*Gen. Rabbah* 19:3).

Another tradition posits that it was Adam who added the further restriction. According to Rabbi Nathan (700-900AD), Adam, after receiving the commandment in its proper form from God, delivered it to Eve adding the restriction prohibiting even touching the tree. His expansion on the limit set by God was done innocently enough; Adam, showing great care and concern for Eve sought additional measures to protect her from harm, as so often one does to protect those they love. Eve subsequently supposed that the embargo, as delivered from the mouth of God, included both eating from *and* touching the tree.

When the Serpent approached Eve (after realizing he could not persuade Adam to partake of the tree), our first mother repeated just what she knew the restriction to be, she told the Serpent what she had heard from Adam. Upon hearing the recounting of Eve, the Serpent realized he had a means to deceive her; saying: "*If you say that God commanded us not to touch the tree, why, I touch it and I do not die! You, too, if you touch it, will not die.*" As in the tradition we discussed before, Eve was no doubt visibly confused, so the deceiver approached the tree more closely and grasped on to it and kicked at it with his feet, shaking it violently until its fruits fell to the earth. Saying again unto her: "*Here, I touched the tree and I did not die, and if you touch it, you will not die.*" He added: "*Here, I eat from the tree and I do not die, and if you eat from it, you will not die*" (*Avot de-Rabbi Nathan* [ed. Schechter], recension A, chap. 1; recension B, chap. 1).

Through both of these traditions, we see that a great truth can be learned in the subtle details found in the Word of God.[40] The Mishnah and the Midrash reveal that it is often our own incorrect perceptions of the world or (despite our best intentions to follow the commandments of God) our over zealousness, which provide the means for our faith to be weakened and for Satan to deceive. Through these interpretations of the story of Satan and Eve, through the addition of "*neither shall ye touch it,*" we see the consequences of the inclination of man to seek to

---

[40] For a thorough presentation of an alternate interpretation of this subject, see Eve's Answer to the Serpent: An Alternative Paradigm for Sin and Some Implications in Theology, P. Wayne Townsend, Scholia et Homiletica, Calvin Theological Journal 33 (1998) pgs. 399-420.

Townsend concludes that the command not to touch the tree was there from the beginning and that it is a precursor to the Levitical law, "*that the Fall and original sin can be understood through the metaphor of uncleanness. Thus, our guilt resides, not first of all in what we do, but in what we are.*" Towsend is incorrect in this conclusion for several reasons: 1) If the command recounted by Eve was original, why didn't the text just say that in Gen. 2:17? 2) Townsend's parallel to the Levitical Law is unwarranted and discounts the law of consumption given to Noah after the flood in the intervening period between the Garden and the Exodus. 3) Townsend knows nothing of our understanding here that the Garden contained nothing unclean for it was entirely a construct within the immortal Physical sphere.

distance himself from sin by adding additional layers of protection. The extreme danger comes when the additions become law and doctrine unto themselves, defining our beliefs and informing our theology. When these additions are challenged and proven wrong, as we expect human constructs to be, our faith, which has now become rooted improperly in these teachings and doctrines of men, can shatter. These are the dangers inherent in the making of additions to the Scriptures and defining the Word of God by interpretation and not by what it says.

In this work, we have seen more than a few of these false constructs and faced many of them head-on. We have noted some that seem foreign to our ears and perhaps even silly as may have been true in the case of the Masoretes changing and forever obscuring the name of God.[41] Through their zeal to protect and obey, they extended the third commandment beyond its original intent to encompassing an effort to prevent even the inadvertent pronunciation of the Holy Name. The end result of their misguided piety is that today no one on Earth knows the true name of the Creator.

The scope of this thesis includes an effort to dispel false interpretations that have been codified as doctrine, with the full force of Scripture, regarding creation itself. We have confronted and resolved conflicts between dogmatic views that are at odds with the World and Universe we see around us. Like Eve, Christians today read into the word of God much that is not there; they accept traditional interpretation as if the interpretation itself was the word of God. When the interpretations, and even our tertiary understandings of those interpretations, are shown to be wrong, a door is opened and challengers to the faith step in to destroy, often with much success.

As we have explored in these pages, today Christianity stands at battlements to defend the faith against science and its precepts. Christians embrace: creationism, young Earth, and the concept of immortal physical constructs transitioning to mortal and back again. They build walls around these beliefs as though they were Scripture and the very Word of God; they truly are not. Christianity holds fast to these errant beliefs defending their teetering wall against the battle rams of things like evolution, geologic evidence of an ancient Earth, and universal death and entropy.[42] When the battlements shake and quiver and begin to crumble and fall, faith can be lost, often irretrievably; this does not have to be so.

---

[41] See The Unutterable Name, beginning on pg. 178.

[42] It is ultimately ironic that the "*battle rams*" so destructive to "*Christian*" beliefs are themselves the very markings of God imprinted on the Universe by His own hand. Those who stand opposed to reading the things found there are themselves denying the Word of God, perhaps in its purest most elegant form: "*For the invisible things of him from the creation of the world are clearly seen, being understood by the things that are made, even his eternal power and Godhead; so that they are without excuse.*" (Rom. 1:20).

As was the case with Eve being held against the tree, we can find ourselves confused because we are standing atop a wall of our own making and not the bulwark of truth that God has revealed. A proper understanding of the Scriptures, as we have discussed them here, tells us where our true fortress stands, impregnable and not far off in the distance. To the protection of its walls we must retreat, there no one can molest.

Eve's addition also presents another consideration; to whom was the commandment "*thou shalt not eat*" given? It is most often assumed that it was given and received by both Adam and Eve, but is this so? [43] Eve's response to Lucifer indicates that she thought the commandment extended to her, but as we have discussed here she may have received the word from Adam; they both may have been mistaken in their belief that Eve was included in the prohibition. Of this possibility we can only surmise; the Father may have spoken directly with Eve or made it clear in some other way that Eve was to be included. We have no doubt that both Adam and Eve experienced the effects of the Fall, but in that, Eve could have been included just like each of us are, vicariously because of Adam. Why even *suggest* such a possibility? Let us reflect on the following: What is the first noticeable consequence of the Fall? When does it occur?

***...we can find ourselves confused because we are standing atop a wall of our own making and not the bulwark of truth that God has revealed.***

"*And the eyes of them both were opened, and they knew that they were naked*" (Gen. 3:7). The Endowment account demonstrates the opening of Eve's eyes through her declaration to Lucifer, "*I know thee now.*" Of significance here is the timing.

Eve does not recognize Lucifer for who he truly is until *after* Adam partakes of the fruit as well. This pivotal transition, the first effects of the Fall, did not occur when Eve ate of the fruit. She did not indicate she knew who Lucifer was when he said to her, "*go and get Adam to partake.*" We will return to this later in our discussion of the Veil of Forgetfulness, but for now it has significance for its relationship to our understanding of the Fall, the temptations of Eve, and her knowledge and intentions. Though it is clear she at least believed she was breaking a law, a commandment of God, we can see how she could have been *beguiled* into questioning her understanding. Is it possible to commit sin just because we think we are, when in fact there is no law in force? This question speaks to our mindset, to our willingness to do what is right, regardless of what we have been told or what others may think.

---

[43] See footnote 12 on page 225 in the section titled, A Choice to be Made.

- Adam and Eve Truly Deceived

The written accounts record the serpent (Satan) as a deceiver, of this there can be no doubt. The Endowment narrative informs that Eve says, "*the serpent beguiled me.*" The deception of Eve is confirmed in the New Testament by the Apostle Paul as well, "*And Adam was not deceived, but the woman being deceived was in the transgression*" (1 Tim. 2:14).[44] Despite these several accounts, many try to deny this actually was the case, "*...The traditional interpretation that Eve was somehow tricked is inaccurate. Such a reading stands in contradiction with the doctrine as taught by living prophets.*"[45]

To those who have received the fullness of the Gospel, there exists a strong desire to maintain a balance in the scales of justice. It is understood punishment can only result when law is intentionally broken, requiring the transgressor to have knowledge of consequences. This, they say, could not have been the case if Adam and Eve were innocent in the Garden. It is reasoned, it simply is not fair for Adam and Eve to be punished if they were deceived. The need, therefore, for denial of tradition is strong. This *necessity* stems, however, from a lack of understanding of the events pertaining to the creation of man and the Garden, the various environs, and the attributes of the three creations and spheres we have detailed here. It is wrong to deny the deception of Eve in the Garden, to do so would undermine so much of what we know to be true. We will explore here why this is so.

The traditional understanding is flawed as well, but its error has a single nexus that is based on the simple misunderstanding of the Fall and its consequent punishment. **The Fall did not result in punishment**. This error (that the consequences of the Fall are punishment) is not scriptural, for nowhere can we find any Scripture that says Adam and Eve were punished by the Fall. Satan is cursed, the ground is cursed for Adam's sake, but of Adam and Eve, no such wording is used. There are consequences associated with the Fall, law was broken, sin was committed in the breaking of that law, but the response from God was not punishment.[46] Consequence is a responsive action that comes through the effect of law. Consequence is necessary, for law dictates that every action has a response. Consequence is not always negative. In many instances, it is manifested as a desired reaction; in the Fall, this was indeed the case. *If the results of the Fall were only consequence and not punishment, then justice is not abused by Eve having been deceived in innocence.*

**The deception of Adam and Eve is a requirement of modern-day revelation.** Any literal reading of the text and the Endowment narrative reveals that Eve was

---

[44] Here Paul does not intend to place blame on Eve for the Fall. In fact, it can be read in the Apostle's words that he believed she was blameless in that she had an excuse in her actions because of the deception of the adversary. Paul records elsewhere that it was the actions of Adam that brought death upon us all, and in doing so he provides no cover of deception for Adam's actions. See 1 Cor. 15:21, 22.

[45] Paradise Lost, Understanding the Symbolic Message of the Fall, Alonzo L. Gaskill, Retoration Studies Foundation Publishing, Provo Utah, 2011, pg. 21.

[46] See our discussion of this concept in, FREE AGENCY "A MATTER OF RISK", pg. 197.

*deceived* and *beguiled*. It is far too naïve to simply pass this off as misdirection or rhetorical effect. The deception of Eve is significant and is supported internally within the sacred texts and in the Endowment. How do we harmonize modern-day, extra-scriptural teachings of apostles and prophets to contrary canonical passages? We do so by fitting them into the landscape of the multiple creations outlined in this book.

**Every last one of us, the sons and daughters of God, begin Temporality in innocence.** Latter-day Saints do not baptize infants, nor accept the false doctrines of imputed guilt and original sin. Justice and continuity of the Plan of Salvation requires the same, if not identical, sequence of experience for all; we know that to be the case with every child born to this Earth, including the infant Jesus born in Bethlehem. **This too must have been the case with Adam and Eve.** This begs the question: At what point in their eternal progression were Adam and Eve innocent and as uniformed as an infant? Were they sent here to the lone and dreary Temporal Earth after their expulsion from the Garden as infants without father and mother to fend for themselves and survive? Or was that stage in their eternal lives accomplished in the paradise of the Garden under the watch and care of Heavenly Father and Mother?

Of Adam's intellectual gifts when he arrived in the Temporal world, President Joseph Fielding Smith observed:

> *"The first man placed on the earth was an intelligent being, created in the image of God. Possessed of wisdom and knowledge, with power to communicate his thoughts and language, both oral and written, which was superior to anything found on the earth today." (Improvement Era, Vol. 22 pg. 466)*

Though it must be said, President Smith incorrectly intended this statement to address Adam's entry to the paradisiacal Garden that he *assumed* was located on this Earth; nonetheless, his proclamation is correct in its description of the attributes of Adam as he arrived after the Fall, on this planet, in the Temporal sphere.

Through our understanding of the three creations of God, we can propose that the innocence stage of Adam and Eve's eternal existence occurred within the confines of the Garden. Our first parents were fully aware and made a deliberate and wholly informed decision to accept the challenges and consequences of the Fall *prior* to being placed in that Garden. After the Fall and their expulsion from the Garden, Adam and Eve were subsequently sent to Temporality as fully functioning, intelligent adults, aware of their actions and eager to pave the way for the rest of us.

This advanced understanding allows us to harmonize the needs of justice by allowing informed choice, deception in innocence, and consequence that is not punishment but a blessing in accord with the predetermined will of God.[47]

---

[47] See Transgression and Sin? later in this work, pg. 278.

- Satan Knew Not the Mind of God

The Moses account records: "*Satan knew not the mind of God.*" Why does this Scripture do so? Some would say that this passage provides the context for teaching that it was Satan who was deceived and not Eve. Hugh Nibley wrote: "*in ancient lore [Eve] is the one who outwits the serpent and trips him up with his own smartness.*"[48] But what *smartness* did Lucifer have if he knew not the mind of God in this thing?

We are told that Satan (Lucifer) knew of the true effects of the eating of the tree of the knowledge of good and evil, for he twisted the truth, if only subtly. Under the inquisition of God, he later exclaimed that he was merely doing, "*which has been done in other worlds.*" Further, God curses Lucifer for what he had done. We are back to the scales of justice here, for Lucifer to have been cursed he must have earned such punishment and that necessitates his actions as being intended and performed with knowledge of the law.

Surely, Lucifer was punished for his intent. Scripture and the Endowment inform us that he intended to deceive Eve into disobeying the will of God; in fact, we are told his exact, deceptive words. Lucifer was not acting behind a veil of forgetfulness; he had enough knowledge of what he was doing to have been punished under the law for them. There can be no doubt of this.

We are certain of Lucifer's intent to deceive. What is not clear is if he was actually successful in achieving what he intended. Did Eve partake in the Garden because she was deceived or because she chose to do so for a greater purpose? We only have her testimony from which to draw our conclusion. Eve told the Father she was beguiled; her veracity in this statement was not questioned. God did not interrogate Eve further, and we would do an injustice to the truth to do so as well. **Eve said she was deceived, and we would do well to believe her.**

So what of the passage in Verse 6 that says, "*Satan knew not the mind of God?*" To put it in complete context, it must be paired with what follows, "*wherefore he sought to destroy the world.*" Further context comes from his statement that he was doing what had been done in other worlds before.

How could what he had done destroy the world? What part of his action had been done in other worlds? And what part of the process that would ensue was known *but not yet* revealed by the Father? These are the three grand questions to be considered here. We do know what Lucifer did do, so we can walk ourselves back from there.

First we must establish the foundation of the Deceiver's knowledge coming into the Garden. Lucifer, the *Shining One*, was a well-informed, well-versed, participant in the Grand Council. He knew of the plan presented by the Father, for our free agency that was to follow. He knew of the plans for the Temporal World that was to come,

---

[48] Old Testament and Related Studies, Hugh Nibley, FARMS, Provo Utah. 1986, pgs. 88-89.

which would allow that *agency* to happen. He was aware of what was to be sin and of the promised Atonement that was planned to compensate for it. This is without question, for Lucifer was an integral part of the debate in the Grand Council through which these things were proposed. Lucifer's opposition and subsequent rebellion against the plan resulted in his being cast down. There is no mention of his loss of any past knowledge that accompanies each of us when we come to the Temporal sphere. It is clear from his revealed actions and statements in the Garden, in the Temporal world immediately following the Fall, and his actions at the meridian of time as recorded in the New Testament, that he remains aware of who he is and what his mission is. He is the "*god of this world*" (2 Cor. 4:4); as such he continues to possess supernal power and knowledge.

With this understanding, we can eliminate any loss of knowledge of things he knew from before. It is a given that Lucifer does not know the mind of the Father, for it is written that the Son of God does not know that either, *in fullness*; "*But of that day and hour knoweth no man, no not the angels of heaven, but my Father only*" (Matthew 24:36). This statement of the Savior's is more than just a remark made in passing; it is reflective of a deeper understanding that is more fully expressed elsewhere in the life of the Savior.

No one can doubt that the Savior was in full comprehension of His Father in all things. Certainly He had full knowledge of the Plan of Restoration and Redemption. He knew of the Fall, the Temporal World, the Spiritual World and the Atonement. But consider the following: "*And in the ninth hour Jesus cried with a loud voice, saying, Eloi, Eloi, lama sabachthani? Which is, being interpreted, My God, my God, why hast thou forsaken me?*" (Mark 15:34). In this verse, we see the very Son of God, the Savior Himself, in the throes of the Atoning Sacrifice, not certain where His Father is, or why He would leave him. The Savior's exclamation reveals that even He did not *know* the mind of God at that pivotal moment.

Is this the kind of *lack of knowledge* spoken of in Moses 4:6? It probably is. Lucifer had general knowledge, very good general knowledge, but not *specific* knowledge of the mind of God. He knew that man must Fall, but was not aware that God intended to use him (Lucifer) as the vehicle of that process.

We are told, *Lucifer attempted to destroy the world*. One way to do that would be to ensure Adam and Eve would live forever in their sin.[49] The story of the Garden tells us that immediately following the Fall, Father and Jehovah *suddenly* returned to the Garden. Perhaps in this unanticipated arrival, Lucifer's full plan was interrupted. What would have happened if it had not been? *What would have been the result if Lucifer had gotten both Adam and Eve, in their fallen state, to have eaten of the tree of life as well?* He would not have had to encourage them to do so, they had

---

[49] Lucifer's strongest desire today was his strongest desire then, "*And because he had fallen from heaven, and had become miserable forever, he sought also the misery of all mankind*" (2 Nephi 2:18). If he could accomplish locking in our first parents to a life of misery with him, he could accomplish his goal; at least he would assure the eternal misery of Adam, his premortal adversary in the Great Rebellion that followed the Grand Council.

permission to eat of that tree, and it seems they already had. Surely in time, in short order, they would have eaten of that tree again "*living forever in [their] sins*;" certainly this situation would have destroyed God's plans for the *world*. The specific "*mind of God*," the knowledge Lucifer did not know, was that God would return to the Garden before Adam and Eve would eat of the tree of life.[50] It appears God was lying in wait just to do so.

Consider this also. Lucifer told Adam and Eve to hide, to avoid the Father; why not tell them to confront him directly in their state of sin? Did Lucifer hope they could avoid detection for a while longer? And what of Lucifer's statement that he was doing the thing that had been done in other worlds? What was that? Eating of the tree of knowledge of good and evil and /or the tree of life? In much the same way he spoke half-truth to Eve, he could in this reply to God be speaking deceitfully, referring to his (hoped for) last half of the menu to be served.

In closing one last parting blow to those who hold that Adam and Eve were more than innocent in the precincts of the Garden before the Fall, that somehow despite Scripture and inspired Endowment teaching to the contrary, they were actually wise, intelligent, and capable of making informed decisions. If that were so, why would they think they could *hide* in the bushes and escape their discovery by God?

-You Have a New World Here

The Endowment account is the only one that shares with us information about Adam's encounter with Lucifer. Though the concept that Lucifer first attempted to convince Adam to partake of the tree was not new when the Endowment account included it,[51] it does not appear in any of the recorded scriptural accounts, ancient or modern, nor is it to be found within the Christian tradition.

In the Endowment, Lucifer appears without introduction and Adam does not ask him who he is. This is instructive in that we can surmise that visitors were common to Adam during his tenure in the Garden. This may have been the case if we allow for Adam to have grown from early childhood to an adult within and/or around the Garden precincts, perhaps much of that length of time without Eve by his side. In contrast, notice that when Lucifer first approaches Eve after giving up on Adam, she immediately asks, "*Who are you?*"

---

[50] The riddle posed by what Satan may or may not have known of the mind of God is entirely a Latter-day Saint theological dilemma. There is no counterpart to be found in the remainder of the Judeo-Christian world. We alone embrace the events of the Grand Council, though it is alluded to in the Greek of the New Testament and is recognized in Jewish commentary of the Old. Latter-day Saint understanding alone teaches that all of us were present in that council, each possessed with sufficient knowledge and the agency to act on it. This teaching lays the groundwork for the context of an adversary who once knew of things of God, who must somehow act contrary to what would be his own desires. We have presented here just one possible scenario that answers this riddle. There is no claim here that this is anything but plausible conjecture.

[51] For example both of the Midrash traditions we just addressed regarding Eve's encounter with Satan, included Satan approaching Eve after failing to convince Adam to partake of the fruit.

When Lucifer first approached Adam and realized that Adam did not have memory of other worlds, he said, "*your eyes are not yet opened, you have forgotten everything.*" Because he is aware that Adam is not yet conscious of his past, Lucifer makes no attempt to infer that he is someone he is not as he would later do with Eve. That Lucifer immediately recognizes Adam's "*eyes are not yet opened*" is yet further evidence that Lucifer still has knowledge of the details of the Plan of Salvation; he straightaway recognized that Adam had "*forgotten everything.*" A *Veil of Forgetfulness* had fallen around Adam and Eve in the Garden, and we have proposed in this thesis that the veil had dropped over the entire Garden, that in fact this veil is what cordoned off the Garden from the rest of Eden.[52]

Because of the veil, both Adam and Eve were in a state of innocence, or specifically a state of forgetfulness. Through frequent, direct, contact, they were aware of whom Father was. However, when the visitor Lucifer appeared to them, he was not recognized. The veil of forgetfulness remained in place until after the Fall. The veil's lifting is first indicated by the utterance of Eve when she declared to Lucifer, "*I know thee now.*"

Of critical significance here is the timing; Eve does not recognize Lucifer for who he truly is until *after* Adam partakes of the fruit as well. The effects of the Fall do not occur when Eve first partakes of the fruit, but only after Adam does. This is in accord with the teaching of Paul that links the death of mankind to Adam and not to Eve, though Paul is fully aware that it was Eve who was deceived and not Adam:

> *"For since by man came death, by man came also the resurrection of the dead. For as in Adam all die, even so in Christ shall all be made alive." (1 Cor. 15:21, 22) "And Adam was not deceived, but the woman being deceived was in the transgression." (1 Tim. 2:14)*

The effects of the Fall are spontaneous, the innocence that accompanied the veil in the Garden disappears without any action on the part of the Father or the Son; there is no summary judgement, the Fall "*just happens.*" This is a further indication that the Fall is not a punishment,[53] but a *consequence*. The Fall did not come about as a result of punitive action, it is not a curse. Note, the marked difference in the case of the costs decreed upon Lucifer. The consequence for Satan is a direct curse; it is an *imposed* sentence. It does not take effect at the time of his action (the tempting and deception of Adam and Eve), but only after the subsequent action of God, after the declaration of His judgement upon Lucifer.

---

[52] A similar veil encloses our Temples on Earth today, though alas incomplete. We attempt in every way to leave the Temporal World beyond the Temple doors when we enter, seeking to shed memory of its sadness and sorrow outside.

[53] "*... if it were possible to eradicate all our culturally induced prejudices... and examine the original Hebrew text of the whole Eden account, we would find a story that actually says very little of what it has throughout the centuries been credited with saying.*" The Redemption of Eve, by Jolene Edmunds Rockwood, in Sisters in Spirit, University of Illinois Press, Chicago, Il. 1992.

**The Fall**

> "*Adam fell that men might be; and men are, that they might have joy.*"

Why? I am completely aware of the growth through experiences to be gained by our tenure of the Temporal sphere. We have discussed that concept at length throughout this work. But why the Fall? Why such an elaborate structure of multiple creations, Restoration, Redemption, and the Atonement? Why all the suffering, cost and loss? Couldn't have, shouldn't have, the progression of the sons and daughters of God been accomplished in a different way? Just the *considering* of these questions is hard enough to fathom, answering them is perhaps impossible; but without addressing them we would rob ourselves of a full understanding of the Fall and the value of the knowledge we can gain from seeking the answers.

God does not react; He proceeds with plan and purpose. Christianity of tradition would have us believe that Adam Fell and changed God's *intended* plan for mankind; that the Atonement was completed in *response* to the Fall, at best, a life preserver kept at the ready, only thrown into the water because it was needed. This understanding limits God's omniscience, at best, only allowing Him enough foresight to allow for the stowing away of a bright orange, floating donut available for use *if* man fell into the sea. Such a concept is absurd and ridiculous.

***When Adam fell, there was no clambering in heaven, for the Fall was part of the plan from the beginning. Our first parents went into the Garden in anticipation of the Fall and with intent and sense of purpose to do so.***

When Adam fell, there was no clambering in heaven, for the Fall was part of the plan from the beginning. Our first parents went into the Garden in anticipation of the Fall and with intent and sense of purpose to do so. They knew of and accepted the challenging consequences of the event while yet in our presence, in the presence of God *before* the Garden was created. With full knowledge of the pitfalls that lay ahead, they placed their faith on the success of the Savior to accomplish the redemption.

Lucifer, having failed in his attempt to prevent the Garden from being established in the first place, became an accomplice in the plan, but for much different reasons than those of our Father. In the Grand Council Lucifer proclaimed, "*I will redeem all mankind, that one soul shall not be lost*" (Moses 4:1). Beginning with his efforts with our first parents, he seeks to prevent the redemption of every one of the sons and daughters of God.

Though Lucifer was involved, he did not bring to pass the Fall. When Paul refers back to the story of Adam and Eve, he does not mention the actions of Satan at all. Paul places the Fall squarely on the shoulders of Adam alone, "*For since by man came death, by man came also the resurrection of the dead. For as in Adam all die, even so in Christ shall all be made alive*" (1 Cor. 15:21, 22). Admittedly, context here drives the content; Paul is associating the actions of a single man in the distant past with their counter, affected by the single man, Christ, of the most recent times. But the point drawn here is that the Apostle is not wrong in assigning blame to Adam alone, for indeed the Fall was affected by his actions and only his actions and cannot be linked to anything Lucifer could have accomplished on his own. Our fate did not even hinge on the actions of Eve, for the consequences of the Fall were not efficacious until Adam partook of the fruit as well.

Our theology identifies two separate and distinct components of the Fall, one spiritual, and the other Physical (Temporal). They are associated with the separate spheres we have addressed in this work and are called, Physical Death and Spiritual Death.

Physical Death comes to all within the Temporal sphere, it is inherent to mortality. Physical Death is not confined to the *living* only, it is an intrinsic quality of everything Temporal, galaxies, stars, planets and the elements themselves are playing out a *lifecycle* that leads to depletion, disorder, decay and finally dissipation. We all die as our mortal existence comes to a close; sometimes the process from birth, *generation*, to death is long, composed of many measurable stages, sometimes it comes quickly, seemingly all-at-once, but eventually it encompasses all that there ever was, or ever will be, in this Universe.

In the context of the three creations we discuss in this work, Physical Death is Temporal in nature, in that it separates the Spirit (embodied intelligence) from the Temporal (physical) tabernacle. With this death comes our entry into the Spirit World, of which we will speak of in depth in a subsequent chapter. Within the Spirit World we await the resurrection, the rejoining of our Temporal elements (in eternal perfection) to that of our Spirit. In the Spirit World, we experience Agency as we did in our pre-existence; it is, however, not "*free*" as it was in mortality but it is limited in its scope of choice, in its ability to increase our understanding, and in its efficacy.

Physical Death is the one spoken of by Paul when he said, "*For as in Adam all die.*" This death is the one that most closely resembles the consequence proscribed by the doctrine of original sin as found in the creeds. Because of Adam we all die, but we do so only *temporally* because we are Temporal. Physical Death affects us all, but as we have seen here, the effects we experience are not because of the imputation of Adam's sin, or guilt; death comes to all things here regardless of sin, its inevitability is because *all* things are Temporal. Paul does not mention sin in this verse or agent men, sons and daughters of God, for that matter; Paul says *all* die, and he means that literally. Death affects everything, whether it is capable of sinning, comprehending, and repenting or not. Physical Death is not the result of sin; it is fundamental to Temporality. Physical death was not an imposed penalty; it is a consequence of the Fall; it is essential to mortality, a by-product of uncertainty which allows for Free Agency and Eternal Progression, both *blessings* from God.

Spiritual Death is something much different than Physical Death. Because we know that our Spirits are eternal and not subject to dissolution, decay, or destruction, Spiritual Death cannot really be a death at all, at least not in the sense we describe it here for life in the Universe.

So what are we to understand Spiritual Death to mean? It is separation from our Father in Heaven; it is not measured by degree and is fully realized as *anything less* than full fellowship with the father and the attainment of all the blessings of God. Spiritual Death has eternal consequences and is a result of our own individual, willful, decisions and actions. Overcoming Spiritual Death is receiving eternal life and eternal increase, becoming co-heirs with Christ. Spiritual Death is a result of disharmony with the will of God, anything less than perfect unison of thought and action with the Father.

Adam died spiritually and his eternal potential was diminished the moment he chose Eve over the Father's will, that is why that instant is called the Fall. Spiritual Death is not, however, a fall from the Grace of God, that remains extended to each of us, but it is a fall away from what each of us can and were intended to be. Adam's sin and spiritual Fall were his own doing, so too are each of ours. Sin is not imputed from father to son.

Because of the consequences of Adam's Fall, we are born mortal and temporal, that makes us carnal and natural; we are predisposed to disharmony with God through imperfect knowledge and the natural selfishness of carnal self-preservation. Spiritual Death affects our Spirits (embodied intelligence). Because of this death, our eternal Spirits are restricted through internal constriction, disallowing Grace to fully sanctify us and preventing us from reaching our full potential as a child of God.

Because there are two separate components of the Fall, so too are there two distinct components of the Atonement: one that is general and universal and one that is individually specific. Latter-day theology refers to these anciently understood concepts as Restoration and Redemption respectively; they are the means to the ultimate end we call Exaltation. Restoration of our Spirits to our resurrected Temporal frame is a singular event, occurring at a place and point in time; Redemption is an eternal progressive process. Though Exaltation represents a final stage of progression, it is an eternal state and as such includes eternal growth of power, glory, and intelligence – what we call eternal increase. All are made possible for us *only* through the singular, *once-and-for-all* sacrifice of Jesus Christ.

The component of the Atonement accomplished by the death and resurrection of Christ reverses the effect of this Physical Death and will allow for *all* to be resurrected, regardless of their state of righteousness. This is the *Restoration* component of the Plan of Happiness spoken of by Alma. Traditional Christianity focuses on the Crucifixion as the empowerment of the Atonement, this is not correct. Physical Death was overcome for all by Christ, not on the cross of Calvary, but by His Resurrection from the tomb. *In death, on the Cross, God became truly man; through the power of the Resurrection, man can become truly God.*

Through the death and resurrection of Christ, all is restored *strictly* by Grace, and this is made manifest by the universality of the restoration, including all, saint and sinner, believer and non-believer, for, *as in Adam all die*, through saving Grace, *shall all be made alive*; there are no exceptions. How could the inanimate, the non-living things that are not sentient, be restored if restoration required anything to be accomplished on the part of the recipient?

*"...we are sanctified through the offering of the body of Jesus Christ once for all."*

(Hebrews 10:10)

The day of Christ's atoning sacrifice began at sundown on the fourteenth day of the month. That evening, following a final meal with His disciples, where He detailed the importance of what He was about to do, they all, save one, traveled on to Gethsemane, on the Mount of Olives, where, before the night was through, the Savior would suffer for all the sins of the World, in prayer, alone in a garden.

Sin is an individual thing. Sin is committed by all who have lived in this Universe, save one, who would, as was His purpose, suffer once for all, "*being in agony ...his sweat as it were great drops of blood falling down to the ground*" (Luke 22:44). In Gethsemane, the Son of God took upon Himself the burden of punishment for all the sin of the world; He was so tortured by it that He was "*strengthened*" by an angel from heaven.[54] Spiritual Death earned by us all through our own individual actions was paid for here by the Savior, alone, without the assistance of the actions of any other,[55] not on the Cross or within the tomb.

Action is required on our part, individually, for this *sufficient Grace* to cleanse us and make us whole. As was the case with Physical Death, overcoming Spiritual Death is something impossible for us to accomplish on our own, but we must be accepting of it, acknowledging the gift.

The accomplishments of the Savior in Gethsemane are the Redemption component of the Plan of Happiness spoken of by Alma. From the very beginning of man's Temporal existence, this path has been understood. The LORD revealed to Adam, "*...by the Spirit ye are justified, and by the blood ye are sanctified*" (Moses 6:60).

---

[54] Luke 22:43.

[55] John 2:19-21; John 10:17-18, these verses specifically state that the Son will do the raising with His own power. Though there are many other verses which state that the Holy Spirit and the Father completed the task, forming a basis for an understanding of the Trinity, that discussion will be left for another place and time. See Making the Son God, pg. 317.

Justification means to be "*declared*" guilt free. The Holy Ghost does that for us as we accept the Savior in our lives and take his name upon us. Sanctification, however, means to be "*made*" Holy. Sanctification can only be given from one who is sovereign and free of guilt himself. We are sanctified by Christ's blood, shed in Gethsemane. Because of His sacrifice in Gethsemane, Spiritual Death is reversed by our coming to Christ and allowing Him to perfect us. We must seek it out, and accept it, living Grace-by-Grace until we are sanctified:

> *"Yea, come unto Christ, and be perfected in him, and deny yourselves of all ungodliness; and if ye shall deny yourselves of all ungodliness, and love God with all your might, mind and strength, then is his grace sufficient for you. That by his grace ye may be perfect in Christ; and if by the grace of God ye are perfect in Christ, ye can in nowise deny the power of God. And again, if ye by the grace of God are perfect in Christ, and deny not his power, then are ye sanctified in Christ by the grace of God, through the shedding of the blood of Christ, which is the covenant of the Father unto the remission of your sins, that ye become holy without spot." (Moroni 10:32,33)*

When we spoke of Spiritual Death, we spoke of internal constriction. By that we mean, that it is imperfection within us which prevents the entry of saving Grace into our Soul. It can be imagined that the Spirit within us, as it was originally fashioned, is flawless and beautiful like a perfectly faceted, clear diamond. In theory, such a diamond would absorb all light that reaches it re-radiating it all away, sharing the beauty of that light entirely with all around, but such is the case only with an unblemished diamond. When we live a life of sin, our surfaces become scratched and fractures begin to appear in our structure. Serious sin results in internal flaws that run deep and threaten to shatter us.

In this metaphor, light represents the power of God that seeks to work in us and through us. The Saving Grace of God is sufficient to restore all perfection to the diamond that we each are. If we can just let go of the sin that fractures, stains and mars us. By holding on to the flaws (*not repenting*), we deny the light entry into our inner parts and prevent it from working there. Light that cannot get in, cannot reflect outward into the world in its brilliant glory; each retained imperfection constricts its power. This constriction is what Moroni speaks of when he says: "*if by the grace of God ye are perfect in Christ, ye can in nowise deny the power of God.*" Only a perfect diamond, our perfect Spirit, can accept and absorb all that the Father has to give, fulfilling its creation in brilliance and beauty.

Spiritual Death is a result of our not allowing the grace of God to fully work within us, for it is the desire of God that it does. Our Father loves us and desires that we return to His full fellowship and achieve our full potential, but though many will fall short of this great blessing, He accepts us all as His children. The Plan of Happiness includes a place for each of us, no matter what flaws we fail to let go. The Savior said, "*in my Father's house are many mansions*"[56] (John 14:2), and within the three degrees of eternal glory, each of us will find a place we can call home.

---

[56] See our discussion on the Three Degrees of Glory in Chapter 9 of this book, pg. 357.

Full fellowship with Father, attainment of all of the blessings of God, becoming co-heirs with Christ, requires that we *allow* Grace to perfect us. Until we do, we cannot fully realize our potential as a son or daughter of God. Agency makes that our choice, however. This full realization of our potential, eternal life, eternal increase, is known as *Exaltation* in our theology. The component of the Atonement accomplished in Gethsemane, in silence, after the Savior told the Father, *thy will be done*,[57] provides for the redemption *and is* the Grace sufficient to perfect us.

In this section, we will address the Fall. As we have done so far in this chapter, we will rely of the book of Moses as our textual guide, as well as Genesis and the Endowment narrative. The Book of Abraham does not include the Fall, or any actions of Adam and Eve within the Garden, but we will find additional resources to draw upon within the Book of Mormon, the Doctrine and Covenants and the Bible.

**The Garden — Moses 4**

14 And they heard the voice of the Lord God, as they were walking in the garden, in the cool of the day; and Adam and his wife went to hide themselves from the presence of the Lord God amongst the trees of the garden.
15 And I, the Lord God, called unto Adam, and said unto him: Where goest thou?
16 And he said: I heard thy voice in the garden, and I was afraid, because I beheld that I was naked, and I hid myself.
17 And I, the Lord God, said unto Adam: Who told thee thou wast naked? Hast thou eaten of the tree whereof I commanded thee that thou shouldst not eat, if so thou shouldst surely die?
18 And the man said: The woman thou gavest me, and commandest that she should remain with me, she gave me of the fruit of the tree and I did eat.
19 And I, the Lord God, said unto the woman: What is this thing which thou hast done? And the woman said: The serpent beguiled me, and I did eat.
20 And I, the Lord God, said unto the serpent: Because thou hast done this thou shalt be cursed above all cattle, and above every beast of the field; upon thy belly shalt thou go, and dust shalt thou eat all the days of thy life;
21 And I will put enmity between thee and the woman, between thy seed and her seed; and he shall bruise thy head, and thou shalt bruise his heel.

---

[57] Matthew. 26:42.

- They Went to Hide Themselves

In the Garden account, we see the two deaths addressed by the Fall, the choice to partake, and the subsequent expulsion (decreed consequence). Adam experienced the effects of the Spiritual Death from the moment he partook of the fruit (his choice). We noted that Spiritual Death is a result of our not allowing the Grace of God to fully work within us. In effect, we, through the internal constraints of pride, or will, or even shame, hide ourselves or deny the presence of God in our lives. And thus here immediately after the Fall, we see this too is Adam's first instinct.

The Endowment account tells us that Lucifer tells our first parents to hide, and the same is true in each of our lives today. For it continues to be the work of Satan to stand in the way of our intimate relationship with God. He does so by encouraging fear and self-doubt. Through a sense of our own unworthiness, Satan seeks to separate us from God, and so it was with Adam, for why did he hide? "*I heard thy voice in the garden, and I was afraid, because I beheld that I was naked, and I hid myself.*" God comes to seek us daily and calls upon us. Do we hide from his presence and deny his call? The story of the Garden is our story as well. We each Fall every single day.

"*I beheld that I was naked, and I hid myself.*" The Hebrew here subtly suggests Adam feels guilt at this time. The Hebrew *erom* (עֵירֹם), translated as naked here, is different from *erummim* (עֲרוּמִּים), which is translated as naked in Genesis 2:25. Both words mean naked, but reflect a different connotation. The difference is very slight and is easily overlooked in the English of the Bible, but the translation does identify it, though almost imperceptibly; Genesis 2:25 includes the phrase, "*and they were not ashamed.*" That the phrase is included in Genesis 2:25 does not have so much to do with what *erummim* means as it does with what the later used *erom* does. This can be understood as follows:

We often use *naked* today to express purity and freshness like that associated with a newborn baby; this is a very joyful and uplifting connotation of the word. We also, however, use naked darkly, negatively, to express shame, exposure, a raw vulnerability, a feeling of weakness, fear and subjugation. *Erummim* means the former; *erom*, here expressed by Adam as a reason for his hiding, means the latter. By selecting *erom* to relate what Adam said to God, the Hebrew scribes intended to express that Adam felt shame, guilt, exposed, and vulnerable after the Fall, feelings he had never experienced before. These are the feelings Satan uses against us all of the time.

When we sin, we stand naked (*erom*) before God. Our shame and guilt drives His Spirit away and our Spirit dies just a little bit more; Spiritual Death is a very real thing. God planned on this circumstance from the beginning. The Hebrew word *kippurim* (כִּפֻּרִים) (Atonement) most literally means *to cover*. In that context, Adam's "*nakedness*," his shame, guilt, and exposure is *covered* by Christ. Christ's Atonement *covers* our nakedness as well; note that it is the *nakedness*, the guilt, and shame, the feelings that prevent us from allowing God's Spirit to dwell within us that is covered, *not* the sin itself.

Verse 15 of the Moses account differs slightly from its Genesis counterpart. Here Moses (through the prophet Joseph) informs us that God did not ask Adam where he was, certainly God knew that, but He asks Adam, *where he was going*? Though God knew the answer to that question as well, we all must answer that very same question for ourselves. God was asking Adam here, *have you decided to leave the Garden and enter into mortality?* Requisite with the conditions and requirements of Agency, that was a decision Adam had to make on his own.

Similarly the answer to the next question asked by God was known to Him, but it was asked because it is important that we each have accountability for our own actions. It is often our nature and perhaps a basic instinct to look for a way to deflect blame, to avoid consequence. When God asks Adam, "*Hast thou eaten of the tree whereof I commanded thee that thou shouldst not eat…?*" Adam replies by deflecting to Eve, who in turn likewise deflects to Lucifer, emphasizing that she had been *beguiled* by him.

The consequences of the Fall were immediate. What was left to be accomplished was the creation of Temporality so that those consequences could fully develop the potential they were intended to.

- I Will Put Enmity between Thee and the Woman

Here we are told of the hatred and disdain that Lucifer will further develop between himself and the Son. The Plan of Salvation (Happiness) includes elements that will empower Lucifer to bruise the heel of Christ, but the same Plan of Salvation includes the Atonement by which Christ will bruise (crush) the adversary's head. The bruising of Christ's heel is manifest in the dominion of the Devil over many of the sons and daughters of God. In the end, however, Christ will be victorious when "*he hath put all enemies under his feet*" (1 Cor. 15:25).

**The Garden — Moses 4**

22 Unto the woman, I, the Lord God, said: I will greatly multiply thy sorrow and thy conception. In sorrow thou shalt bring forth children, and thy desire shall be to thy husband, and he shall rule over thee.
23 And unto Adam, I, the Lord God, said: Because thou hast hearkened unto the voice of thy wife, and hast eaten of the fruit of the tree of which I commanded thee, saying—Thou shalt not eat of it, cursed shall be the ground for thy sake; in sorrow shalt thou eat of it all the days of thy life.
24 Thorns also, and thistles shall it bring forth to thee, and thou shalt eat the herb of the field.
25 By the sweat of thy face shalt thou eat bread, until thou shalt return unto the ground—for thou shalt surely die—for out of it wast thou taken: for dust thou wast, and unto dust shalt thou return.

- Cursed Shall be the Ground

These *consequence* verses are replete with symbolism and prophetic parallels. They detail what was yet to lie ahead in the Temporal World, not only for Adam and Eve, but for each of us and the Son of God Himself. It is important for us to realize that the Temporal sphere is about binding. Perhaps nowhere else is that binding more evident than through the shared experiences discussed here.

Effects native to the Temporal sphere bind Spirit to flesh, they bind Man to God, and the actions of the Savior within the Temporal sphere bind God to Man. The binding is accomplished both through covenant and shared experience. Here we see consequences of the Fall listed for both Adam and Eve. Each of the consequences would be experienced by the Savior as well. For Adam's sake, the ground is *cursed* (v23); for our sake, Christ became cursed for us, "*Christ hath redeemed us from the curse of the law, being made a curse for us: for it is written, cursed is everyone that hangeth on a tree*"[58] (Gal. 3:13). Eve would have pain in her conception (the travail, *labor* of childbirth) (v22); Christ also, "*shall see of the travail of His soul*" (Isa. 53:11). Adam would suffer because of thorns (v24) and so too the Savior, "*And they clothed him in purple, and plaited a crown of thorns, and put it about his head*" (Mark 15:17). Sorrow entered the world because of the Fall (v22, 23); and for the sins of each of us, Christ, in Gethsemane experienced great sorrow, "*...My soul is exceeding sorrowful, even unto death*" (Matt. 26:38). Because of sin, Adam would sweat (v25); because of the sins of Adam and his posterity, Christ sweat, "*great drops of blood falling down to the ground*" (Luke 22:44). Eve would be in subjection to Adam, *he [Adam] shall rule over thee* (v22); and Christ subjected Himself to the Father "*...the Son also himself be subject unto him that put all things under him*" (1 Cor. 15:28).

For all the parallels to be found here in the consequence verses, the contrast in roles played in the Plan of Happiness by Adam and Eve in the Garden and the Son of God on Calvary stands out and should be emphasized. Through the Fall, Adam brought death into the world both spiritually and Physically (Temporally) "*as by one man sin entered into the world, and death by sin; and so death passed upon all men*" (Rom. 5:12); Christ died on the cross and suffered in Gethsemane to undo that, "*...he surely must die that salvation may come; ... he dieth, to bring to pass the resurrection of the dead*" (Hel. 14:15). Helaman further goes on to teach of the distinction between the two deaths as he declares,"*...this death bringeth to pass the resurrection, and redeemeth all mankind from the first death—that spiritual death; for all mankind, by the fall of Adam being cut off from the presence of the Lord, are considered as dead...*" (Hel.14:16). The Fall brought to pass the "*first death*;" the *second death* followed after the creation of the Temporal sphere and the Temporal death that was attendant to it.

Because of the Fall, the consequences spoken of in these verses (a Fallen, thistle bearing Earth, child bearing and its travails) are pronounced, but they are as yet still

---

[58] See also Deuteronomy 21:22, 23.

future. It will be noted that they are separate, detached from the Fall itself which was immediate and had already taken place. This distinction is important, for it unlinks the creation of Temporality and its associated attributes, only some of which are the consequences noted here, from the effects of sin, *willful disharmony with the will of God*, the "*first*" death as spoken of by Helaman.

This Universe we inhabit, what it does to us, and provides for us, are not the effects of sin, but they are gifts from God, the means for which we can return to Him; Temporality is not a prison.

Of interest is Verse 23 of the Moses translation and its parallel, Verse 17 of Genesis, Chapter 3, they include: "*Because thou hast hearkened unto the voice of thy wife.*" Though this phrase is no longer found in the Endowment narrative, it once was. The phrase cannot be dismissed as a corruption of the original Hebrew text for it was included in the Endowment and still remains in the restored Moses translation. Indeed in the Endowment we are told that it was the very act of considering the logic and needs of Eve that led Adam to partake of the fruit that was forbidden. The Endowment narrative records his submission when Adam says, "*I see that this must be.*" Adam was noble and considerate in this decision, by joining Eve, he gave up his life, for the sake of another, the most Christ-like thing any man can do. Through this unselfish personal sacrifice, giving up immortal life in paradise, for expulsion to a hostile world and certain death, Adam became the first of many types of Christ to be found in the pages of the Old Testament.

| **The Garden — Moses 4** |
|---|
| 26 And Adam called his wife's name Eve, because she was the mother of all living; for thus have I, the Lord God, called the first of all women, which are many. |

## *The Mother of All Living*

The Endowment narrative instructs us that Adam names Eve before the entry of this first pair into the Garden, "*because she is the mother of all living.*" After which Elohiym confirms "*that is right.*" The timing of this event within the Temple is enigmatic, for in the written accounts, both ancient and modern do not depict the naming of Eve by Adam until after the Fall. Both Genesis and Moses imply that the naming was in fact after our first parents are in the Temporal sphere, maybe even after she first bore a child.

This *later* timing of the naming in the written accounts is consistent with our understanding of why Adam and Eve partook of the fruit. The Endowment informs us that Eve asked Adam if he intended to keep all the commandments of God, and if he did, how could he multiply and replenish the Earth if she was to be expelled, "*and you will be a lone man in the Garden.*" In placing the naming of Eve after the

Fall, the written accounts, by implication, confirm our understanding that prior to the Fall, within the Garden, Adam and Eve could not conceive and have children:

> *"And now, behold, if Adam had not transgressed he would not have fallen, but he would have remained in the garden of Eden. And all things which were created must have remained in the same state in which they were after they were created; and they must have remained forever, and had no end. And they would have had no children; wherefore they would have remained in a state of innocence, having no joy, for they knew no misery; doing no good, for they knew no sin." (2 Nephi 2:22, 23)*

Reproduction, the ability to conceive, is not merely associated with the Fall in time, it, in fact, required it. It appears to require the Temporal sphere as well, at least the physical bodies that we attain here. The Temporal body (in its perfected, immortal, resurrected form) is needed for eternal increase as well; we will explore this concept in further detail later in this work.[59]

Eve was created female from the outset and had the attribute of gender from the moment of the creation of her Spirit. She was a woman, identified as a mother, before she had the ability to procreate, before that was even an available option to her. Gender, both man and woman is a fixed, immutable, attribute of the eternal Physical sphere.

The Hebrew translated as Eve is *hawwah* (חַוָּה); it is a derivative of *hayyim* (חַיִּים), the word for life. In the Genesis account, Eve is only mentioned by name twice and so *hawwah* only appears two times in the entire Hebrew Bible. In full context, Adam correctly identifies the woman, his wife, Eve, as not only the mother of all living, but the *source* of life itself in the Temporal sphere.[60] In this regard, Eve's actions are critical. It is Eve who initiates the events of the Fall. Though we have argued she was indeed innocent in the Garden and was truly beguiled by Satan, for all of our inspired records tell us so, we also know that she was not without knowledge and the strong desire of a woman (and those of a potential mother) to nurture and care for a family. Satan played upon these instincts to encourage her disobedience of one commandment, in the attainment of what to her was no doubt a higher calling.

We have noted here, that the Scriptures provide some reason to infer that Eve's entry into the Garden was accomplished because she would precipitate the Fall when it appeared Adam perhaps would not. To this point, to the execution and enactment of the essential requirements of the Fall, Hugh Nibley stated somewhat satirically, "*Eve is the first on the scene, not Adam, who woke up only long enough to turn over and fall asleep again. ...in all that follows she takes the initiative,*

---

[59] See Eternal Procreation, pg. 521.

[60] There is another interesting side note regarding the name of Eve. The language of ancient Sumer spoken and written in the third millennium BC was a development of symbiosis between Sumerians and Akkadians both Semitic in origin. In the earliest common script of Sumer, the marking for the word life, which later evolved into the Hebrew *hayyim*, was the same marking used to signify the word rib.

*pursuing the search for ever greater light and knowledge while Adam cautiously holds back. Who is the wiser for that?"*[61] We here, second that finding.

| **The Garden — Moses 4** |
|---|
| 27 Unto Adam, and also unto his wife, did I, the Lord God, make coats of skins, and clothed them. |

- Coats of Skins

Upon informing Adam and Eve of the things that were ahead of them in the Temporal sphere, God began to provision them by making for them "*coats of skins*." Previously, Adam and Eve had made for themselves coverings of leaves.

That the LORD God would make the *coats of the skins* of animals is significant for two reasons. First, it is the *earliest* indication of the entry of death itself (*not just the warning or promise of it*) into the World and in fact within the Garden. Also note that God uses the skins of animals, not some form of textile to clothe Adam and Eve; this represents tacit approval from God himself of the killing of animals for the use of man. This speaks loudly to those who would have us understand otherwise. Secondly, though without a doubt very important, is what these skins represent and where and when Adam and Eve receive them.

The "*coats of skins*" spoken of are symbolic representations of Adam and Eve receiving their *mortal* bodies of *Flesh and Bone and Blood*. This embodiment is in recognition of the Temporal destiny the two would now face and a necessary attribute of their beings for the accomplishment of the things that were yet to occur to them in the Garden. It is important to fully appreciate the fact that bodies of *Flesh and Bone* were had by Adam and Eve while still in the Physical sphere within the Garden. Those bodies were held by the first man and woman in that place in both an immortal, *perfect* state, and a *Temporal* state (now with blood) after the Fall. More on this important understanding will be discussed in Chapter 9.[62]

After the Fall, Adam and Eve made aprons of leaves, "*And they sewed fig leaves together and made themselves aprons*" (Moses 4:13). In the ancient Middle-East, young Semitic children, both boys and girls, were typically clothed in long, loose, shirt-like frocks, bare underneath. As the young children aged and reached maturity, they would begin to wear a loincloth or "*apron*" underneath the frock. Wearing a loincloth connoted they had reached the age of maturity, no longer an innocent child, they had reached adulthood; their underparts, now capable of reproduction, were covered, protected, and hidden from view. By donning aprons in the Garden,

---

[61] From: Hugh Nibley, Patriarchy and Matriarchy (first delivered on February 1, 1980, to the annual women's conference at Brigham Young University).

[62] See Our Temple Garments, pg. 379.

the story symbolically infers that Adam and Eve had newly received the power of procreation.

Additionally, aprons were also a symbol of priesthood authority. The High Priests of Levitical Israel wore an *ephod* (apron) as a symbol of their priestly responsibilities and authority. The Endowment account tells us that Satan wore an apron that represented his Priesthood authority; it also tells us that the Father introduced the Levitical Priesthood to Adam while he was still in the Garden after the Fall. The aprons spoken of in the Garden account are full of symbolic meaning at many levels and remain symbolic in our Temple experience today.

| **The Garden — Moses 4** |
| --- |
| 28 And I, the Lord God, said unto mine Only Begotten: Behold, the man is become as one of us to know good and evil; and now lest he put forth his hand and partake also of the tree of life, and eat and live forever,<br>29 Therefore I, the Lord God, will send him forth from the Garden of Eden, to till the ground from whence he was taken;<br>30 For as I, the Lord God, liveth, even so my words cannot return void, for as they go forth out of my mouth they must be fulfilled.<br>31 So I drove out the man, and I placed at the east of the Garden of Eden, cherubim and a flaming sword, which turned every way to keep the way of the tree of life. |

## *Man Has Become As One of Us*

So what has changed? What was accomplished by the Fall? We are aware of the consequences; we each deal with them every day, but what was gained by the Fall? What was worth so great a price? "*To know good and evil.*" This attribute sets the sons and daughters of God apart from all the rest of creation. By it, we who kept our first estate and entered into the mortality of the Temporal sphere, "*become*" as God.

Please note the tense, man was not *to* become, but "*is* become." The attainment of the ability to distinguish good from evil was *immediate* with the Fall; knowing what evil *was*, would be another thing all together, and speaks back to our original set of questions posed about the Fall and the elaborate structure of a creation in three parts.

Evil (in its full realization) and depravity do not exist in the Spiritual sphere or within the Physical; there pain, suffering and death are at best just concepts, distant and vague, not experienced and intimately known. Knowledge leads to wisdom, it is not wisdom itself. Wisdom comes from successful application of knowledge; it even comes more powerfully through the *unsuccessful* application of knowledge.

Wisdom cannot be attained without application, and application is not possible without a means of its exercise. *That exercise is what the Temporal sphere is for.*

The Temporal sphere, the Universe we live in today, is governed by law; law that is known and understood by every particle within it. Every Hadron and Lepton, Proton and Electron, Muon and Gluon follows the law, filling "*the measure of its creation,*" transgressing "*not the law*" (D&C 88:25). We understand these laws as the Laws of Physics today. Newton's famously discovered law, "*for every force there is an equal and opposite reaction,*" governs the mechanics of the world we live in and describes the way we each willfully move. But God's laws govern the mechanics of our will itself. Lehi famously said, "*For it must needs be, that there is an opposition in all things*" (2 Nephi 2:11). This is a description, though lacking the broader lexicon of modern language, of the *Laws of Consequence*. Just as we gain physical strength, coordination and dexterity through the application of force applied by our muscles against an opposing force of inertial matter, we gain wisdom by application of our knowledge and will against opposing forces. The application of the physical law of opposing forces builds great structures, grows crops, and skilled surgeons use it to mend and heal the sick. Likewise, through the application of moral law, we gain moral and spiritual strength, *wisdom*, exercising our will, consequence by consequence, until we have gained full knowledge of the results of each action.

So here with the attainment of the ability to know good and evil, Adam and Eve have become like God. The attainment of knowledge was not immediate, but would require experience; what was immediate was the choice that was made to gain that experience and its consequence. Knowledge of evil would require the creation of the uncertain Temporal sphere; only after having known of evil would they be able to recognize the pure joy of good. This is the answer to the question: why the Fall?

Within the Physical sphere, our embodied Spirits can exercise will but with eternal consequence. So that we may fully appreciate the good we all knew well, the Temporal sphere was created for us to experience the depths of evil in a manner that does not persist through eternity.

In the end, after death, through the Atonement, our bodies are healed and restored to the glory they once held in the Physical sphere. The redemptive power of Jesus Christ restores us to the perfection we need to once again enjoy the intimacy of our Father in Heaven. This is the only way *wisdom* can be attained; it could not have been gained in any other way. We experience suffering and pain at great cost that we might obtain wisdom, that we might have joy.

- Wisdom is God Himself

Man became like God himself when he gained the power to know good and evil; this is *wisdom*. The word, *know* here is in the familiar Biblical context that demands intimacy and oneness, as a husband *knows* a wife. Adam and Eve would come to truly know good and evil in this way gaining true wisdom because of it. It is in this way that they truly became like God.

In the opening verses of his Gospel, the Apostle John elaborates upon what was in his time a well-established Wisdom (Logos) Theology. It is fascinating to see how Verse 22 of Genesis, Chapter 3 and its parallel verse in the fourth Chapter of Moses relate the divine attribute of *wisdom* as known by John with the concept of "*knowledge of good and evil.*"

Wisdom Theology first began to appear in Jewish writings of the Second Temple era (first and second centuries BC), as well as the writings of the celebrated Jewish philosopher, Philo of Alexandria (c. 25-50 BC). Wisdom Theology sought to personify the many mentions of wisdom found in the Hebrew Bible as a means of reconciling an immutable God of Greek philosophy with the moveable God of body, parts, and passions found in the pages of Scripture. Philo enlarged Wisdom Theology in the century before the rise of the Christian era. In fact, Philo was the one who first called this "*wisdom*" (חָכְמָה) (*chokmah*) of the Hebrew by the similar Greek term, *Logos* (λόγος), in the process authoring what has come to be known as Logos Theology.

Indeed Philo's Logos (*Wisdom*) Theology was broadly accepted in the then, much Hellenized Jewish world, defining Jewish thought and religious understanding of their God. Logos (*Wisdom*) Theology no doubt would have continued to dominate Jewish religious thought, had it not been so thoroughly adopted and amplified by John to lay the foundation of the Christian deification of Jesus Christ. Once the Logos was so broadly associated with Jesus of Nazareth, Logos Theology died out in the Jewish world and it became the basis for Christianity's understanding of God.[63]

---

[63] Note that Philo's Logos Theology sought to identify wisdom and Logos as divine; in fact, it insisted on it. This was obviously a step towards a divided and subordinate Godhead. That this theology was so readily accepted by the Jews speaks to the fact that they were not monotheistic and did not become so until they were faced with the plurality of the Christian Godhead comprised of both Father and Son (see below). As Christianity developed and built upon the Logos Theology, it came to be recognized as a heresy within Judaism, the acceptance of which by Jews was blamed for the destruction of Jerusalem and the Temple and the Diaspora. The belief in "*two powers in heaven*" by both Jew and Christian led to the adoption of the 19th Benediction late in the first-century. Read daily within the synagogue, the *Birkat ha-Minim* (םינימה תכרב "blessing/curse on the heretics") was meant to identify and separate out for expulsion from the congregation those Jews who proclaimed a separate power in creation and Christians proclaiming a distinction in God himself. The reading of the *Birkat ha-Minim* "*curse on the heretics*" initiated the separation of Christian and Jewish religions. Prior to this time, it would be proper to identify Christians as a sect of the Jews. Note: This is significant testimony and witness of early Christians teaching and believing in a distinction between the Father and Son.

Deuteronomy 6:4 "*...which makes a statement about the unity and the indivisibility of God, ...(though it makes sense in a later Jewish context as a polemic against Christianity). The verse* [Echad] *makes not a quantitative argument (about the number of deities) but a qualitative one, about the nature of the relationship between God and Israel. Almost certainly the original force of the verse, as the medieval Jewish exegetes (Rashbam and Ibn Ezra) recognized, was to demand that Israel show exclusive loyalty to our God, YHWH—but not to deny the existence of other gods! In this way, it assumes the same perspective of the Decalogue, which by prohibiting the worship of other gods, presupposes their existence... Once true monotheism became the norm in the Second Temple period... Second Temple readers and translators of the Shema were thus forced to read this and similar passages in a way that made them consistent with the normative monotheism.*" The Jewish Study Bible, Featuring the Jewish

The *wisdom* passages drawn upon so heavily by Philo and later by the Apostle John to support each of their nascent theological thoughts were found in many locations in the Kethuvim and the Jewish apocrypha: such as, Proverbs, Sirach, and the Wisdom of Solomon and Baruch. These works often describe "*Wisdom*" (Greek, *Sophia*, σοφία) as a distinct substance and more correctly a reality, entity, unto itself. Philo was first to refer to the *Logos* (λόγος) as the "*first-born*" or first "*created being,*" (Deus. Imm., 31). The Apostle John immediately recognized this *Logos* and *Wisdom* to be the Christ, the Son of God. John drew upon all the many references in Hebrew literature to wisdom to declare that Jesus Christ was divine and define His eternal role as the Son of God. In the literature known to both John and Philo, Wisdom is ascribed attributes similar to those possessed by the *Logos* (λόγος), and *The Word* of John's prologue. For example:

-Wisdom is the "*fashioner of all things.*" (Proverbs 8:26-31; Wisdom of Solomon 7:22).
-Wisdom was with God in the beginning. (Proverbs 8:22-31; Sirach 24:9; Wisdom of Solomon 6:22).
-Wisdom is the one who gives eternal life. (Proverbs 8:35; Wisdom of Solomon 8:13)
-Wisdom came into the world and dwelt among us. (Proverbs 8:30–31; Sirach 24:4-8; Wisdom of Solomon 8:1; Baruch 3:36–4:1).
-Wisdom is connected with light. (Proverbs 8:35; Wisdom of Solomon 6:12; 7:25–26, 29–30)
-Wisdom possesses glory as one of its attributes. (Wisdom of Solomon 7:25; 9:11; Sirach 14:27; 24:1-2).
-Wisdom speaks as one having authority and reserves the power to judge. (Proverbs 1:24-32).
-Wisdom was rejected by those to whom she came to bring salvation. (Proverbs 1:20–30; Baruch 3:12; 1 Enoch 42:2).

In the ancient verses of the Garden account, we see a direct relationship between knowledge, *wisdom*, *logos*, and divinity, millennia, before the concept would appear in the Kethuvim and the Jewish apocrypha listed here. These relationships were established well before they coalesced as doctrine and religious thought in the minds of Philo and John, and the budding Christian church.

Knowledge of good and evil is a divine attribute and is not sin. It must be noted that the word "*sin*", *chata* (חָטָא) (to miss, go wrong) has still not been used to this point in our readings of the creation stories, the Garden stories, or the Endowment. Any interpretation of the Garden passages as a tale of *original sin* must have been a later emendation.

What we do have here is a proclamation that Adam and Eve had changed through the Fall, that they had progressed and had become like deity, perhaps not in total,

---

Publication Society, Tanakh Translation, Berlin. Adele and Brettler, Marc Zvi, editors, Oxford University Press 2004, pg. 380.

not yet, but in part. Those that would hold as truth any concept that defines man, the sons and daughters of God, as something base and not worthy of the consideration of God, before the Grace of God can work a salvific affect in the Soul, have to deal directly with Moses, Chapter 4, Verse 28 (Gen. 3:22).

- Lest He Put Forth His Hand and Live Forever

We have already discussed the consequences that would ensue had Adam partaken again of the Tree of Life after having Fallen and that it was possible that this might have been the intent of Satan all along. Here the verse informs us that death is *attendant* to the Fall, a *necessary consequence*. Consistent with our reading here, we understand that death is a consequence and not an imposed punishment; as such, God sought to *protect man* by preventing him from *living forever* in a state that would forever preclude him from the precincts of the Father.

There are many risks involved with the Plan of Happiness, many things that could result in permanent separation from the Father. Here, we see certainly one of the most immediate threats to the plan and perhaps the most deadly. For if Adam or Eve had eaten of that tree again, *living forever in [their] sins,* God's plan for the *world* would have been destroyed before it had even started.

- My Words Cannot Return Void

"*For as I, the Lord God, liveth, even so my words cannot return void, for as they go forth out of my mouth they must be fulfilled.*" The Moses translation here provides words that are not found in any other account. They tell us that God's Word cannot be broken. By them we understand that the demands of consistency on the Word of God comprise and empower His justice. Adam and Eve are to be driven from the Garden because The LORD God said that was what He would do, not because of anger, reproach or retribution. Adam and Eve were told death would be a consequence of the eating of the tree of knowledge of good and evil; *thus it was said, and thus it would be*. As eternal embodied Spirits, physical death for Adam and Eve would not be possible; but within a Temporal sphere, within a Temporal body, it could. Simply put, what does punishment have to do with Adam and Eve's estrangement from God and their entry into the lone and dreary world? *Absolutely Nothing*...

- The Father Speaks No More

"*...and I placed at the east of the Garden of Eden, cherubim and a flaming sword, which turned every way to keep the way of the tree of life.*" That Cherubim are placed at the East of the Garden is symbolic of the sealing up of the pathway to God. Man in his Fallen state, within the Temporal sphere, would no longer have direct access to the Father but could only come to Him through the Son. After this point in the Garden story, the Father, *Elohiym*, ceases to have direct contact with the sons and daughters of God, only appearing again from time-to-time throughout the rest of history to testify of the nature and authority of Jehovah, His son.

God is symbolically oriented away from us to the East, His glory is represented by the rising Sun. The Garden was placed in the East of Eden to be near Him, now in the East, nearest His precincts, Cherubim are placed to guard the way. At this point in the story, the veil is fully lowered and closed; just like the curtain within the Temple separated the Ark of the Covenant (the seat of God) within the Holy of Holies from the rest of the world.

The withdrawal of the Father is a necessary requirement of His continued creative process, this time creating the Temporal world through tzimtzum[64] (contraction, withdrawal). Through God's withdrawal, *tzimtzum* increases and imperfection is introduced. Because of *tzimtzum*, through its creation by subtraction, the Temporal world is Fallen, less than the perfect whole.

## *Transgression and Sin?*

The Fall came with great consequence, many struggle with this. Their concern stems from a need for divine justice and fairness and the logical disconnect of punishment for a choice made either in complete innocence or an informed decision that was made in absolute alliance with the will and purpose of God. How could Adam and Eve be punished so harshly, as it appears, in either of these two cases? To resolve this apparent logical paradox, attempts have been made to mitigate the actions of Adam by calling them a *transgression* and defining that as something *less* than a sin. These attempts are at once both ignorant and unnecessary.

God is perfection, and sin is sin. No matter what it is called, sin, transgression, mistake, error, blunder or simple gaffe, intentional or not, *it is* something less than the perfection associated with God. It is all an abomination to Him and separates us (*the perpetrator*) from His presence, period.

God is no respecter of persons, and He does not *qualify* sin by degree. There is no *almost*, to sin; in the same way a woman cannot be almost pregnant, a person is either all the way in, or a person is not. It is just whistling Dixie (so-to-speak) to term Adam's Fall as a transgression as if that were somehow *less* than a sin. The use of the word transgression in our shared Scripture and proclamations never intended to infer that there was a qualitative hierarchy to sin. The word transgression in the Scriptures was never intended to be used in that manner.

The paradox exposed here (if Adam was innocent in the Garden, why did he face such dire consequence) is completely resolved when we understand that the choices made by Adam and Eve in the Garden were indeed made in the complete innocence demanded by the Scripture and the Endowment, *and at the same time*, with complete knowledge of the choice they made. How could this dichotomy be so?

Our first parents experienced consequences because they both existed as sentient, informed beings before their placement in the Garden. They possessed a state of full

---

[64] See <u>CREATION BY SUBTRACTION</u> in this work, pg. 13.

awareness *at*, *after*, and possibly *before* the Grand Council, an event that transpired *before* the Garden was planted and they were placed within it. They were informed of the purpose of the Garden, voluntarily accepted the task, and entered into the Garden aware of the consequences and possible dangers of being the first man and woman to populate the Temporal sphere. Like a spaceship, the Garden was the vessel to get them to a new world, not that new world itself.

Before entering the Garden, this first couple was fully knowledgeable of the laws and will of God; they knew of the risks and the rewards, the consequences of disobedience and the possibility of progression and eternal increase. They were aware they would have to sin to progress, that the ability to choose made possible here would necessarily make us imperfect and separate us from our Father, but they fully understood that it was God's plan that we do so. So here we have described consequence for chosen imperfection and full knowledge of those consequences beforehand.

With the desire to accomplish the Fall, Adam and Eve gallantly stepped up to the task and were placed and formed within the Garden in a complete state of innocence and forgetfulness. The veil was necessary, for their experiences had to be truly independent of any certain knowledge. Here we have the complete innocence spoken of in the Scriptures. There is no injustice.

Additionally, for those who struggle with the expression of *justice* in the Fall, we must understand that there is no punishment associated with the consequences of the Fall for Adam and Eve, the effects they experience, and we as well, are just consequences. These consequences are no more an *imposed* penalty than the consequences experienced by an auto racer who runs out of fuel in a race or damages his racecar when he hits the wall; they are the price to be paid for the thrill of the ride. Quite contrary to punishment, God has put in place a safety net, *the Atonement*, to make us whole again in the end.

- Adam's Path to The Fallen World

Creedal Christian dogma presents the justice paradox in full: Adam was created ex-nihilo in a state of utter innocence with the exception of just one prohibitive command, to not partake of the fruit of the tree of the knowledge of good and evil. The creator of *all* things (including the adversary) allows Adam, in his innocence, to be deceived by that adversary who is a supernal being and more powerful and more intelligent than Adam. The result must have been expected, the disobedience of the prohibition. Because of Adam's disobedience, God's plans for Adam were disrupted and Adam would have to be punished; not only Adam, but all of his posterity, indeed all of the Universe would be punished as well.

The creeds provide no explanation; there was nothing to be gained in any of this, tremendous cost, with no benefit.

God could have prevented the Fall in so many ways and repaired the rift right at Adam without infecting the remainder of creation. Why create the adversary and allow evil? Why place the tree before Adam? Why give the disobedience of just one man the power to affect the whole Universe? Why? Why? Why? Indeed, without answers to these questions, the story does not make any sense at all. The creator, at best, appears unthoughtful and unreasonable, worse he seems incompetent and malevolent. Something is clearly missing from our understanding here, but this is the whole of the situation as taught by the Christian church for nearly two millennia.

| **Original Sin/Freedom and Choice** | |
|---|---|
| **Innocence**<br>**Deception**<br>**Disobedience**<br>**Punishment** | **Knowledge**<br>**Informed Choice**<br>**Purpose**<br>**Fulfillment of:**<br>***God's Will***<br>***God's Plan***<br>**Blessing** |
| **Atonement**<br>**Restoration** | |

Table 7-1 *Competing Processes of the Fall.*

Alternatively, Latter-day Saint doctrine has *evolved* to provide answers to these questions and make sense of the Fall. This doctrine begins with a purpose for the Fall and informs us that the Fall was planned and in accordance with God's plan for the progression of us, His children. The Fall was God's intention from the beginning. We are told that the Fall is the means to an exaltation. That, because of the Fall, we can progress, despite the natural state it imposes. Recognizing that the Fall was the plan of God all along and providing a purpose for it, this modern doctrine provides an answer to why the Fall, and justifies its extreme cost. However, it still does not provide relief as to the matter of justice. Only our understanding here can do that.

The antithetical views of Adam's path to mortality are shown in *Table 7-1*. Both cannot be right, or are they? To answer this question, let us consider what we know in the context of multiple creations of God.

-God Blessed the Broken Road

I used the word "*evolved*" above with intention; for the Latter-day doctrine, as it is commonly understood today, has moved away from what was originally taught. It parses the actions of Adam in the Garden and now describes his choice as a *transgression* and not a sin, specifically identifying that as something less than sin. This tact was not taken until recent decades and is in ideological conflict with teachings found in the Scriptures and in the Endowment.

The doctrine has become muddled[65] because it attempts to rationalize ideas that appear to be in conflict:

---

[65] If we find the word of God in conflict with doctrine, in multiple places, and in varied ways, then the doctrine is false and needs to be revisited and corrected. Doctrine not supported by scripture and the Endowment narrative is no more than blind dogma. Those that support dogma over the revealed word of God, do not strengthen the Church nor the Gospel, but are a weakness. They are no different than the teachers of errant concepts taught by "*orthodox*" Christianity.

-Adam was innocent in the Garden, "*...if Adam had not transgressed he would not have fallen, but he would have remained in the garden...wherefore he would have remained in a state of innocence*" (2 Nephi 2:22, 23).

-Adam had to have been informed of the law because he suffered a penalty "*he that knoweth not good from evil is blameless*" (Alma 29:5).

-Adam did what he had to do to fulfill God's plan, "*But behold, all things have been done in the wisdom of him who knoweth all things. Adam fell that men might be; and men are, that they might have joy*" (2Nephi 2:24, 25).

-Adam (*and Eve*) was deceived in the Garden, "*And the woman said: The serpent beguiled me, and I did eat*" (Moses 4:19, Endowment Narrative) "*...He became Satan, yea even the devil, the father of all lies, to deceive and to blind men...*" (Moses 4:4). "*The serpent beguiled me*" (Eve, Gen. 3:13, Moses 4:19, 2 Cor. 11:3, Endowment Narrative).

As this section started, we said no matter what it is called, sin is sin; anything less than the perfection associated with God, is abomination and will separate us from the presence of our Father. Calling Adam's action a *transgression*, does not justify it. Intentional or not, with knowledge or in innocence, the action resulted in the Fall. Consider the following in understanding how sin and transgression are related:

> *"Sin is a transgression of divine law, as made known through the conscience or by revelation. A man sins when he violates his conscience, going contrary to light and knowledge—not the light and knowledge that has come to his neighbor, but that which has come to himself. He sins when he does the opposite of what he knows to be right. Up to that point he only blunders. One may suffer painful consequences for only blundering, but he cannot commit sin unless he knows better than to do the thing in which the sin consists. One must have a conscience before he can violate it. 'Where there is no law given there is no punishment...no condemnation.' He that knoweth not good from evil is blameless." (Saturday Night Thoughts, Elder Orson F. Whitney, Salt Lake City, UT (1921), pgs. 239, 240)*

Despite what is an obvious equivalency of terms "*sin is a transgression*" indicated in the very first sentence of Elder Whitney's statement, he goes on to describe the parameters of sin, expressly indicating that sin requires the full knowledge of the committer and willful intent to do wrong; the sinner must have a conscious, he must know better. What this statement does not do is delineate between transgression and sin, as it does so clearly with sin and a mere blunder. *Elder Whitney clearly understands sin and transgression to be the same thing.*

We understand the concept of an age of accountability; in some very real respects, Adam in a (*state of innocence*) while in the Garden precincts, never reached that age until after the Fall. The Endowment informs us that when Adam woke in the Garden, "*he became as a little child.*" We understand that children under the age of accountability are not answerable to God for their sins, but does that mean they do not suffer consequences for their actions? Certainly we know they do. Is it possible then to sin in innocence? President Brigham Young taught, "*If I do wrong knowingly, then I sin. When this people can live and never do a wrong knowingly, if they should sin in their ignorance, God will freely forgive that sin, if they are ready to repent and refrain from it in the future.*"[66] *President Young understood it to be possible to sin in ignorance.*

Note here as well, the conditions of forgiveness for an *ignorant transgression* are identical to those required of us for willful disobedience, they are, repentance and refraining from returning to the old way.

| The Path of Adam and Eve |
| --- |
| **Knowledge** |
| **Informed Choice** |
| **Purpose** |
| **Fulfillment of:** |
| ***God's Will*** |
| ***God's Plan*** |
| **Innocence** |
| **Deception** |
| **Disobedience** |
| **Blessing** |
| **Atonement** |
| **Restoration** |

Table 7-2 *Correct Understanding of the Fall.*

Sin is sin; disobedience is disobedience, no matter what name we may choose to give to it. There is no gradation; the result of disharmony with the will of God (*and that is what sin is*) is the same, regardless of intent, degree, or what is perceived by us to be the seriousness of the violation. The words *degree* and *perception* here are *most critical* to our understanding, for in their use we err.

God is perfection; He lives in a perfect sphere. Anything short of perfection, no matter how small the flaw, is cut off from that perfection and is doomed to remain cutoff until the flaw is covered by the Atonement. Repentance and refraining is not penitence; it is not an imposed fine, but it is a *requirement* of restoration. Without a change of mind and heart, we remain imperfect. Forgiveness covers the past; the future is ours to deal with. If we do the action again, we are separated from the perfection anew.

The apparent paradox regarding the actions of Adam and Eve in the Garden, the effects of their disobedience and God's reaction to it, has left many disturbed and confused for centuries. It need not be resolved by redefining sin and denying sacred truth regarding the innocence of life within the Garden. Eve does not need to be proven the intellectual superior of Satan. This conflict of precepts can be resolved by properly sequencing what is known of the, events, decisions, knowledge, responses, and effects told within the three creative spheres of the Spiritual, Physical and Temporal (*Table 7-2*). Doing so correctly illuminates the truth of what has become known as the Fall, shedding new light on the concepts of Original Sin, the punishment of mankind, and the purposes of our Temporal existence and the Atonement.

---

[66] Brigham Young, in Journal of Discourses, Vol. 9, pgs. 220-221.

- Teaching by Example – a Common Path for All

It is common practice to identify the story of Adam and Eve in the Garden as a type for the existence of man, a metaphor for the life of each of us in mortality. What is often lost in this treatment of the story is that Adam and Eve were actual persons experiencing life, its challenges and purpose just as we do. In fact, it was required that they do so. In that context (the reverse of what is usually applied, considering Adam and Eve must have experienced their existence as we do, instead of we experiencing as they), we can see that many of the necessary steps in life have been left out of the traditional telling of the Garden story.

Through the filling in of those gaps, allowing them to occur in the separate created spheres, it will be shown that the two conflicting tracks for the Fall, one of deception in innocence, the other of informed choice, transgression, and blessing are actually both part of the same path forming one continual road leading to God. It was the path taken by Adam and Eve, and it is the same path we each walk along today.

*Adam and Eve had knowledge of the challenges of disobedience that would transpire in the Garden and accepted those challenges and consequences willingly and with intent before they were placed in the Garden.* They understood what was to be gained and fully comprehended the risk before accepting their call to enter the Temporal sphere. "*The Lord had warned Adam and Eve of the hard battle with earth conditions if they chose to eat of the tree of knowledge of good and evil. He would not subject his son and daughter to hardship and the death of their bodies unless it be of their own choice. They must choose for themselves. They chose wisely, in accord with the heavenly love for others*" (Elder John A. Widtsoe).[67] In this our first parents were no different than each and every one of us, *the sons and daughters of God.*

They, as we, were embodied Spirits in the Physical sphere before a veil of forgetfulness was lowered over them as they entered the Garden and its immediate precincts. *Attending the Grand Council, Adam and Eve understood the purpose of the plan and made an informed decision to follow it, in fulfillment of God's will.* This would require the Fall, and mortality. They were the first to Fall, but each of us do so as well when we leave the precincts of Heaven behind us to come to Earth.

*Adam and Eve were born in innocence in the Garden and grew in knowledge and stature there.* Like each of us on Earth, they "*became as a little child,*" learning through experience until they were deceived, lulled, and coerced by the adversary to disobey the will of God. Their story is our story.

*Disobedience requires punishment, and punishment was meted out for the one individual who knew enough to be accountable within the Garden; Satan was cursed and driven away from the presence of God.* Just as a little child is not held

---

[67] *Evidences and Reconciliations*, Salt Lake City, UT (1960), Vol. 1, pg. 194.

accountable for sin or breaking the law, but suffers the consequences of those actions *just-the-same*, Adam and Eve were not punished but were faced with the consequences of the Fall. We here, through the Atonement, its redemption, and restoration, find ourselves not *punished* for our sins, but live in a Fallen world experiencing the consequences of illness and death, pain and sorrow, and all of the horrors that unfettered evil can bring.

The consequences of disobedience are severe, the risks attendant with the Fall are great, but with those conditions comes great reward. *In temporality Adam and Eve were given the blessings of procreation and a truly dynamic "free" agency.* Allowing them to not only obey law, but understand it, compose and craft it, and experience it. Their reward is our reward as well.

The effects of the Fall are severe and thoroughly separate us from the presence of our Father in Heaven; there can be no accomplishment of the natural man sufficient enough to bridge the chasm that divides us. The road home would forever remain impassible if not for the Grace of God. *The Atonement of Christ makes it possible for Adam and Eve and each of us to return to the home of our Spirits to the God who breathed life into us.*

- Original Sin Discredited

The orthodox concept of original sin is deemed to be in error in light of our concept of understanding as detailed herein. The reality of the second Physical (spiritual) Creation, one independent of the Temporal state men are found in today, disconnects each of our individual sins from the sins of our first ancestor Adam.

Apologetic descriptions of the orthodox Christian view of the Fall encompass an understanding of a "*Fortunate Fall.*" In this reading, however, the Fall is only "*fortunate,*" in that the ultimate sin of infinite hubris committed by our first parents brought about the wonderful gift of the Savior's infinite grace. The *celestial solution*, the incarnation and Atonement of Christ was a reaction to the Fall, a necessary response to the unplanned disobedience and open rebellion of man. The Fall is understood to be a disaster that results in the ultimate gift of divine restoration from God, *but it is first, above all else, a disaster* beyond the will and plan of our Father.

Conversely with the full understanding we have developed here, the Fall is shown to be in accord with the divine will and intent of a loving, purposeful God from the outset of the Physical Creation and subsequent creation of the Garden. Through the Fall, Because of the mortal incarnation of man and all that we each can learn from it, the ultimate potential of the sons and daughters of God is achievable.

> *"And now, behold, if Adam had not transgressed he would not have fallen, but he would have remained in the garden of Eden. And all things which were created must have remained in the same state in which they were after they were created; and they must have remained forever, and had no end. And they would have had no children; wherefore they would have remained in a state of innocence, having no joy, for they knew no misery; doing no good, for they knew no sin. But behold, all things have been done in the wisdom of him who*

> *knoweth all things.* ***Adam fell that men might be; and men are, that they might have joy.*** *" (2 Nephi 2:22-25)*

Through our proper understanding of the things of God, we find: 1) A separate and distinct Physical Creation and the placement of the Garden there within it, not here in the Temporal sphere; 2) Our first parents' foreknowledge of their placement in and of the challenges to be met in the Garden; 3) The purpose of life; we find ourselves directed to a firm knowledge of the truth.

Adam and Eve, indeed all of mankind had gained instead of suffered loss through the effects of the Fall. The Fall had transpired according to *"the wisdom of him who knoweth all things." (2 Nephi 2:24)*. In our first parents' own words:

> ***"Blessed be the name of God, for because of my transgression my eyes are opened, and in this life I shall have joy, and again in the flesh I shall see God.*** *And Eve, his wife, heard all these things and was glad, saying: Were it not for our transgression we never should have had seed, and never should have known good and evil, and the joy of our redemption, and the eternal life which God giveth unto all the obedient." (Moses 5:10, 11)*

Our reading of the Scriptures and understanding of the Endowment here develops (beyond all previous bounds) an understanding of a truly physical corporeality to things eternal. Nowhere is this more literally understood than in the story of the Garden. Through our reading, we have firmly planted tangible and physical things of the Garden in the realm previously described as *spiritual*, the immortal Physical sphere that existed and exists outside of the Temporal Universe we inhabit today. In the Garden, we have envisioned plants and animals, Adam and Eve, and (far more descriptive of the attributes present there) the commonplace presence of Lucifer, the Father, and the Son.

Within the Garden, Adam and Eve enjoyed the company and presence of the Father in a way that seemed natural to them. His presence was in no way atypical to them; He walked among them "*in the cool of the day*." The presence of the Father, and the Son and Lucifer there alongside Adam and Eve in the Garden did not appear supernatural, for the Garden itself was within the supernal sphere. The Father, Son and Lucifer appear to our physically embodied first parents just as they are, corporeal, for they truly are.

"*Wherefore, men are free according to the flesh; and all things are given them which are expedient unto man. And they are free to choose liberty and eternal life.*"

(2 Nephi 2:27)

**That Which Has Been Done (Then Again Once More)**

Within this work, we assert that Lucifer and his followers were *cast down* at the outset of the Great War in Heaven and not after its conclusion, as we affirmed that no such conflict could occur in the perfect precincts that attend the presence of the Father.[68] We even suggested that Michael led the righteous forces because the perfection of the Son prevented Him as well from approaching the fray. Our understanding here that Temporality itself was formed to facilitate conditions where discord is possible (a condition we know was only allowed because of the withdrawal of the presence of the Father) testifies of the truthfulness of our understanding that the war must have been fought away from the habitations of God.

To where then are the rebellious cast? This most enigmatic of questions is further exacerbated by our understanding of the three created spheres; specifically, our understanding here that the *transitory*, Temporal sphere had not yet been created at the point when the rebellion began.

The generally accepted traditional teaching that the Adversary and his followers were cast down to a Fallen Earth cannot be supported by the details of this work and the chronology they support. So what are we to understand of this?

The answer is to be found in several seemingly unrelated teachings from the Endowment narrative and the Genesis 1 accounts. Let us lay the structure of our understanding here carefully, introducing each of the several teachings individually, in a sequence that develops an eternal perspective of things within the Physical sphere.

First, returning to the verses in Genesis, Chapter 2 that tell of the creation of the Physical sphere:

> "¶*These are the generations of the heavens and of the earth when they were created, in the day that the Lord God made the earth and the heavens, And every plant of the field before it was in the earth, and every herb of the field before it grew: for the Lord God had not caused it to rain upon the earth, and there was not a man to till the ground.*" *(Gen. 2:4, 5)*

These two verses have formed an awkward pause and even a diversion in the seamlessness of the Biblical creation story and the subsequent Garden account, but they tell an important part of the story that is otherwise overlooked, that of the creation of the Physical sphere.

Likewise, the Endowment narrative also pauses at the same point in the chronological sequence between the creation of the (this time) Physical sphere and

---

68 See our discussion, - Michael and His Angels, pg. 353.

the telling of the Garden story. This interregnum is equally important because it puts solid ground beneath our feet on which several key truths can find support.

In the pause, we are witness to the "*birth*" of Adam, "*then let us go down and form man in our own likeness and image*;" we witness the introduction of Eve, "*Adam, here is a woman whom we have formed and whom we give unto you,*" and we witness the blameless state of the first man for, "*having forgotten all, [he] will have become a little child.*" Throughout all of this, Adam and Eve are not yet in the Garden for God says: "*we will place you in the garden,*" and later He commands, "*Jehovah, introduce Adam and Eve to the Garden which we have prepared for them.*" There is great significance in this detail. By it we are informed that both Adam and Eve are innocent and fully embodied within the Physical sphere before their entry into the Garden; we are informed along with them, that throughout it all they have been on an Earth, organized for them.

Innocence here does not imply that they were uniformed. It does speak to the fact that a veil of forgetfulness had enshrouded them already, a fact emphasized in the narrative by the reiteration of the plan to them by the Father, "*we will there command you and Eve to multiply and replenish the earth, that you may have joy.*"

Somewhere then, on the surface of a beautiful, eternally perfect, planet called Earth stood Adam and Eve, beyond the confines of a Garden planted there, "*eastward in Eden,*" waiting to be introduced therein. This Earth, we are told, was the one we observed being created throughout the previous five days of the Endowment narrative. It is the one that is the Physical orb reflected in the very presence of the Temporal Earth we inhabit today, in the exact same way we each are merely an image of the perfect Physical being we are in the Physical sphere. We are aware this "*Earth*" is not the only such world in the Physical sphere, because at the outset of the creative process displayed, we are informed God said: "*yonder is matter unorganized, go ye down and organize it into a world <u>like unto the worlds we have heretofor formed.</u>*"

The creative process described in the Endowment narrative, like its companion creation (that of the Temporal sphere), formed not only the entire sphere, but also individual and numerous worlds (lands or Earths) within it; many of them inhabited long ago before this Universe was created, before the event we call the Big Bang. These numerous orbs are home to countless worthy Physically Embodied Spirits, all of which were created at the same time as Adam and Eve; "*I, the Lord God, had created <u>all</u> the children of men; and not yet a man to till the ground; for in heaven created I them; and there was not yet flesh upon the earth*" (Moses 3:5). Within the Physical sphere, are also to be found many worlds that are home to those that had followed Lucifer into rebellion.

Let me explain.

The Physical sphere is the sphere from which we came, but it is also the same sphere to which we will return and carry on our eternal existence. We understand the Physical sphere to be the stage on which much of our pre-existent lives played out, but do not let the fact be lost that it was created to be our home throughout an eternal future.

The Physical sphere from its *initial* creation holds countless planetary spheres found in regions of vastness that were then, are now, and forever will be, organized and segregated into three realms: the Telestial, Terrestrial, and (at least) some elements that are Celestial. The Great War in Heaven was waged in the confines of this same Physical sphere and Lucifer and those that chose to follow him find their home even now, in this sphere, [69] in the border wastelands there, on very many worlds and locales, in the realm called Outer Darkness. In those darkened environs to which they were cast, evil dwells and grows, as the deceiver introduces more maleficence to them each day.

The Temporal sphere exists entirely within the confines of the Physical sphere, as if in a bubble, formed in one selected region in the Telestial realm within the vacuum created by the withdrawal of the presence of God. Enclosed in this expanding "*void*" of Temporality, are mere images of greater things found in the Physical sphere, simple copies of existent things that are eternal. Temporality is limited and will evaporate in time, leaving nothing but its memory behind and eternal Souls made better through the experiences and agency it afforded.

This sphere, the Physical, also holds the Spirit World brimming with embodied Spirits who have already experienced their time in Mortality and whom are anxiously awaiting the promised resurrection. (*Figure 7-3*) Within this Physical sphere Adam and Eve stood as innocents on their orb, in the *Terrestrial* region, during this brief period after the sixth day of creation ensued and before the Garden story begins.

With a firm understanding of the geography of this dynamic eternal sphere, let us continue. At the time immediately preceding Adam's entry into the Garden, the Physical sphere was teeming with embodied Spirits all created by God. Those that had kept their "*first estate*" waited anxiously for the Fall and the experiences that would ensue in Temporality. Those that kept not their first estate waited as well, but for different reasons. It was the task of Lucifer in the Garden to expose Adam to evil and trap him in its clutches eternally, as is the case with Lucifer and all who followed him. Descriptively, and to our point here, this plan of Satan reveals something of the nature of the Physical sphere and confirms our conclusions here. When pressed by the Father about his actions in the Garden, Lucifer in defense replies, very enigmatically: "*I have been doing that which has been done in other worlds.*"

These words of the deceiver are most often understood to refer to previous rounds of Gardens and their separate inhabitants being tempted and Falling, but this is not so. Here we understand Lucifer to be referring to the worlds to which he and his followers were cast, at the outset of the Great War in Heaven. On those orbs the

---

[69] See Michael and His Angels, pg. 353.

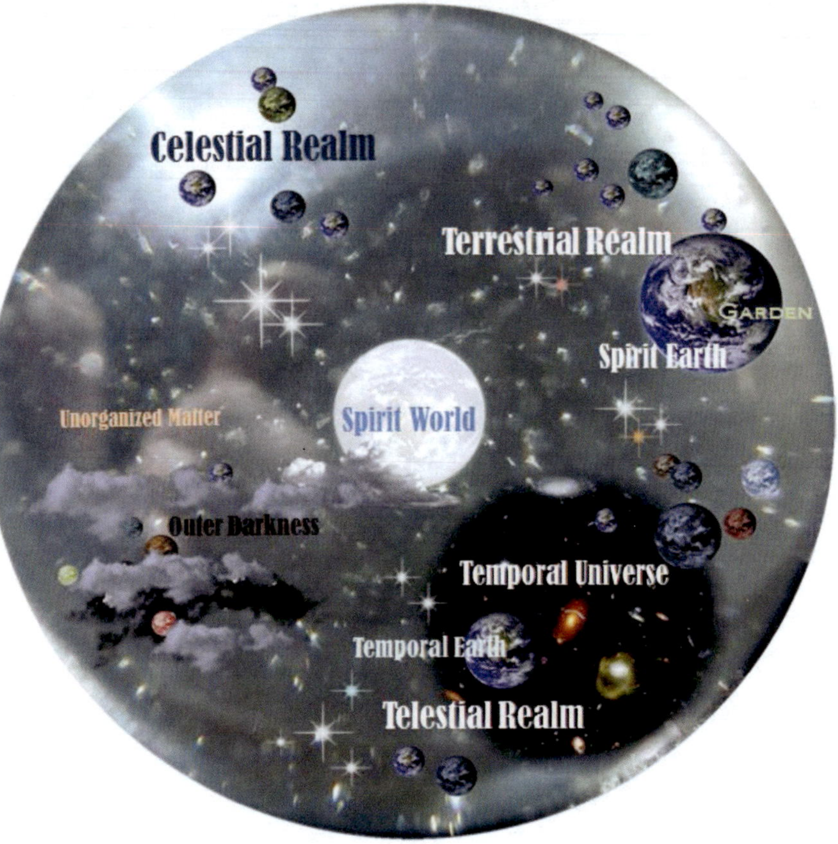


Figure 7-3 *The Physical Sphere and its Environs.*[70]

---

[70] I am extremely hesitant to include such an illustration because, at best, any such representation is inadequate. It is impossible in two dimensions to adequately detail the three dimension realty of the Temporal sphere. It is immeasurably more difficult to describe the added dimensionality of the overlay of the Physical sphere on top of temporality. It is wholly improper to define the separate spheres and realms contained in the Physical sphere as places, confined to location in any terms that could be illustrated, indeed it is misrepresentation to understand them as bounded at all. Additionally, the concepts I have shared here in this work feed into the unique knowledge of the individual readers, developing in their own minds images and understandings that are much different than mine. What we are left with here then, is an entirely inadequate representation of the Physical sphere as I understand it. This is a representation of the relationship of its parts only, not a factual display of any of its elements. Here it can be understood that the three degrees of Glory, the Spirit World and Outer Darkness are all subsets of the Physical Creation. The entire Temporal Sphere is too a subset of the Physical, a *temporary effect* within the Physical created through the subtraction of the presence of God. Though impossible to illustrate here, I believe the Temporal Sphere shares the same "*space*" as the Physical sphere, overlaying its components like our Temporal bodies overlay our embodied Spirits. As such everything we are in this Universe is in effect everything we are in the Physical, just less so, more constrained, and decidedly less perfect. In light of this context, the illustration shows two separate Earths, one Spiritual and one Temporal, when in fact they are both individual manifestations of the same thing: one dying, changing and decaying (much different today than it once was), the other pristine, eternal and perfect. Also, though the Spirit World is depicted as a separate sphere in the middle of the picture, it too overlays Temporality completely,

knowledge of evil was no doubt known, introduced to the unfortunate inhabitants by the "*shining one*" himself through his prideful willfulness. In his response to the Father here, Lucifer, as he so often does, distorts the truth by misrepresenting it. Though he was being *selectively* honest when he said that what he was doing in the Garden was what he had already "*done in other worlds*,"[71] he neglected to include the fact that it was *he* who had introduced evil to the inhabitants (*his followers*) on many worlds before, as he waged, and even then continued to wage, war on the Father's plan. Lucifer truthfully spoke, for by introducing Adam and Eve to evil in the Garden, he truly was doing what he had done on other worlds before.

Here in the passages of the pause, we see truth contained. We are told through allusion of the state of the Created Spirits awaiting their Temporal existence on extant worlds in the Physical sphere. Through the recorded words of the Deceiver spoken in the Garden, we know of the location of the rebellious, of where they were cast and of their fate in the darkness.

We have no need to imagine a repeating chain of the events depicted in the Garden, of multiple Adams and Gardens; the Endowment narrative does not infer that at all.

I would be remiss here if I did not state that the record of the Garden account, as is true with so very much of revelation, is truly not constrained to a single coherent narrative or any one place or time. The story not only represents the very real events critical to the Plan of Salvation, truly actions carried out *once* by one Adam at the outset of the path which lead to the Temporal sphere, but it also is a representation of the history of each of us, the sons and daughters of God.

Through the story of Adam and Eve, we see our story told as well. Like they, we each stood at the Grand Council and accepted the Father's plan; we fought in the struggle of the Great War in Heaven. We each became an embodied Spirit and were fully informed of the consequences beforehand; we chose to enter Temporality in innocence, as little children, to Fall and experience the wages of sin and disobedience. We are, like our first parents, recipients of a Temporal body and are cast out from perfect precincts to a Fallen world. The Deceiver tempts each of us, and like Adam and Eve, we place our trust in a redeeming Savior to atone and restore us to the presence of our Eternal Father – to the home from which we came. We must remember that in the eyes of God all things are before him at once, that He cannot separate our beginnings from our ends, because those concepts have no meaning in eternity. The creation and Garden stories told are at once linear in time and wonderfully circular with an eternal perspective.

---

including the Temporal Earth. The Spirit World not only surrounds us, but it is within us, and can be thought of including all that is portrayed here within its bounds. For what it is worth, with all of its insufficiencies, I hope you enjoy the picture I see.

[71] The "*worlds*" spoken of here by Lucifer need not represent fully organized spheres but very well may have been cobbled together habitations out along the fringes of the Physical sphere, an eternal refugee camp where darkness reigns. Perhaps "worlds" as used here refers to orbs or the "*earth*" (*matter*) of the Physical sphere, but more likely it was to the vastness of the open heavens within the Physical realm that the great armies massed and joined in struggle, truly a realm that could be called *outer darkness*.

# PART 3

# THE SPIRITUAL CREATION

*"Man was also in the beginning with God. Intelligence, or the light of truth, was not created or made, neither indeed can be."*

*D&C 93:29*

CHAPTER 8

# THE REALM OF OUR ETERNAL FATHER

## Spiritual Creation

In Chapter 1, we discussed that for much of the last 2000 years, scientific and philosophical thought determined the Universe to be without a beginning; as such it was understood to be eternal in nature. That it was eternal had not been determined by any empirical evidence, but was deduced by logic. The logic of this unavoidable truth was so sound and seemingly inescapable that the eternal nature of the heavens had been accepted theologically as well, despite scriptural pronouncements that would indicate otherwise. Remember that early in the last century, Elder B.H. Roberts wrote:

> *"... we hold the universe to be self-existent and eternal. Duration, space, matter, force, mind – each infinite after its kind – are its prime and included factors. The universe is comprised of these.*
>
> *Such a universe can have no first cause, since that would imply a time when there was no cause, and there has been and can be no such time. Causation is eternal, and in the eternal and infinite universe there has been operation always Eternal Cause."*[1]

In the context intended by Elder Roberts (a discussion of the Universe around us, our Temporal habitation), his logic is misapplied. The logic of an eternal construct requiring an eternal cause remains without fault and is unassailable, but it does not apply to the Temporal nor to the Physical sphere, for they were both created with purpose and are not without beginning. Elder Roberts, and indeed all of the best minds of the world in the opening decades of the 20th century, assumed the Universe we inhabit to be eternal, static and unchanging; in this they were all wrong. There is,

---

[1] The Truth, The Way, The Life, pg. 260; (B.H. Roberts, edited by Stan Larson: Smith Research Associates, 1994).

however, an eternal sphere, one which subsists through causation that is eternal. The construct of a creation process in three parts as detailed in this work, restores the integrity of the logic of Eternal Cause, while providing for Temporal and Eternal created realities which fulfill all of the requirements of the revealed word of God.

The Scriptures clearly describe a beginning, a single point of time in the distant past when all things were created.[2] In the last century, scientific thought finally caught up with this notion when evidence was discovered for a beginning to the Universe and the Big Bang theory became its standard model. We have ascribed these, the scriptural and scientific beginnings to the same event, the creation of the Temporal Universe and all things within it. This was, we said, initiated through the precept of "*first cause*" (creation of an uncertain world in which free agency could be expressed and experienced, was itself, a resultant of the Fall).

We have introduced an understanding of a creation before that of the Temporal, we called and described it as Physical. This creation is detailed in the Endowment narrative, and through it, an eternal, tangible sphere was created, one without an end but with a beginning. We understand that this too was initiated through the precept of "*first cause*" (embodiment of the primordial intelligences, allowing for eternal progression). These two creations, however, are not of the "*self-existent*" and "*eternal*" described by the logic of Brother Roberts and the many that went before him, they are not the foundation of that which exists because of "*Eternal Cause*."

Within this chapter, we will finally discuss the eternal as expressed by the traditional use of the word. This sphere, the Spiritual sphere, is without beginning and without end. We are not describing here an eternal round of Universes, a never ending chain of new beginnings following a like chain of repeated endings. We are also not describing a *multiverse* or anything like it. For the Spiritual sphere truly is eternal, like the Father Himself, and through Him, it too is self-existent. Of what then do we speak when we talk of the *Spiritual Creation*?

What we are discussing here is the creation and state of things *within* the Spiritual sphere, *not the creation of the sphere itself*. In both of our previous discussions of the Physical and the Temporal, the creative processes we described began with the making of the spheres which included the creation of things within them by the same processes; this will not be true in our discussion of the Spiritual sphere.

The Spiritual Creation we will discuss and detail is not, and should not be understood to be, the creation or beginning of the Spiritual sphere itself. Of its beginning, as well as that of Father God, we make no attempt to explain or qualify. Within this work it is our understanding that Father, and the sphere He inhabits, extend forever into an eternal past.

---

[2] The opening words of the Hebrew Bible, *Beresheet bara Elohiym*, literally means, at the head, God created. "*At the head*" (בְּרֵאשִׁית) refers to the beginning of time. As the Christian era began, the Apostle John opened his testimonial gospel with the Greek words, "*en arche en ho logos*." John, in his use of "*en arche*" (ἐν ἀρχῇ) also spoke of the most ancient of times.

Because we have no direct experience with the Spiritual sphere,[3] we can only discuss that which is known of it through revelation. We understand explicitly that things *spiritual in nature* are things unembodied; they are without a discernable physicality or tangibility. The spiritual things we will discuss in this chapter were created at some point in the distant past by our Father God. The creation of these things, the first of which was the Son, Jehovah, Christ, "*who is in the image of the invisible God, the firstborn of every creature*" (Col. 1:15) took place within the Spiritual sphere in accordance with the nature of things there.

***Throughout this work, we have assumed a definition of eternal that includes something which has no end but does not address a beginning. We understand by this an eternal nature for things that are in fact created.***

Throughout this work, we have assumed a definition of eternal that includes something which has no end but does not address a beginning. We understand by this an eternal nature for things that are in fact created. For example, it is hoped that through the Atonement, the saved will live forever with Father in Heaven. What is forever if not eternity? In this context, the Temporal is not eternal because we foresee it to have an end. The Physical is eternal, for though it was created at some point in the past, it will have no end. Likewise we understand that within the Spiritual sphere many things (individual intelligences) were *created* but though they have a beginning, they are eternal for they will have no end.

Christ, Jehovah, *YHWH* of the Old Testament, is a created being, the *firstborn*, yet He is eternal and can rightfully say, "*I am the Alpha and Omega, the beginning and the end*" (Rev. 21:6), for He *was* first and *will be* last. Of the Father and the Spiritual sphere we have no knowledge of a beginning, and despite many who have speculated on such a beginning, we will not do so here. Within this work, we describe and conclude that the Father extends into eternity in both directions of time, into an eternal past with no beginning AND an eternal future that will have no end.[4]

---

[3] The circumstances of Joseph Smith's first vision aside, the experiences of the Brother of Jared and Moses appear to be the only recorded instances where God allowed the viewing of His true spiritual being. In both cases, the viewing was enigmatic and supported through the supernal powers of the Holy Spirit: "*Behold, this body, which ye now behold, is the body of my spirit*" (Ether 3:16), "*...there shall no man see me and live*" (Exodus 33:20).

[4] Note here that we did not include mention of the Spiritual Sphere as having no beginning for we cannot conclude that to be true. We understand an eternal God to be eternally creating, progressing and

Very little has been revealed about the creations within the Spiritual sphere. We know that within the Spiritual sphere there existed Intelligences before they were embodied and that there also was unorganized matter. The scriptural record includes passages from: Revelation, Abraham and the Doctrine and Covenants. The Endowment narrative describes this realm as well for it was the backdrop of the creative process that formed the Physical sphere described therein. We will examine all of these sources here and include, when appropriate, additional teaching both ancient and modern.

**The Spirit World — Abraham 3**

19 And the Lord said unto me: These two facts do exist, that there are two spirits, one being more intelligent than the other; there shall be another more intelligent than they; I am the Lord thy God, I am more intelligent than they all.

-Intelligence, the Core of Our Existence

As with the first verse of Genesis, this verse covers a *lot* of ground; in just a few words very much is said. First, let us deal with both Spirit and Intelligence. Note the context here; God is not saying that Spirit and Intelligence are the same thing. In this verse intelligence is an adjective, describing a quality of Spirit. The concept of independent intelligences, entities unto themselves, does not first appear until three verses later, in Verse 22.

Intelligence is fundamental to our very Being; it is what makes us, us. Intelligence is what is in us that *is* of God. Everything else, our Physical eternal body, and our Temporal body, that which made our Intelligence first a Spirit, then a Living Soul respectively, that which will comprise eventually our Immortal Soul will have been *added* to that intelligence. We are intelligence at our core. It is proper, though perhaps confusing to the ear to describe us, man, as Spirit, Soul, and Intelligence. In our hierophantic teachings, these terms are often intertwined; of this we must remain aware.

Though it is a founding belief for our country, the precept that "*all men are created equal*" may not be true. This verse, however, does not decide that question, for it speaks of the intelligence of Spirits, which through their independent action and experience varies significantly as the quality of intelligence is added upon. It *may* be

increasing, as such He may well have *created* that which we know to be purely spiritual and that the spiritual itself may have *become* a state to which God evolved through eternal progression from some other unknown previous state. We do not accept or hold the view that God is immutable, we understand Him to be forever good. "*For behold this is my work and my glory—to bring to pass the immortality and eternal life of man*" (Moses 1:39); in this He will be *forever faithful*.

true that each individual intelligence was created with an equal allotment, but it is *not necessarily so*. By the time we get to Verse 22, this point is moot, as it is readily apparent that already before the embodiment of the Physical Creation the Intelligences are differentiated, some qualified as "*noble and great*." We are not told if this differentiation is as a result of individual effort and advancement. What is clear here, however, is that to whatever height of intelligence a Spirit may attain, God will remain more intelligent than all. Since we are told, "*The glory of God is intelligence*" (D&C 93:36), we can be certain that none of us will ever exceed or equal Him in Glory, and from the Scriptures we are assured that no one being ever has.[5]

- Intelligence is Eternal

> *"Intelligence is eternal, and exists upon a self-existing principle. It is a spirit[6] from age to age and there is no creation about it." (Joseph Smith, King Follett Sermon, April, 1844.)*

The self-existing principle spoken of here is the "*Eternal Cause*" of Elder Roberts' logical argument. It is the "*law irrevocably decreed in heaven before the foundations of the world, upon which all blessings are predicated*" (D&C 130:20). God has a desire to create and that desire is a blessing for us, His created. As is true with us, the pure desire to create comes from love. Procreation is a gift; it is associated in the eternities with eternal increase, and *eternal increase is eternal cause*.

Though we were created, that is, *though our intelligences were created*, the substance of our intelligence comes from God; it is indeed a part of Him and as such it is eternal. Of this subject, the Prophet Joseph explained: "*Element had an existence from the time God had. The pure principles of element are principles which can never be destroyed; they may be organized and reorganized; but not destroyed. They had no beginning, and can have no end*"[7] (The Contributor, vol. 4, pg. 257). The Prophet further went on to expound the same teaching as follows:

> *"The intelligence of spirits had no beginning; neither will it have an end." (History of the Church, 6:311) (Joseph Smith, Mill. Star, Vol. 23, pg. 262.)*

---

[5] This fact alone must be considered and used for context with every utterance spoken about Man becoming like God and God having once been like man. Within this work we will make no such claims to man ever becoming equal with the Father in glory or sovereignty. We feel no need here to imagine an endless chain of Gods stretching into the past. Our God is eternal without beginning and without end, more intelligent and more glorious than all.

[6] "*A spirit from age to age*"— not *spirit from age to age*; but "*a spirit*;" that is *an entity, a person, an individual*. The Prophet's statement here could well be taken as an interpretation of, "*Man was also in the beginning with God. Intelligence, or the light of truth, was not created or made, neither indeed can be*" (D&C 93:29).

[7] "*There is no such thing as immaterial matter. All spirit is matter, but it is more fine or pure, and can only be discerned by purer eyes; We cannot see it; but when our bodies are purified we shall see that it is all matter*" (D&C 131:7, 8).

Intelligence is the light of truth and independent in action. "*Man was also in the beginning with God. Intelligence, or the light of truth, was not created or made, neither indeed can be. All truth is independent in that sphere in which God has placed it, to act for itself, as all intelligence also; otherwise there is no existence*" (D&C 93:29, 30). Of the Christ, the Apostle John testified, He, "*...was the true Light, which lighteth every man that cometh into the world*" (John 1:9). Because we are intelligence, Man was in the beginning with God. Our independence began in the Spiritual sphere long ago. We, however, were not the first to be born, as John said, our light comes from yet another.

Christ, Jehovah was the firstborn of the Intelligences; "*And now, verily I say unto you, I was in the beginning with the Father, and am the Firstborn;...Ye were also in the beginning with the Father; that which is Spirit, even the Spirit of truth*" (D&C 93:21, 23). Because the Savior was the firstborn, we understand that all of our intelligences were created through Him; "*All things were made by him; and without him was not any thing made that was made*" (John 1:3).

These verses and teachings affirm not only the co-eternity of the Son, but that of the Spirits (Intelligences) of men.

**The Spirit World — Abraham 3**

22 Now the Lord had shown unto me, Abraham, the intelligences that were organized before the world was; and among all these there were many of the noble and great ones;
23 And God saw these souls that they were good, and he stood in the midst of them, and he said: These I will make my rulers; for he stood among those that were spirits, and he saw that they were good; and he said unto me: Abraham, thou art one of them; thou wast chosen before thou wast born.

- The Intelligences were Organized

Abraham was shown the intelligences and was told they were "*organized (perhaps meaning formed) before the world was*." These verses are somewhat enigmatic because Abraham variously calls those he saw, Intelligences, Spirits and Souls. Our understanding of a Soul precludes this scene from a time before the world was. Spirits properly fit into the Physical sphere, while Intelligences alone could be applied to each of us in any of our many manifestations in the three spheres of creation. Abraham saw that individuals had in some way set themselves apart from others, becoming "*great and noble*." Abraham among others was chosen by God to be a ruler before he was born.

There can be no doubt that Michael, as an independent intelligence, before he became the embodied, immortal, Adam in the Garden, within the Physical sphere, had distinguished himself among the intelligences in such a way that he was included as co-creator with the Son in the forming of the Physical sphere. This single detail alone provides the strongest evidence of an existence of individual entities, sentient and agent before the creation of the Physical sphere.

***Michael existed, distinguished himself, co-created as an independent being, before the Physical Sphere was created, before he was given a physical body and made an eternal Spirit by God the Father and Jehovah, in the sixth period of the Physical Creation.***

Let this fact not be lost. We began our description of the three creations with the Temporal, where we are today, intentionally with this purpose in mind, to lead to an understanding that there is indeed a third creation, long ago in the past, before the forming of the eternal things of the Physical sphere. We detailed the six days of the Temporal Creation and made a case that the Genesis 1 account described the forming of our Universe today and all of the natural processes within it. We developed an understanding that the creation of the eternal Physical sphere was a separate event preceding the Big Bang, that all things were embodied Physically (created spiritually) before they were made naturally. We detailed the six creative periods of the Endowment account which describe the creation of this eternal Physical sphere and spoke of the Garden placed within it.

In the sequencing of these discussions we were deliberate, for that brings us directly to Michael and his role of co-creator of the Physical. His existence, in a time before the time before time, is marked and without question. Michael existed, distinguished himself, co-created as an independent being, before the Physical sphere was created, before he was given a physical body and made an eternal Spirit by God, in the sixth period of the Physical Creation. That we are *equal* with Adam in species and order of created being is also unquestionable, placing our creation, existence, along with his, at a time and place before the creation of the Physical sphere and the embodiment of our intelligences, which formed our Spirits.

President Romney indicates that the organization of Spirits included first a division into groupings of male and female; gender, he espoused, is a fundamental attribute of intelligence: "*In origin, man is a son of God. The spirits of men 'are begotten sons and daughters of God.' Through that birth process, self-existing intelligence was organized into individual spirit beings*" (Marion G. Romney, Conference Report, Ensign, Nov. 1978, 14). President Romney seems to allude to the point in time when self-existing intelligence is organized into individual entities (Spirit beings). If we read this in that context, we discover a critical insight. Though we expect gender to be associated with our Spirits as part of our spiritual embodiment, we understand here, through the words of President Romney, that gender appears to be more fundamental to whom we are than even that.[8,9] Though we can understand our gender finding root within our minds, our thoughts, what we readily associate with our intelligence, we have no hope of understanding how both male and female intelligence emanates from the Father without there being a Mother as well, or at the very least a female component in this unique creative process. Why would intelligences, not capable of procreation, be created from their inception as both male and female if gender is not a critical attribute of our individual eternal existence?

**The Spirit World — Abraham 3**

24 And there stood one among them that was like unto God, and he said unto those who were with him: We will go down, for there is space there, and we will take of these materials, and we will make an earth whereon these may dwell;

- One Like Unto God

The "*one among them like unto God*" is the Son of God, Jesus Christ, who is recorded as describing what we recognize to be a component of the great Plan of Happiness. That this was seen and recorded by Abraham makes it the earliest recorded mention of the Grand Council available. The Savior here previews, in words familiar to the Endowment, the making of a dwelling place for man. Because Abraham has freely used Spirit, Soul and Intelligence interchangeably, we are unfortunately left without a clear picture of which creative event is spoken of here. Though the words used are similar to those of the Endowment, we cannot be certain that the making of a dwelling place is that of the Physical sphere. Therefore, on the evidence found in this verse alone, it remains unclear when the Grand Council took place and whether it was in the Spiritual or the nascent Physical sphere.

---

[8] Christ himself affirmed gender to be a fundamental attribute of our existence; it is not a manifestation of cultural influences, or a result of natural, Temporal conditions, but is an attribute of Spirit, as shown here, even an attribute of our intelligence. When the Pharisees enquired of the Savior about marriage and divorce, He replied, "*But from the beginning of the creation God made them male and female*" (Mark 10:6).

[9] Gender is fundamental to what we each are. However, we must understand that here in Temporality there are many factors at play which alter and distort our true makeup, what was intended for each of us by God. The uncertainty at play in the natural world can have deleterious effects on our genetic makeup, deforming or developing our bodies in many ways including those most readily apparent as gender. Layered upon these natural, harmful, causes can also sometimes be found in the effects of the Adversary and sin in our lives, transforming individuals psychologically through wrong choices and unnatural desires.

**The Spirit World — Abraham 3**

25 And we will prove them herewith, to see if they will do all things whatsoever the Lord their God shall command them;

- We Will Prove Them

Herewith is a very descriptive word providing a context not always understood; it can alternately mean, *along with this*, or, *by this means*. This text was rendered through inspiration directly into English by the Prophet Joseph; so we do not have a comparison to the original language to provide further insight and clarity. It remains unclear how we are to understand this statement. Would the creation of a realm for man to dwell be the means of his testing, or would a separate testing be employed in addition to the process of creating a dwelling place? The latter interpretation fits more in the context of the creation of the Physical sphere, the subsequent planting of the Garden and the test of obedience to be given therein. The former describes the Temporal sphere which by nature, through its creation, is uncertain and fallen, producing conditions that test men directly.

**The Spirit World — Abraham 3**

26 And they who keep their first estate shall be added upon; and they who keep not their first estate shall not have glory in the same kingdom with those who keep their first estate; and they who keep their second estate shall have glory added upon their heads for ever and ever.

- The First and Second Estates

The Eternal spheres were not without challenges that tested the bounds of Agency. At some point, after the Grand Council, a choice had to be made and conflict ensued. Those that chose the Plan of the Father and followed Michael in support of the Atonement of the Son are those who kept their *first estate* (their first inheritance). Notice here that it is said those that chose to follow Lucifer, *shall not have glory in the same kingdom...* What does this mean? It remains unclear where and when the Grand Council occurred. The answer to this significantly impacts our view of the first and second estates and the state of development accomplished for Lucifer and his following.

We *can* imagine the Grand Council occurring in the Spiritual sphere, before the Intelligences were embodied with the creation of the Physical sphere. Such a chronological sequence requires the Intelligences to be independent and agent, capable of understanding and making what has to be the most important decision in their eternal existence. This chronology would support an understanding that Lucifer and all others at the time did not have any form of body (tangible) at all. This would be in keeping with a *non-scriptural* Latter-day understanding that those who followed Lucifer and Lucifer himself were denied physical bodies. (It is, however, assumed by that understanding, that the bodies *denied* are those Temporal bodies we attain here in this Universe. In this work, we understand a separate Physical sphere and *embodied* Spirits within it.) This chronology also aligns with our understanding that Intelligences were capable of growth and choice allowing some to distinguish themselves from others. If the war in Heaven took place in the Spiritual sphere, its forces were led by Michael before his embodiment, before he had become Adam. This scenario conforms to Scripture and common Latter-day teaching. But as we discuss in this work, this scenario is not without a conundrum. For a war and discord to take place in the perfect precincts of the Spiritual, would mean that such activity would of necessity transpire in the presence of God.

In light of the three creations we discuss in this work, there is another chronology to be considered, one that would not require direct conflict in the Pavilions of God. Perhaps the Grand Council took place in the Physical sphere after it was created but before the Garden was planted. In this scenario, all who participated in the Council would have been embodied with eternal Physical bodies, including Lucifer who stood in opposition to the Plan of the Father. The war would have been accomplished by forces embodied. The righteous forces were led by Michael, meaning that his renaming within the Garden is associated with it being a Temple and in recognition of him stepping into a representation of the Temporal world.[10] This chronology also conforms to Scripture and Latter-day teaching, but it requires us to accept that Satan and his followers are embodied, that the keeping of the first estate is rewarded in some way other than the receipt of a Physical body. Somehow then, the Temporal body we realize here in this Universe carries an advantage over the Physical body we previously attained in the eternal Physical sphere.

The results of the Grand Council, wherever and whenever it took place, were not to the liking of Lucifer: "*And the second was angry, and kept not his first estate; and, at that day, many followed after him*" (Abr. 3:28). He and his followers were denied the attainment of Glory that others would achieve as a result. We understand Glory to be intelligence, the attainment of which comes through experience and the exercise of free will. Lucifer and his followers are denied the experience of Temporality. This may be difficult to comprehend, but consider this:

Though Lucifer, *satan* (שָׂטָן), is the God of this world, and dwells here, He is not empowered as we are. His followers seek to possess flesh in any form (including the bodies of swine), seeking its experiences, attempting to exercise "*free*" will. Lucifer was cursed in the Garden, but he retains the power to bruise the heel of man. However, the flesh of Temporality, worn by the sons and daughters of God, as offspring of Eve, our mother, empowers each of us to crush the Adversary's head. Through the denial of free will attained in Temporality and without the breath of life, the *Ruach*, which connects us intimately with the Father, those that kept not their first estate are deprived of experience, the increase of intelligence, and the Glory of God that attends it.

---

[10] Much more on this subject will be presented in the next chapter, Day Breaks On An Eternal Morning, pg. 357.

## Intelligence

It is our express understanding in this work that the entities that comprise each of us were strictly intelligence, in the Spiritual sphere, before the creation of the Physical sphere, before they were embodied. We will discuss in detail just what an embodied Spirit is in a subsequent chapter,[11] but it is comprised of both a physical, eternal, tangible body which encloses a self-existing, independent, and *agent* intelligence.

In Latter-day contexts, *intelligence* is usually interpreted in one of two ways. Intelligence(s), used as a noun, frequently occurs in Latter-day Scripture and teaching, always referring to a form held by the premortal sons and daughters of God in the most distant past. In its noun form, Intelligence represents the literal offspring of our Heavenly Father as individual beings prior to their Spirit birth (embodiment). These individual Intelligences are not just young and childlike, but experienced, noble and great as well. As an adjective, intelligence is a *quality* and a *thing* to be *acquired*; it is light and truth, "*the true Light, which lighteth every man that cometh into the world*" (John 1:9). The Scriptures also speak of intelligence as the primordial substance (element) that existed in and with God before our individual Intelligences (the noun) were created. Intelligence, *as an attribute*, has always existed.

The scriptural source for the concept of *intelligences* as independent entities is the third chapter of the Book of Abraham; it is first discussed there in Verse 22: "*...the Lord had shown unto me, Abraham, the intelligences that were organized before the world was.*"

Some of the characteristics and qualities of intelligence, that of our individual being and as an attribute to be acquired and added upon, are listed and expounded upon here:

- The LORD showed Abraham "*the intelligences that were organized before the world was*" (Abr. 3:22). The individual intelligences that we are, before we were embodied by the creation of the Physical sphere, were organized, as individuals and in familial, tribal and social groups. The twelve tribes of Israel are a familial association (organization, grouping) of the intelligences that extends back to these earliest times. This does not, however, refer to a Temporal genetic association; children from the tribe of Ephraim can be born to Earthly parents who are not. Our tribal affiliations represent something spiritual in nature that extends from this ancient period.

By this we are also to understand that the individual, sentient and agent beings called intelligences were created and formed before the creation of the Physical and Temporal spheres.

---

[11] See <u>The Nature of the Family of God</u>, pg. 499.

- All intelligence is "*independent in that sphere in which God has placed it, to act for itself*" (D&C 93:30). Intelligence is agent. By its nature, it must act. It does so individually and independently as soon as it acquires its uniqueness, its singular existence. The *Intelligences*, as individual entities, have been agent and independent since their inception in the Spiritual sphere. They have remained agent within both of the subsequent Physical and Temporal spheres.

- "*Intelligence cleaveth unto intelligence*" (D&C 88:40). This is the basis for love; bonding one *Intelligence* to another is a fundamental attribute, a primary *law of intelligence*. The work of Lucifer is to bind independent intelligences to him by severing the bonds we have with others. It is intelligence that causes a righteous man to truly "*cleave unto his wife*" clinging and sticking to her until, "*they shall be one flesh*" (Gen. 2:24).[12] Intelligence, as an individual entity, or as an exchangeable, acquirable, attribute, *attracts*, drawing one thing to another. Intelligence is Glory; as such the Glory of God draws us to Him and that same Glory is love itself.

This Law is manifest in the Laws of physics as well; Gravity perhaps is a manifestation of this Law in the Universe. The Law of Intelligence, its attribute of *cleaving* to itself, may allude to an answer to the enigmatic question: What is the source of the *force* of Gravity? We note here our previous discussion of Dark Matter in the Universe and propose the following set of facts: Dark Matter, though undetectable by any means, reveals itself through the attractive gravitational force it contributes within the Universe. All Spirit is refined and pure matter, undetectable by us in Temporality. Spirit is embodied (tangible, material) Intelligence; intelligence attracts intelligence. *If the source of gravitational attraction emanates from this attribute of intelligence, then it is fundamentally different than the other forces of nature which derive their power from material element itself.* This disconnect perhaps explains the resistance the Law of Gravity presents to its unification with the other known forces of nature.

- Intelligence "*was not created or made, neither indeed can be*" (D&C 93:29). God is eternal; He is without beginning. Intelligence is *His quality*, a divine attribute, and always has been. We are from Him, created by Him, through His giving of Himself. Though we are independent of Him, what we are is the substance of Him, forever a part of Him. Every experience we acquire is one that is shared with Him, adding equally to our intelligence and to His.

> *"...if there be two spirits, and one shall be more intelligent than the other, yet these two spirits, notwithstanding one is more intelligent than the other, have no beginning; they existed before, they shall have no end, they shall exist for they are gnolaum, or eternal." (Abraham 3:18)*

---

[12] See -She Shall be Called Woman, pg. 234.

- "*The glory of God is intelligence, or in other words, light and truth*" (D&C 93:36). It is also God's work and His glory to bring to pass the eternal life of man; "*For behold, this is my work and my glory—to bring to pass the immortality and eternal life of man*" (Moses 1:39). When these two statements are added together, we are to understand that we are intelligence itself. We are made of Glory.

- Intelligence is light and truth, Truth is fundamental to our being. "*Light and truth forsake the evil one*" (D&C 93:37). At our core we object to and turn away from the adversary. The *light and truth* within us combines with the *Neshama* we are endowed with to empower our inherent sense of right and wrong, our conscience. It its raw, unembodied form, Intelligence cannot rebel against God. This truth informs us that the war that followed the Grand Council occurred after intelligence was added to, after the embodiment of those intelligences in the Physical sphere.

- Intelligence acquired "*in this life will rise with us in the resurrection*" (D&C 130:18). Intelligence is a noun, but is also an adjective in that it not only exists as an entity, sentient and distinct, but it is also a quality, an attribute, and attainable, something that can be acquired and added upon. Intelligence is not fixed; it is dynamic and grows through experience with time.

Think of the Earth we have described in this work, accreting, growing in mass and size, gaining in gravitational influence, beauty and diversity with every added particle or meteorite. We are much like the forming Earth as we acquire knowledge through experience. Sometimes the acquisition is shocking, devastating, like the hit of a large asteroid, leaving destruction in its wake and altering forever the course of our life. Most often, however, the acquisition of knowledge goes unnoticed, like the thousands of flashes of light, high in the sky, unseen by gazers below. All of it, however, adds to what we are, day-by-day growing in beauty and wonder, eventually changing us, like the Earth, from something barren, to someone capable of producing life, sustaining it, and causing it to flourish.

Though often thought to be lost because we can forget, intelligence is not lost; it is added upon with every moment that we live. It defines our character making us who we are. "*And if a person gains more knowledge and intelligence in this life through his diligence and obedience than another, he will have so much the advantage in the world to come*" (D&C 130:19).

- The LORD rules over all the intelligences; "*...I rule in the heavens above, and in the earth beneath, in all wisdom and prudence, over all the intelligences...*" (Abr. 3:21). God is intelligence, and we are from Him, part of Him, and an extension of Him. He is and forever will be greater than we, the sum of who we *all* are and much more. Whatever height of intelligence we may attain as an independent intelligent entity, God will remain more intelligent than all. It is implicit in any proper understanding of Latter-day doctrine that none of us will ever equal, let alone exceed, our Father in Glory and we must express that no one being or any grouping of beings ever has.[13]

***It is implicit in any proper understanding of Latter-day doctrine that none of us will ever equal, let alone exceed our Father in Glory...***

- Describing an Intelligence

What is an Intelligence? What does it look like or represent? Other than an understanding that it is somehow existent without an embodiment, we know no further. Our intelligences we affirm are eternal, having no end, but were created. They had "*no beginning*" because their entire substance was once *of the Father* and as such are eternal in form with Him, uncreated and self-existent.

With words that still hold true today, in 1936, then still *Elder* Joseph Fielding Smith cautioned:

> *"Some of our writers have endeavored to explain what an intelligence is, but to do so is futile, for we have never been given any insight into this matter beyond what the Lord has fragmentarily revealed. We know, however, that there is something called intelligence which always existed. It is the real eternal part of man, which was not created or made. This intelligence combined with the spirit constitutes a spiritual identity or individual. No formal pronouncements have been made by the leading councils of the Church to clarify what additional meanings and attributes may be assigned to the word "intelligences," beyond that which identifies intelligences as spirit children of God."*[14]

---

[13] Attempts by Latter-day teachers to logically aver that our God has a history much like our own, that he too was created and had a Father God before Him, must confront the reality expressed here. We do not entertain notions that suggest we will eventually become equal unto Him in any measurable way. How is it possible then to think of even one tier of father before and above ours, "*I am the Lord thy God, I am more intelligent than they all,*" let alone an endless chain of them? It is illogical, futile and even foolish to assert such concepts and certainly to entertain such thoughts as representing valid doctrine. See our previous discussion -Intelligence the Core of Our Existence as well, beginning on page 296.

[14] Here Elder Smith uses the word spirit differently than we have defined it in this work. In regards to this subject, consider the following from Elder B.H. Roberts:

Elder Smith here points out a truth that we must not ignore. Of the *nature* of an *intelligence*, our Scriptures are silent. And we must remain so here.

---

*"It is often the case that misconceptions arise through a careless use of words, and through using words interchangeably, without regard to shades of differences that attach to them; and this in the scriptures as in other writings. Indeed, this fault is more frequent in the scriptures than in any other writings, for the reason that, for the most part, they are composed by men who did not aim at scientific exactness in the use of words. They were not equal to such precision in the use of language, in the first place; and in the second, they depended more upon general tenor of what they wrote for making truth apparent than upon technical precision in a choice of words; ideas, not niceness of expression, was the burden of their souls; thought, not its dress. Hence, in scripture, and I might say especially in modern scripture, a lack of careful or precise choice of words, a large dependence upon the general tenor of what is written to convey the truth, a wide range in using words interchangeably that are not always exact equivalents, are characteristics. Thus the expressions, 'Kingdom of God,' 'Kingdom of Heaven,' 'the Whole Family in Heaven,' 'the Church,' 'the Church of Christ,' 'the Church of God,' are often used interchangeably for the Church of Christ when they are not always equivalents; so, too, are used the terms 'Spirit of God,' and 'Holy Ghost;' 'Spirit of Christ,' and 'the Holy Ghost;' 'Spirit,' and 'Soul;' 'intelligencies,[sic]' and 'spirits;' 'spirits,' and 'angels.' I mention this in passing, because I believe many of the differences of opinion and much of the confusion of ideas that exist arise out of our not recognizing, or our not remembering these facts. Hereafter let the student be on his guard in relation to the use of the words 'intelligencies,[sic]' 'spirits,' 'soul,' 'mind,' etc.; and he will find his way out of many a difficulty."* (Seventy's Course on Theology, pgs. 183, 184).

As an aside to this observation and instruction of Elder Roberts, I would like to add that I prefer a hierarchy of expectation in the receipt of revelation and teaching as it comes to us by God and subsequently his ministers. God's communication to man is perfect, limited only by the ability of the receiver to understand. In my view it is common for prophets to receive and record what they are incapable of understanding; in this they often misrepresent, or do not fully appreciate, the totality of the divine message themselves when they expound upon it. Thus it would have been impossible for Moses or Abraham to have completely detailed the facts presented in this book, as they relate to the science of the creative events revealed to them by God, for the understanding of such was unknown to them. They, as diligent prophets, recorded faithfully what they saw and were told, in the most perfect way known to them, what could only be fully understood now, many thousands of years later; such is the way of prophecy, and the teaching of God. I find no error in the receipt of scripture from God to man. Some, but little error, can be found in its transcription into different languages, and over time, in its original language, as originally received, it remains highly descriptive and remarkably precise. The error Elder Roberts speaks of, I would agree, is completely due to man, inherent in annotation, interpretation and in the expounding of scripture. I too agree, in the modern-day these effects are much more apparent, but this is because of the proliferation of the literate, those who can and do write, and the ease in printing and reprinting an overabundance of commentary. As we have discovered in this work, the Word of God, is and always has been, more than sufficient and precise enough to relate and discover the truths of the mysteries of God.

## Creation Out of Nothing?

It would seem that theology suffers the same ultimate stop that physics does, a beginning without a first cause. Though, on the surface, this dilemma may seem to vex both disciplines equally, in fact it does not. Theology can without fault or concern of lost principles, turn to the supernal and evoke eternity; science (however often its practitioners attempt to do so) cannot. At the point where Scientists do insert infinity (∞)[15] into their equations, they leave the realm of science behind and evoke, if not in fact welcome, the metaphysical. In this they seek to have their metaphorical cake and eat it too.

The history of theological thought is not without error of logic in this concern however. Theologians, for unnecessary reasons, framed a cosmological concept that in the end too was fatally flawed. In an effort to ensure absolute divinity and superiority for the monotheistic God of Christianity, second and third century religious thinkers sought to bolster His position against the pantheon by declaring that their God created where there was absolutely nothing. This was novel, for nowhere else had any god been recorded as doing so. The doctrine of creation *ex nihilo* Latin (*from nothing*) signaled a fundamental change in theological understanding of both God and the Universe.

***The Creator of the Universe, who had given up so much to be born a man and die as one of us, was no longer imagined to be so.***

Though never expressed or enforced by decree, or deliberately formulated in creed, the concept of *ex nihilo* is fundamental Christian dogma today. In its original framing as a theological precept, it was meant to ensure an understanding that God was greater than, and transcended all of creation. But with that came a dire consequence that thoroughly divided the Creator from His creation. By insisting God created *all from nothing*, *ex nihilo* meant God did not share of His substance when He created; the Universe and we are of an entirely different *nature* (Greek, *ontos*, ὄντος) than that of the Father.

The rift that only begins here runs much deeper; our alienation from God is profound. Because of *ex nihilo*, our intelligence, physical body, the Spirit they combine to form, indeed we as living Souls are all a product of creation (in our contextual understanding, Temporality) and have *nothing* in common with the Father. The ontological divide is complete; we can share no kinship with the divine.

---

15 The infinity symbol, sometimes called the lemniscate.

The change was stunning. In less than 200 years, the intimacy of God with his creation so fully evoked by the literal incarnation of the Son of God was completely forgotten and replaced with a philosophical construct of man borrowed from pagan Greek thought. Christ who literally became man in the flesh, was now fully God but of a substance so utterly foreign to human nature as to be incomprehensible to man. The consequences of *ex nihilo* do not stop there; the divide is much more expansive.

As the concept of creation *ex nihilo* fixed, a radical dualism engrained theological thought. Christians began to accept, as true, a universe that existed on two separate planes, not just in quality of existence, but in every substantive, measurable and detectable way. God was not only of a substance that was foreign to us, but so too was everything associated with him. The things of God are incomprehensible to us, of a different *order* than the Universe around us.

These new concepts provided fertile ground for heretical ideas to take root. Docetic, Substitutional, and Adoptionist teachers expounded that God and the things of His experience were so far removed from ours that the incarnation of Christ was somehow not real, that it never happened or it only merely *appeared* to be so. Though the heresies were condemned at the time, many of the concepts were absorbed by the developing Christian religion and remain today buried in the theosophies that support creedal thought.

What began as a right-minded attempt to elevate and separate the Christian God above pagan gods of stone and bronze, ended up raising Him so far above us as to separate Him from man entirely. Ironically, this great error was accomplished at a time when the footsteps of the incarnate Son of God could still be heard echoing in the air. The Creator of the Universe, who had given up so much to be born a man and die as one of us, was no longer imagined to be so.

In many respects, the monster of separation between the philosophies of man (the disciplines of science included) and any belief in God or a creator, was birthed in the doctrine of *ex nihilo*, teaching us, if nothing else, to be ever vigilant and careful in watching what we may wish for.

## *What Nothing Means*

In some contexts nothing means a lot, such is the case here.

Within this work, we developed an understanding that the Temporal Universe around us was formed and is comprised of things that existed before. God created it through subtraction; what we are and what we experience in Temporality is somehow less than the perfection from which it came. We also understand the Physical sphere to have been created from "*unorganized matter.*" The stars and worlds there were formed from materials chaotically filling an expanse, matter that existed from a time before Jehovah and Michael organized and created with it at the direction of our Father. Joseph Smith taught us, "*Element had an existence from the time God Had*" (History of the Church, vol. 6, pg. 308). By this we understand that it is eternal; intelligence is eternal as well, both emanating and finding their source in

God. If the Universe was created *ex nihilo*, then none of what we understand here is possible. For if there was *nothing* before, there in fact can be no concept of before.

*Ex nihilo* creation establishes a logical paradox and presents an entrapping conundrum that engulfs all reason. Consider what nothing means. Imagine what that might look like, an expanse, a void, darkness? Now remember that traditional theology (the same that developed the concept of *ex nihilo*) debated and understands that God has a "*substance.*" Since God is eternal, logic demands that *substance* is eternal as well. Such eternal substance must be something, it is decidedly not *nothing*. Traditional theologies also understand that God fills the vastness of space with His presence, that in fact there is *no place where He is not*. If He is substance and He is everywhere, where then existed nothingness from which He created *ex nihilo*?

It is important to address here that sectarian Christianity today still clings to the philosophies and scientific understandings of the world when the creeds were formed. Of those ideas, God has said they, "*were an abomination in his sight*," that they "*were all corrupt*," because they are the "*doctrines... of men*" and "*deny the power*"[16] of God.

Theology is the study of the things of God. Any study embarked upon by man is constrained by man's understanding and his familiarity with the objects, the truths, and the experiences of the Universe around him. Any derived doctrine begins with human precepts and develops concepts that are described through the realization of things known.

The creeds, and the theologies that came from them, were man's best attempt at the time to explain the Godly in Temporal terms, for they could do no other. Those terms of necessity expressed the Universe, matter, energy, and their relationships in ways that are known today to be wrong. This set up a fatal cascade that has crippled current sectarian theological thought.

The Bible does not speak at all about the *substance* of God[17] and neither do the creeds. At best, the creeds can only start with defining what the Temporal and natural are, then set the Godly and the spiritual apart as something other than that. However, in this, the creeds were captive to what was known of the Universe in the third, fourth and fifth centuries. What we today know the Universe to be, is something far different than what was known then. Had the creeds been composed and debated today, against the backdrop of modern science, our understanding of quantum and relativistic physics, modern chemistry, biology and microbiology, there is no doubt they would have developed much different theological concepts.

---

[16] *Joseph Smith History* 1:19

[17] The Bible does, however, enigmatically describe faith (an intangible) as a *substance* (Heb. 11:1). Joseph Smith taught that faith "*...is the principle of action in all intelligent beings*" (Lectures of Faith, Lecture First, 9). In the context of the law that *Intelligence cleaveth to Intelligence* as discussed in this work, we can understand that faith does move things and when intelligence is embodied, as it is, that motion relocates substance.

The Creeds were influenced as well by ancient notions of government, law, monarchy and judgement; each of these concepts leaving their marks on our traditional theology.

Like the addition of "*neither shall ye touch it*" to God's prohibition surrounding the tree of knowledge of good and evil,[18] the creeds themselves were codified and revered as though they came from the mouth of God Himself; such is not the case. They take up for doctrine things that never were in the Word of God as delivered.

Christian theists today define their religion, establish its boundaries, and espouse its core doctrine, as much from the tenets of the creeds as they do the Scriptures themselves. The result is, theology, today, still seeks to describe God in terms that lost currency some 1600 years ago. How could the authors of the creeds describe the substance of God, when they did not truly understand the qualities of Temporal substance? The answer is *they didn't*. They were errant then, and the creeds they developed are errant still.

It must be remembered, against the current of a Christian world that refuses to do so, that the creeds are not Scripture, they are not the Word of God; they represent merely, the best attempt of fallen men to describe the indescribable. Only a fool would profess as truth today the anatomy of man based on what was known 2000 years ago. No sane person would enter into surgery with a surgeon who described the human body in those ancient terms, of course not. We understand that the body is a work of art, intelligently designed and developed through processes set in place by God in the beginning. We accept that the best minds of ancient Greece and the early Christian world were no less capable than ours are today, they were no less spiritually aware, but the information of the world around them was much less than ours and as such their understanding of medicine and the human body was incomplete, wrong and even deadly.

Such is also true of theology. The Word of God is also a work of art, intelligently designed and delivered to man through processes set in place by God. As delivered, the Word, like the human body, is to be studied and understood. What one generation discovers from its truth is a result of their understanding of the things of the world around them, and as such can be wholly lacking,[19] though not by any fault of their own.

Theology today must accept this truth and return to the Word of God for understanding, applying what we know today to it for enlightenment. Espousing the creeds as doctrine, judging and interpreting Scripture from premises established by

---

[18] See our previous discussion, <u>Neither Shall Ye Touch It</u>, on pg. 251 in this work.

[19] The scriptures and the ability of the Holy Ghost to reveal its truths and meaning to man are wholly sufficient for Salvation in any given age or situation. That is the true purpose of scripture, the providing of a foundation upon which the Word, through the Spirit, enters our lives, expands our knowledge and understandings, and brings us to the faith. Scripture, and our understanding of it, is a means to an end. It is designed to lead and point in the direction of God, much more than it is to disclose all of His knowledge and the truth of all His workings.

them is as harmful and deadly to the Soul as an ancient surgeon would be to a patient in a modern operating room.

When *ex nihilo* created us from nothing, we became nothing, certainly not something worthy of the atoning sacrifice of the veritable Son of God. Because *Ex nihilo, Nihil Fit* (Out of Nothing, Nothing Comes), the doctrine separated and divided us from our Father in Heaven; as such it played directly into the hands of the deceiver, the accuser. By creating all *from nothing*, God did not, could not, share of His substance with the created, and as a result we are not of Him.

***When ex nihilo created us from nothing, we became nothing, certainly not something worthy of the atoning sacrifice of the veritable Son of God.***

*Satan* (שָׂטָן) has been trying to separate us, the sons and daughters of God, from the Father since a time before the Grand Council, and from the Garden when Lucifer told our first parents there, "*Father will see your nakedness, Quick! Hide!*" The doctrine of *Ex nihilo*, completed the divide, separating us in fact from a family that is in truth our own, the very family of our Father in Heaven.

## *Returning Home, the Family of God*

Latter-day doctrine overturns the Christian theosophy of creation *ex nihilo* and all it derives; it restores the Creator to His creature with the intimacy of eternal kinship. We are literal sons and daughters of God, created long ago before the Temporal things of this Universe were created. We are not only in God's image here on Earth, but the material of our Spirit and the Intelligence that joins with it, are ancient, from a time long before.

Here we will learn that such beliefs are not new, they have been held as certain before, and the memory of that understanding has been lost over time. Platonic philosophy that did its part to redefine the personal God of Christianity into one without *body, parts and passions*, gave way to Aristotelian philosophy, which worked to put away any conceptualization of a pre-existent world. In time, what had been common belief was decreed heretical.

As the Christian era dawned, both Judaic and Greek traditions commonly accepted as fact, the ante-mortal existence of man. Varied sources include (among a broader list of Jewish writings from the first and early second century AD), the *Apocalypse of Abraham* and the *Prayer of Joseph*. These two books from the pseudepigrapha describe the premortal existence of the patriarchs and those who were of the tribes to be born of Abraham, those who were *yet to be born* as the chosen people of God.

The first known listing of the New Testament Canon, included the Wisdom of Solomon. In its pages, Solomon confirms his belief in the pre-existence of his Soul as he records: "*As a child I was naturally gifted, and a good soul*[20] *fell to my lot; or rather, being good, I entered an undefiled body.*" The Gospels themselves record similar thinking of the disciples, "*And his disciples asked him, saying, Master, who did sin, this man or his parents, that he was born blind?*" (John 9:2). In both of these recordings, Solomon and the LORD's disciples, express an understanding that one's pre-mortal, independent, actions as an *agent Spirit*, affect one's standing and placement at birth and within mortality. (It must be noted that the Savior's answer to his disciples, that the man and his parents did not sin in the pre-existence, does not include anything in the way of a refutation of such a concept.)

Though belief in the pre-mortal existence of Souls was widespread (if not universally accepted); the pages of Scripture offered little to support it, and less still a description of how it was meant to be understood. In this vacuum of knowledge, and as Christianity developed, so too did the discussion of the origin of the Soul (Spirit) of man. Through that process, a belief in the pre-existence endured.

At the outset of the third century, Origen (185–254 AD) (considered by many to be the church's first theologian) believed in and taught of the pre-existence of Souls. To him it was the only explanation for the existence of the good and evil beings that populated the Universe and the heavens above.

Origen reasoned that God would not, indeed could not, be the author of evil. God did not create evil, nor did He impose an inherent predisposition towards evil onto the things He created. Origen pressed that since man and nature are agent and free, the current status of a Fallen nature must be consequential to their own actions and considerations from a time past. Origen said: "*Some sinned deeply and became daemons, others less and became angels; others still less and became archangels; and thus each in turn received the reward for his individual sin. But there remained some souls who had not sinned so greatly as to become daemons, nor on the other hand so very lightly as to become angels. God therefore made the present world and bound the soul to the body as a punishment.*"[21]

Over two hundred years later (early in the fifth century), Augustine (354–430 AD) and Jerome (347? – 420 AD) still grappled with the concept of the pre-mortality of man. Jerome called the problem of the pre-existence of the Soul, "*one of the greatest with which the church has to deal.*"[22] Augustine also discussed his uncertainty of the origin of the Soul; he wrote, [the] "*obscurity of this darkest question, simply does not admit of a clear solution.*" Though Augustine confessed,

---

[20] Note that Solomon here is describing the "*anatomy*" of a Soul as the combination of Temporal body and a pre-existent "*good,*" thus agent, Spirit being. As such, he confirms the concept of our being as we have described it in this work.

[21] Origen, *De Principiis* 1.8, Translated by G.W. Butterworth, in *Origen on First Principles, Being Koetschau's Text of the De Principiis* (New York: Harper & Row, 1966) pg. 67.

[22] Letter to Marcellinus and Anapsychia, *Jerome letters and Select works*, Edited by Phillip Schaff, Peabody, Maryland, Hendrickson, 1999.

"*It would be rash to affirm any of these,*" he listed four views about the origin of what he termed, Souls, two of which admit to the concept of pre-existence: *(1) Souls come into being through propagation* [Traducianism]*; (2) Souls are individually created for each person at the time of their birth (conception)* [creationism]*; (3) Souls are pre-existing at some location and are sent by God into the bodies of those who are born* ["sent" pre-existence]*; (4) Souls sink into temporal bodies by their own choice* ["fallen" pre-existence].[23]

To Augustine, the first two modes of generation on this list came with frustrating complications.

The theological concept that human parents created the Spirit concurrently with the Temporal body at conception, that the Spirit is imputed from one to another in the same manner as original sin, is called Traducianism. It is problematic because it seems to impart almost divine like qualities (the creation of a Spirit) to the material, natural man. Traducianism is also inconsistent with an attribute Spirit is thought to have. Spirit is supposed to be fundamental, elemental, incapable of division, Traducianism infers otherwise.

Creationism teaches that at conception, God creates the Spirit, placing it within the new life just beginning to form in the womb. This concept, however, seems to make God complicit in the every contemptible act that can be associated with procreation, rape, incest and the like. Today it also involves the Creator in, in vitro, and ex vivo pregnancies asking, would God provide a Spirit for the forming embryo if these technical procedures are not condoned by Him, as is taught by many churches?

Creationism also begs yet another heavy question, one that strikes right at the heart of the doctrine of Original Sin. How can a Spirit, so recently made by the Father, be so utterly corrupted through age old actions of Adam the instant it is placed into the flesh by God's own hand? It must be noted, that Augustine, if however reluctantly, settled on this view as the mode he favored most. Perhaps because he did, more than any other reason, even though it is hobbled by the question posed here, it is the view held by the majority of Sectarian Christians today. Origen and Augustine's decisions aside, the question of the origin of the Spirit remained unanswered for centuries leaving subsequent Christians without a definitive answer.

It must be remembered that this discussion included not only the state of man before mortality but extended to include uncertainty about his fate in the future as well. For if man was not eternal, that is created both spiritually and temporally at inception, then it seemed at least possible, if not reasonable, that the Spirit could be destroyed as well. That this was considered is evidenced by the fact that the Catholic Church would not recognize the immortality of the Soul, that it will survive death, continuing in possession of an endless conscious existence, for almost another thousand years when it did so at the Fifth Council of the Lateran (1512-1517 AD).

---

[23] Augustine, On Free Choice of the Will, translated by Thomas Williams Indianapolis, Hackett 1993.

This belief, however, is not universally held even to this day, and it is still not considered a necessary principle of Christian faith by many protestant ecumenical councils, because several "*Christian*" denominations and many distinguished teachers expound a belief in (Annihilationism) the absolute destruction of the Soul.[24]

The Restoration of the Gospel brought with it the several understandings necessary to resolve this question once and for all. First we were told that the pure material of the Universe is eternal. Joseph Smith taught: "*Element had an existence from the time God Had. The Pure principles of element are principles that can never be destroyed; they may be organized and reorganized; but not destroyed. They had no beginning, and can have no end*" (The Contributor, vol. 4, pg. 257).

With this the *Restoration* directly refuted the errant concept of creation *ex nihilo.* God had an eternal substrate from which to create. This material is the invisible matter from which the things visible consist: "*For by Him all things were created, both in the heavens and on earth, visible and invisible*" (Col. 1:16). The LORD revealed that the eternal substrate is fine and pure, that it cannot be seen by Temporal man; "*There is no such thing as immaterial matter. All spirit is matter, but it is more fine or pure, and can only be discerned by purer eyes; We cannot see it; but when our bodies are purified we shall see that it is all matter*" (D&C 131:7, 8). Through this singular revelation and with the same divine breath, God disclosed the Spirit is of this same matter, confirming it has a structure and physicality to it in terms that can be understood by man. Through the Restoration, we are once again told that Spirit is element, restoring the original meaning of the word *pneuma* (πνεῦμα) used by John in his gospel to describe God, "*God is a spirit...*" (John 4:24).[25]

Joseph also restored a truth that was known to Abraham and others long ago but had been completely lost, that man is at his core an Intelligence, sentient and independent, from a time long ago and once at home in the precincts and presence of our Father in Heaven. Joseph taught that this intelligence was eternal and immortal: "*The intelligence of spirits had no beginning, neither will it have an end.*"[26] What Joseph says about the intelligence of man is profound. It is most always misconstrued; certainly in all cases outside of the church, but also in almost every case when it is discussed within. This is the exact context and understanding that ties us directly to our Father in Heaven and cannot be left misunderstood. We will clarify here.

---

[24] The doctrine of *Annihilationism* is considered to be true in various forms by (among many) Seventh-day Adventists, Jehovah's Witnesses, and some who follow the Anglican Communion.

[25] See our previous discussion of this subject, WHAT IS SPIRIT?, pg. 172.

[26] King Follett Sermon, *(Teachings of the Prophet Joseph Smith, Joseph Fielding Smith, comp., Salt Lake City: Deseret Book, 1976, pg. 353).*

Joseph said: "*....Intelligence is eternal and exists upon a self-existent principle. It is a spirit*[27] *from age to age and there is no creation about it... The first principles of a man are self-existent with God.*"[28] The prophet does not here imply or suggest that the Intelligences, the independent, individual Intelligences you and I are, were not created, and were not begotten by our Heavenly Father, for that would be in direct conflict with the word of God. What Joseph does say here is, that *intelligence* is eternal, that it is self-existent; these (eternal, self-existence) *are* the attributes of God. God is also defined as ultimate glory; it is a foundational attribute of Him, "*The glory of God is intelligence...*" (D&C 93:36). Intelligence is eternal in Him, emanates from and is sourced from Him. *An Intelligence* (what you and I fundamentally are) is solely comprised of *intelligence*, parted from God.

***Because the restoration details that we are principally, first, above and before all else, intelligence, our familial relationship to our Father in Heaven, His parentage of us, could not be more intimate.***

Joseph, breathlessly, states that the ***first principle(s)*** of each of us (meaning man) is literally the primordial "***stuff***" which is self-existent with God (intelligence).

*Think of what that means....*

We are living Souls on our way to becoming immortal souls. We are comprised of three principle elements Temporal matter, Physical (Spiritual) matter, and Intelligence. These are our *principles.* Intelligence joins with Physical matter to form Spirit; Spirit joins with Temporal matter to become a Living Soul. Our *first principle* then is Intelligence, and that Intelligence (Joseph tells us) is self-existent with God. Intelligence is not only from God, it *is* God. Because the Restoration details that we are principally, first, *above and before all else*, Intelligence, our familial relationship to our Father in Heaven, His parentage of us, could not be more intimate.

What started millennia ago as a sensible, though a misguided, attempt to *sanctify* God in theological terms, ended up ripping, we, His children, from His very family. Through the Restoration and the truths that accompanied it, we are truly *restored* to our proper place as the literal sons and daughters of God.

---

[27] Ibid. Here, as we have discussed before, the Prophet freely interchanges the disparate words "*spirit*" and "*intelligence.*"

[28] Ibid., pg. 354

## The Firstborn of Every Creature

Creation *ex nihilo* had another detrimental effect on critical Christian doctrine. When *ex nihilo* insisted that all created things were of a different nature (substance) than the Creator, any understandings that allowed for the Son to be created (a creation of the Father) meant that He was not equal with God. The resultant theosophical gap had to be closed. The very many recorded sayings of the Son Himself, declaring that He was not the Father, but merely His Son, had to be explained away. The gap didn't close all at once, and it didn't close neatly; it left some ragged and frayed edges in the process. To seal up the breach, *Docetists*, *Adoptionists*, *Substitutionalists*, *Patripassianists* among others, each took their turn at bat. The thought processes were not sequential either, some running concurrently at times, some leading or trailing others significantly in consideration. Very much at odds with one another, the varied courses developed antagonists and protagonist, who debated in some cases for centuries. In the end, despite great efforts to do so, and contrary to what some claim today, no resolution was found expressing one universally accepted view.

### *Making the Son God*

Often great error can come from honest, right minded attempts to understand the un-understandable. I am confident you may be familiar with the term, *not one iota of difference*, but do you know where it came from? What is its origin? The source for this term is to be found in just two words at the center of a debate that raged centuries ago and to great dismay still rages today.

In an obscure town in 325 AD, just one letter, the Greek letter *iota* (I) was critical. There a debate raged for days between learned bishops over just two little words that differed from one another in spelling only by the inclusion of the one letter *iota* in one word and its absence in the other. This seemingly small difference turned out to be *all the difference in the world*. The two words: *homoousius* (ὁμοούσιος) and *homoiusius* (ὁμοιούσιος) though similar in appearance, in fact represent concepts that are much different when placed in the context of the relationship of the Son to the Father.

*Homoousius* means of a *similar* substance; *Homoiusius* means of the *same* substance. When the words were applied to the Savior, the difference meant that the Savior either *was*, or *was not* God. Faced with this question, the bishops at Nicaea properly decided that Jesus was in fact God, because if He wasn't, the Atonement could not be infinite as it indeed must be. The result of their decision, however, started a cascade of events that grew into the establishment of the doctrine of the Trinity.

- The New Math

Trinitarianism is a concept that is *not* taught in the New Testament; many a Sectarian Christian would argue against this assertion, but they cannot do so convincingly. The Catholic Encyclopedia states, "*among the Apostolic Fathers, there had been nothing even remotely approaching such a mentality or perspective.*"[29] It must be remembered, the Apostolic Fathers are those whose very hands penned the New Testament, those that were taught by the Savior or themselves taught by those who had been taught by Him. The Anchor Bible Dictionary, edited and composed by Protestant Christians affirms as well, "*one does not find in the N*[ew] *T*[estament] *the Trinitarian paradox of the coexistence of the Father, Son, and Spirit within a divine unity.*"[30] These statements are indisputable.

Henry Bettenson in his iconic book, The Early Christian Fathers, affirms "*'subordinationism', it is true was pre-Nicene orthodoxy.*"[31] Likewise, Martin Werner writes: "*every significant theologian of the church, in the pre-Nicene period, had actually represented a Subordinationist Christology.*"[32] And one more for good measure, consider the following as well from R.P.C Hanson: "*There is no theologian in the Eastern or the Western Church before the outbreak of the Arian Controversy [in the fourth century], who does not in some sense regard the Son as subordinate to the Father.*"[33] Subordinationism is itself a belief that the three members of the Godhead are separate and distinct, of significant different orders of glory. Despite its universal acceptance, by *every* early church father, as a proper representation of the relationship of the Father to the Son, and of them both to the Holy Ghost, Subordinationism was anathematized through the very councils that conjured up the Trinity. *It cannot be truthfully argued by anyone that the Doctrine of the Trinity is original to the Faith.* So how did we get here from there?

Before we get a little deeper into the concept of the Trinity itself, we must understand that today we are not really where some Christian advocates want to say we are. Despite claims to the contrary, the Trinity, as it was codified by decree so long ago, is *far* from being universally accepted and understood by Christians today. Less than half of those who sit in Protestant church pews on Sunday today believe the doctrine of the Trinity as it is codified in the Nicene Creed. There exists a distinct disconnect in the minds of Christians between what is preached, what is professed, what is understood, and what is believed. A very recent study[34] of

---

[29] The New Catholic Encyclopedia (Vol. XIV, 1967, pg. 299). "*The formulation 'one God in three Persons' was not solidly established, certainly not fully assimilated into Christian life and its profession of faith, prior to the end of the 4th century. But it is precisely this formulation that has first claim to the title the Trinitarian dogma. Among the Apostolic Fathers, there had been nothing even remotely approaching such a mentality or perspective.*"

[30] Anchor Bible Dictionary, (Vol. 2, God, Yale University Press, 2007, pg. 1055).

[31] The Early Christian Fathers, Henry Bettenson (Oxford University Press, New York, 1956, pg. 239).

[32] The Formation of Christian Dogma, An Historical Study of its Problems, Martin Werner, Translated by S.G.F Brandon, Harper & Brothers, New York, 1957.

[33] The Search for the Christian Doctrine of God, R.P.C. Hanson, Baker Publishing Group, 2008. Richard Patrick Crosland Hanson (1916-1988 AD) Bishop of Clogher in the Church of Ireland was Lightfoot Professor of Divinity, University of Durham, professor of theology, University of Nottingham, and professor of historical and contemporary theology, University of Manchester. He is noted for his outstanding work on the evolution of Christian theology during the years between the Council of Nicaea (325 AD) and that of Constantinople (381 AD) and the decades before.

[34] A survey of 3,000 people conducted by Lifeway Research, commissioned by Ligonier Ministries as reported in the Federalist, October 10, 2016. This was not a trick question, and is revealing to the core.

Americans revealed that although seven in ten respondents confirmed their belief in the Trinity (The Father, Son and Holy Spirit are three Persons, but one God) and six in ten agreed Jesus is both human and divine, more than half indicated that they believed Christ to be created affirming he was, "*the first and greatest being created by God.*"

That this is the case is telling; it speaks of the utter nonsense that is wrapped up in the dogmatic belief intrinsic in Trinitarian doctrine. The same disconnect that exists today between what is thought commonly and what is decreed by theologians, existed the first time this subject was officially debated. To understand why that was so then and remains so today, let us review briefly the history that led to the formation of the doctrine of the Trinity.

We have already discussed the vacuum of knowledge that existed in the blossoming Christian world in the second century as those who had been taught by the Savior and the next generation, those who were taught by them, passed away. With their departure, a dearth of understanding ensued that facilitated a variety of interpretations of the, then still evolving, cannon of the New Testament. In the second century after the advent of the Savior, four primary conceptualizations describing the form the Son of God held during His ministry on Earth developed. Docetic, Substitutional, Adoptionist and Patripassianist Christology's began to be accepted, each competing to be the view accepted by the nascent orthodoxy. A brief summary of each of these theological concepts is provided here:

-Docetism, from the Greek δοκεῖν/δόκησις, *dokeīn* (*to seem*) *dókēsis* (*apparition, phantom*) is the belief that Jesus only *seemed* to be human, and that His human form was completely an illusion. Christ was apparitional, only a sensation within the mind of His observers. His historical and bodily existence, His human form, was a facade without any true substance. Docetism represents perhaps the deepest penetration of the consequential errors stemming from the concept of *ex nihilo* into Christology.

Docetists taught that everything material, physical, and substantive was so debase and degenerate that anything divine would not, *could not*, interact with it. The things of God (whatever that could possibly mean) were enveloped in a sphere beyond that of ours; so much so, that the Divine could not be associated with the human, Christ was not who and what He appeared to be at all. In modern terms, the ministry of the Savior was the ultimate special effect, an illusion for the ages.

---

This study shows that it is disingenuous of Clerics to espouse that Christian salvation hinges on a professed belief in the doctrine of the Trinity, when what the professed belief is actually understood to be by those holding it is ambiguous and without strict definition. Those that seek to exclude whole swaths of people, who espouse Christ to be the Savior of the world and the Son of God, from Christianity because they refuse to publicly affirm this creedal fallacy, must turn inward to their own adherents, within their own congregations, to the more than fifty percent among them who, despite their confession, do not believe the precepts of the Trinity themselves.

Bishop Serapion of Antioch (197–203 AD) first used the word dokētaí, (δοκηταί), (illusionists) when he referred to a Christian community at Rhosus (modern Arsuz in Southwestern Turkey) that denied the humanity of Jesus. It is entirely improper to talk about docetic Christologies in a way that would suggest it had a single definition; it has not. Docetism itself was as vaporous and inexpressible as the God it espoused. There were hundreds of variations in its form and its Christologies. Because *ex nihilo* and Docetism both stemmed from the infusion of Greek understanding of the supernal into Christian thought, elements, or more properly shadings, of these philosophies found their way into all of the subsequent Christian understandings of God and Man, into the discussions, debates, contentions, into the creeds and eventually into the dogmatic beliefs of Christianity in the modern age.

Docetism in all of its divergent varieties was the most extreme of the four alternate Christologies. As such, it was universally condemned by the Early Christian Fathers even though its precepts, *ex nihilo,* and the complete separation of the creator from the created, still remain foundational to traditional Christianity.

By establishing an alternate reality that prevented God from entering the Temporal sphere, there existed no division between the Son, *who did not actually exist*, and that of the Father. The *substance* of God and the creation of the Son was not an issue; to the Docetist, Christ (an illusion) had no discoverable substance.

Oddly, Docetism through its insistence that God the Father projected a semblance of Himself into the minds of the humans who experienced the Christ, both elevated Jesus all the way to God and completely subordinated the Christ, His reality, into absolute nothingness.

- Adoptionism, also sometimes referred to as dynamic monarchianism, teaches that Jesus was merely a man (though some variations curiously admit, He was still somehow born of a virgin), that He lived a life like any other man but was most pious. Because of "*his sinless devotion to the will of God,*"[35] He was adopted by God as His Son, variably, either at baptism, the resurrection, or at the ascension. Through adoption, the man Jesus was empowered to become the "*High Priest over the House of God*."

It is uncertain when this Christology first began to be understood, some place its appearance as early as the decade after the death and resurrection of the Savior. There are evidences of this to be found in the earliest copies of the Gospel of Mark (c 65 AD), which *curiously* do not include the phrase "*Son of God,*" evidently an Adoptionistic omission in the text. A work that once held status within the cannon, the Shepherd of Hermas (2nd Century AD), describes Jesus as a "*virtuous man,*" "*filled with the Holy Spirit,*" and "*adopted as the Son.*" There are variations of this Christology that elevated the relationship of the man Jesus with the Father to

---

[35] From Epiphanius' account of the Ebionites in his compendium of Heresies the Panarion or, *Medicine Chest* (written c. 375 AD). *The Panarion of Epiphanius of Salamis*, Frank Williams, Translator (E.J. Brill, Leiden).

something more than adoption, drawing somewhat on the *modalism* of Patripassianism and describing it as a "*union*" with God.

Adoptionism is, of itself, consistent with a subordinationist theology and, in its insisting that Jesus was just a man, created as such, but subsequently adopted by God, it demands that He be of a lesser order than that of the Father.

-Substitutionalism. Like Adoptionist, Substitutionalists believed that Jesus was merely man, made just of flesh and blood, fully mortal, the literal son of Mary and Joseph. Jesus was not born of a virgin, and He was not born with a divine nature. Jesus was notable only through his exemplary righteousness and wisdom.

***An obvious flaw in all three of the Christologies of Docetism, Adoptionism and Substitutionalism, was the lack of salvific power in the Atonement, for all three removed Deity from the sacrifice on the cross at Calvary.***

Like the Adoptionists (those who marked the baptism of Jesus as the adoption point), Substitutionalists considered the baptism by John a pivotal moment. Instead of adoption, however, Jesus in receiving the Spirit upon Him, was instilled with a share of the Godhead; at that moment the divine Christ began to indwell him. At that point the man Jesus in a sense slept and the indwelling Christ took over, ministering, teaching, and working miracles. When Jesus was about to suffer, the Christ withdrew and the man, who was born of Mary by Joseph, died alone on the cross a mere mortal. Adoptionists pointed to the lighting of the Spirit, descending from heaven like a dove, as the point of reception of the divine Christ, and the cry of Jesus, "*Eloi, Eloi, lama sabachthani*" ("*My God, my God, why has thou forsaken me"*) (Mark 15:34), as the moment when the Christ withdrew.

Substitutionalism was very often criticized by the early apologists as a docetic teaching, and indeed it was, for it too taught that Deity only *appeared* to be in fleshly form on the Earth. Like the Docetist, Substitutionalists believed God was insubstantial and could not have a substance, for Substance was created. Substitutionalism, like Adoptionism seems to have a very early origin, for it also denied, in fact never addressed, the virgin birth, pegging its formation to a time before the nativity narration of the Gospel of Luke was widely circulated.

-Patripassianism. An obvious flaw in all three of the Christologies of Docetism, Adoptionism and Substitutionalism, was the lack of salvific power in the Atonement, for all three removed Deity from the sacrifice on the cross at Calvary.

Recognizing and addressing this theological shortfall, the last of the four major alternate Christologies developed and gained relevancy early in the third century. Patripassianism (from the Latin patri- "*father*" and passio "*suffering*") received its name because this Christology taught that the Father became incarnate in Christ, and because of that He had suffered the complete extent of the Passion.

Also known as modalism or perhaps incorrectly as modal monarchism, Patripassianism espoused that God the Father, Jesus Christ, and the Holy Spirit are not three distinct personages within the Godhead, but are merely three different perceived modes (approaches) or characteristics of a single being. This Christology defined a *monadic* God; there is no actual substantive difference between the three modes of existence. This resolved the creation problem by making the Son the same (in the most literal way) as the Father Himself. Patripassionism, in its primary form, placed God the Father on the cross when it denied that the Son was separate and distinct from Him. There are variations to this Christology, for example, though some Monarchians were accused of teaching that the Father died on the cross as Patripassionist, many expressed an Adoptionist view as well, in both it appears the Monarchians envisioned a Godhead that was unified in such a way that allowed the Father to *actually experience and suffer* the torment of the Son.

Patripassianism did not share the subordinationist understandings that were consistently apparent within the other three alternate Christologies we have addressed here, and this fact drew its condemnation by Origen and others. Though Patripassianism was condemned by the church, its modalism is inherent if not semantically evident in the Trinitarianism expressed in the Athanasian Creed. Patripassianism, its popularity at Rome, and its ability to describe the efficacy of the atoning sacrifice of Christ on the cross, positioned it above the other heretical Christologies for adoption into the nascent doctrine of the Trinity. As such, it laid the footing for the eventual contest between Arius and Athanasius that was to follow.

Each of these four schools vied for preeminence and found expressions of their views within the emerging and evolving cannon. It must be remembered that the texts of the New Testament were still developing in this period. Each of these theological factions were themselves reproducing the Gospels and the letters of Paul and the other Apostles. The tradition of infallibility of Scripture and the practice of trying to preserve it without change was still decades, if not centuries, yet future. It was common for each faction to record, within their communities, versions of the text that varied from those copied by others; adding or subtracting an article here and there, doing the same with an adverb or pronoun within the text was common. Less common, but not rare, could be found insertions or deletions of entire phrases. All of these changes altered the meaning of the written word in favor of, more often in counter to, one of these four views, supporting the interpretations of one Christology over that of another.

The changes were not for the most part falsehearted, intending to mislead or redirect, but were primarily made as a means of commentary, amendment and instruction. What was done then differed not much at all from what is done today each Sabbath over uncountable pulpits by a teacher or expounder of Christian doctrine. Each preacher, interpreting anew what is written in Scripture, making it express what he or she knows it to *truly* mean, as opposed to what it actually says. In such a person's mind nothing harmful is being done, after all they are merely relating the true message of the Scriptures, *for it could not have originally intended to mean anything else*... (Think back to our previous discussion of the debate between Calvinism and Arminianism.)

The "*error*" of course in this interpretive situation is always owned by *another*; the most recent hand is not changing anything but is merely *correcting* or *clarifying* a text that is insufficient in some way. Today this same process continues through the multiplication of scholarly translations of the Bible (NIV, American Standard, KJV, New Living Translation, NKJV and New American Bible among them) and within the production of these versions into other languages; each time, men, on their own accord, interpreting into a new text what they *know* to be true, *providing insight* or *correcting* to restore something lost.

Such was also true in the second and early third centuries; however compounded, for there was no establishment of a traditional text, no way to determine originality of its form or its reading, No way to determine one alternate reading over another. That was eventually accomplished through the institution of an *orthodox* view, by which all texts could be compared.

Orthodoxy itself was established through empirical means, by the popularization of thought, or by the sheer power of individual or governmental influence. By the end of the second century and into the third, Patripassianism, or Modal Monarchism as it is also known, had gained the most currency and indeed was the view held most in favor within the church in the precincts of Rome and its associated sees, though it was not the one held by most of those who believed in the Christ.

That there were many copies and versions of the early texts cannot be denied, for after some 1800 years, examples of the many versions of the texts remain. Each showing signs of the influence of these four theological movements on the written word.[36]

These Christologies, their effects on the developing cannon and indeed the content of the cannon itself, are important to the history of the church, the establishment of the orthodox view, the creeds, and what became traditional Christianity. To the common Christian however, to those who lived in the second and third centuries, their importance and yet potential impact were entirely unknown. The debates, disputations and quarrels over these contending Christologies were for the elites, the philosophers, those few who could read and write.

---

[36] For a thorough discussion and understanding of this subject, see *The Orthodox Corruption of Scripture: The Effect of Early Christological Controversies on the Text of the New Testament*, Bart D. Ehrman, Oxford University Press, 1996.

In the first and second, and into the third centuries and beyond, the common Christian continued to believe what they believed through faith and because of the feelings that sprung from that faith, the Spirit revealing to them what was true. They were like the Christian faithful today, identified by the previously discussed survey, those who believe at their core, something far different than what has been codified by creed. The Early Christians believed what they had been told by their elders and bishops, what they had personally witnessed in a time before there was a written form of Scripture, before the learned attempted to interpret the "*true*" meaning of the written word. Those that had walked in the footsteps of Christ when He preached in Palestine, those who had been converted by the Gospel when they actually heard it preached unto them by Peter and Paul, those who learned from their own Christian mothers and fathers, knew the Christ to be the literal Son of God, God who walked among men on Earth, the Savior of the world, doing the work His Father, our Father, who had sent Him to Earth to do it. To the common Christian and to the vast majority of the Bishops that ministered to them, this was a Subordinationist view of the Godhead.

- Subordinationism and the Nicene Creed

*Ex nihilo* had imposed an understanding that Christ could not in any way be created, and the four Christologies we discussed each found a way to describe Christ as something, in fact, other than a creation of the Father. They all did so in spite of Paul's own testimony of the fact: "*And He is the image of the invisible God, the first-born of all creation*" (Col. 1:15). As the third century dawned, the distinction between the created and the creator continued to expand, and this put pressure on theologians to define the Son. We already discussed the elevation of the Son to the status of the Father as it was espoused by the Patripassianist; this view, however, was in conflict with the Subordinationist teachings that had been taught in the Church since the days when Christians heard the voice of the LORD and the twelve Apostles themselves. Christ taught: "*...My Father is greater than I*" (John 14:28) and again, "*...but to sit on my right hand, and on my left is not mine to give, but it shall be given to them for whom it is prepared of my Father*" (Matt. 20:23). Clearly the Christology taught by Christ and that recorded by the Apostles expressed a subordinate view of the Godhead.

In direct response, and with full consideration of the leanings of the evolving church in his day, Origen wrote in his commentary of the Gospel of Matthew:

> *"Those who entertain false notions about christ under pretence of doing him honour are not to be thought of as 'for' him: such are they who confuse the conception of the Father and Son and suppose that the Father and Son are one in individual being and only admit distinctions of function in the identical subject." (Comm. in Matthaeum, xvii. 14)*[37]

---

[37] The Early Christian Fathers, Henry Bettenson (Oxford University Press, New York, 1956) pg. 233.

Origen clearly taught that God the Father and His Son Jesus Christ are separate and distinct. He expresses that those who were then espousing Christologies that suggested the Son and the Father were the same (Patripassionists), were teaching false notions (doctrine) and in doing so are not "*for*" him. It is not possible to imagine a more direct and pointed condemnation of such an understanding. Note here what Origen attributes the practice to, those who seek to do "*him honour*;" such is also true of those who espouse the Trinitarian doctrine today.

Origen understood that the Father exceeded the Son in knowledge and intelligence:

> *"If the Father comprehends [i.e. holds in his grasp] all, and the Son is included in that 'all', then the Father comprehends the Son. But some will raise the question whether the only-begotten Son's knowledge of the Father is identical with the Father's self-knowledge, and will demonstrate that the saying 'The father who sent me is greater than I' is true in all respects. So that in respect of knowledge the Father is known by himself more fully and clearly and completely that he is known by the Son." (De Principiis, IV.35) (Greek in Justinian, Ep. Ad Menam [Mansi, 1x. 525])*[38]

His teaching of the subordination of the Godhead was clear and without equivocation, the Son is as far removed in excellency from the Father as we are from the Son; *"We say that the Son and the Holy Spirit excel all created beings to a degree which admits of no comparison, and are themselves excelled by the Father to the same or even greater degree" (Comm. in Ionnem, xiii. 25).*[39] Origen was not alone in his understanding, as we have already stated, Subordinationism was the view held by virtually *all* of the ante-Nicene Bishops, and indeed all of those from whom we have extant writings.

***Of the Son Tertullian wrote, He "comes into being," and "there was a time when the Son did not exist."***

Tertullian (c.155-c.240 AD) wrote a treatise to the Modal Monarchian, Praxeas, in which he sought to express *distinction* in the Godhead, in a subordinate way, in opposition to Praxeas' Patripassianism. Ironically, in his apology, Tertullian introduced in a Christian context the terms "*substance*" and "*person*" *for the first time in recorded history*, terms that would be later used at Nicaea in framing the Trinitarian doctrine.

Tertullian's *Trinity*, however, is markedly different than the one eventually drawn up by the council in Nicaea. Tertullian did not divide the creator God from His

[38] Ibid, pg. 233.
[39] Ibid, pg. 233.

creation, and Christ though divine and efficacious in achieving an infinite Atonement, is subordinate to the Father for He was created by Him. Of the Son Tertullian wrote, He "*comes into being,*" and "*there was a time when the Son did not exist.*" Don't be confused by these statements; Tertullian is unambiguous. He means plainly that the Son is a creation of God. In his words you cannot find the convolutions of the Nicene Creed; there is *nothing* here of anything like "*begotten and not made;*" that idea simply did not exist in Tertullian's mind.

It cannot be clearer that the concept of a created Son of God, separate and distinct from the Father was *original* to Christianity. Those that hold and declare that the early church fathers only appeared to be subordinationist, because they lacked the terminology, because they had not yet *formulated* the words and concepts of the trinity, are absolutely wrong. Men like Paul, Origen and Tertullian understood *exactly* what they believed and taught. They were *completely aware* of the concepts behind the false Trinitarian creed and their consequences on theological thought. They stood against those concepts and warned of their consequences long before those same concepts were "*declared*" orthodoxy.

Within a few years of the passing of both Origen and Tertullian, another was born that would take up the defense of Subordinationism against the greatest challenge to the original Christology of the church. Arius (256-336 AD) would face, head on, the ever growing challenges from Greek philosophy as it began to be entrenched in the West, at Rome and within its outlying allies.

As one of the two major opponents in the great theological controversy that came to bear his name, Arius would stand to defend the doctrine of a subordinate Godhead, the belief held by the vast majority of the bishops of the church, against a new Christology, a non-Biblical construct of man. Arius' position and teachings would be opposed for nearly a century by a volatile and determined Athanasius who espoused the concepts that would eventually become the Nicaean Creed.

Throughout the fourth century, the original doctrine of the church faced great forces that sought to change it, forces that would eventually win. The great conflict began quietly in the mannered teaching of a simple presbyter in North Africa and would find its way to an unlikely spot near the crossroads of the East, in the Southern part of modern Turkey, at Nicaea. Those geographical crossroads turned out to be the site of perhaps the most important historical junction for the Christian church as well, for the events that transpired there, so long ago, reverberate throughout the church today.

- The Great Arian Controversy

History is to be found in books after the memories and those who remembered it are gone. Because such is the case, those who own and write the books have the power to mold the past. Because this is true, the Christian world will find that its books today define the doctrine of a subordinate Godhead as heresy, as something that offended the *original* Trinitarian belief, but in fact the exact opposite was true. A study of the actual events surrounding this critical time in Christian history is revealing and instructive. Its story is told primarily in that of two men, but includes much more.

Athanasius was born in Alexandria about 295 AD. As a young man, Alexander, the Bishop, made him his assistant, and in 319 AD, he was ordained a deacon. When he stood against Arius at Nicaea, he did so at 27 years of age, in the role of Alexander's secretary. Three years after Nicaea, Alexander died and Athanasius replaced his mentor in the see. His life was one of constant struggle against the church and with a string of Emperors; he is known as "*Athanasius Contra Mundum*," Latin for, *Athanasius against the world.* A fitting appellation for the man, for it bespeaks of the reality that what Athanasius espoused, the foundational precepts of Trinitarianism, was not orthodoxy, was not the original teaching of the church, but a new, aberrant, theology altogether.

Arius, was a popular and gifted rector of an outlying Alexandrian parish named Baucalis. Although his character has been thoroughly assaulted in the intervening years, the honor and affection shown to him (almost universally) by hundreds of his contemporary bishops, speaks differently of the man. Arius was conservative, and he struggled against the mixing of Christian theology with Greek thought. He was a student of Lucian before his service in the precincts of Alexandria.

It is all but impossible to determine what was actually taught by Arius (what his doctrinal stances truly were) for after Nicaea, in an effort to still the clamor, the Emperor ordered, "*if any writing by Arius should be found, it should be handed over to the flames*." To effect this course, Constantine further ordered, "*if someone should be discovered to have hidden a writing composed by Arius, ...his penalty shall be death*." With such an order of destruction in place, it is understandable that little of what was written by Arius himself is to be found today. What is known of Arius' arguments indicate that he espoused: God the Father was supreme[40] and that Christ the Son was the first, oldest and most beloved of the Father's creations (*ex-nihilo had an impact on Arius; for he believed that Christ was created from nothing*).[41] Arius taught that the Son was pre-existent, the primal creation of the

---

[40] "*We acknowledge One God, alone Ingenerate, alone Everlasting, alone Unbegun, alone True, alone having Immortality, alone Wise, alone Good, alone Sovereign; Judge, Governor, and Providence of all, unalterable and unchangeable, just and good, God of Law and Prophets and New Testament*" From a letter written by Arius to Alexander, Bishop of Alexandria, c. 320AD as recorded in Athanasius' *De Synodis.*

[41] "*For He is above Him, as being His God and before Him. But if the terms 'from Him,' and 'from the womb,' and 'I came forth from the Father, and I am come' (Rom. xi. 36; Ps. cx. 3; John xvi. 28), be understood by some to mean as if a part of Him, one in essence or as an issue, then the Father is according to them compounded and divisible and alterable and material, and, as far as their belief goes, has the circumstances of a body, Who is the Incorporeal God.*" Ibid. This perhaps more than anything else was the sticking point at Nicaea, for by this Christ was like the rest of creation, not made of the substance of God. Any consideration of the outcome of Nicaea, of the debate over *homoousios* (ὁμοούσιος) will show that indeed it was this issue that was the heart of the conflict. Arius, perhaps, was the first to logically press *ex nihilo* to its ultimate conclusion that anything created by a God who creates from nothing, is not of the essence of the Father. A syllogism that is often attributed to Arius, shows the effects of *ex nihilo* in this conversation: "*If the Father begat the Son, he that was begotten had a beginning of existence: and from this it is evident, that there was a time when the Son was not. It*

Father, His very first accomplishment.[42] Arius affirmed that only God was without a beginning and He alone was eternal and infinite.[43] Arius taught that though the Son was created, all other things were created through Him.[44] To Arius only the Son was directly created by God, making the Son the *Only-Begotten* of the Father.[45] Arius taught that Christ was the Son in the truest sense, there was a time when the Son did not exist, and He must have come into existence after the Father.[46] Christ was independently agent having His own free will.[47] Since there was a time when Christ was not, He could not be infinite. In true Subordinationist fashion, Arius concluded, the divinity of the Father was greater than that of the Son and he declared the Son was under the Father, not co-equal or co-eternal with Him.

These beliefs were completely in line with the, by then, almost universally espoused Subordinationist view of the Godhead. What Arius added, it seems, was his understanding that a created Christ, because of *ex nihilo*, would of necessity be not of the same substance as God.

Alexander, it must be noted was a severe Patripassianist, and he along with a few of his associates were at this time beginning to formulate the understanding of a Modalistic Godhead that represented a single God, in three parts, somehow equal in every way, the same, and yet separate. Arius, who taught differently in his own see could not be tolerated, and in 321 AD, Arius was deposed by Alexander, his bishop. The new formulation of the Godhead was adamantly championed by Alexander's young assistant, Athanasius.

By the time Arius was deposed, his teachings and word of the controversy created by Alexander's attempt to suppress them, had spread far beyond the Alexandrian see. Arius was not alone in his beliefs; in fact, he was supported by virtually all of the bishops in the East as well as those in the extended West. After his deposition, Arius fled to Caesarea and to his friend there, the Bishop Eusebius. Eusebius was more than sympathetic and promptly restored Arius, eliciting great ire from Alexander who wrote that Eusebius had overstepped his bounds and had no ecclesiastical authority to do so.

---

*therefore necessarily follows that he had his substance from nothing."* Ibid. Nicaea, began a process that ended in Constantinople which resolved the paradox of Arius in the wrong way. Instead of dropping *Ex Nihilo*, they dropped the truth that the Son was created.

[42] *"who begat an Only-begotten Son before eternal times..."* Ibid.

[43] *"And God, being the cause of all things, is Unbegun and altogether Sole,"* Ibid.

[44] *"...Only-begotten Son before eternal times, through whom He has made both the ages and the universe;"* Ibid.

[45] *"but the Son being begotten apart from time by the Father, and being created and founded before ages, was not before His generation, but being begotten apart from time before all things, alone was made to subsist by the Father."* Ibid.

[46] *"For He is not eternal or co-eternal or co-unoriginate with the Father, nor has He His being together with the Father, as some speak of relations, introducing two ingenerate beginnings, but God is before all things as being Monad and Beginning of all. Wherefore also He is before the Son;"* Ibid.

[47] *"...and begat Him, not in semblance, but in truth; and that He made Him subsist at His own will, unalterable and unchangeable; perfect creature of God,"* Ibid.

From Caesarea, Arius reached out to other bishops writing to them seeking their backing; in return he received much in the way of encouragement. The support was not lightly given either; there was a deep bond between Arius and most of those to which he wrote. Many of them, like Arius himself, had been taught by the same famous teacher of the previous generation, Lucian of Antioch. To bring to mind their common bond, Arius addressed some of his letters "*Dear Fellow-Lucianist.*"

What Arius believed and taught largely comes to us through the writings and records of Alexander and others who eventually opposed Arius. It is rare that we can find in Arius' own words what he knew to be true; fortunately, there exists some correspondence written between Arius and other bishops in the years leading up to Nicaea. In one of those letters, free of the rhetoric that has, over the centuries, clouded this issue, Arius affirms his faith as follows:

> *"...God, who is without a beginning, existed before the Son... that the Son is not unbegotten, nor in any manner a part of the unbegotten, or of any matter subject to him; but in will and design he existed before all times and ages, perfect God, the only begotten, unchangeable; and that he existed not, before he was begotten, or created, or determined, or established, for he was not unbegotten."*[48]

Further, Arius goes on to say:

> *"We are persecuted, because we have said that the Son has a beginning..."*

This single line from Arius is telling, for it affirms that his persecution from Alexander was not about an aberrant view of Christ held by Arius, but it was about Arius not accepting the nascent trinity as espoused by the Patripassianist Alexander. The Trinity, for the first time in Christian history, expressed that the Son's existence extended eternally into the past with that of the Father; somehow, it taught the Son was "*begotten and not made.*"

One of the bishops to whom Arius wrote from Caesarea was yet another Eusebius, this one the bishop of Nicomedia. The Nicomedian Eusebius was the bishop of the Emperor Constantine, and as such, was very influential in the eventual convening of the Council at Nicaea. Eusebius was also the close friend of the Empress Constantia, Constantine's sister.

In response to the letter he received from Arius, Eusebius also sent letters to other bishops. In one to Paulinus, of Tyre, he spoke of the dim view he and Arius shared for the new *Platonic order* of the Godhead, the Nicomedian Eusebius urged

---

[48] The Ecclesiastical History of Eusebius Pamphilus, Appended by A Historical View of the Council of Nice; With a Translation of Documents, by Rev. Isaac Boyle, D.D. pg. 40. On this page and the one preceding it, the Rev. Boyle provides a letter from Arius to Eusebius of Nicomedia, where the former laments the persecution that has befallen him and others who "*for the sake of truth*" *have endured it. Arius correctly states that his view is held by "all the bishops of the East... with the exception only of Philogonius, Hellanicus and Macarius,*" whom Arius classes as, "*heretical men, and uninstructed in the faith.*"

Paulinus to join with him and Arius in actively suppressing it. Of the errant, developing, Trinitarian doctrine, Eusebius wrote:

> *"We have never heard, my Lord, of two beings unbegotten, nor of one divided into two; nor have we learnt or believed that he could suffer any thing corporeal, but that there is one unbegotten, and another truly from him, and not made of his substance, by no means partaking of his nature, nor being of his substance, but altogether differing in nature and in power, yet made in the perfect likeness and nature and power of his creator... And we say this, not from our own reasonings, but instructed by the scriptures."*[49]

Eusebius says the doctrines of the church, as he understood them to be, as he knew them to always have been, were "*instructed by the Scriptures*;" in this he is poking at the fact that the Trinitarian belief is a sole construct of the mind of men, *reasoned*, not inspired. Of this fact, spoken from one so close to the doctrine's origination, we would be wise to take notice.

- Nicaea and its Aftermath

At the promptings of Eusebius Bishop of Nicomedia, the Emperor Constantine dispatched his closest Christian advisor Hosius, Bishop of Cordoba to Alexandria with a letter from him urging Alexander and Arius to reconcile. When that mission failed, Constantine called for an Oecumenical Council in 325 AD. The council was to be held at Nicaea (modern day Iznik in Turkey) and would be largely made up of bishops from the East very much Subordinists and aligned with Arius' thinking. Of the approximately three hundred bishops in attendance, only five were from the West and aligned with Athanasius.

***Bishops, many of whom because of their faith were disfigured and maimed, some halt and even without limb from cruel punishments pronounced upon them by more than a few emperors, now all stood in front of another emperor to clarify that same faith***

The bishops there presented an unusual sight, one that had previously been unimagined. Bishops, many of whom, because of their faith, were disfigured and maimed, some halt and even without limb from cruel punishments pronounced upon them by more than a few Emperors, now all stood in front of another Emperor to

---

[49] From, The Ecclesiastical History of Eusebius Pamphilus, Appended by A Historical View of the Council of Nice; With a Translation of Documents, by Rev. Isaac Boyle, D.D. pg. 40.

clarify that same faith, that it may be decreed the official faith of the land. What we know of the council is largely due to the recording of another Eusebius, the Bishop of Caesarea.

The bishops were called to resolve the dispute that had flamed between Arius and Alexander over the distribution and substance of the members of the Godhead. Athanasius (who attended the Council as secretary to Bishop Alexander, who was in failing health) could not present, though he was responsible for much of what was presented for the minority and nascent Western view of the Godhead. A view which would become the foundation of the Trinitarian doctrine we know today.

Arius, on the other hand, spoke for himself and the Subordinationist view that had long been accepted. It would be centuries still before the title Trinitarian would be universally applied to what Athanasius advocated at Nicaea. But for that time, and for the decades that were to immediately follow, the concept he backed was termed Athanasian because of him. Likewise what Arius advocated was not called Subordinationism but was termed Arian and Arianism. It appears in fact that the view Arius advocated at Nicaea represented a specific view of Christ that varied, if however slightly, from that usually understood to be Subordinationist today.

History, as tended by the orthodox and Trinitarian church, has not been kind to Arius and has shoveled on perhaps many untruths about what he espoused. So Arianism, as it has come to be known, incorporated more than the simple subordination, as we have detailed in this work (for to Arius, Christ was created from nothing).

The council of church bishops at Nicaea sought to draw upon the greatest skills of mankind, intelligently they reasoned, diplomatically they compromised, lovingly they sought to mend torn reputations and restore peace and tranquility. In the end, they were marginally successful, far more so than should be expected. However, in the final analysis, this was a human endeavor, corrupted by human weakness. Men of God sought for a human understanding of things heavenly. In this they achieved all that they were capable of. The Council of Nicaea used Earthly terms to create an enigmatic God that on the surface looked to be grand but lacked any of the truly divine attributes of simplicity found in the truth. God is not empowered by a decree of men. Men are empowered by the truth bestowed upon them by God.

Those gathered at Nicaea (it is thought at the suggestion of Bishop Hosius) embraced the term *homoousios* (ὁμοούσιος), '*of the same substance,*' to describe the joining together of the Father and the Son, that they both were equally divine. It must be remembered that it was the *substanc*e from which the Son was made, that the original conflict with Arius and Alexander was about. In that original context, the *same* substance, *homoousios*, was not presented in the way it is described by theology today; God and Christ were not the *same thing* in identity, but were composed of the same "*stuff*," of the same order of material, both divine. This fact can be understood by Eusebius' own words to his congregation at home. In describing the way the substance of God was expressed by the Athanasians, Eusebius said:

> *"They admitted that the words "of the substance" signified that the son was of the Father, but not as a part of the Father. We thought it well to assent to this explanation, as conveying the pious doctrine, that the son was of the Father; but not, however, a part of the Father. We therefore agreed to this opinion; nor did we reject the word consubstantial, having in view the promotion of peace, and being anxious to avoid a departure from the right belief. For the same reason we approved also of the words "begotten, not made," since the word made, they said, was common to the other creatures which were made by the Son, and to which he has nothing similar; and that therefore he is not made like those who were created by himself, but is of a more excellent substance than any created being."*[50]

The Eusebiuses, *both of them*, and the rest of the Arians, understood this to mean that the Son was made of the same materials as the Father. This simply expressed that what that substance was, was much different than what Temporality is composed of. It did not speak at all of the subordinate relationship of the Son to the Father, but formulated that the Son was God. To the Arians (Subordinationists), Christ was begotten, created, but not fashioned at His creation of the things Temporality is composed of.[51] We must not overlook this fact: *All who came to the Council a Subordinationist, by far the vast majority, left the Council a Subordinationist, and remained so the rest of their lives; as did most of the church in their generation and for many more generations to come.*

Keep this in mind, Subordinationism was not the heretical view at Nicaea; Trinitarianism represented a new way of describing God, the Father and the Son. Those in that day who were called Arian, the Subordinationists, accepted the Nicene formula as it was presented through their interpretation of its words and, "*having in view the promotion of peace, and being anxious to avoid a departure from the right belief.*" The *right belief* they intended to uphold was the Subordinationist view, the one they presented there at the outset of the council:

> *"We believe in one God, the Father Almighty, Maker of all things, visible and invisible; and in one Lord Jesus Christ, the Word of God, God of God, light of light, life of life, the only begotten Son, **the first born of every creature**, begotten of the Father before all ages, by whom all things were made; who for our salvation was made flesh and conversed among men; who suffered, and rose again the third day, and ascended to the Father, and will come again with glory to judge the living and the dead. We also believe in one Holy Spirit; believing every one of these to be and to subsist, the Father truly the Father, the Son truly the Son, and the Holy Spirit truly the Holy Spirit; as our Lord, when he sent his disciples to preach, said, "Go, and teach all nations, baptizing them in the name of the Father, and of the Son, and of the Holy Ghost."*[52]

---

[50] See Letter from Eusebius Pamphilus to the Church of Cesarea in the Appendix, pg. 548.

[51] In a very real sense, Nicaea rejected *Ex Nihilo*, at least as it applied to the created Son; in this Arius lost.

[52] In his letter to his home see after the Council at Nicaea, Eusebius of Caesarea expressed the view of the majority of the Bishops, that they had been successful in defending their belief in a subordinate

The words chosen at Nicaea, particularly the one word, *homoousios* (ὁμοούσιος) (consubstantial), as it was understood by the bishops there, did not change the subordinate relationship of the Son to the Father, but defined in some small way what God was made of, of something other than what made up man (Temporality as we have described it in this work). *Homoousios* did not decide whether the Son was created or not, just that he was of the "*stuff*" of the Father. They rejected the *ex-nihilo* creation of the Son, espoused by Arius, but not his Subordinationist view of the Godhead. Their view, by-the-way is in complete harmony with the view we describe in this work. It is the same view taught by Joseph Smith.

The great debate between Athanasius and Arius did not end with the Council of Nicaea. Many bishops — in the East the vast majority — still openly or secretly sided with Arius; *i.e., they considered Christ the Son of God, subordinate to the Father and not eternal for he was both created and begotten.*

The Emperor Constantine himself, after having such an active part in formulating the decree and banishing Arius, invited him to a personal interview in 331 AD. Through the interview, Constantine could find no heresy in Arius, and recommended his restoration and the other Arians to their churches. To this Athanasius protested loudly, and Eusebius of Nicomedia responded by obtaining from Constantine an order for his recall. Athanasius refused the recall, and in 335AD, the Nicomedian Eusebius convened a Council at Tyre that directed Athanasius' deposition. In response, the Emperor sent Athanasius into exile in Gaul where he remained for the next two years.

Arius visited the Emperor Constantine for a second time and professed adherence to the Nicene decree. *It is certain that Arius could find the same values in the decree that both Eusebiuses and the rest had found earlier when at the council they found it proper enough to accede to.* Perhaps Arius was prompted to make the concessions by those who believed in him and sought to see him reconciled. Whatever the case, after this second meeting, Constantine bade Alexander, Patriarch of Constantinople, receive Arius into communion.[53] Though Arius did not live to receive this

---

Godhead, an *Origenist* view. The text of the entire letter is provided in the appendix at the end of this work.

[53] This restoration has not gone without notice of many anti-Arian church historians who, in an effort to cast aspersion and in some way further their cause, have at times repeated, a no doubt mythical story of the miraculous and painful demise of one whom, for his beliefs, they deem worthy of death. The ecclesiastical historian Socrates is one of many who have related the story. He says, on the Saturday before his restoration "*divine retribution overtook his daring criminality*" and "*a terror seized him,*" and he "*almost immediately died*" (From *Ecclesiastical History*, By Socrates, I, 37-8). The complete statement of Socrates as rendered by William Durant is provided as follows: "*It was then Saturday, and Arius was expecting to assemble on the day following; but divine retribution overtook his daring criminality. For going out from the imperial palace ... and approaching the porphyry pillar in the Forum of Constantine, a terror seized him, accompanied by the violent relaxation of his bowels. ... Together with the evacuations his bowels protruded, followed by a copious hemorrhage, and the descent of the smaller intestine; moreover, portions of his spleen and his liver were eliminated in the effusion of blood, so that he almost immediately died.*" The Story of Civilization, The Age of Faith, by Will Durant (New York, New York. Simon and Schuster, 1950, pg. 7). It is wholly inconsistent with the history of God's dealings with man to assume there is any truth to this story. Any consideration of this man's actions and

communion, his rapprochement with the Emperor and the Roman church was made complete when on his deathbed, a mere year later, Constantine received the rites of baptism from Arius' friend, defender counselor, and fellow Arian, Eusebius, Bishop of Nicomedia.

At the death of his father, Constantius II ascended to the throne in the East of the now divided Empire. To Constantine, matters of religion, though mildly amusing, were an important factor in the control of the state and a means of culling public approval. For his son, such matters were of greater importance. Constantius II took theology more seriously than his father had and made his own inquiry into the nature of the Godhead; subsequently he adopted the Arian view and felt a moral obligation to ensure it was accepted by all of the empire. Church Councils, called and dominated by the new Emperor, re-established Arianism (Subordinationism) by affirming that the LORD Jesus Christ was in the *likeness* of His Father and not consubstantial. This period saw the beginning of the great divide between the Church in the West and the Church in the East. From this point forward, the struggle for doctrinal hegemony ceased to be decided by theists, through adherence to Scripture and what had been revealed, but by power struggles within the Empire.

Athanasius took advantage of the death of Constantine to return to his see in Alexandria which had remained vacant since his disposition. At the behest of the Nicomedian Eusebius, Athanasius was again expelled in 339 AD, this time Constantius replaced him, through appointment, with the Arian bishop, George of Cappadocia. Athanasius was forced into exile for a second time, this time in Rome, at the courts of Constantius' brother, Constans, Emperor of the West. Constans, fearing division in the Empire, sought reconciliation on behalf of Athanasius with his brother Constantius and achieved the restoration communion of Athanasius in 346 AD. Four years later, Constans was murdered and Constantius became sole Emperor. That Subordinationist Christology remained the *orthodox* view after Nicaea continued to be evidenced, when yet again, another Council, the Western Council of Milan, condemned Athanasius (a third time in 15 years) for his continued disturbance of the faith.

Athanasius would continue this cycle of exile and restoration for the rest of his life, being exiled two more times, by two more Emperors, before his death in 373 AD. The fifty years that followed Nicaea were witness to not only the volatile history of Athanasius, with his checkered cycles of elevation and exile, in the shadow of many Emperors, but they also saw an unparalleled parade of Councils and their formulations, wherein creed, and its counter-creed, dividing manifesto and fought for compromise, fell one-after-another in a reeling and disorienting series.

---

life work, compared with others of whom we have record, e.g. Hitler, Borgia, Herod, Nimrod, Nebuchadnezzer and the priests of Baal and Molech, will show that punishment of Arius for his actions, while allowing such as these to live on, could only be the action of an unjust God. Arius, it seems, was more probably in fact brutally murdered, before he could be restored to full fellowship in the church. Someone who could do that to an 80 year old man, who spent a life in service to God, will receive his own judgement, beyond that to be found on the pages of any history book.

Arianism, in the form expressed by its opponents as a radical departure from accepted Christian doctrine, in fact never did exist. Arius did not himself adhere to the exaggerated claims recorded to have been presented in the polarized atmosphere of the debates at Nicaea. As far as Arianism deviating from accepted doctrine is concerned, it followed more closely to the substitutionalist form of faith held by the majority of the church and its bishops than can possibly be imagined.

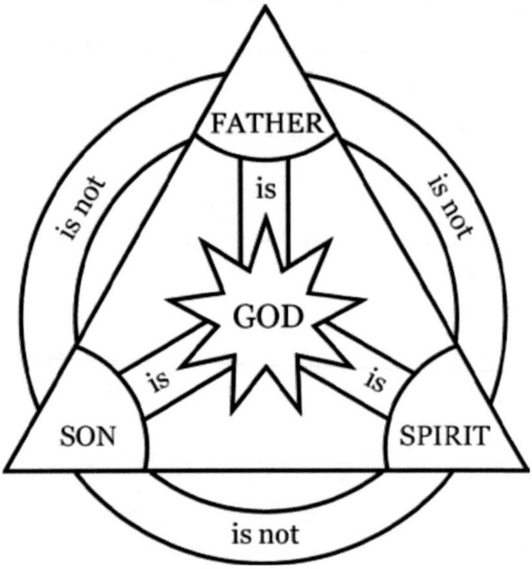

Figure 8-1 *The Trinity.*

Arius found support in the writings of Origen, Lucian and Dionysius of Alexandria. In many of his passages, Origen stressed the subordination of the Son as did many other early writers, as we have discussed here. Arius stood confident in his doctrine because behind him were these great men, years of Christian tradition, and the many passages of Scripture which he could quote as proof-texts of the true doctrine. As a representation of the faith, the tenets of Arian thought, in moderation, always had, and would continue to be for some time to come, the form of Christianity that represented the norm.

The Council of Nicaea (the *Great Council*) and its attendant decree, did not resolve any issue. Though the Athanasian decree of Nicaea remained intact and was never rescinded, anyone who chose to (a majority of the Church including the Emperor and those Emperors that followed after him) read it as a declaration of Arian thought.

Trinitarianism did not rule the day; everyone but Bishop Alexander, his secretary Athanasius, and the few bishops who came with them, walked away from Nicaea as they had arrived – Subordinationists. It would be another fifty plus years before the Trinitarian concept that was thought to be agreed upon at Nicaea would take its final form (*Figure 8-1*) at another Council, held at Constantinople, in 381 AD.

Signers of the decree at Nicaea and those they represented took from the great council exactly what they brought to it; their faith had not shifted. Constantine and his successors remained Arian. Most of the bishops remained Arian. Eusebius of Caesarea, whose eyewitness account of the council is virtually the only extant account of the grand council, went to Nicaea an Arian and remained one upon leaving. Eusebius' letter home, written just after the council, expresses this fact clearly.[54]

Athanasian precepts were only truly adopted in the metropolitan areas of the West on the Italian Peninsula and its surrounds and then only after the Nicene creed was reworked in 381 AD. That they have become the dominant Christian precepts today, is a result of a sequence of events that worked together to separate the West from the East and establish a sense of protectionism in Rome.

Faced with dynamic cultural changes, Christianity over the centuries drew upon wrong interpretations of by now distant events and an ancient decree to establish orthodoxy in a vacuum of leadership that had filled with ignorance, corruption, political pressure and greed. At best, from the Athanasian point of view, the decree resulted, through the Emperor's willingness to compromise, in some shift of political power from the prelates in the East to those in the West. That eventuality served the Emperor well in stemming concerns in the West that the power of the empire was shifting eastward toward Constantinople, but the shift would later lead to the acceptance of the Trinitarian doctrine as well. A debate that began at Nicaea, though called not entirely for theological reasons, became wholly politicized, driven by the needs of the Empire over those of the faithful. As it progressed into the next generations, the final outcome would be driven by war, conquering armies, and the expediency of power as Rome fell.

After 381, at the close of the fourth century, in Rome and the regions round about, Arianism, though far from dead, had been delivered a severe blow. Here, along the Appian Way, in the shadows of the old Roman temples, the remnants of the culture of *Summus Deus*[55] enforced an interpretation of the Savior's life and ministry that favored an Athanasian world view. Rome indeed was becoming Athanasian in thought. Though it seemed, for now at least, to be politically overwhelmed in Italy, Arianism was about to win a peculiar victory among those who would dominate history in the coming centuries, a victory that in the end would cost Subordinationism the war.

---

[54] A complete copy of Eusebius' letter written to his home see, immediately following the council at Nicaea, is provided in the Appendix at the end of this work.

[55] About the time the future Emperor Constantine was born (c 270 AD), a most interesting new cult, Sol Invictus (the Unconquered Sun) began to be heavily promoted by the Emperor, Aurelian (214-275 AD). Sun worship, not unusual among pagans, was in this case very different because of its Monotheistic nature. Cultists worshipped a single unrecognizable spirit who is supreme, detached from human and worldly affairs, yet all of the Universe was organized and is controlled by his actions alone. This spirit was symbolically identified by the sun, a symbol in which this spirit somehow manifested itself. Under Aurelian rule (270-275 AD), this cult was organized with great splendor. The temple of the Sun that he built and dedicated in 270AD at Rome was one of the most beautiful structures the world had ever seen. Roman coinage under Aurelian bore the inscription —The Sun is the Lord of the Roman Empire. Additionally, this cult was in complete harmony with another Monotheistic belief system rapidly finding favor and that was at this same time ever gaining followers in high places throughout the empire, the cult of Summus Deus (the God who is supreme.)

Christianity had first been carried northward to the barbarian Teutonic tribes by Roman captives taken in the Gothic invasions of Asia Minor in the third century. The "*apostle*" Ulfilas (311? -381 AD) was the descendant of one of these Christian captives taken from Cappadocia. Born and raised among the Goths who lived in the North beyond the Danube, he was consecrated as their bishop about 341 AD, by the (now familiar to us) Arian prelate, Eusebius of Nicomedia.

When the Gothic chieftain Athanaric persecuted the Christians in his dominions, Ulfilas obtained permission from the Arian Constantius II to bring the Gothic Christians across the Danube into Thrace. To instruct and multiply his converts, Ulfilas patiently translated, from the Greek into Gothic, all of the Bible except the Books of Kings, which he omitted as dangerously martial, and as the Goths had yet no written language, he composed a Gothic alphabet based upon the Greek (it is that same Gothic alphabet that is used in the printing of this book today). Ulfilas' Bible was the first literary work in any Teutonic tongue. The devoted and virtuous life of Ulfilas generated among the Goths such confidence in his wisdom and integrity that they accepted his Arian Christianity without question.

As other *barbarians* received their Christianity in the fourth and fifth centuries from the Goths, nearly all the invaders of the Empire in the period of its decline were Arians (Subordinationists), and the new kingdoms established by them in the Balkans, Gaul, Spain, Italy and Africa were officially and popularly Arian, believing in a subordinate, non-Trinitarian Godhead.

Conquerors and conquered differed by only an iota in their faith: The now "*orthodox*" conquered peoples at Rome and its surrounds (after 381 AD) officially held Christ to be identical in being, *homoousios* (ὁμοούσιος); the Arian *invaders* considered that same Christ only similar in being, *Homoiosios* (ὁμοιούσιος), with God the Father. That difference, however, became vital in the politics of the fifth and sixth centuries as the balance of power shifted yet again.

By this chance concatenation of events, Arianism held its ground as the belief of the majority until minds were forced to change at the point of a sword, first when the "*orthodox*" Franks overthrew the Visigoths in Gaul, followed by the conquering of Vandal Africa and Gothic Italy by Belisarius, and finally when Recared changed the faith of the Visigoths remaining in Spain. Through war and bloodshed, under the trampling feet of wave-after-wave of conquering army, Christianity became ever more orthodox (Athanasian) and ever less true.

After years of debate and incessant rancor, the plain and precious things of God became the muddled doctrines of man. Who God is, who His Son is, how they relate to each other, and their combined relationships to the Holy Ghost, are matters of keen importance. With the guidance of the Holy Ghost, these are concepts easy to understand by anyone willing to hear. It has always been the intent of God for His children to know the truth concerning these things. In the decades of the fourth century, the number of saints who were listening to, and in turn sharing, the simple truths of the gospel was continually in decline.

The activities of the prelates who should have been guided by the Spirit had in fact driven it away. Unwilling to listen, they were fast losing their chance to hear. As the century drew to a close, the long-suffering and patience of our Father in Heaven was wearing thin, and within thirty years, an age that had been prophesied a thousand years before would begin.

To the Prophet Amos the LORD had revealed the image of a time when He would withdraw the fullness of His Spirit; He said:

> *"Behold, the days come, saith the Lord God, that I will send a famine in the land, not a famine of bread, nor a thirst for water, but of hearing the words of the LORD: And they shall wander from sea to sea, and from the north even to the east, they shall run to and fro to seek the word of the LORD, and shall not find it." (Amos 8:11, 12)*

In the world surrounding the Mediterranean, Christianity would never be the same. Leaders who received their authority from the LORD and His Apostles, and those they had appointed, would no longer be found. Church doctrine would be decided by a council of politically minded leaders instead of being formed by what is said in the Scriptures or revealed by the Spirit. In the new world, half a globe away, history was following a similar course. In 425 AD, somewhere in America, the prophet Moroni (under the direction of the Holy Ghost) hid up a book, which had been abridged by his father but finished by him, and divinely purposed to be brought forth in the last days. Here the same deafening silence was about to fall.

As the shadow of the LORD passed over the Earth, the Light of Christ darkened and dimmed. History began to record the medieval centuries and Mankind started to fall soundly to sleep; the Dark Ages were upon them. When the early Christian church with her Saints, martyrs, and subordinate Godhead would awake in the eventual morning she would be changed and perhaps even unrecognizable.

The Dark Ages, however, were never as dark as many would suppose. Night never came. The Sun never fully set but remained low on the horizon, like an arctic summer, dimly lighting the landscape while never rising overhead. Because the Spirit God continued to inspire great Christians like Aquinas, Francis, Dominic, the *transcendent* John and others; in the enduring twilight there was always the illumination of the faithful to be found, if, however, obscured.[56] When the remnant of the Church would arise at the dawn of the renaissance, she would, however, be in need of reform, in disarray, corrupted and Trinitarian in belief. Is it any wonder why the LORD was more than ready to supply her with what she needed – inspired men of God ready to serve and minister to those willing to listen: Luther and Tyndale, Wycliff and Wesley and Williams – and then a young boy named Joseph Smith Jr.

---

[56] For a fascinating read detailing the paradigm shift in the Mormon understanding of *The Great Apostasy* narrative, read: Standing Apart: Mormon Historical Consciousness and the Concept of Apostasy (Edited by Miranda Wilcox and John D. Young, 2014, Oxford University Press, New York).

## *The Church of the Firstborn*

Those who insist on the uncreated, creedal Christ do so in direct opposition to what is plainly taught by the Scriptures and in conflict with what was believed and taught by the Christian church in its earliest times; they recast the definitions of words to fit their interpretation. One of the words and concepts that fall to this errant parsing is, *Firstborn*. To the Trinitarian, Christ was not truly born; His incarnation aside, He has always existed. Firstborn to them then, must fit within a different context, one that supports it as a title, or rank, something more reflective of His position or preeminence, His Lordship over all of creation.

The problem with "*fitting*" words into preconceived notions is that doing so, one most often gets it wrong. Interpreting *Firstborn* in the context of Christ's Lordship and not properly, with the primacy of His creation or divine birth as it truly represents, opens the door to problems more vexing than those its adherents hope to avoid.

In a monarchy, a Prince (heir to the throne) has the right conferred through his position as firstborn of the kingdom, this does not exclude the King from having other sons or even that the prince may truly be the firstborn son (refer to Abraham's sons, Ishmael and Isaac). Such a Prince, with consideration given to his abilities and attributes, may not even be the most qualified, or capable, to rule the kingdom. The Prince may not be the wisest or oldest of the royal family or even of among the King's subjects. No one understands the Prince to be of a different substance than others who inhabit the kingdom.

This descriptive analogy, is imperfect and breaks down, but is illustrative nonetheless at describing the many detrimental consequences that arise from forcing an interpretation of a regal title onto the theological concept of *Firstborn*.

The proper place to receive direction for interpretation is within the Biblical texts themselves. The Savior said, "*Search the scriptures; for in them ye think ye have eternal life: and they are they which testify of me*" (John 5:39). With this in mind, we should apply what the Scriptures reveal through the many Bible stories that teach of the relationship of a firstborn. In *every* Type of Christ found in the Old Testament, Joseph, Moses, Isaac and Jacob, the same facts exist. That those facts appear at all is significant; that they appear *every time* means they are not to be ignored and are there for a very important purpose.

Every one of these *types* receive the rights of the *Firstborn* though they were in fact NOT the first son to be born, each had an older brother. *This illustrates that these rights are earned by individual merit through the selections and election of the Father, not patriarchal order and succession*; such is true of the Melchizedek Priesthood, the priesthood after the order of the Son of God. The Levitical Priesthood was patriarchal in order, conferred on men through birth, not through direct calling and election from God.

Each of these *types* had a brother who contended with him for the position of *Firstborn*, struggled with him for the birthright, the glory of position next to the father, the respect of his siblings, and the blessings that accompanied the position. *This speaks to the contention between Lucifer and Jehovah.*

Further consider just one of these *types*, Joseph. He was favored of his Father and by God; he was sent to seek out his brethren who had become lost to his Father. His brothers cast him into a pit, declaring him to be dead when he was not. In Egypt he served faithfully but was falsely accused and was cast into prison. There he ministered to captive Souls, both the innocent and the guilty, and was given the keys to the prison by its keeper. He was recognized as a Prophet (an interpreter of dreams). He became a savior in Egypt (the Biblical representation of the world) becoming it's de facto ruler, though not recognized as so by the kingdom's inhabitants (his law was the law that saved the land). He literally provided the bread of life to his brethren and to the entire world (Egypt). In the end, he revealed himself to his brethren, not as their ruler, not as someone who was *other than they*, but *he revealed that he was their literal brother*.

***The parallels shown in the events of the lives of Joseph and Christ are striking and not without divine intent.***

The parallels shown in the events of the lives of Joseph and Christ are striking and not without divine intent. A capstone is placed upon the divinity of this message by Jesus Christ Himself. Consider the following: After declaring, "*here am I*," Joseph was sent by his father to Shechem to look after the wellbeing of his brothers; to complete the parallelism, Christ, after his baptism by John and subsequent brief visit to Jerusalem, began His mortal ministry to His brothers and sisters by publicly declaring for the first time, "*I... am he*" to the Samaritan woman at Jacob's well, in Shechem…

There can be no doubt, these *Types of Christ* in the Scriptures testify of Him, prophetically anticipate His ministry, and detail truths regarding who He is, of His relationship to His Father and His relationship to we, His brothers and sisters. In hermeneutics (Biblical interpretation), context is everything, and with what we have shown here in mind, there can be no other context to affix Paul's teaching: "*And He is the image of the invisible God, the first-born of all creation*" (Col. 1:15). Jehovah (Christ) was the first to be born and by, and through, His virtue, He is the *Firstborn* as well, sovereign and divine.

## *Christ is Created*

Our understanding here is that Jesus Christ is the same who was called Jehovah before the incarnation; He is the God of Abraham and Jacob and Isaac. It is He who said to Moses, "*I am.*" Jesus Christ who died on the Cross at Calvary, was buried in the tomb and raised again the third day, is the Creator of the Temporal Universe we inhabit. That Christ is the same, who joined with the Father to create our individual, distinct, eternal, physical bodies; by such, He earned the right to be called the "*father of spirits*" (Heb. 12:9). As Jehovah, that same *He* took direction from the Father and directed Michael as co-creator in the forming of the Physical sphere. He is the Word, the Light of Men, and the Divine Logos; "*the same was in the beginning with God. All things were made by him; and without him was not anything made that was made*" (John 1:2, 3). All of this being said, understood and thoroughly accepted, we also affirm that Christ, Jehovah, was and is the first-born of the Family of God, *created* by the Father a very long time ago. In this belief we are also joined by the Apostle Paul who wrote of Christ:

> *"Who is the image of the invisible God, the first-born of every creature. For by him were all things created, that are in heaven, and that are in earth, visible and invisible, whether they be thrones, or dominions or principalities or powers; all things were created by him and for him: And he is before all things, and by him all things consist. And he is the head of the body, the church: who is the beginning, the firstborn from the dead; that in all things he might have preeminence. For it pleased the Father's that in him should all fullness dwell." (Col. 1:15-19)*

To the Colossian Saints, Paul began his description of Christ's relationship to His Father with the phrase, "*Who is the image of the invisible God.*" The word "*image*," used here is the Greek, *eikon* (εἰκὼν), which means *copy* or *likeness*; it is not used to express that one thing is the *same* thing as something else, it is not an equivalency. Paul could have said Christ was the Father, but he did not; in fact, he *never* does and neither does the Savior. An object cannot be an image of itself; an image implies, in fact demands, the existence of another.

In the context used here, the word encompasses much more than a resemblance or an exemplification. It would be improper and misleading to suggest that Paul meant to portray Christ as an image, like that which shows in a mirror, just a manifestation of the singular entity it represents; that would be Docetism. A reflection in a mirror is still not *identical* to that which it is an image of; it is separate, unique with its own attributes, completely alike, but in appearance only without life of its own.

Though our interpretation of what Paul meant by *image* is strongly supported by the text itself, it remains just an interpretation; what is needed is his personal testimony of the context itself. Fortunately for us, the statement we have been evaluating is just the first half of a single sentence. In the last half, following the comma, Paul explains exactly what he understands the image of Christ to be. To Paul, Christ is not the *invisible God*, but he is a *creation* of that invisible God, separate and distinct from God; Paul uses the Greek word, *ktiseos* (κτίσεως). *Ktiseos* specifically speaks to something created where nothing had existed before, an original creation, not a reformulation, change of appearance, or alternate manifestation. It is nothing at all even remotely similar to a third personage *piece* of one single, uncreated, always-existing, eternal God. To further put *image* in its proper context here, the one intended by Paul, we can hearken back to Genesis 1, man is a created being, no one can doubt that, or will, yet man, like Christ, is also created in the *image* of that same invisible God; there is no mistaking that.

We do not accept that God cannot be divided, nor do we accept the concept of *ex nihilo*, these we affirm are not taught in the Scriptures but are theoretical constructs of man. Our God is neither immutable nor impassible; He lovingly created us out of Himself and He weeps with us when we do.

What this means is the Greek philosophers had it all wrong, creator and creature can be of the same substance which is in fact what we see happen every day all around us. Substance is akin to kind; it does not relate directly to form or function. Everything in Temporality is of the same substance, the same elements from the periodic table (the vast majority of it from just the first 16 of those elements listed). That substance, however, comprises an innumerable variety of form, all with different characteristics and attributes. The architect and steelworker that creates a bridge over a vast stretch of water is made up of the same iron and oxygen and hydrogen; the creator, the created and the expanse are all the same thing. So it is with God and ourselves His creation, one much greater than the other; one with attributes of Deity, the other with attributes far less capable, but of substance, we are the same.

Sure, all of this makes sense when viewed from the perspective of today, but to the philosophers and theists of 2000 years ago, such an understanding of the world was unfathomable. The Greek world that birthed the Christian era was made up of only four elements: Earth, Fire, Water and Air. There was a hierarchy of elemental spotlessness; Earth, from which we were formed, was the most corrupt and base. Air, of which Spirit consisted, was pure and near perfect; it could catch sail, bend branch and make leaves dance; it brought life with a newborn's first breath and took it with the dying's last; it was ethereal, mysterious and unknown. To Plato and those that thought like him, God created the elements and as such was not *of* them. He was unmoving and without change. He neither had a body or parts, because that could only be composed of element which was corrupt and corruptible. God looked down upon war and peace, joy and misery, and to the philosophers did not intervene; therefore, He was without passion. Naturally, they framed their concepts of God, the created, and the relationships between the two in terms they understood and so too (eventually) did the new Christians.

- When was the Son Embodied?

What is required for Godhood? To answer any query as to when the Son was embodied, it is first necessary to answer what is required of God. In providing an answer to this primary question, Latter-day Saint hierophantic teaching, most often espoused in popular exposition and not repeated officially, differs widely from what the Scriptures describe.

It is common to hear teaching that to be fully God one *must* be embodied. Popular explication often goes on to declare, *for this reason we came to Temporality to receive a body.* Everything we have discussed here debates this assumption. The purpose of Temporality is far removed from that (though we understand that embodiment remains a peripheral component of the purpose of the Temporal).

Without question God our Father and His Son, Jesus Christ are now embodied. We have discussed here that the pre-incarnate Christ, as Jehovah, was embodied in his interaction with ancient prophets, including Moses, the brother of Jared and others. We understand that the Father and the Son were embodied when they walked in the Garden of Eden with Adam and Eve. The *embodiment* of the Father and the Son is a reality, truly ancient in effect. A Platonic view of an immutable God would derive that a God who *is* embodied *always has been* embodied, but is this the case? And is embodiment truly a requirement of Deity?

As to when the Father may have become embodied or even if there ever was a time when He was without one, the revealed word of God is completely silent. We cannot and will not speculate here. As to the time of Jehovah's embodiment we are no closer to understanding, except that we understand that there was a time when He was begotten and that before that, He did not exist as a separate entity.
Any doctrine that requires the embodiment of God runs into not one, but several logical stops when it is related to other doctrines and the revealed Word. The first of those logical stops is most apparent; the Holy Ghost is declared to be God, the third member of a triune Godhead. Yet, we are told, the Holy Ghost is unembodied and must remain so, "*The Father has a body of flesh and bones as tangible as a man's; the Son also; but the Holy Ghost has not a body of flesh and bones, but is a personage of Spirit. Were it not so, the Holy Ghost could not dwell in us*" (D&C 130:22). We could develop an understanding here that the Holy Ghost does have some form of a body that is other than "*Flesh and Bone*" but that would make the Holy Ghost embodied in such a way that is different from the Father and the Son *and* not align with what is revealed here about why the Holy Ghost is without body.[57]

The second logical stop is less apparent and only reveals itself when we consider another known attribute of God – the ability to create. We are told that Michael co-created with Jehovah all things in the Physical sphere; Michael accomplished this before he was embodied in *Flesh and Bone* as Adam. This stop is not secure and fast in that there are extenuating circumstances that are themselves, perhaps, revealing. The unembodied Michael is co-creator; he is joined and directed by Jehovah, and Jehovah is the one who receives direct instructions from the Father. This speaks of a significant difference between Jehovah and Michael that may include the fact that Jehovah was embodied and Michael was not. However, if we are to accept that only Deity can create, and allow that Michael was "*creating*" and did so unembodied,

---

[57] Refer to our discussion in Chapter 10 of this book, The Holy Ghost, pg. 417 for a detailed understanding of the actual embodiment of the Holy Ghost.

then we have to resolve the conflict that arises from Michael displaying qualities of deity while yet unembodied.

Perhaps finally, and of extreme importance, is our understanding of the Son and His relationship to us before the creation of the Physical sphere. Latter-day theology understands Jehovah, the Firstborn to be our elder brother; He is of the same order of creation as we, but first and superior. If we declare that He, Jehovah, was God from the beginning, as indeed we must do *for Scripture demands it*, and insist that to be God he has to be embodied, then our simple understanding of Him as our elder brother may be wrong, for if He was embodied at the *moment* of his primal inception, He never would have existed as an unembodied intelligence as we were. This understanding infers His creation is different than ours in substantive ways, ways substantive enough for Him to perhaps be of an entirely different order of creature. *We can be fine with this concept and still hold true to all that we have discussed in this work, but we must recognize the significance of this consequence that stems from an insistence that deity has always had to be embodied to be Deity.*

***If we declare that He, Jehovah, was God from the beginning, as indeed we must do, for scripture demands it, ...then our simple understanding of Him as our elder brother may be wrong.***

Some have found a further logical stop in the incarnation of Christ, arguing that His incarnation was a second embodiment, a reincarnation, since He was embodied as He appeared to the Old Testament period prophets. This view is incorrect. Christ was not reincarnated when His mortal body began to develop in Mary's womb. That He was embodied prior to this in pre-mortality, requires us to understand that the embodiment that was witnessed by Moses and the brother of Jared, was of a different form, character and nature (*Flesh and Bone*) than that He had assumed on Earth (*Flesh and Bone and Blood*). We will discuss this later in this work in Chapter 10.[58]

Examining what we have discussed here brackets a period in time for the event that resulted in Christ's Physical embodiment, if however loosely and tenuously established that timing is. We have suggested that Jehovah had a period of existence as an unembodied intelligence just as we, placing His embodiment (a physical and immortal embodiment, as a personage of Spirit) sometime *after* the

---

[58] See our discussion, The Physical Nature of the Godhead, pg. 412.

creation of His individual intelligence. We discussed that when Jehovah created the Physical sphere, He was different in nature than his co-creator Michael; we suggested that this difference may have included His embodiment. Though we cannot be certain, because what has been revealed to us is woefully incomplete, our time frame for the embodiment of the Son then is placed at some point within the Spiritual sphere *before* the Physical was created. This alludes to a conclusion of sorts that the Son is an embodied being of a different order than we, if for no other reason than we had to wait for the creation of the Physical sphere to be embodied, and it appears He did not.

- The Divine Feminine

We have seen here that it is apparent that gender may be an attribute of intelligence. "*In origin, man is a son of God. The spirits of men 'are begotten sons and daughters of God.' Through that birth process, self-existing intelligence was organized into individual spirit beings*" (Marion G. Romney, Conference Report, Ensign, Nov. 1978, 14). As such, gender is truly a basic and fundamental quality of the divine within us. But does that mean that gender is also an attribute of God? Logic, it would seem, demands so, for we, both male and female are made in the likeness of God. We have almost universally spoken of God as our Father, a male; it can and has been argued that this is nothing more than a representation, reflecting the internal bias present in millennia of male dominated cultures. Though there are those who speak of a Mother God, and have been those who have done so for millennia, it has almost always been done with a marked lack of conviction, as if spoken only counter-culturally, in terms rarely ever more than just suggestive of the possibility.

In contrast to those who speak of mere possibility, Hebrew Kabbalist tradition speaks of the Spirit of God, the "*Eyn Sof*" (אֵין סוֹף),[59] as encompassing wholly the Father and a Mother, decidedly assigning a feminine attribute to the Godhead. The *Eyn Sof* is a wholly Hebrew triune godhead, including not a divine son like the Christian Trinity, but composed of a Father, Mother and Spirit. The *Eyn Sof*, is combined into a unified singular entity, in a manner not at all unlike that imagined by the traditional Christian Trinity.[60]

Latter-day expressions of a belief in Heavenly Mother stem from our understanding of our own families in an eternal construct. Though little is revealed of its particulars, we understand that eternal increase means husbands and wives (sealed together through the power of the Holy Order of the Priesthood of God) will experience eternal increase that is thought to include such a sealed pair generating

---

[59] The Word, pg. 33; William S Eldridge; 2010

[60] Even though the Jewish Sages considered Eyn Sof (Ain Soph) indefinable, they did most universally agree that Eyn Sof was a Triune Godhead including a Father and Mother, having both masculine and feminine attributes. Judaism based this assertion on the Genesis 1 account: "¶*And God said, Let us make man in our image, after our likeness: and let them have dominion over the fish of the sea, and over the fowl of the air, and over the cattle, and over all the earth, and over every creeping thing that creepeth upon the earth. So God created man in his own image, in the image of God created he him; male and female created he them.*" (Genesis 1:26, 27)

Spirit children of their own. In a revelation recorded in July of 1843, but no doubt received, at least in part, as much as 12 years earlier, the Prophet Joseph detailed the rights and privileges assured to those who enter into the *New and Everlasting Covenant of the Priesthood*:

> *"And again, verily I say unto you, if a man marry a wife by my word, which is my law, and by the new and everlasting covenant, and it is sealed unto them by the Holy Spirit of promise, ... they shall pass by the angels, and the gods, ... to their exaltation and glory in all things, as hath been sealed upon their heads, which glory shall be a fulness and a continuation of the seeds forever and ever." (D&C 132:19, 20)*

It has been logically inferred that the blessing of "*glory*" which is a, "*fulness and a continuation of the seeds forever and ever*" pronounced on worthy male and female pairs is an emulation of the process our Heavenly Father must have accomplished with an eternal companion of His own, at least in the distant past, if not even now and continually. That *eternal companion* of our Father, is understood to be Heavenly Mother. It is this reasoning alone that engenders the Latter-day conceptualization of the Mother God.

There is no presented revelation to describe how "*a continuation of the seeds forever and ever*" would effectually be accomplished. Though some have speculated both within the church and without, with various intentions, that this generation is an extension and continuation of a sexual relationship between a husband and a wife, imitating in eternity what is performed in the Temporal sphere. This, however, does not appear to be so, for we have proposed and shown here the creation of our Intelligences proceeds from subdivision of pure intelligence from God himself. We have also detailed the subsequent embodiment of those Intelligences (the creating of our Spirits) was a task performed by the Father (Elohiym) and the Son (Jehovah). In both these creative tasks, there is no understanding of a *mother* involved. If eternal creation completed by both male and female pairs in the eternities to come is indeed the expected course, it would not parallel any of the processes we have so far described in this work pertaining to the generation of our Intelligences and our eternal Physical bodies.

This being said and understood, consideration should be given to the possibilities suggested by the Feminine in God and in our Intelligences and within the Hebrew definitions of the Name *YHWH*; these are of significance even if we are not yet to understand why.

## Debate in the Grand Council, War in Heaven

The LORD instructed the patriarch Abraham regarding the premortal experiences of all who have been, or ever will be upon the Earth. Among those events was the Council in Heaven, at which the Father's Plan of Salvation for His children was discussed. Disagreement at the Council resulted in contention and conflict called the War in Heaven

It is not clear when the Great War in Heaven occurred, though it is directly associated with the Grand Council and follows it, most probably very soon after.

It is interesting to note that Jewish scholars interpret many passages of the Old Testament as providing a description of a *Grand Council* in Heaven, among them are: 1 Kings 22:19-22; Isaiah, Chapter 6, and Job, Chapters 1-2. We have previously discussed the "*Let us*" and "*as one of us*" phrases of Genesis 1:26 and 3:22, as they are interpreted by this same tradition. A commentary for Genesis 1:26 can be found in The Jewish Study Bible,[61] which reads: *"The plural construction (Let us...) most likely reflects a setting in the Grand Council. God the King announces the proposed course of action to his cabinet of subordinate deities, though He alone retains the power of decision...."* Here, one of the most respected sources of Jewish exposition continues on with this quoted exegesis to speak of debate in that same council, regarding the creation of man, Temporality and evil; "*The midrash manifests considerable uneasiness with the plan to create something so capable of evil as human beings are*;" "*God told his ministering angels only of 'the way of righteousness' and hid from them 'the way of the wicked'*"'[62] and finally, "*...while the angels were debating the proposal among themselves, God took the matter in hand. 'Why are you debating? ' he asked them. 'Man has already been created!'*"'[63]

We must note here, that in this discussion, indeed in all of our discussions of creation so far which pertain to events that occur before the introduction of the Temporal realm, we are hampered by our want to place one thing before the other. We feel pressed to fulfill our need for a chronological element, *that* which we sense as *Time*. In discussing things spiritual, a sense of time may be wholly inadequate and misleading. Time, the concept of a before and after, does not comfortably fit within the definition of the eternal world that exists in our minds, but this may only be an enigma because of our limited understanding. Consider the following:

The Scriptures are replete with references comparing the time frame of us (man) to that of God. It is important not to overlook that there is in fact a *comparison* of time frames discussed. I don't think this fact can be taken lightly. It is clear that no attempt is made to claim that the realm of God is without a reference frame that

---

[61] *The Jewish Study Bible, Featuring the Jewish Publication Society, Tanakh Translation* (Berlin. Adele and Brettler, Marc Zvi, editors, Oxford University Press 2004), pg. 14.

[62] Genesis Rabbah 8.4

[63] Genesis Rabbah 8.5

could in some way look like a time arrow, pointing to the future, from a point in the past. The Creator himself identifies as both Alpha and Omega. Implicit in that claim is the human concept of one thing after another, a bracketing of things between a beginning and an end. It may be true that in this the Creator is just speaking in terms that mortal man can understand, but it is equally true that our concept of omnipresence is hindered by our being mortal. The idea of a God that is without time is a work of sentient constructs, it conforms to Platonic logic, separate and apart from truths revealed or not revealed in Scripture.[64]

Although the *Grand Council* is often spoken of as a single event, what has been revealed to us almost certainly describe actions that occurred over a span of time and in multiple proceedings. Our all-knowing Father in Heaven knew the course that had to be completed from the outset, so it must be understood that the Plan, its development, was not a work of committee, or anything of that kind. That we were independent Intelligences, already agent, demanded that we became informed so that we could freely choose. We are told that at these occasions, Jehovah and others were foreordained to the work they would accomplish in the Physical and Temporal spheres; "*Even before they were born, they, with many others, received their first lessons in the world of spirits and were prepared to come forth in the due time of the Lord to labor in his vineyard for the salvation of the souls of men*" (D&C 138:56). Understanding these processes, those of foreordination and the detailing of an already formulated Plan in the pre-existence, and how they relate to *what we can* and *what we perceive we can* do in Temporality is instructive.

At the *Grand Council(s)*, we were informed of the Plan and were told that a time would be appointed when we could accept or reject it in our lives, through a simple, up-or-down vote. We did not develop the Plan, nor contribute to its development, for in the presence of the Father, that is not possible.[65] Temporality affords us unique opportunities to not only choose between *what* is right and wrong, but we can *develop the concepts* of right and wrong, write our own rules, watch them play out, experience the results and learn what right and wrong means. These things are not possible in the presence of the Father. Learning through trial and error is a dirty, muddy process, the precincts of the Father are ordered, pristine and perfect; anything less than that is an abomination and as such is barred from His presence.

---

[64] As an example of conflicting premises relating to God and time, consider this: If God does not exist in time, before and after having no meaning to Him, how can God track our progression (e.g. unrepentant and penitent), account for it and actually progress Himself?

[65] It has been said, "*Within the Mormon tradition there are very few well defined and uniquely Mormon theories about Atonement.*" Within this work, we have developed one of them, perhaps the *only* one defined, illuminated and in harmony with all that we know of the Universe today. There is another theory that has achieved some acceptance, "*Intelligence Theory*," first proposed by Cleon Skousen. Intelligence Theory, advocates a process that describes a god who holds power through popular acceptance of independent intelligences. To retain his position of popular approval, this imagined god must remain firm and just in appearance, exacting law and demanding a price be paid for disobedience (thus the Atonement). This theory is wholly unacceptable in that it denies the intrinsic sovereignty of God and replaces it with a similitude of political power held through support of the masses.

The Plan was perfect as it was presented, any deviation from it through amendment or adjustment could not be considered by the Father. It is important to note here that though the Plan is perfect, the execution of it, however, requires the imperfect. To bridge the gap between the two disparate worlds, a Redeemer/Justifier/Sanctifier was needed. The Gospel was established, and prophets were identified and called to be born and to work in the Temporal sphere at critical times, providing waypoints in a raging chaotic sea, by which we, who are lost, can navigate our way back home.

The Plan, and the truly "*free*" agency it provided for, required the redeeming of a Fallen Universe (and all that inhabit it) from the Physical Death brought about by the Fall of Adam. The Plan provided for a restoration (at great cost) of the sons and daughters of God from the Spiritual Death that results from sin. The time came when, following the principles of agency, one who would accept the burden required, had to freely choose to do so. The moment was not without a significant amount of drama and discord.

**The Spirit World — Abraham 3**

27 And the Lord said: Whom shall I send? And one answered like unto the Son of Man: Here am I, send me. And another answered and said: Here am I, send me. And the Lord said: I will send the first.
28 And the second was angry, and kept not his first estate; and, at that day, many followed after him.

The perfection of the precincts of God began to be challenged when a usurper openly opposed the Plan and sought to draw others with him. We have already discussed the Adversary, *satan* (שָׂטָן), and his actions at the time of the *Grand Council.*[66] The Scriptures themselves, in two separate places, focus in on this specific and critical series of events when instruction became debate, then broke out into struggle and open rebellion. It was no doubt after long periods of explanation, and teaching in which the Plan was expounded and justification was given for the risks, that the Father asked, "*Whom shall I send [as the Redeemer]?*" The first to answer was Jehovah, Jesus Christ, *the great I AM*; agreeing to follow the Father's plan He spoke, "*Here am I, send me.*"

Abraham and Moses both record the same event similarly. Abraham is quoted in the blocked text above and Moses' account of the same incident, though provided before, is repeated again here:

> *"That Satan, whom thou hast commanded in the name of mine Only Begotten, is the same which was from the beginning, and he came before me, saying—Behold, here am I, send me, I will be thy son, and I will redeem all mankind, that one soul shall not be lost, and surely I will do it; wherefore give me thine*

---

[66] See SEND ME beginning on page 204 of this work.

> *honor. But, behold, my Beloved Son, which was my Beloved and Chosen from the beginning, said unto me—Father, thy will be done, and the glory be thine forever. Wherefore, because that Satan rebelled against me, and sought to destroy the agency of man, which I, the Lord God, had given him, and also, that I should give unto him mine own power; by the power of mine Only Begotten, I caused that he should be cast down; And he became Satan, yea, even the devil, the father of all lies, to deceive and to blind men, and to lead them captive at his will, even as many as would not hearken unto my voice." (Moses 4:1-4)*

| **The Spirit World — Abraham 3** |
| --- |
| 28 And the second was angry, and kept not his first estate; and, at that day, many followed after him. |

- And At That Day Many Followed

When the one who would be called Lucifer was rejected, his anger kindled a fire of rebellion in him, "*he became Satan, yea, even the devil, the father of all lies, to deceive and to blind men, and to lead them captive at his will, even as many as would not hearken unto my voice.*" As his rage grew, he led many away after him.

> *"for behold, the devil... rebelled against me, saying give me thine honor, which is my power; and also the third part of the hosts of heaven turned he away from me because of their agency;" (D&C 29:36)*

> *"And there appeared another sign in heaven; and behold, a great red dragon, having seven heads and ten horns, and seven crowns upon his heads. And his tail drew the third part of the stars of heaven, and did cast them to the earth. ...Neither was there place found in heaven for the great dragon, who was cast out; that old serpent called the devil, and also called Satan, which deceiveth the whole world; he was cast out into the earth; and his angels were cast out with him." (JST Revelation 12:4, 8)*

Here for the first time, we are introduced to another synonym for Lucifer, the Great Dragon, or Great Red Dragon. We already delineated the significance of the names of the deceiver and adversary;[67] Lucifer, Satan and the Devil were shown to be equivalent. We also detailed that this same one was known as *nachash* (נָחָשׁ), the Serpent, or through a surprising derivative of this word, the "*Shining One.*" John draws on the same imagery this time animating the characterization further to that of a seven headed, ten horned dragon. The heads and horns viewed by John have eschatological significance confirming the works of this *satan* (שָׂטָן) will continue to the last days.

Satan is cast out, but along with him, fall also a third part of the Intelligences that were at the *Council(s)*. However, a *third part* does not actually mean 1/3 or 33%. In

---

[67] See our discussion titled, The Serpent, beginning on page 245 of this book.

the Biblical Hebrew, a *third part* means a portion, defining the concept of *some* as opposed to *most*. It is more than a few, but may not be as much as 1/3, though it could have been more.

- The Anointed Cherub

It is curious that two stood to say "*send me,*" as if they both had full expectation of being equal to the task. How could this be so, when we know that the task could only be accomplished by the *Firstborn*, the Only Begotten of God. Satan proposed that he could "*redeem all mankind, that one soul shall not be lost*" (Moses 4:1). We can assume this was believed to be a reasonable promise; why would those who chose to follow him have fathomed doing so, if he stood on an empty, unachievable platform? Clearly there is more to be understood here.

Without doubt, Lucifer, before the rebellion, was positioned with great honor among the Intelligences in the Spiritual sphere. The Prophet Ezekiel describes his pre-existent prominence as follows:

> *"Thou hast been in Eden the garden of God; every precious stone was thy covering, the sardius, topaz, and the diamond, the beryl, the onyx, and the jasper, the sapphire, the emerald, and the carbuncle, and gold: the workmanship of thy tabrets and of thy pipes was prepared in thee in the day that thou wast created. Thou art the anointed cherub that covereth; and I have set thee so: thou wast upon the holy mountain of God; thou hast walked up and down in the midst of the stones of fire. Thou wast perfect in thy ways from the day that thou wast created, till iniquity was found in thee. By the multitude of thy merchandise they have filled the midst of thee with violence, and thou hast sinned: therefore I will cast thee as profane out of the mountain of God: and I will destroy thee, O covering cherub, from the midst of the stones of fire. Thine heart was lifted up because of thy beauty, thou hast corrupted thy wisdom by reason of thy brightness: I will cast thee to the ground, I will lay thee before kings, that they may behold thee. Thou hast defiled thy sanctuaries by the multitude of thine iniquities, by the iniquity of thy traffick; therefore will I bring forth a fire from the midst of thee, it shall devour thee, and I will bring thee to ashes upon the earth in the sight of all them that behold thee. All they that know thee among the people shall be astonished at thee: thou shalt be a terror, and never shalt thou be any more." (Ezekiel 28:13-19)*

Lucifer was created with special significance; the emblems of his glory were *in* him when he was created; "*the workmanship of thy tabrets and of thy pipes was prepared in thee in the day that thou wast created.*" Thus he was made that way by God. Ezekiel confirms this when he says Lucifer was anointed by God, specifically the, "*anointed cherub that covereth, and I have set thee so.*"

The choice of the word, *covereth,* here, *hasowkek* (הַסּוֹכֵךְ) in the original Hebrew, hearkens back to the Atonement; we will remember that we have already shown that the word *kippur* (לְכַפֵּר), Atonement, also means "*to cover.*" Relating the Fallen one to a title that is in effect, *one who atones* is high praise indeed and bespeaks of the immense potential once possessed by Lucifer, indeed he was a "*Shining One.*"

The authority associated with Lucifer in the times of the *Council* is further enhanced by God's vision of him as viewed by Ezekiel. In that vision, the breastplate of the one who is cast down holds nine precious stones, they are; "*the sardius, topaz, and the diamond, the beryl, the onyx, and the jasper, the sapphire, the emerald, and the carbuncle.*" Compare these instructions the LORD gave to Moses for making the breastplate of the High Priest in Israel:

> "*¶And thou shalt make the breastplate of judgment with cunning work; after the work of the ephod thou shalt make it; of gold, of blue, and of purple, and of scarlet, and of fine twined linen, shalt thou make it. Foursquare it shall be being doubled; a span shall be the length thereof, and a span shall be the breadth thereof. And thou shalt set in it settings of stones, even four rows of stones: the first row shall be a sardius, a topaz, and a carbuncle: this shall be the first row. And the second row shall be an emerald, a sapphire, and a diamond. And the third row a ligure, an agate, and an amethyst. And the fourth row a beryl, and an onyx, and a jasper: they shall be set in gold in their inclosings. And the stones shall be with the names of the children of Israel, twelve, according to their names, like the engravings of a signet; every one with his name shall they be according to the twelve tribes." (Exodus 28:15-21)*

It is clear that the coverings of Lucifer are identical to the breastplate of Aaron, curiously missing the stones of the third row, "*a ligure, an agate, and an amethyst.*" Perhaps this speaks of a priesthood of a lesser order, or one even that is somehow incomplete.[68] When this same Fallen one, Lucifer, appears in the Garden, he is asked of his garments, particularly of his apron, to which he replies that, "*it is an emblem of my power and priesthoods.*"

Of Lucifer, the Fallen one, God proclaims, "*Thou wast perfect in thy ways from the day that thou wast created, till iniquity was found in thee.*" The precincts of God are perfect and within them discord, contention and rebellion cannot exist. All of these are abominable and corrupt in the eyes of God. Referring again to Lucifer, the "*Shining One*," God declares: "*Thine heart was lifted up because of thy beauty, thou hast corrupted thy wisdom by reason of thy brightness.*"

Isaiah declares that it is those that, "*walketh righteously, and speaketh uprightly*," who "*will dwell on high*," "*with the devouring fire*" and "*everlasting burnings*" (Isa. 33:14, 15). And here in vision Ezekiel sees Lucifer being cast from, "*the midst of the stones of fire.*" Because of Lucifer, God says: "*thou hast sinned: therefore I will cast thee as profane out of the mountain of God: and I will destroy thee, O covering cherub, from the midst of the stones of fire.*" Isaiah, similarly witnessed of this fall of one who was so great, both asking and answering, "*How art thou fallen from heaven, O Lucifer, son of the morning! how art thou cut down to the ground, which didst weaken the nations! For thou hast said in thine heart, I will*

---

[68] Each of the twelve stones on the breastplate of Aaron are each associated with one of the twelve tribes of Israel. The breastplate of Lucifer only contains nine. Perhaps this signifies a lesser kingdom or dominion, one represented by the number nine instead of twelve.

*ascend into heaven, I will exalt my throne above the stars of God: I will sit also upon the mount of the congregation, in the sides of the north.*" (Isa. 14:12, 13).

- Michael and His Angels

With his annunciation, "*And there was war in heaven: Michael and his angels fought against the dragon*" (Revelation 12:7). John places Michael in the Great War in Heaven. John identifies Michael as a leader of significance for the just cause in the struggle. Daniel identifies Michael as "*the great prince which standeth for the children of thy people,*" (Dan. 12:1) who are the *people of God.*

Ancient sources often conflate the identities of Michael and Melchizedek; it is clear the two were identified as the same by many. One source (an extremely fragmentary Dead Sea Scroll manuscript, 11Q13) originating with the first and second century BC inhabitants of the community at Qumran, records Melchizedek as the divine being who allies the Angels against Belial (the Devil) and his followers. This telling, however, may more properly be referencing the *Priesthood* Michael held and not the man Melchizedek; "*priests of the Most High, after the order of Melchizedek, which was after the order of Enoch, which was after the order of the Only Begotten Son*" (D&C 76:57). It is appropriate that Michael would be associated with the Melchizedek Priesthood,[69] as he led the struggle to ensure the mission of the Savior could be accomplished.

That Michael fought in the great confrontation does not speak at all to what form his being took at the time, nor does it speak of ours. That *he* was the leader of the warring angels and *not* the Father or the Son does speak of something that is perhaps very significant and on reflection obvious.

As we have discussed sin and discord cannot exist in the presence of God, so we must understand that once the rebellion started and the war ensued, it was conducted away from the Father, beyond His precincts and even from those of the pre-existent Son. Because of this we find that it was not the Father, or Jehovah who lead the righteous forces, but Michael; "*And there was war in heaven: Michael and his angels fought against the dragon; and the dragon fought and his angels, And prevailed not; neither was their place found any more in heaven. And the great dragon was cast out, that old serpent the Devil, and Satan, which deceiveth the whole world: he was cast out into the earth, and his angels were cast out with him*" (Revelation 12: 7-9).

Lucifer and his followers were cast down at the outset of the war and not after its conclusion. This begs the question as to where that place of exile may have been,

---

[69] In consideration of what we have discovered about the coverings of Lucifer and the Breastplate of Aaron, it would be very interesting to discover what any vestment of Michael may have looked like in the Physical sphere.

perhaps someplace alluded to in the words of Lucifer in the Garden, "*I have been doing that which has been done in other worlds.*"[70]

> ***The original choice presented to the sons and daughters of God was not to choose the plan of the Father over that of the adversary, but it was to accept the challenges of Temporality or to remain as they were in perfect harmony with God.***

It is impossible to imagine contention and war invading the perfect precincts of the Father in the Spiritual sphere, so we must envision it having occurred elsewhere beyond the Holy Pavilions of our Father in the Spiritual sphere. We understand here that the war occurred after the creation of the Physical sphere and the embodiment that created the Spirits of Adam and Eve and ours as well.[71]

Perhaps it was the creation of the Physical sphere that led to the conditions that allowed for conflict and discord. Perhaps it was to an Earth, or the "*earth*" (unorganized *matter*) of the Physical sphere to which Satan and his followers were cast. It may have been on the plains of a heavenly orb or in the vast open heavens of the Physical realm that the great armies massed and joined in struggle.

What we can of certainty say here is, the Great War in Heaven followed the Grand Council and occurred before Michael disappears in the Endowment narrative and reappears as Adam in the Garden.

In time, the conflict (within the Physical sphere or in that of the borders of the Spiritual) came to a close, and those that chose to follow God's plan were at peace

---

[70] See our previous discussion, That Which Has Been Done (Then Again Once More), pg. 286.

[71] This important doctrinal consideration was addressed almost 90 years ago by five apostles who objected to B.H. Roberts' inference to Intelligences rebelling against God. They wrote: "Those who rebelled in the world of spirits were begotten spirits, who, if they had remained faithful, were prepared to come into this mortal world. The revelation which speaks of intelligence [D&C 93:30] says: 'man was in the beginning with God.' (When was this beginning?) Then this thought follows: 'Intelligence, or the light of truth, was not created or made, neither indeed can be.' Again we are taught that 'Light and truth' –intelligence– 'forsaketh that evil one.' [D&C 93:37] This being true, and treating intelligence as an entity, [as we do within this work] then that entity cannot rebel against light and truth, for it would rebel against itself." (*References and comments in brackets added*) (Report of **DOCTRINAL POINTS QUESTIONED BY THE COMMITTEE** (Elders; George Albert Smith, David O. McKay, Joseph Fielding Smith, Stephen L. Richards, and Melvin J. Ballard, all then Apostles) appointed by the First Presidency to read the manuscript of B.H. Roberts' The Truth, The Way, The Life, submitted 15 May 1930.)

again and happy; "*the morning stars sang together, and all the sons of God shouted for joy...*" (Job 38:7).

We should note here that the original *choice* presented to the sons and daughters of God was *not to choose the Plan of the Father over that of the Adversary*, but it was to accept the challenges of Temporality or to remain as they were in perfect harmony with God, though without the blessings of eternal progression. This fact is almost always overlooked. Understanding this concept, however, provides substantial contexts on which to hang those of foreordination, election and other conceptualizations that place what *appears* to be predestination in direct opposition to free will. We understand here that all of these conflicts are resolved through choices freely made and actions that followed those choices within the pre-temporal spheres.

By accepting the challenges faced by Temporality, risking the consequences attendant with it, those that would come to be called the sons and daughters of God, covenanted with the Father that by entering into the uncertainty of the Temporal world and enduring there to the end, they would receive increased knowledge and power and would continue on their progressive journey to become like the Father, co-heirs to His kingdom with Jehovah, the very Son of God.

- The Great War Continues

> *"And the dragon stood before the woman which was delivered, ready to devour her child after it was born. And the woman fled into the wilderness, where she had a place prepared of God, that they should feed her there a thousand two hundred and threescore years. And there was war in heaven; Michael and his angels fought against the dragon; and the dragon and his angels fought against Michael; And the dragon prevailed not against Michael, neither the child, nor the woman which was the church of God, who had been delivered of her pains, and brought forth the kingdom of our God and his Christ." (JST Revelation 12:4-7)*

The account of the War in Heaven found in the Book of Revelation and virtually all of the other ancient sources speak eschatologically of the confrontation, so we understand that it still rages on; this should be of no surprise to any Christian. Some see in the symbolic imagery of the war's telling events yet future, of a great conflict yet to occur. Many have placed this great clash in the future, as occurring at the end of time, within *the last week of the tenth jubilee*.

It is perhaps fitting that the timing of this event is so enigmatic, for indeed it is both past and future; it is present as well, continual and constant. As such, what we have described as the Great War in Heaven was not really a *war* but an opening battle in an enduring struggle that will continue to rage between forces of good and evil until in the end, Christ is victorious and the Father shall "*put all enemies under his feet*" (1 Cor. 15:25).

## CHAPTER 9

# DAY BREAKS ON AN ETERNAL MORNING

### The Three Degrees of Glory

It may not have occurred to you that what we have been discussing in this work (three separate creations or spheres) perhaps forms the basis for the Three Degrees of Glory which are so familiar to Latter-day Saints. Perhaps you have surmised that the three separate creations provide the substructure for the context of the three personage Godhead, as well as our three part individual being, *Intelligence*, *eternal Physical body, and Temporal body*, which together comprise our living Soul? Such a correlate of threes is not coincidental and must be considered significant.

If you have made a *direct* connection between the three creations we have detailed here with the three Eternal Kingdoms, you have come to a conclusion that is not supported by this work. The Spiritual sphere is not the Celestial Kingdom, neither is the Physical sphere the Terrestrial Kingdom. The Eternal Telestial Kingdom is not what we have described here as the Temporal sphere. Such an equivalency is not my intention, and we must correct such thinking here.

In *Figure 1-2* we have shown the Temporal to be finite in duration and entirely a subset of the Physical, which had a beginning but will have no end. Both the Physical and Temporal are subsets of the Spiritual, which is eternal having no *knowable* beginning and which will have no end. We understand the Telestial, Terrestrial and Celestial Kingdoms to be eternal realms, wholly within the Physical sphere,[1] itself something a part of a greater Spiritual sphere, the beginning of which we will not speculate on at all in this work.

---

[1] See Figure 7-3 *The Physical Sphere and its Environs*, pg. 289 and footnote 70 on that same page.

Let us take a moment to reiterate an important understanding. In the last chapter, we spoke at length of the Spiritual Creation. We understand explicitly that things spiritual in nature, things unembodied, without a discernable physicality or tangibility, were created at some point in the distant past by Father God, He whom we call *Elohiym*. The first of these things created in the Spiritual sphere was the Son, Jehovah, Christ; "*in the beginning was the Word, and the Word was with God*" (John 1:1). He was created in accordance with the nature of things there. The Spiritual Creation we have detailed in this work is not to be understood as the creation, beginning, of the Spiritual sphere itself, but by it we describe *things* that are within the Spiritual sphere. We have not made any attempt to describe a beginning for the Spiritual sphere or a beginning for the Father. Within this work, Father and the sphere He inhabits, extend forever into an eternal past, without consideration or speculation of a time before.

Throughout this work we have defined the eternal as that which includes some things which have a beginning but do not have an end. We understand by this an eternal nature for things that are in fact created. For example, it is hoped that through the Atonement the *saved* will live forever with Father in Heaven. *What is forever if not eternity?* The three Heavenly Kingdoms detailed in this chapter are to be understood as such eternal constructs, something, the beginning of which we know, but which will have no end.[2]

There is similarity between the Three Degrees of Glory and the three creations in number, hierarchy and gradation of influence and association with the Father, but the connection should stop there and not be carried further. To reiterate, the Spiritual Creation we discuss here does not represent that which has been called the Celestial Kingdom. The Physical Creation should not be understood as representing merely the Terrestrial Kingdom or the confines of the Garden in Eden. The Temporal sphere we inhabit today is not the Telestial Kingdom nor should we expect it to be transformed in any way to become so.

## *The Kingdoms of God*

Ancient scriptural references to the three Kingdoms of God are sparse and, in fact, non-existent in the Book of Mormon and The Pearl of Great Price, but we do have strong ancient testimony that the few Biblical verses describe a context that associates with our modern understanding.

The Gospel of John records a teaching of the Savior that details the coming Spiritual World. The Savior said, "*Let not your heart be troubled: ye believe in God, believe also in me. In my Father's house are many mansions: if it were not so I would have told you. I go to prepare a place for you*" (John 14:1, 2). Latter-day Saints understand

---

[2] Later in this chapter, we will make a case for an understanding that the Celestial Kingdom is not so much a location or place, but represents an intimacy with the Father and describes a state of eternal increase and progression.

*many mansions* to refer to the *three Kingdoms of Glory*; in this it turns out we are not alone.

Early in the second century, Irenaeus preserved a tradition he variously attributes to "*one of the ancients*," "*the divine elder*," "*a certain elder*," "*one who was before us*," "*one better than we are*," and so on. There is little doubt that Irenaeus considers the *elder*, the *one who was before* and *better than we are* to be the Apostle John. There is much evidence to suggest that this *tradition of the elders* was facilitated to Irenaeus, himself Bishop of Lugdunum, Gaul, Circa 200 AD, through Polycarp[3] and Papias,[4] both disciples of John and bishops themselves. Tradition holds that Papias transcribed the gospel of John, as John, *while carried away in the spirit*, dictated it to him. Irenaeus recorded that "*the elders*" (Polycarp, Papias, and John) elaborated on what the Savior's teaching of the mansions meant. Irenaeus wrote:

> *"As the elders say, then will those who have been deemed worthy of an abode in heaven go there, while others will enjoy the delight of Paradise, and still others will possess the brightness of the city, for in every place the Savior will be seen, to the degree that those who see him are worthy. They say, moreover, that this is the distinction between the dwelling of those who bring forth a hundredfold, and those who bring forth sixtyfold, and those who bring forth thirtyfold: [Matt. 13:8,23] the first will be taken up into the heavens, and the second will dwell in Paradise, and the third will inhabit the city. FOR THIS REASON THEREFORE, OUR LORD HAS SAID, "IN MY FATHER'S HOUSE ARE MANY ROOMS," for all things are of God, who gives to all their appropriate dwelling.... The elders, the disciples of the apostles, say that this is the order and arrangement of those who are being saved, and that they advance by such steps, and ascend through the Spirit to the Son, and through the Son to the Father, the Son finally yielding his work to the Father, as it is also said by the apostle: "For he must reign until he puts all enemies under his feet." (Against Heresies 5:36 1-2)*

The context here is clear and unambiguous; the mansions are not individual dwellings, but compasses, the abode of multitudes, segregated by worthiness in not two, or four, or seven, but <u>*three*</u> degrees. There is order and advancement of those being saved; this is *progression*. In the *Elder's* account, those that have departed mortality *ascend* from the presence of the Spirit, to that of the Son and finally to the precincts of the Father. There can be no more precise correlation found between any teaching and the Latter-day concept of the *Three Degrees of Glory* than that found here. This teaching is direct from the Apostle John who heard the Savior voice the very words. John most likely asked for, and no doubt received, clarification and additional understanding beyond what his Gospel records. Here we have an obscure, ancient recording that describes in *exact* detail progression in the eternal sphere

---

[3] Polycarp (Bishop of Smyrna 69-155 AD). Jerome recorded that Polycarp was a disciple of John the Evangelist and that John had personally ordained him bishop in Smyrna.

[4] Papias (Bishop of Hierapolis, 70-163 AD). Papias is the author of the five volume, <u>Exposition of the Sayings of the Lord</u>. It is unfortunate that this work is lost except for excerpts found in the writings of Irenaeus and Eusebius. It remains, however, a very important source for early oral traditions of Christianity and of the Gospels.

through *Three Degrees of Glory* and varied association with the persons of Deity. This schema, taught anciently, precisely describes what Joseph Smith taught most recently.

The same apostolic authority held by John was held by Paul, who made reference to *Three Degrees of Glory* as well, describing them in terms of heavenly objects in the sky. Though Paul here is indirect in reference, putting his words in the context we just discussed, leads us to understand Paul to be discussing the same three Heavenly Kingdoms:

> *"There are also celestial bodies, and bodies terrestrial: but the glory of the celestial is one, and the glory of the terrestrial is another. There is one glory of the sun, and another glory of the moon, and another glory of the stars: for one star differeth from another star in glory." (1 Cor. 15: 40, 41)*

Paul's teaching here seems a bit confusing because his intent was not to describe the Kingdoms but the differences between two things, our current Temporal bodies and our future resurrected bodies. Because of this, he starts by comparing the differences between just two glories then mentions the three. What can be understood here is the degrees of Glory differ much from one another.

Paul here refers to the transition from the Temporal sphere to the Spirit World; "*So also is the resurrection of the dead... It is sown a natural body; it is raised a spiritual body*" (1 Cor. 15:42, 44). In passing let us, like Paul, deviate briefly from our discussion of the Kingdoms of Heaven to discuss something of the bodies that will enclose us there. For the state of our existence in the Temporal sphere, Paul uses the phrase, Natural Body, Greek, *psychikon soma* (ψυχικόν σῶμα); we have encountered and discussed this description before. To Paul, our current bodies are "*natural*," he uses the Greek word *psychikos* (ψυχικός), which is the adjective form of the word psyche or Soul. This is in accordance with our understanding of a *living Soul*, an eternal embodied Intelligence enclosed in a Temporal body. Of the body of greater Glory, the one we will inhabit in the Spirit World, Paul uses the Greek word *pneumatikon* (πνευματικόν), the adjective form of *pneuma*, translated as *Spirit* in the KJV; the Apostle Paul displays the same understanding of an eternal physical embodiment as Origen did.[5]

Because Paul is using the degrees of Glory to illustrate the differences of the two bodies, Temporal and Eternal (our Physical), this verse does not name all three of the Kingdoms, only two.

We must make two observations; first Paul's descriptive use of the Kingdoms here infers that his hearers know of them, a teaching that became lost in short order. Secondly, Paul describes differences in the bodies obtained in the resurrection, along with Kingdoms of varying Glory in which we will live; he describes differences in the Glory of such bodies, not all attaining the same.

---

[5] See our previous discussion, WHAT IS SPIRIT?, pg. 172.

## *Celestial*

The Celestial Kingdom is the paramount of the three Kingdoms of Glory. Those who attain this Kingdom will abide forever in the presence of the Father and His Son, "*They are they who are the church of the Firstborn*" (D&C 76:54). It is misleading and improper to describe this Kingdom as a location or a place, for it properly represents an intimacy with the Father and includes a state of *eternal increase* and progression; as such, it must be borderless and unconstrained. Those who inherit the Celestial Kingdom achieve *Exaltation*.

The Celestial Kingdom is the place prepared for those who have *"...received the testimony of Jesus, and believed on his name..."* and have been *"...overcome by faith, and are sealed by the Holy Spirit of promise, which the Father sheds forth..."* (D&C 76:51, 53). To inherit this gift and realize this Kingdom, we must follow the counsel of Moroni who said:

> *"Yea, come unto Christ, and be perfected in him, and deny yourselves of all ungodliness, and if ye shall deny yourselves of all ungodliness, and love God with all your might, mind and strength, then is his grace sufficient for you, that by his grace ye may be perfect in Christ; and if by the grace of God ye are perfect in Christ, ye can in nowise deny the power of God." (Moroni 10:32)*

The Celestial Kingdom is about *perfection* and *sanctification*; these are gifts from God freely given and not earned; "*... if ye by the grace of God are perfect in Christ, and deny not his power, then ye are sanctified in Christ by the grace of God, through the shedding of the blood of Christ, which is in covenant of the father unto the remission of your sins, that ye may become holy, without spot*" (Moroni 10:33). We understand through these words of Moroni that *coming* to Christ works to perfect us and that perfection is a process which requires our remaining open to receiving. We express that openness through our willingness to obey and follow Christ in baptism and other ordinances, learning, "*...precept upon precept; line upon line, line upon line; here a little, and there a little*" (Isaiah 28:10). Obedience, however, does not earn the gift, but is an outward expression of an inner desire. God *gives us* what we desire; He does not *deny us* anything.

"*For I, the Lord, will judge all men according to their works, according to the desire of their hearts*" (D&C 137:9). Attainment of Glory in the Kingdoms of God are not as much about worthiness and reward nor about punishment as it is about receiving the desires of one's heart, the attainment of true happiness and wellbeing. We attain in eternity what we seek and are most comfortable with. *Exaltation is not earned or given as reward*; such thoughts need to be curtailed. The Prophet Joseph said, "*I also beheld that all children who die before they arrive at the years of accountability are saved in the celestial kingdom of heaven*" (D&C 137:10). Consider what this implies. These will attain the *Celestial* without effort or even intent to do good. The Celestial is the *default* state of Glory, *something to be lost, not won.*

The Celestial Kingdom is also about eternal progression, and we find from another of the Prophet Joseph's revelations that within the Celestial Kingdom itself are to be found three more degrees. To fully achieve eternal increase (eternally building upon family relationships), we must enter into "*the new and everlasting covenant of marriage*" (D&C 131:2), accepting that covenant and remaining true to its requirements. This is Temple marriage and sealing. The everlasting covenant of marriage is a *Priesthood Order*. Through the Temple, the efforts of our Spirits in the Spirit world and in accordance with the principles of eternal progression, all who are worthy and desire to enter into the new and everlasting covenant of marriage will have that opportunity, whether in this life or our eternal life to come. In fact, it must be understood, that this covenant and the increase it allows is the true definition of eternal life. It is what we actually came here to achieve; it is the reason for all the cost, the loss and suffering. It *is* the prize that was *worth the risk* of the Fall.

The Celestial Kingdom then represents a state of existence not unlike that which the sons and daughters of God experienced before coming to the Temporal sphere; the enhancements being that of procreation, facilitated by a resurrected body, experience with the effects of evil, and eternal increase, which differs from eternal progression.

## *Terrestrial*

Having presented a view of the Celestial Kingdom, we will now turn to that of the Terrestrial. First we must understand that there is a qualitative difference in the physical form of those who inherit this Kingdom of Glory from that of those who attain the Celestial; "*Wherefore, they are bodies terrestrial, and not bodies celestial, and differ in glory as the moon differs from the sun*" (D&C 76:78). What this difference represents is not known but is significant; it is no doubt associated with an ability to procreate and have eternal increase.[6]

Those who inherit Terrestrial Glory "*...receive of the presence of the Son, but not of the fulness of the Father*" (D&C 76:77). This informs us of the relationship between the Terrestrial and Celestial. None can doubt that the Son has inherited all that the Father has, that He enjoys the fullness of the Father, eternal increase and has a Celestial body (or at least will have), yet here we are told that the Terrestrial receive His presence, indicating there is no physical obstruction to the differing physical forms interacting, and indeed we should not expect that to be the case. That the Son's presence can be experienced in the Terrestrial is clear, and by inference so too can that of the Father and, by extension, any other being from the Celestial Kingdom. Those within the Terrestrial Kingdom may be visited by the Father and others from His precinct; the Scripture quoted here does not exclude this reading as

---

[6] Remember we have identified in this work that the "*Glory of God is intelligence*" (D&C 93:36); our Glorified states then may be reflective of our individual attainment of intelligence as measured against the precepts of God.

it properly states only, that the Terrestrial does not receive of the "*fulness*" of the Father. This we understand to be true and consistent.

We have come to understand that the attainment of Glory in the Kingdoms of God is the receiving of the desires of one's heart and that God gives us what we desire. He does not deny us anything; so how can the attainment of the Terrestrial Kingdom be a realization of true happiness for those who dwell there? Who are these sons and daughters of God? What is different about their desires and expectations?

We are told, that among others, those in the Terrestrial Kingdom include "*honorable*" people "*...who were blinded by the craftiness of men*" (D&C 76:75) and "*...they who are not valiant in the testimony of Jesus*" (D&C 76:79). By these statements we are to understand that the vast majority of living Souls in Temporality would end up in the Terrestrial Kingdom.

***We have come to understand that the attainment of glory in the Kingdoms of God is the receiving of the desires of one's heart and that God gives us what we desire. He does not deny us anything…***

The Celestial Kingdom comes with many gifts, but with that comes much responsibility: "*For unto whomsoever much is given, of him shall much be required...*" (Luke 12:48). Such responsibility is not for everyone, indeed perhaps most. Though we are burdened by the Natural and Temporal here, the nature of our Soul probably does not fall far from the proverbial tree near which our Spirits dwell. Much can be said by just looking at what valiant means: brave, bold, intrepid, fearless, and courageous. Not all exhibit these traits here, those that do will not often be found "*blinded by the craftiness of men*" either.

There is another group here, curiously, "*...these are they who died without the law*" (D&C 76:72). How this group would differ from those "*children who die before they arrive at the years of accountability* [who] *are saved in the celestial kingdom of heaven*" (D&C 137:10) is entirely unclear. We have a firm understanding; "*where there is no law given there is no punishment; and where there is no punishment there is no condemnation; ...the mercies of the Holy One of Israel have claim upon them*" (2 Nephi 9:25). So this group in attendance within the Terrestrial is not a result of punishment or condemnation. In fact, we have made clear within this wo, that, in the eternities (after the judgement and resurrection), punishment and condemnation are inoperable principles. They are in use here, only in Temporality, as facilitators of change and correction. It is therefore possible, and I propose here probable, that this group chose in the pre-existence to come to mortality without law, choosing to avoid a path through the dangers of mortality, accepting beforehand a life without exceptional challenges and an eternity without resultant increase, taking in effect a *safe bet*.

Within the Terrestrial we will also find two other groups, those: "*Who received not the testimony of Jesus in the flesh, but afterwards received it*" (D&C 76:74) and those "*spirits of men kept in prison, whom the Son visited and preached the gospel unto them*" (D&C 76:73). Of these two groups what is meant by "*received not*" is more accurately expressed as *rejected*, and although it is not specifically mentioned in Verse 73, it must be inferred that these are those who *accepted* the Gospel after it was preached, *not just those who merely heard it*. Here we see the process of repentance at work in the Spirit World, for these two groups within the Terrestrial Kingdom clearly did not repent of their ways while in the flesh. Within the group spoken of in Verse 74, we may hopefully find the unrepentant malefactor who hung beside Jesus on Golgotha.

We must consider something else here, eternal progression as it pertains to these Heavenly Kingdoms. Consider those of whom we are told, "*...received not the testimony of Jesus in the flesh, but afterwards received it.*" What are we to conclude from this? That they are permanently, eternally confined to the Terrestrial?

Before answering, consider some other groups, those "*without the law*" and those that were "*not valiant*." Now consider what we do by proxy, sealing eternal families together within Temple walls. We fully understand what we do here on Earth must be ratified by actions in Heaven, that at best we perform *images* of things eternal in Temples on Earth today. But the concept is valid; why seal an individual from one of these Terrestrial groupings to those within the Celestial Sphere if the sealing is not ever to be efficacious in precincts beyond the veil? What would your eternal life be like if you were denied companionship and intimacy of a child who was forever confined to places beyond your sphere?[7] Indeed for the Plan of Happiness to be what it is intended to be, progression too must be eternal. The states of these Spirits, their classification and assignments described here, are initial and need not remain eternally fixed.

## *Telestial*

Despite what uninformed detractors often say, Latter-day doctrine is not exclusive, but it is very inclusive. It allows for virtually all who have lived on the Earth to attain Heaven, perhaps fully realizing for the first time in Christendom the words of Paul, "*For since by man came death, by man came also the resurrection of the dead. For as in Adam all die, even so in Christ shall <u>all be made alive</u>*" (1 Cor. 15:21, 22). By this we believe that to be made *alive* means so much more than just being embodied to stand at the Judgement Bar only to be utterly destroyed or sent to an ever burning lake of fire if deemed unworthy of the Son's *Saving Grace*. But we

---

[7] We have described in the book a structure and relationship to the Kingdoms of Heaven that allows interaction of those who attain a "*higher Glory*" with those who do not; see the last paragraph on pg. 362. We then expect relationships to maintain efficacy between beings in the separate spheres.

believe that the *Infinite Atonement* is sufficient for all, that Adam's transgression brought death to us all, and Christ's Atonement restores equally *all* who have entered into the Fallen state.

The last time a similar doctrine was taught was at the outset of the third century BC. The great Christian teacher/scholar Origen taught of a *universal salvation* using the Greek term *apocatastasis* (ἀποκατάστασις), which when rendered into English reads restitution or reconstitution. *Apocatastasis* only appears once in the Bible, in the New Testament, in a sermon delivered by Peter; "*...he shall send Jesus Christ, ...Whom the heaven must receive until the times of (ἀποκατάστασις) restitution of all things...*" (Acts 3:20-21). For Origen, and perhaps for Peter, there was an expectation that the Salvation provided through the Atonement of Christ would apply to all.

Origen believed in the pre-existence of our Souls, so to him *apocatastasis* not only meant restitution, but quite literally the reconstitution (restoration) of things to a state in which they were before. Origen taught that so great were God's powers of *apocatastasis* that in due course even Satan himself would be saved. Latter-day Saint doctrine does not espouse a truly universal *apocatastasis*; for we believe there are those who possibly may never be saved;[8] they are the sons and perhaps daughters of Perdition. We will discuss them again later in this work.

Within the Telestial Kingdom are those who "*... are thrust down to hell*" (D&C 76:84) and later "*...redeemed from the devil...*" at "*...the last resurrection...*" after "*...Christ the Lamb...*" has "*...finished his work*" (D&C 76:85). These are they "*...who deny not the Holy Spirit*" (D&C 76:83). Those who inhabit the Telestial Kingdom, "*...received not the gospel of Christ, neither the testimony of Jesus*" (D&C 76:82) and this after having had the Gospel presented to them throughout their Temporal existence and subsequent stay in the Spirit World. As before, "*received not*" means *rejected*. It is hard to fathom such intransigence.

Our understanding here is that the Telestial Kingdom will occupy the space that, before the last resurrection and final judgement, was the Spirit World. Those who receive this Glory will not experience much of a change from their existence in the Spirit World they inhabited before their judgement. This Kingdom most closely resembles the Garden *after* the Fall before the creation of the Temporal sphere. The Telestial is an idyllic, beautiful, environ looking and feeling just like our Temporal Universe does today, but eternal, without strife, pain, sorrow or fear; the Glory of the Telestial "*surpasses all understanding...*" (D&C 76:89). There will be found planets, stars and galaxies all in perfect motion and time. Like the Garden, the Telestial Kingdom is separated from the precincts of the Father and those who experience with him full fellowship and eternal increase. This locale fits entirely the

---

[8] Eternity is a very long time, and given that we are all, including Satan, the sons and daughters of our Heavenly Father, it is conceivable (very much so) that an all knowing, all powerful, all loving Father would succeed in His efforts to redeem and progress all of His children. In such a scenario true *apocatastasis* is imaginable, if however remote and distant that achievement may seem to be.

description of the Heaven of creedal Christianity and inheriting such an afterlife could be deemed salvation by any traditional (orthodox) definition.

- Perdition, *A Special Place in Hell*

With Three Kingdoms and all the inclusiveness engendered by the Grace of God, there are some who will still reject His favor and deny His sovereignty; for these few there can be found no place in the Glorified Kingdoms of Heaven. These express open rebellion to the Will of the Father, after having known of it intimately: "*...those who know my [God's] power, and have been made partakers thereof, and suffered themselves through the power of the devil to be overcome, and to deny the truth and defy my [God's] power*" (D&C 76:31). Of these *sons [perhaps daughters] of perdition*, God has said, "*...there is no forgiveness in this world nor in the world to come–* " (D&C 76:34) and "*... it had been better for them never to have been born*" (D&C 76:32).

The sons of perdition are "*...vessels of wrath, doomed to suffer the wrath of God...*" (D&C 76:33) for they "*...denied the Holy Spirit after having received it, and having denied the Only Begotten Son of the Father, having crucified him unto themselves and put him to open shame*" (D&C 76:35), they are the "*...only ones on whom the second death shall have any power, ...the only ones who shall not be redeemed*" (D&C 76:37, 38).

The environs of Perdition are described as "*...the lake of fire and brimstone, the abode of "...the devil and his angels*" (D&C 76:36). Other scriptural and Latter-day teachings describe this lake as a metaphorical representation of the state of mind of those driven away from the Heavenly realm. We are then left uncertain as to what and where Perdition is, and for certain we are not told how it fits into the eternal realms of God.

## The Temple, a Bridge Between the Spheres

A Temple is for teaching and covenant making. For the faithful, it represents both a righteous goal and another world. But most importantly, it forms and maintains a connection between two spheres. Here on Earth, Temples link the mortality of the Temporal to the immortal Physical sphere. A Temple "*...is a place of connection in history... You participate in ordinances and make covenants and you are allowed to reach back to your ancestors and you make covenants that will bless your descendants*" (Pres. Henry B. Eyring). We enter Temples with the purpose of sealing together eternally paired Souls and seek also to seamlessly connect the family of God.

Latter-day Saints are familiar with the concept of binding/sealing within Temple walls, and we are fully aware of the thinning of the Veil to be found there. For us, on this side of the Veil, Temples are a place of spiritual refuge and attainment in the *Fallen* and *Natural* world. Within the Temple we look upwards to Heaven and feel Heaven reaching down to us.

***Within the Temple we look upwards to Heaven and feel Heaven reaching down to us.***

It is fitting that we learn of the Garden and Adam and Eve in Temples on Earth, each dedicated to instruct us in things of the Physical (spiritual) world. Through the Endowment, we learn that the Garden was a special place in Eden (the Physical sphere), truly a Temple of God[9] in which the divide between the Physical and the Temporal was so thinned that it left the spheres touching. Today our Temples (The House of the LORD on Earth) are hallowed places where the divide between things *natural* and *spiritual* is thinned as well.

Through the story of Adam and Eve, we are taught sacred truths, providing us with knowledge about why we are here and where we are going; we are given the keys we will need to cross over to the other side. The story of the Garden informs us that Adam and Eve were taught in the Garden Temple of who they were and where they were going; they too were given the necessary tools they would need to cross the same divide, in their case from the Physical to the Temporal. The Garden story tells us that Adam and Eve were sealed in marriage within the confines of a Temple in the same manner as we are here on Earth; there is beauty and comfort to be found in the continuity. The Garden Temple represents a true bridge between the eternal Physical world of our Spirits and the Natural Temporal world we inhabit today.

---

[9] See footnote 1 on pg. 212.

Moses, Chapter 3, Verse 9 reads: "*And out of the ground made I, the Lord God, to grow every tree, naturally, that is pleasant to the sight of man; and man could behold it. And it became also a living soul. For it was spiritual in the day that I created it; for it remaineth in the sphere in which I, God, created it,*" here *naturally* (natural) is associated with "*living soul*" and in this, we can establish definition. A *living soul* has a natural, and in this context, a Fallen state, with attributes and qualities of the Temporal, something fundamentally different than what is to be found in the Physical sphere.

Later in this work in Chapter 11, What is Man, we will detail what it means to be a *living soul*, but for now we must simply know that it is defined as an embodied Intelligence (a Spirit) enclosed within a mortal (*or resurrected* immortal) fleshly (Temporal) tabernacle. That may sound complex, but it really isn't. We are intimately familiar with the full realization of this concept; it is simply what we are.

Each of us as a son or daughter of God (in Temporality) is a three part construct of intelligence within an immortal physical body, inside a *Flesh and Bone and Blood* Temporal frame; this is the natural body of Paul; "*...There is a natural body, and there is a spiritual body. And so it is written, the first man Adam was made a living soul...*" (1 Cor. 15:44, 45).

When we discuss the embodiment of Intelligence to form *Spirit* and the subsequent enclosure of these immortal, physical, tangible, Spirits within Temporal bodies to become living Souls, we must remember that the Temporal is in fact a subtraction from a more perfect pre-existing Physical sphere. Additionally, we understand that the Physical sphere includes life *in spiritual form* of all manner, plants and animals, all with varying degree of intelligence forming their embodied Spirits. Inanimate constructs of Spirit as well, or what has become common to term inanimate, exist: rock and water, particle and element, star and planet, all possessing an eternal Spiritual physicality imbued with the proper ration of intelligence for it to carry out the measure of its creation. In Temporality all these spiritual forms are overlaid with a Temporal twin that is both Fallen and mortal,[10] suffering the effects of entropy and its inherent decay and eventual disintegration. All of these Temporal constructs, Spirit within the Fallen, is what is termed *Natural.*

Throughout this work, we have struggled to remain consistent in referring to the Physical sphere and things within it as Physical; in this endeavor we have not always been successful. The realm of the Physical sphere and all it contains has been called by so many, for so long, spiritual and Spirit, it has become a habit that is difficult to shake. This issue will be compounded in this section as we begin to discuss the three degrees (or Kingdoms) of Glory and the way station Spirit World that awaits all of us before the Resurrection. In this regard, we will understand that the Telestial, Terrestrial and Celestial Kingdoms are each an eternal realm within

[10] See Appendix, Envisioning the Temporal Reality, pg. 555.

the Sphere we have called the Physical. The Spirit World, though not eternal in construct, for it serves a purpose only until the Judgement and Resurrection, is wholly confined within the Physical sphere as well, for as long as the Spirit World remains a waiting area.[11]

## *Connecting the Past, Present and Future*

We discussed what is *said* in the Endowment in detail in our chapter about the Garden and the Fall, but the Endowment liturgy reveals much more than what is spoken. The structure of the narrative, its presentation and order, reveal many truths about: the three Creations, the Three Degrees of Glory, the Priesthood, and the purposes of the Garden. It reveals much about the Temporal sphere and the Spirit World as well. Here we will examine these details of the Endowment liturgy itself with an eye for what they teach us.

Figure 9-1 *Symbolism of the Temple.* (Stained glass art pane from the new Nauvoo temple.)

In the Endowment liturgy, we symbolically *progress* as we follow Adam and Eve from the Garden into the Fallen *lone and dreary world*. From there we move to the Terrestrial and on to the Celestial. Along the way we make covenants, are informed of Law and receive explanation of the Priesthood. In every place, with associated movement and action, and through the instructions we receive, we are educated of things Eternal.

We will begin our discussion here "*symbolically*" (*Figure 9-1*) (*Table 9-1, Page 375*) in the Garden with our first parents.

- Garden

We have laid the groundwork for our understanding that the Garden represents a Temple in the Physical sphere; as such, it is a link to the Temporal sphere. Within the Garden, Adam and Eve were prepared for their release into the Telestial world; they were empowered with the ability to choose, and when they Fell, they were subsequently taught what they needed to know to survive in Temporality.

---

[11] It very well may be the case that the Spirit World will become the Telestial Kingdom after the Judgement. If that is the case, then it always has been and truly is an eternal realm.

In the Endowment narrative, while our first parents are still within the Garden after the Fall, they, and we likewise, are informed of the Aaronic Priesthood. The Endowment identifies two "*stages*" of this priesthood, and it will identify two stages for the Melchizedek as well. It may not be proper to call them *stages*, but we will see that it is significant that both Priesthoods have these two readily identifiable separate components. While yet within the Garden, the first *stage* of the Aaronic Priesthood is given and we are told that it accompanies the *Law of Obedience and Sacrifice*. The significance of this informs us, as it did Adam, of the purpose of our Temporal existence. The Law defines and delineates the two primary elements of the Temporal sphere: Obedience to Law, because the Temporal must be governed to remain uncertain, with bounds set so experience may be gained and *Sacrifice* can justify. The Pearl of Great Price provides us with divine insight into this concept:

> *"For by the water ye keep the commandment; by the Spirit ye are justified, and by the blood ye are sanctified" (Moses 6:60)*[12]

Sacrifice is obedience to the law, keeping the commandment, allowing the Spirit to justify (proclaim or declare) that we are free from guilt. Sanctification comes not through the blood of those animals sacrificed, but through the blood of the Savior spilt in the processes of the Atonement. The sacrifice needed then is twofold, one part is that of the Savior's, and the other part is that of ours which we accomplish through obedience to the commandments (this is to be taught later when we receive the second stage of the Aaronic Priesthood in another place at another time).

The Law given encompasses obedience and sacrifice, the two essential elements required in the Atonement and the essential components of the Plan of Happiness. We are instructed in the Endowment that this Plan was explained and accepted through covenant before our first parents entered the Temporal sphere. By application of analogy, we know, and it is confirmed, that we also were told of the Plan, its sacrifice, and we covenanted with our Father to accept the challenges of mortality before we came to this sphere as well. This was done by each of us in association with the Grand Council convened long before this world was.

It is important to note that the Garden is not in the Telestial world, but it is still Terrestrial in nature, walled off from the precincts of the Father but truly Physical (Eternal). It is in fact a Terrestrial Temple separate from the Physical sphere around it. Our Temples on Earth today are Telestial Temples walled off from the Fallen, mortal, world around them. Because of its special nature as a Temple (a bridge between two spheres), the Garden allows certain conditions to exist that cannot exist elsewhere in the Physical sphere.

Signifying that the Garden is a Temple, we find Michael being renamed Adam when he enters the Garden. In like manner, we are given a new name when we enter

---

[12] We will discuss in detail later in this work, that "*blood,*" or what is signified by it (death) represents the corrupting component of the *Flesh and Bone and Blood* bodies of the Temporal sphere. The *shedding* of blood associated with all of the ancient sacrifices was meant to focus attention to the fact the we all must literally "*shed*" blood (overcome death, cast it away) to become mortal again.

the Temple here on Earth. Note: we are not informed of Eve's previous name in the Endowment Narrative. Also, Adam and Eve both obtain from the Father "*coats of skins*" that represent their receipt of bodies of an entirely new nature. More will be mentioned about all of these concepts later in this treatise.

Next, we follow Adam and Eve as they are driven out into the "*Lone and Dreary World*." This same locale is subsequently also identified as the Telestial world. Here the Endowment continues to instruct us.

-Earth = Telestial World

First, we must clarify that it is clear the "*Lone and Dreary World*" to which Adam and Eve were driven is the same Temporal mortal Earth we inhabit today. Though it is called the Telestial World, it cannot be, nor should we expect it is intended to be considered the immortal, eternal, Telestial Kingdom we have already spoken of with detail in this chapter. As we discussed earlier, the Telestial Kingdom may in fact encompass a spiritual Earth, spoken of by John the Revelator, "*And I saw a new heaven and a new earth for the first heaven and the first earth were passed away...*" (Rev. 21:1). The "*new*" Earth John saw in vision may in fact be the one already existing in the Physical sphere or even a number of such orbs there. We will explore later in the work what John understood "*new*" to actually represent.

Satan is the god of this world (he proclaims, *now is the great day of my power*). He reigns over all of the Earth. Earth, as spoken of by Lucifer, is the substance of Temporality itself, matter, by which he means to infer that his reign encompasses the entire Temporal Universe. We understand this to be true for we have discussed that the Adversary is both cause and the driver of opposition in the Earth. The Endowment shows him at work as an adversary, a *satan* (שָׂטָן) in the *Lone and Dreary World*.

Within the "*Lone and Dreary World*" (in Temporality), we are given the second of the two *stages* of the Aaronic Priesthood, and we are put under covenant to observe its associated law: The *Law of the Gospel*. This Law governs the principles of repentance and faith which, through the Atonement, allow the blood of Christ to cover[13] our guilt and sanctify us. Sanctification means to *make Holy* and is different from justification (obtained through the first law given, obedience and sacrifice), which is merely the *declaration* that we are free from guilt. The second *stage* of the Aaronic Priesthood represents the office of Priest today, the High Priest of old Israel during its temple period. A Priest today has the authority to baptize, the ordinance that prepares each of us to become a member of the Church of God; baptism is a spiritual cleansing. Israel's High Priests were the only men allowed to enter into the Holy of Holies (the presence of God) and this they could do only on the behalf of all of Israel once a year on the day of Atonement when they completed an ordinance that cleansed Israel of its collective sin.

---

[13] See They Went to Hide Themselves in this work, pg. 267.

The Aaronic Priesthood is the *preparatory* Priesthood, and its two *stages*, along with their associated laws and principles, are preparatory in nature. Stage one, and the Law of Obedience and Sacrifice given to Adam in the Garden, prepared him to understand what the Temporal world would hold and informed him of its purpose. The second stage confers upon the priesthood holder the keys required to justify through the cleansing of baptism and the subsequent entry into the presence of the LORD that cleansing allows because of sanctification gained through the Atonement. Cleansing, purification, repentance and, specifically, faith are all principles of the Temporal (Telestial) world. The Creator, Jehovah himself, would have to take on *Flesh and Bone and Blood* and be born into and die within the Temporal sphere to "*fulfill the law*" of the Gospel, completing the circle of repentance, faith, restoration and redemption. The structure and ordered presentation of these Priesthoods in the Endowment narrative within the Garden, and after entry into the *"Lone and Dreary World,"* emphasize the Temporal nature of the Aaronic Priesthood and its associated ordinances and principles.

Our symbolic entry into the *"Lone and Dreary World*" of the Endowment liturgy, is marked by the donning of robes. We are told these are the robes of the Priesthood; their significance, however, has much greater meaning than this. They in fact represent our Spirits being clothed in a Temporal body of *Flesh and Bone and Blood* (as do the *Coats of Skins* received by Adam and Eve). Through the Endowment, we are to understand that the Aaronic Priesthood is associated with, *and perhaps depends upon*, the Temporal body. The attainment of the Priesthood and the donning of the robes (symbolizing our Temporal Bodies) empower Adam and each of us to command Lucifer, just as Peter, James and John do.

- Terrestrial (Temple)

Unlike the Aaronic Priesthood, which we have seen is Temporal, and tied to our Temporal bodies (the robes on the left side), the Melchizedek is eternal; we are merely practicing its use here. In the Temporal, we are blessed to exercise the Melchizedek Priesthood through faith in accordance with the will of God and empowered by the Son. To emphasize the eternal nature of the Melchizedek Priesthood,[14] the Endowment instructs us to move the robes from the left to the right side in *preparation* for us to officiate in the ordinances of this Priesthood. By changing the robes to the right side, we are symbolically shedding our Temporal bodies and taking on our *resurrected* bodies. This we do *before* entry into the Terrestrial World which is itself symbolized by our moving into the Terrestrial room within the Temple[15] prior to receiving the Melchizedek Priesthood and its associated laws and covenants.

---

[14] The Melchizedek Priesthood functions in whole within the eternal Physical and Spiritual spheres, and the Endowment liturgical sequencing illustrates that it only truly can be exercised there. Our receipt of it in Temporality is in recognition of foreordination to it in the eternal realms (Alma 13:1-4). We are blessed to preview its power and authority here in Temporality but must be cognizant of the fact that we are only exercising its authority in part, in similitude of its full exercise and efficacy in Eternity.

[15] In some Temples the Endowment participants do not move from room to room, but the room itself is transformed in some way to acknowledge the transition from the Temporal to the Terrestrial.

The detail and sequencing is critical here. It is important for us to recognize that the Temples we enter here on Earth are special places in Temporality. Within their walls, we experience images of the Telestial, Terrestrial and Celestial, but they are just images. After detailing the creation of the Physical sphere, the Endowment presentation turns to the Garden story and symbolically places us there. The Garden (it will be remembered) is itself a Temple, one within the Physical sphere representing the unfallen Telestial World.

The Temples we enter here on Earth are in the Temporal; they are Fallen representations of the Physical (spiritual), and within them (through the Endowment liturgy), we symbolically progress in stages from the Garden to the Telestial (Temporal) "*Lone and Dreary World*," then (skipping the Spirit World) we symbolically advance to the Terrestrial followed by the Celestial. Within the Temple, we never symbolically enter the Spirit World, because while we are yet clothed in our Temporal bodies (robes on the left side) we are placed into our resurrected bodies (robes on the right side) before being given the Melchizedek Priesthood and enter directly as resurrected beings into the Terrestrial world. What then of the Spirit World? We will see that, in a sense, within the Temple we are there all along.

The Endowment opens with the creation of the Physical sphere itself. The Physical sphere encloses the compasses in which we all will live out the entirety of our remaining eternal existence, the three Kingdoms of Glory: the Telestial, the Terrestrial and the Celestial. Within the Physical sphere can also be found, Perdition, the Spirit World and the Garden. There is a common link between the Garden, the Temporal "*Lone and Dreary World*," the Spirit World and the Telestial Kingdom. We propose here more than just a common link between these four, but a realization that they are all the same *place* observed in several individual manifestations, as if we are viewing separate paintings of the same landscape scene, completed in each of the four seasons.[16]

With the robes on the right side, in our resurrected bodies, within the Terrestrial world, we are given the two stages of the Melchizedek Priesthood.[17] The

---

[16] The Garden before the Fall is an image of the future Telestial Kingdom. After the Fall, the Garden displays the characteristics of the Spirit World; it is the Garden that most closely resembles the Spirit World to which we all will go before the resurrection. The "*Lone and Dreary World*" is the Temporal Universe, our home today; it is the *Temporal twin* of the Spirit World and the Fallen Garden. We all go to the Spirit World to await the resurrection and judgment, returning again to conditions similar to those found in the Garden after the Fall. The resurrection is a restoration; it reunites us to immortal elements of our Temporal bodies and renews the Heaven "*And I saw a new heaven... for the first heaven... [had]passed away...*" (Rev. 21:1). After the resurrection, the Spirit World completes a transformation to its final expression as the eternal Telestial Kingdom. (Imagine the entire Universe raised to a state similar to that known in our temples on Earth today — an immortal realization of the Temporal sphere.) The Garden, the "*Lone and Dreary World,*" the Spirit World and the Telestial Kingdom, each in order, are manifestations of the same place in their own times.

[17] The Melchizedek Priesthood has "*the privilege of receiving the mysteries of the kingdom of heaven, to have the heavens opened to them, to commune with the general assembly and the church of the Firstborn, and to enjoy the communion and presence of God the Father, and Jesus the mediator of the new covenant*" (D&C 107:19).

Melchizedek holds "*the power and authority of the higher ...Priesthood...*" and "*hold[s] the keys of all the spiritual blessings...*" (D&C 107:18). The Aaronic or Levitical is the lesser Priesthood and is set apart, through its office of Bishop, "*unto the ministering of temporal things*" (D&C 107:68, 73). The first stage of the Melchizedek is associated with the binding and sealing powers of this higher Priesthood through its association with the *Law of Chastity*.

It is important to note that chastity is associated with this eternal Priesthood and also with the eternal precincts of the Terrestrial and subsequent Celestial Kingdoms. This is significant because, families are eternal, procreation is eternal, and the power of the sealing ordinance is eternal in nature. Chastity is a requirement of the sealing ordinances performed by the Melchizedek Priesthood. It must be noted here that chastity as a law gains its importance in securing the binding among those it binds as well as ensuring the Priesthood remains Holy. In this respect, both those who perform eternal marriages and those who are bound in them are under the Law of Chastity and must hold the Melchizedek Priesthood and conform to its covenants.

The Melchizedek Priesthood is that which functions in the Eternal Kingdoms. It functions here in the Temporal world symbolically and in similitude of its true powers confirmed and carried out in the realms of the Physical and Spiritual. Through it, the righteous have power to seal in Heaven and on Earth, but all things eternal are secured and cemented in the Eternal spheres. The first *stage* of the Melchizedek Priesthood associated with the *Law of Chastity* is given and functions here in Temporality because procreation first functions here as well.

The second *stage* of the Melchizedek Priesthood is accompanied by the *Law of Consecration.* This Law is fundamentally different than the other laws in that it does not have negative consequences applied to it, only immense potential. We covenant that we do consecrate ourselves, our time, our talents, and all with which we have been blessed or will be blessed to the Church and the Kingdom of God. It is important to understand that this law can only partially be fulfilled by each of us while on Earth and is only capable of realization in its full extent in the Eternal Kingdoms.

Throughout the Endowment, each of the covenants we enter into are associated with a token, a sign and its name. The token and sign for this last covenant, the one associated with the second *stage* of the Melchizedek Priesthood, is not given to us immediately but reserved for later in the Endowment, when we reach the veil at the entry point to the Celestial World. This is significant and illustrates that the second *stage* of the Melchizedek Priesthood is associated with exaltation and eternal increase in the Celestial World. It is fitting that the covenant is sealed by the voice of God directly and is not done so until our entry into the Celestial World, for only there are we truly capable of offering up the consecration of *ALL* the blessings our Father has to give us.

It must be noted that with our symbolic entry into the Terrestrial World in the Endowment, we have no longer been in the company of the *Adversary*; Satan last spoke to us back in the *"Lone and Dreary World"* before we received the power of the Priesthood, which when righteously applied, gives us the power to rebuke him.

The satan, *ha-satan* (הַשָּׂטָן) is limited to the Temporal now; "*upon thy belly shalt thou go, and dust shalt thou eat all the days of thy life*" (Moses 4:20) and does not have the power to molest us after the resurrection within the eternal Kingdoms of God; consider how this truth applies to the Spirit World.

After having received the two *stages* of the Melchizedek Priesthood and clothed in our resurrected bodies, the sequencing of the Endowment liturgy now readies us for presentation at the Veil and introduction to the Father.

**-Garden**

-Preparatory for release into the Telestial World, the Garden is a link (Temple) to the Temporal sphere.

**-<u>First</u> Aaronic Priesthood** The Law of Sacrifice [Justification].

-This Principle/Plan is taught and accepted through Covenant, *before* entry into the Telestial (Temporal) World.

-It is a Terrestrial Temple in the Celestial Sphere.

-Michael is given a new name (Just as we are given a new name).

-Adam and Eve are clothed in "*coats of skins*" (Garments), bodies of *Flesh and Bone* first, then *Flesh and Bone and Blood.*

**-Earth=Telestial**

-Satan is the god of this world.

**-<u>Second</u> Aaronic Priesthood** The Law of the Gospel, Repentance and Faith [Sanctification].

The precincts of God (Holy of Holies) can only be approached by the High Priest through the Atonement.

-Clothed on the left side (Symbolizing our Temporal bodies, like Adam).

**-Terrestrial (Temple)**

-Preparatory for entry into the Spirit World to come.

-Melchizedek Priesthood given (Practicing Priesthood for what is to come in the Spirit World, power to bind in Heaven and on Earth).

**<u>First</u> Melchizedek Priesthood** (Ordinances on Earth) The Law of Chastity. Because procreation is introduced, Melchizedek Priesthood has powers to bind families together.

**<u>Second</u> Melchizedek Priesthood** (Power in Heaven) The Law of Consecration. Only partially possible here on Earth, but fully required in the Spirit World and beyond.

-Clothed on the right side. (Christ on the right hand of God; symbolizing our departure from the Spirit World and our resurrected bodies, like that of Christ).

-No More Satan in the Endowment

*(Satan does not have influence in the Spirit World).*

**-Celestial (Through the Veil)**

-Symbolizes our return to the precincts of the Father (Something beyond that of the Spirit World).

-Entry is granted only by the Father Himself.

-The Name is only spoken by the Father.

Table 9-1 *Symbolism in the Endowment Liturgy.*

- Celestial (Through the Veil)

Passage through the Veil symbolizes our return to the precincts of the Father. This reveals truth about the separation of the varied degrees of Glory. For without (not yet passing through) the veil, we are *symbolically* fully ordained with Priesthood authority; representing Adam at the veil, we are introduced as "*faithful in all things*," we have resurrected bodies (robes on the right side), yet still there remains a *veil* between us and the Father in the Celestial World. This is representative of the same set of conditions in the Garden before the Fall, with its veil in place, separating Adam from the rest of the Physical World.

Entry into the Celestial world is granted only by the Father. It is important for us to be aware that the Celestial Kingdom is not so much a place, as it represents an intimacy with the Father and a state of eternal increase and progression.

The name associated with the *Law of Consecration* is given only by the Father from the precincts of the Celestial World (beyond the Veil); this is significant in that it speaks to an understanding that the *Law of Consecration* is eternal,[18] can only be fully lived and expressed in terms of eternal increase and progression and that eternally everything we have, or will ever have, are blessings from and owed to the Father.

- The New Name

Admission to the mysteries of the Endowment narrative is accompanied by the receipt of a new name given to each participant. What is the significance of the new name, and what does it mean?

When we name a child at birth today, we often choose a name for the way it sounds, one we have affection for, or one we associate with someone or something special. In contrast, within the Scriptures, we find names frequently had deep significance, some carefully chosen with purpose to have meaning. The meanings are often of such significance that they become ingrained into our language itself. Often, it seems, it is difficult to be sure if the name was given to the individual due to its prior meaning, or if it came to mean what it does through direct association with the characteristics of the person to whom it was given.

There are occasions when God provides a new name for an individual; Abram, Sarai and Jacob are among them. Abram became Abraham, "*father of a multitude*" or "*father of many nations*," signifying God's covenant with Him and his seed. Sarai became Sarah, "*princess*," because of her position as the first mother of those under the Covenant. God renamed Jacob, Israel, "*prince of God*" or also "*he who struggles*

---

[18] An important evidence of the eternal nature of this Law, is that within the Temple, we are only asked to covenant that we "*accept*" this law, we are not asked to "*observe and keep*" this Law in the manner that was done three previous times with the Law of Obedience and Sacrifice, the Law of the Gospel, and the Law of Chastity. The Law of Consecration and its attendant second stage of the Melchizedek Priesthood are not in full affect (binding) in the Temporal Sphere.

*with God*" after wrestling with the patriarch all night long. We have already discussed in this work the extreme significance of Christ renaming Simon as Peter on the rocks at Caesarea Philippi.[19]

In these most legendary name change circumstances, God is expressing the significance of His relationship or expectations for the renamed (honoring their spiritual accomplishments, prophesying of their spiritual potential, or esteeming the bestowal of His blessings on them). We often find ourselves incapable of imagining ourselves as God sees us, and through the renaming of these legendary figures, God's foreknowledge challenged the perceptions of what any thought could be.

When God changed Abram's name to "*Abraham*" (*father of many nations*), the ability of an almost one hundred year old, childless man, to obtain what his new name implied must have seemed absurd, but God knows our full potential and truly knows who we each are. God named these sons and daughters with full knowledge that the meanings of these appellations were already true, and then He proceeded to "*worketh in*"[20] them to make each who they were eternally meant to be.

Our receipt of the new name in the Temple may *seem* similar to these instances, yet it is fundamentally *much different*. The new name is associated with a single scriptural reference found in the Book of Revelation. Though referenced in D&C Section 130 (v 10), the verse mostly goes unnoticed by Latter-day Saints; it reads:

> *"He that hath an ear, let him hear what the Spirit saith unto the churches; To him that overcometh will I give to eat of the hidden manna, and will give him a white stone, and in the stone a new name written, which no man knoweth saving he that receiveth it." (Rev. 2:17)*

What does this enigmatic verse mean? And why do we get the "*new*" name?

-The new name of the Book of Revelation is engraved on a stone, that it is, is a direct reference to the stones found on the breastplate of Aaron and the subsequent High Priests of Israel. God directed that a breastplate be created containing the names of the 12 tribes engraved on 12 precious stones.[21] This breastplate was accompanied by the Urim and Thummim.

-This revelation was received by John while on Patmos; he had been placed in exile there following Roman judgement against him for teaching of Christ. Having been judged then, John was familiar with the Roman practice of awarding to a person on trial a white stone if acquitted and a black stone if condemned. In the vision, the stone bearing the new name is white.

-John was in the Greek world at the time as well; athletes at the Greek games, as their prize, won white stones upon which were written their names.

---

[19] See Upon This Rock, and pg. 133.

[20] Philippians 2:13.

[21] Exodus 28:15-30.

The white stone seen by John in vision then, bore evidence that the recipient was *free from guilt*, and *one who had achieved a winner's prize* by enduring to the end. Its religious significance directly relates to the Atonement, which empowers us through the Priesthood of God to come directly to him, acting as our shield or breastplate, our personal armor of God, for it bears a stone with our name upon it.

The Revelator described the name as "*new*" (Greek, *kainon*, καινὸν). John's use of this word, and not *neos* (νέος), means he had something beyond out-of-the-box-brand-new in mind; in the original language, the context describes something much different.

***Our "new" name isn't new to us at all; it is our true name, the one by which our eternal Spirit was called; it is our Heavenly name, the name God gave us, the one we have had since our inception and the one that we will again have through eternity to come.***

*Kainon* means more than just new, it denotes something that is unused or unaccustomed; it does not signify something that did not exist before but something that has a different form and quality than it once had. To John, the Greek word *kainon* (translated as *new* in the KJV) had a meaning that most properly fits *renewed* as we would use it today; John used *Kainon* elsewhere as follows: "*And I saw a new heaven and a new earth: for the first heaven*[22] *and the first earth were passed away*" (Rev. 21:1). In both of these verses, John envisions a restoration to an eternal Glory held once before. In this sense, the *new* name written in stone isn't a *new* name at all, but a very *old* one restored to the holder of the stone when the holder assumes a different quality or regains a quality they once had.

In one very real sense, the new name represents our entry into covenant with the LORD; in this case, it mirrors the renaming of Abraham and Jacob. Our new name also confirms our restoration to our original family, the family of God. We left that

[22] This verse may provide tangential evidence that the Spirit World becomes the Telestial Kingdom after the Resurrection. Though we understand there to be finally three eternal Kingdoms, the purpose of the Spirit World as a way station for our Spirits before the Resurrection and the Judgement exists only for a limited duration of time. As the Spirit World is our first abode in the Heavenly precincts we will inhabit after mortality, it may very well be understood to be the "*first heaven.*" In this case, John saw in vision the renewal of both the Temporal Earth and the Spirit World. In the case of the *new* Earth, what John saw was the Glory of the pre-existing physical orb called Earth, or perhaps many like it, in the Physical sphere.

family to come to Temporality, and through the ordinances of the Temple, we begin our return to intimacy with the Father. The new name represents our becoming again who God created us to be. It represents in eternal precincts what baptism represents here – rebirth, being born again.

The truest relationship of the new name with precedent, however, is with another name given upon an entry into a Temple, a Temple built by the hand of God Himself. Our receipt represents the new name given to Adam in the Garden.

Our "new" name isn't new to us at all; it is our true name, the one by which our eternal Spirit was called; it is our Heavenly name, the name God gave us, the one we have had since our inception and the one that we will again have through eternity to come. The name we have here in Temporality is not ours, it was given to us by our parents, perhaps dictated by culture; it represents our Earthly existence. Through the Endowment, we recognize that this name, our *given* name is linked forever with our true "*new*" name, attached to it, and adding upon.

As a matter of course, like so many of the things we receive in Temple, the new name given to us there is symbolic. The name we are given in the Temple merely *represents* our true new name; *it is not* our new name itself. The name we are given for our own Endowment is assigned by date to everyone who enters the Temple that day. The name is assigned from a list, one each for either male or female, that changes each day, until when the end is reached, it returns to the beginning and runs through the same list all over again. As is true with so much of our Temple experience, it is not so much *what we do* as it is, *that we do*. The new name we are given in the Temple, what that name actually is, is not the important thing. What is important is what the name *represents* and that we accept it by covenant and revere it as though it was the one we will receive in Heaven – *written in stone.*

- Our Temple Garments

There is an extremely significant component to the Garden story that is mirrored within the Temples on Earth; one, despite its significance, that is most often entirely overlooked.

Upon our initial entry into the precincts of the Earthly Temple, we are washed and anointed (a purification) in preparation for each of us to be clothed in the Temple Garment. We are clothed in the Garment prior to our receiving our New Name. Later in the Endowment narrative, we are told that the Garment represents "*the coat of skins spoken of.*" Earlier versions of the Endowment narrative included "*you have received the garment, also your new name*," directly relating the Garment and the new name to one another. The Garment and the New Name both represent an important transition in the *being* of man. They both mark the receipt of a body of *Flesh and Bone.*

The Garden accounts describe God clothing our first parents in "*coats of skins*" after having found them hiding in the Garden after the Fall. The Endowment narrative elaborates by equating the "*coat(s) of skins*" with the Garment to "*cover*" our "*nakedness.*" We have previously reviewed the symbolic context of "*nakedness*" as shame, guilt and exposure. The effects of these things are covered by Christ's Atonement which *covers* them.[23]

The Temple ordinances, through their equating of the "*coats of skins*" to our Garment and the New Name, not only associates them through symbolism to our Fallen bodies of *Flesh and Bone and Blood*, but they also represent the immortal body of Adam before the Fall. They then represent the Temporal component of our three part being in both its iterations (Fallen and Resurrected), mortal and immortal. With this understanding in mind, let us see what we can learn from it.

Immediately after providing "*coats of skins*" to our first parents, God utters the following revealing statement: "*Behold the man has become as one of us…*" (Moses 4:28). By that we are to understand that *Flesh and Bone* truly completed the form and creation of man in the image of both the Father and the Son. And by that we know the Father and the Son are Physically embodied in *Flesh and Bone* as well.

In this context Adam and Eve, who were pure and clean before the Fall, received their *Flesh and Bone* (after the Fall bodies with *Blood* as well) bodies within the Garden Temple symbolized for us today by the Garment and anciently by what we are told is a *like* Garment, the "*coats of skins*" given to our first parents.[24]

The initiatory is the first Temple ordinance we receive in the liturgy; it is preparatory to our receiving the knowledge of the Endowment. We are clothed with the Garment at the initiatory (today symbolically clothed, though in times past

---

[23] See our previous discussion, - They Went to Hide Themselves, pg. 267.

[24] At this point, I want to make a point of differentiation to clarify. Previously we discussed our understanding that the "*coats of skins*" made by the Father and given to Adam and Eve represented a phase transition in their being, and symbolized the first pair's enclosure in bodies of *Flesh and Bone and Blood* (- Coats of Skins pg. 272). The Garments we wear, we are told in the Endowment, represent these same "*coats of skins,*" but there is more to the representation than just that.

For the "*living,*" the Temple experience *begins* with the initiatory and the clothing of the patron in the Garment immediately before the receipt of the new name. I propose here that the timing of these ordinances is significant and represents the initial embodiment of Michael in a body of *flesh and bone* in the Garden in conjunction with his receipt of his new name, Adam. In this context, Adam (as well as Eve) was created in the Garden as a *flesh and bone* image of what was later to come in Temporality. He, like all other things in the Garden, was immortal but occupied a special place in the eternities, a place that is a *template* of the Fallen world. Adam's being, in its state before the Fall, was an eternal body of *flesh and bone* reflecting our future state as resurrected beings. After the Fall, yet still within the Garden, God clothed the first man in a "*coat of skins,*" which symbolize the transfiguration of Adam's immortal being to a Temporal one precursory to his entering the Temporal sphere.

Beyond placing its timing at the point of entry of Adam to the Garden, we are not informed of any detail of the creation of our first father's body of *flesh and bone*.

It is a fact that the Priesthood of God, its authority and power, is equally shared on Earth by both a man and woman sealed within the Temple. It is fitting then that this singularly most important event in the process of progression for the sons and daughters of God, the attainment of a body of *flesh and bone*, is memorialized in the Temple by Priesthood ordinances performed by both men and women. In fact this is the only place and time on Earth where women perform Priesthood ordinances. I propose here that this is not a reflection of practicality or convenience or for moral necessity but it is in fact done in remembrance of the initial creation of our *flesh and bone* bodies so long ago completed in the Garden by both Heavenly Father and Mother; see Hearing the Softest of Voices, Our Heavenly Mother, pg. 449.

actually clothed); the timing and sequencing of this (immediately upon entry and before the Endowment of knowledge) is significant in that it correlates with the entry of Adam into the Garden and when he first received his *Flesh and Bone* body. We are told when the Garment (*Flesh and Bone*) is placed upon us: "*inasmuch as*" we "*do not defile it*" and are "*true and faithful*" to our "*covenants,*" the Garment will be "*a protection*" to us against the "*destroyer*" (the *ha-satan*, הַשָּׂטָן) until we have finished our work on Earth. *Think of what that means...*

Put in its proper context, the Garment represents our Temporal body; it then (Temporality itself) provides a protection to us against the *Adversary*, and in fact everything we are told about the *satan* (שָׂטָן) and our relationship to him here on Earth confirms that we have been given power over him and can be shielded here from his actions.

The *covenants* referred to in the initiatory are those we accepted in the pre-existence, before we came to the mortal world. If we abide by them, Temporality itself protects us from the deteriorating effects of sin in the eternities beyond. Temporality and the *Flesh and Bone and Blood*[25] we receive here, provide a protective environment whereby we can experience the effects of sin in an uncertain world.[26] Adam and we (symbolically through the receipt of the Garment and in actuality because we are mortal) can experience true "*free*" agency and, through the Atonement, receive sanctification and restoration.

Further, after the Fall (for the sons and daughters of God), the body of *Flesh and Bone* is transformed through the addition of blood. This transformation (however symbolic the addition of blood in fact is) is significant, in that death now becomes inherent to our nature. The *blood* (death) of Temporality absorbs, shields and separates our Spirits from the sin of the world, buffering its destructive impact, gathering it up and taking it away at our passing. The Atonement of Jesus Christ removes us from the sin absorbed by that blood, which death allows us to shed. The salvific powers of the Son's sacrifice wash away those sins, purifying and sanctifying us in the reverse order symbolized by the initiatory and our being clothed in the Garment (representing the *coats of skins*) within the Temple.

The symbolism is powerful and intrinsic to every sacrifice offered properly through Covenant with God since Adam. Through the shedding of blood (casting away our Temporal nature at death), we can return to the presence and companionship of the Father. We understand this sphere and the receipt of a mortal (Fallen) body not to be

---

[25] In a following chapter in this work, What Is Man, pg. 469, we will address the significance of blood in the Temporal bodies we receive in this Universe.

[26] The wearing of the Garment by Latter-day Saints is often the focus of ridicule and misunderstanding. Frankly the true meaning of the Garment is unknown by many, if not most Latter-Day Saints. They are not "*magic*" in any way. The protection they offer is not from falling rocks of screaming bullets; they do not keep disease away or alter the laws of physics and biology in their environs. But what they represent, our mortality, the covenants we made to come here, their safeguards, and the protective effects of the Atonement in empowering Free Agency in this Universe without permanent deleterious effects in the Eternities, is nothing short of *magical...*

a punishment but a blessing, one that allows us not only to become more like God in form and substance, but in experience and wisdom as well without permanent injury inflicted on our Spirits.

Through all we have learned here about our nature and the symbolism of the Garment, we must not overlook the truths to be gleaned from a realization that Adam and Eve, within the Garden Temple, were at some point clothed not only in a Physical body (that of an embodied intelligence, a "*spirit*"), but they also had a body of *Flesh and Bone* (and subsequently of blood also) before entering the Temporal sphere.

The Temple Garment symbolizes Michael's receipt of a body of *Flesh and Bone* marked with his receipt of a new name (Adam) at his initial entry into the Garden and its precincts. We receive the Garment and the new name *immediately* upon our initial entry into the Temple. After his Fall, Adam is given a *Coat of Skins*, symbolizing that death had fallen upon him and the world; his body was transformed from that of *Flesh and Bone,* to *Flesh and Bone and Blood*, symbolized in the Temple by our donning robes on the left shoulder. When the Fallen Temporal body is restored to its perfected state through the Resurrection, we symbolically recognize that transition by moving our robes to the right shoulder.

- Decoding the Lecture at the Veil[27]

Adam is not God. The most glaring evidence of this fact comes from the Endowment narrative itself. We are told that Michael becomes Adam and as such we have Michael and Adam in three separate spheres existing as a fully independent, separate and distinct individual alongside the Father himself. In the Spiritual sphere, Jehovah and Michael co-create the Physical sphere at the direction of Elohiym the Father, with Jehovah very obviously an intermediary. The Endowment is explicit in detail when it informs us that Michael became Adam in the Garden, and it further details Adam in direct conversation with the Father, one obviously inferior and responding to and receiving instruction from the other. These several exchanges occurred both before and after Adam's Fall. After the Fall, it is clearly shown that Adam was expelled from the Garden to the Telestial World (the Temporal Universe), and while he and Eve are there, we find Elohiym continuing to inculcate Jehovah in interaction with Adam, directing that Peter, James and John, observe and report, instruct and advise.

Any understanding that Heavenly Father could indeed be Adam would require us to envision and construct a being as foreign to our common sense understanding as the Trinitarian (three beings in one) concept of Deity we so vehemently object to.

---

[27] In February of 1877 Brigham Young introduced a lecture that was to be delivered at the Veil in the newly dedicated St. George Temple. That lecture detailed President Young's understanding that Adam was God. The Prophet had taught the concept before, as early as 1855, but the teaching was not received by way of revelation, nor was it ever canonized in any form. It is entirely unclear if the lecture was actually incorporated into the endowment ceremony outright, or even if it was ever included in any temple ceremony other that at St. George, if even there, and for how long.

## Foreordination and Agency in the Pre-Existence

Agency, choice, the foreknowledge of God and predestination of man are concepts that to some seem at best incongruent. To most they are completely irreconcilable, and indeed they are, despite some philosophical attempts to make them compatible. That they may seem mismatched is not because of the notions themselves, the notions are not the root of the discord; but it is a constrained world view that imposes limits on the activities of God, and by association, those of man, within a construct of space and time that frames the conflict. Our view, as we have developed it here in this work, lifts these restrictions and provides an eternal continuum in which the actions and Will of God and man play out.

For millennia the mind of man has struggled to understand a world that allowed for a fully sovereign, all-powerful, all-knowing God and an independently agent and accountable mankind. When these disparate principles collide, philosophical conflicts arise. At the discordant crossroads, many false teachers can be found debating errant and even blasphemous philosophies of men. We have discussed some of these at length already in this work.[28] Our understanding of the three separate creations broadens the canvas of experience, action and result, freeing the confines placed upon these notions by time, within a beginning and an end, that only appears to be firmly in place, but actually does not exist at all. Here, in this section, we will discuss the concepts of foreordination as it relates to free agency and uncertainty in the Temporal sphere.

- Agency and Sovereignty, Dividing a Christian World

In upstate New York during the year preceding 1820, a young, questioning Joseph Smith became very much aware of a theological struggle that began far away and well over a millennia before. In his own personal history Joseph records:

> "*...there was in the place where we lived an unusual excitement on the subject of religion... Indeed, the whole district of country seemed affected by it, and great multitudes united themselves to the different religious parties, which created no small stir and division amongst the people, some crying, 'Lo, here!' and others, 'Lo, there!' Some were contending for the Methodist faith, some for the Presbyterian, and some for the Baptist...*" *(Joseph Smith History 1:5)*

Joseph's mother and three of his siblings joined the Presbyterian church and perhaps because his father displayed a reluctance towards *organized* religious affiliation, the mind of young Joseph "*was called up to serious reflection and great uneasiness*," as he sought to determine what was right and what was wrong:

> *"In the midst of this war of words and tumult of opinions, I often said to myself: What is to be done? Who of all these parties are right; or, are they all*

[28] See our discussions, The Root of All Evil and God of Our Fathers, earlier in this work, pgs. 118 and 17, respectively.

*wrong together? If any one of them be right, which is it, and how shall I know it?" (Joseph Smith History 1:10)*

Through the details of his testimony, it is clear that what young Joseph had been exposed to was the playing out of a debate that had come to full bloom just 200 years earlier but had its roots over a thousand years before that. Joseph's statement, "*Some were contending for the Methodist faith, some for the Presbyterian, and some for the Baptist...*" reveals that the young man was thrown headlong into a debate between Arminians and Calvinists, one which continues to this day.

The Calvinist–Arminian rift, in its modern form, began early in the 17th century in the Netherlands with theological disputes between followers of John Calvin (Jehan Cauvin, 1509-1564 AD) and Jacobus Arminius (Jokob Hermanszoon, 1560-1609 AD). It continues among Protestants today following the interpretations of Scripture codified and espoused by these two men. The divide centers on vastly disparate views driven by the false precept of original sin regarding the very fundamental doctrines of the extent of salvation, the universality of the Atonement, and the predestination and depravity of man.

This great rift is wholly formed around either the acceptance or denial of a singular reasoning: *Mankind is so debased by original sin that he cannot on his own even accept the offer of salvation, therefore salvation comes to the individual through a direct call and action from God completed by the Holy Spirit; God chooses for himself who will be saved, leaving all others, the vast majority of man, to eternal damnation because of Adam's lone actions in the Garden.* Calvinists believe this to be true, Arminians deny it, preferring a theology instead that instructs: *God extends his mercy and salvation to all, desiring that all be saved. It is up to man to freely choose this gift from God through individual faith and belief in the Atoning sacrifice of the Savior Jesus Christ.*

The contention that Joseph witnessed was severe, "*...for a scene of great confusion and bad feeling ensued—priest contending against priest, and convert against convert*" (Joseph Smith History 1:6), and this commotion was reflective of the debate since its inception. The dispute that troubled Joseph traced its origin to clashes that occurred between 1609 and 1618, known as the Quinquarticular Controversy, but the seeds of the contention can be found planted in a time much more distant in the past.

Late in the fourth century AD, a young monk from Britain named Pelagius (circa 360-418 AD) traveled to Rome and was sickened by the thoughtless, undisciplined behavior within the churches there. Remember, that to many, Christianity was liberating, the elevation of man to an object worthy of the love of God to the extent that He would die for men's Souls, as taught in the early Christian church, developed in some a libertine enthusiasm. It unshackled Christians from burdensome temple worship, sacrifice and obedience to law. One can easily imagine congregations that appeared as carefree and unconventional as the Haight Ashbury community of young people in San Francisco did in the mid-to-late 1960's. For this reason alone, contemporary Jewish leaders could feel justified in their common description of the sect of the Nazarenes (Christians) as the *Kingdom of Arrogance.*[29] Pelagius sought to correct this lack of holiness by preaching the Gospel beginning correctly with justification through faith *alone* (though often ascribed as a creation of Martin Luther, Pelagius first added the word *alone* to the faith phrases of Paul) but finishing it with a demand of God requiring human effort and morality.

Pelagius accounted value to men's Souls and found in them worth and divinity. In principle he opposed the despondent and cynical view of human nature espoused by Augustine of Hippo (Aurelius Augustinus, Bishop of Hippo Regius, 391-430 AD). In time Pelagius' followers went beyond the bounds of their teacher and removed justification through faith, codifying instead a doctrine of salvation that was contingent upon the works of morality, known through time today as Pelagianism. We can only surmise their actual teachings though because what we know of it survives only in the writings of its two strongest opponents – Augustine and Jerome.

In response to what the teachings of Pelagius had become, Augustine, no doubt informed by the Greek philosophies espoused by Platonism and Manichaeism, developed, and expanded upon, the (errant) doctrines of original sin (which Pelagius denied) and predestination. Augustine reasoned; *man is so depraved that he is incapable of even accepting the offer of salvation — it must be God who chooses for himself individuals to bring to salvation.* This conclusion of Augustine drives the doctrines of *limited atonement* and *irresistible grace*, which were adopted by classic Calvinism. In time Pelagianism moderated, becoming much more akin to what Pelagius had first taught, but still standing in contrast to the Augustinian view declaring simply that, *man's faith came from man and was an act of free will, unassisted by previous internal grace.*

---

[29] *BIRKAT HA-MINIM* – (בִּרְכַּת הַמִּינִים) (*against heretics*). The "*benediction concerning heretics*" is a statutory prayer of the daily *Amidah* that invokes divine wrath upon and against heretical sects in general: the *min* (plural *minim*), the *malkhut zadon* "*kingdom of arrogance*," and specifically Judeo-Christians, the *nozerim*. The intent of this prayer was the exclusion of these minim from participating in the synagogue. Beginning in the first century AD, Rabban Gamaliel II required the hazzan to recite this prayer in public worship three times a day with a response from the congregants confirming it. Gamaliel did not apply this regulation to any other benediction; it read: *"For the apostates let there be no hope. And let the kingdom of arrogance be speedily uprooted in our days. Let the noẓerim and the minim be destroyed in a moment. And let them be blotted out of the Book of Life and not be inscribed together with the righteous. Blessed art thou, O Lord, who humblest the arrogant."* By refusing to confirm the curse through the voicing of *Amen* at the prayers conclusion, Christian congregants would be exposed, marking themselves for expulsion from the Synagogue (See John 9:22, 12:42 and 16:2). At issue here is the identification of the Christians as the "kingdom of arrogance." This benediction dates to the second Temple period, and although it is unclear what its early readings were (no doubt not including the *nozerim* originally), the "*kingdom of arrogance*" has been thought to apply in turn to both the Greek and Roman empires who ruled over the Jews. By the time of the early Christian era, however, it no doubt applied to Christians who were arrogant in their belief that by faith they could be saved by a God who had become man like unto themselves.

Official Creedal resolution of this matter came in 531 AD, with the Council of Orange. The Council of Orange adopted what has become known as a *semi-Augustinian* theology. The council decreed that faith is an act of free will but is still a gift, stemming from the Grace of God, enlightening the human mind and enabling belief (this as it turns out is very much in line with how Latter-day Saint doctrine would describe faith). The Council of Orange adamantly denied predestination, stating, "*We not only do not believe that any are foreordained to evil by the power of God, but even state with utter abhorrence that if there are those who want to believe so evil a thing, they are anathema.*"

- Protestant Division

At the outset of the 17th century, the Calvinist reformers used the canons of the Council of Orange to provide a historical pretext supporting their recent constructions of the depravity of man through the effects of the Fall and original sin claiming that *depravity*, as fundamental doctrine, had been taught much earlier in the church. The council of Orange was also used by Arminian theologians to affirm their view for it opposed irresistible grace and strict predestination.

What began as a purely theological debate (The Quinquarticular Controversy), became an open struggle during the period of 1609 to 1618, having serious political implications in the Netherlands. Quinquarticular (i.e. "*having to do with five points*") refers to points of contention raised by the Arminian party through its publication in 1610, of the five articles of Remonstrance. These articles were rejected by a Synod convened in Dort (1618-1619) and published in a canon. The Dort canon provided responses to the articles of remonstrance; they are commonly referred to as the Five Points of Calvinism and provide the basis for Calvinist doctrine today.

The disposition of the Quinquarticular controversies ended any hope of unification among the nascent Protestant movement, cementing the divide between Lutherans and Calvinists, and within the Netherlands, resulted in the imposition of an edict which transformed the Arminian movement into a separate, persecuted church.

Today Protestant Christians are affiliated with many dissimilar partitions that can be grouped into two camps identified as Arminian and Calvinist. Both camps originated in the Reformation and are of the Reformed Movement, but because the rift between their beliefs have loomed so large, those who hold the Arminian view are now rarely identified as Reformed Christians. Arminian churches include, among others, such diverse congregations as: Lutheran, Baptist, Methodist, Mennonite, Amish, Seventh-day Adventist, Wesleyan and Nazarene. Calvinist churches largely identify today as Reformed, and include: Presbyterian, Anglican, Reformed Baptist and United Church of Christ.

Because of the Council of Orange, the Roman Catholic Church is semi-Augustinian in theology. Eastern Orthodox Churches, and the Church of Jesus Christ of Latter-day Saints affirm theologies that more closely align with Arminianism, drawing upon elements from both the semi-Pelagian and semi-Augustinian theological stances.

In an attempt to present the appearance of a *unified* faith, which commonly supports the exactness of written Scripture that clearly delivers the inerrant word of God, this rift is often glossed over by Christians. However, the depth of disagreement about who God is and how He deals with His creation is stunning, representing two different Gods entirely.[30] That two groups can read the same Bible and come to such different understandings, speaks directly to the errancy of the doctrine of "*sola scriptura*" (Scripture Only).

***The character and purpose of the God of Calvinism and Arminianism is as different and conflicting as can be; they are as dissimilar as Christ is from Krishna, Zeus and the Demiurge.***

Latter-day Saints understand that a Christian, through faith, develops a personal relationship with God, our common Father. We believe that such a relationship involves intimacy, and association with Him. This closeness delivers a clear and indisputable understanding of His divine character and purpose. In contrast, the character and purpose of the God as understood by Calvinism and Arminianism is as different and conflicting as can be; they are as dissimilar as Christ is from Krishna, Zeus and the Demiurge.

How can such divergent concepts develop among the "*faithful?*" It is possible only because many choose to rely on the reason of man applying those understandings to the Scriptures without accepting the guidance of the Spirit in interpreting the Word of God. Christians who espouse the inerrancy of Scripture, relying on it alone for guidance from God and the attainment of Salvation, must account for its allowing the faithful to develop such discordant theologies. They must provide a mechanism, attested to in the Scriptures themselves, that explains how Scripture can remain unaffected in transmission by the effects of open and continual discord.

---

[30] It is important to understand the severity of the Calvinist point of view. It is errant to think of their interpretations as being insignificant from ours. Consider the following brief collection of quotes generated in response to just a single observation from a Christian who testified that *God Loves All* as a point of illustration: *bracketed* comments were not spoken but their inferences were included in the discussion. "*God does love all, just not all the same. They* [non-Calvinists] *expect God to be in a way almost subhuman; man does not love all in the same way, why do they expect God to do so?*" "*God does love them, He does not wipe them out the moment they come into existence, He could rightly and justly do so, but He brings the rain and allows the sun to shine upon them, He shows them many great gifts and tender mercies...* [Before he condemns them to the eternal, torturous hell, by not calling them to seek salvation, though He could do so if He chose to] *He loves them, but not with the same love He shows for the elect, whom He brings to regeneration and faith in God.*" "*This Baptist said that anyone who calls upon the name of the Lord will be saved..... Every Calvinist will agree and say the same thing... Except Jesus says, no one can do that!* [Only God can cause him to want to do so]." Such thinking represents a horribly distorted understanding of our common Father in Heaven.

At the base of this theological dispute, are the concepts of God's sovereignty and His ability to choose juxtaposed to that of man's *agency*. For the Calvinist, who seeks to uphold God's complete control in every action in space and time, *salvation* is of necessity a result of the action of God alone. He calls only those He chooses, leading them to Him, instilling the very desire to do good in them and separating them out from those to whom He chooses not to save. Man does not have a thing to do with any of it.

I don't need to expound upon the *many* problematic logical incongruities this thinking imposes; we will just discuss two here.

If God chooses to save just a few, it must have been His intent to do so from the beginning. Why then create all of those Souls, the overwhelming majority, of which He damns from the outset, with no *hope* at all for salvation? Where is the Glory in consigning to hell, and declaring wicked and unprofitable, most of His own creation, when it was *He who made them so*?

Calvinists insist upon the complete sovereignty of God, that man is not agent;[31] we are condemned by original sin and saved by grace through choice and action of God alone. That reasoning, however, must extend in reverse as well; Adam's disobedience and Fall then, were entirely God's doing, causing original sin to begin with. What Glory then is in the gift of Salvation? God is responsible for evil. In fact some Calvinists go so far as to claim God created evil so that His Glory would be manifest when He conquered it. *Not an impressive challenge really, when you think about it, claiming victory over something of your own making and completely within your control.*

These are the views of the Presbyterian Church that Joseph's mother and his siblings had joined, but apparently not those of Joseph, "*...my mind became somewhat partial to the Methodist sect...*" (Joseph Smith History 1:8). The Methodists, to which Joseph had been drawn, were Arminian and confirmed God's call to all and an infinite Atonement through Jesus Christ for all that freely come unto Him in faith with a desire to follow Him. To those who hold the Arminian point of view, God's Glory is defined by His unrestricted, abiding love for all of His children equally.

---

[31] To be transparent a great many Calvinists are Compatibilist, a philosophical construct that allows for predetermination and full sovereignty of God, yet somehow man is truly agent and free to choose for himself. In this hypothesis, God both chooses and calls the saved to Salvation, for without God's "*effectual call*" man has absolutely no capability of coming to God on his own. A man thus called is changed into something that desires to do good and that desire is a gift from God, not inherent in man. The obverse is the sticky wicket, those to whom God does not extend the call continue in their sinful ways because they *freely choose* to. The disconnect here is that they were clearly created with inherent original sin and could not choose to do good, for that desire is not in them. The logic of Compatibilism is completely assailable because free will and effectual call are *absolutely* incompatible.

- Light for a Darkened World

Faced with these disparate theologies and what appears to be differing thoughts on these matters in the home, Joseph turned to God in prayer. His response was the seminal moment for the Church in the Latter-days. In a quiet grove of trees, unremarkable in nature and appearance, a stand we however call sacred today, the Father and the Son made an appearance to the young man. After a time and with fervent desire, Joseph voiced his question to Deity:

> *"...I asked the Personages who stood above me in the light, which of all the sects was right (for at this time it had never entered into my heart that all were wrong)—and which I should join. I was answered that I must join none of them, for they were all wrong; and the Personage who addressed me said that all their creeds were an abomination in his sight; that those professors were all corrupt; that: 'they draw near to me with their lips, but their hearts are far from me, they teach for doctrines the commandments of men, having a form of godliness, but they deny the power thereof.'" (Joseph Smith History 1:18-19)*

As young Joseph grew and became the Prophet, much would be revealed to him. Through his translation of the Book of Mormon and the receipt of direct revelation, the answer to his first question, one that spawned from the dispute about predestination, would be given time and time again. Joseph would deliver to the world a theology that resolved the centuries long struggle, affirming a universal love of God for all His children and properly focusing all things on the singular purpose of God and His creation; "*For behold, this is my work and my glory—to bring to pass the immortality and eternal life of man*" (Moses 1:39). By that we understand our Father means *ALL* men, every son and daughter, not just a chosen few.

- Foreordination

Calvinists will draw heavily upon certain Scriptures to bolster their argument for a people *chosen* by God through an *elective* call. In this they are much like the Sadducees of old. For the Sadducee, the history of the Jews is one of selection. Beginning with Shem through Eber, and Abraham, then through Isaac and Jacob, each generation was distinct because it was chosen by God to be so. The Jewish Priesthood, originally conferred on Aaron (in their view), was passed from father to son as well.

Sadducees felt they were born, *selected by God*, to rule, judge and preside in Israel. Patrilineal descent is not a path of restriction; however, it is one of diversity and dilution through volume as each generation was born, father having sons and eventual grandsons, each having more sons and grandsons of their own. In recognition of this, the Levitical Priesthood and the patents of its authority narrowed within the family line over time, beginning with the sons of Aaron and narrowing to those of Zadok only, until narrowing again today to those surnamed Cohen. The Sadducee was divinely chosen, through no action of their own, through birth, chosen by God without regard to their own righteous endeavors. The Sadducees had enormous scriptural support, as well as thousands of years of tradition to support their views and claims. The Calvinist view of those chosen and elected by God is in principle no different than this.

In comparison, the Pharisees were much more Arminian, though not entirely so. Their election was an unearned gift from God as well, through birth. The Pharisees had to be of Levi, but not necessarily from the lineages of Aaron or Phineas or Zadok. By the second Temple period, with over 1800 years of generations in between, almost everyone in Israel could draw a line through their family tree to Levi, so there was, in effect, a *General Call* in place among the brethren in the land. Israel, however, was chosen above the rest of the gentile world. As with the Arminian view, the Pharisee had to freely choose God. Pharisaical authority came not just by birth, but through study and command of the Scriptures and through obedience to, and performance of, the requirements of Law with exactness – all acts of free will. As was the case with the Sadducees, the Pharisees could, and would, point to Scripture to support their authority. Certainly none were better at understanding it and espousing it than these forbears of the rabbinic rights.

The problem of course is both the Sadducees and the Pharisees had it wrong, and so also do the Calvinists and Arminians of today; "*…they draw near to me with their lips, but their hearts are far from me, they teach for doctrines the commandments of men, having a form of godliness, but they deny the power thereof.*" What then of Scripture that refers to a *chosen people*[32] and prophets chosen by God[33] before they were born? Is the Priesthood foreordained? Who are the elect? Our understanding here in this work, through the structure of three spheres, allows these things for a few, allows God to call and choose, without demanding Him to abandon and discard the many.

Perhaps Joseph's first glimpse into the true meaning of the enigmatic Biblical passages, that had been so fiercely debated, came as he translated the words of Alma, as they were recorded by Mormon. On this subject those words said:

> *"And this is the manner after which they were ordained--being called and prepared from the foundation of the world according to the foreknowledge of God, on account of their exceeding faith and good works; in the first place being left to choose good or evil; therefore they having chosen good, and exercising exceedingly great faith, are called with a holy calling, yea, with that holy calling which was prepared with, and according to, a preparatory redemption for such. And thus they have been called to this holy calling on account of their faith, while others would reject the Spirit of God on account of the hardness of their hearts and blindness of their minds, while, if it had not been for this they might have had as great privilege as their brethren." (Alma 13:3-4)*

---

[32] See Romans, Chapter 9 and Deuteronomy 7:6.

[33] The Prophet Jeremiah, we are told, was foreordained to his calling: "*Before I formed thee in the belly I knew thee; and before thou camest forth out of the womb I sanctified thee, and I ordained thee a prophet unto the nations*" (Jeremiah. 1:5); so too was John the Baptist foreordained, his mission was to prepare the way of the LORD (see Isaiah 40:3; Luke 1:13-17; 1 Nephi 10:7-10).

Alma explains here that, "*from the foundation of the world*" some were called "*according to the foreknowledge of God.*" This was not done through arbitrary means, at the sole pleasure of God, but in respect for the agency and actions of those "*being left to choose good or evil*" and "*on account of their exceeding faith and good works.*" Alma had an understanding of a premortal existence fully in line with ours here. In the Physical sphere, before this Universe and the Earth we stand on was created, many of the children of God distinguished themselves "*exercising exceedingly great faith*" and were "*foreordained*" (elected and chosen) to fulfill the purposes of God in times *then* yet future.

Later Joseph would uncover the words of Abraham, who had seen in vision a time before the creation of the world as well, a time even more distant in the past, even before the Physical sphere existed. The patriarch's vision revealed the noble and great of the Spiritual sphere being chosen for a purpose:

> *"Now the Lord had shown unto me, Abraham, the intelligences that were organized before the world was; and among all these there were many of the noble and great ones; And God saw these souls that they were good, and he stood in the midst of them, and he said: These I will make my rulers; for he stood among those that were spirits, and he saw that they were good; and he said unto me: Abraham, thou art one of them; thou wast chosen before thou wast born." (Abraham 3:22, 23)*

Joseph F. Smith would see in vision this same event. It is recorded in the Doctrine and Covenants as follows:

> *"I observed that they were also among the noble and great ones who were chosen in the beginning to be rulers in the Church of God. Even before they were born, they, with many others, received their first lessons in the world of spirits and were prepared to come forth in the due time of the Lord to labor in his vineyard for the salvation of the souls of men." (D&C 138:55, 56)*

In all of these instances, God provides an understanding of election and foreordination that is in complete harmony with His sovereignty, Law, and His purpose. Foreordination, as it is described here, does not supersede agency, but it is a result of the exercise of it. By eliminating illusory boundaries of time, imposed by our limited Temporal vision, we can view an expansive continuum of existence before and after our time on Earth. A continuum which allows us to choose, fail, advance and achieve throughout the expanse of eternity. We all had agency in pre-mortality. Those having chosen good, exercising much faith there, were sent to Temporality with blessings and opportunities being called to do Holy things. Joseph Smith said, "*Every man who has a calling to minister to the inhabitants of the world was ordained to that very purpose in the Grand Council of heaven before this world was*" (Teachings of the Prophet Joseph Smith, pg. 365). Though we are called and our actions in the pre-mortal spheres may have warranted foreordination, we are not predestined.

The multiple spheres of creation detailed in this work allow us to differentiate between a factual understanding of foreordination and the mistaken theosophy of predestination.[34] We understand God knows "*all things, for all things are present before [his] eyes*" (D&C 38:1-2). Because of this and His knowledge of our past history, He can anticipate our choices.

God's foreknowledge does not predestine us, but it forewarns Him. God does not impose His will on us, we freely choose. Though God knows our future potential, it is because of our past actions that we are foreordained and blessed to help bring about His purpose. Foreordination also manifests itself in another way that more closely resembles that of predestination, but in concert with free agency we see it not to be so. We understand that God in accordance with our pre-mortal desires establishes for each of us, "*bounds of habitation*" (Acts 17:26). This means that the times, places, and circumstances of birth into mortality for our individual Spirits are appointed, to best allow for the fulfilling of our purpose, mission and assignment in Temporality. Our nation, the age of history, and our familial associations are fixed in this way setting the "*bounds*" discussed by Paul, but allowing for us to be fully free agent within them.

Our free agency here in Temporality permits our choice to accomplish Holy orders, or fail to meet their expectations.[35] We are not all born equally, it seems, but we are equally loved by our Father, equally on a path to perfection, and all chosen (at some point, here, or after)[36] to accept and receive Salvation, perfection and exaltation. God's plan is perfect, our agency is complete, and our Father is patient, knowing the beginning from the end. We may stumble today, fall short of the rewards of Grace at this time, but His Grace is sufficient and He knows it will cause us to succeed if we will allow it to strengthen us.

---

[34] Augustine and later Calvin interpreted the word "*predestinate,*" Greek, proōrisen, προώρισεν, in Romans 8:29-30 and Ephesians 1:4-5 dictated a divine precausation. For them and Calvinists of today, God is the ultimate causal agent of any action in man which is always and ever only an effect. The teaching of Joseph Smith rejected this interpretation offering the doctrine of foreordination as detailed here, believing instead that Paul intended to convey that we are foreordained to divine sonship through Christ, that the Plan of Happiness was set in the pre-existence to affect the redemption of all the Souls of men.

[35] It is fully understood that there are many who were foreordained before the world to a greater state than they have prepared themselves for or may achieve here. An example of these can be found in the Apostles Peter, James and John. Through what we have discussed here and the Endowment narrative, we know they were foreordained to the Apostleship, yet the Gospels tell us that all three were happily engaged as simple fishermen before Christ stepped up to them along the shores of Galilee calling them out of their boats. Not a one was at that time aware of their foreordination or the lofty heights of their potential and purpose in the plan of God. We are told, "*Behold, there are many called, but few are chosen*" (D&C 121:34). By this we can understand that despite being among the noble and great, those whom the Father would make chosen leaders, there are some who may fail to fulfill here all that their calling intended, e.g. *King David*. We must remember that the achievements we may have been foreordained to, include paths we each may take in the eternal spheres and kingdoms to come. It appears then that "*now*" may not necessarily be the times that try men's Souls.

[36] Elder B. H. Roberts, of the Seventy, taught that all, at some point either sooner or later, will manifest, "*the strength of that intelligence and nobility to which their spirits had attained in the heavenly kingdom before they took bodies upon earth.*"

Because of errant understanding held by those who contended over the human constructs of predestination and limited salvation, the boy Joseph was led to question. When he took his question to the Eternal God, he was answered. It would take years for the Latter-day doctrine of foreordination to be fully exposed, but in the instant of initial communication, in direct response to the entreaty of a young man who was searching, the Son of God replied boldly and without ambiguity:

> "*...they draw near to me with their lips, but their hearts are far from me, they teach for doctrines the commandments of men, having a form of godliness, but they deny the power thereof.*"

What is the Power of God? It is His abiding love; such as is expressed in the incarnation of the Son, His sacrifice and Atonement for *all*, "*For as in Adam all die, even so in Christ shall all be made alive*" (1 Cor. 15:21, 22).

Salvation through the Atonement is for *all* men, denying this fact is *denying the power of God*. Christian churches and traditions that communicate a limited Atonement, for a chosen few, are teaching the false doctrines of men shrouded in "*a form of godliness*," hiding behind an incorrect interpretation of Scripture. A heart, one espousing to be Christian or not, that does not recognize all of us as God's children, that does not know that He seeks and died for each of us, and that He searches for everyone lost, weeping with us in our pain, is a heart that is "*far from*" the heart of the Father.

We here within this work affirm that it is the Work and Glory of God to bring to pass the eternal life and Salvation of *all* of the sons and daughters of God. It is His desire that not one child is lost. This He does while allowing and fully enabling our free agency in the process.

His *calling of the elect* was completed in the pre-existence; His choice was made there. Because all of the sons and daughters of God chose to accept His Plan, we all come to Temporality through His Grace, experiencing here the results of that decision freely made. All experiences are possible here, no matter what we freely choose, because He made it possible and empowers it. Because we lived before, experienced progression and accomplishment individually, some of us completed processes that set us apart from others before this Earth was formed, including ordination and receipt of priesthood office. As such, when we receive an office here in Temporality, we are merely restored to an office we were ordained to in a time before.

It is entirely consistent, within this *continuum* of existence, for there to be a people chosen to experience things here that are different than the experiences and expectations of others. With an understanding of agent activity in a Spirit World to follow, an interpretation of scriptural passages that suggest some may not receive, *or are even intended to receive*, the gift of Salvation in this life remains consistent with a God who desires to seek after and save all born into mortality.

Our understanding of the risks and rewards of agency, made possible by an uncertain Temporal Universe in which God creates by His withdrawal, establishes a superior theology that harmonizes seemingly conflicting scriptural passages.

*We undeniably worship a fully sovereign God who is at once all-powerful and a personal Father who weeps when we do. That is the God of the Bible, the God of the Universe, the God every true Christian will come to know.*

## Revealing the Spirit World

We have described here the three Kingdoms of Glory and perdition as the destinations of our Spirits in the Resurrection after the Judgement. And we have implied that these places are initial habitations and indeed for some may not be the last stop on their eternal journey. The Plan of Happiness as we understand it in the context of the continuum of eternal experience we discussed in the previous section of this work must provide for the equalizing of experience for the multitudes of Souls who lived out an infinite array of histories in a completely uncertain Temporality. The Spirit World provides for this equalization.

We understand that following the death of our Temporal bodies, our Spirits are separated and subsequently manifested in the eternal Physical sphere as embodied intelligences. They are immediately eternal, but because they have not been reunited with the resurrected Temporal body, they are not yet what is understood to be immortal. The purpose of the Spirit World is not just that of a way station to the Resurrection and Judgement, but it exists to provide an opportunity for many, who because of varied experience in this life did not have equal exposure to right and wrong, to accept the *Law of Obedience and Sacrifice* and the *Law of the Gospel*, to exercise faith and repent.

In the next few pages, we will examine the Spirit World, focusing on the state of the Spirits there and the opportunities the Spirit World provides them. We will observe the Spirit World from the different viewpoints of the Bible, the Book of Mormon and through Latter-day hierophantic teaching. We will uncover contrasting terms used to describe this same Heavenly realm and will learn much from the dichotomy. Our discussion will not immediately turn to Hell, for the Spirit World does not contain such a place.

### *The Biblical Account*

Within the pages of the Bible, the Spirit World is called both *paradise* and *prison*, and we find these terms used alternately. In his first epistle, Peter spoke of the Spirit World: "*Jesus, while his body was yet in the tomb, went to prison and preached to spirits some of which had lived disobedient mortal lives in the days of Noah*" (1 Peter 3:18-20). Luke calls the same place visited by the Christ "*paradise*" as he recounted the LORD's words to the repentant but *not-yet-baptized*, *not-yet-confirmed*, thief who hung to His right on the cross; "*Verily I say unto thee, To day shalt thou be with me in paradise*" (Luke 23:42, 43). These verses are confirmed in their accuracy through Joseph Smith's Inspired Version of the Bible. The verses in Luke and Peter are corrected for minor grammatical errors but left intact as to the particulars of context and place names. In the account in Peter, the Prophet Joseph did not correct the reference to prison. Additionally, Joseph adds the statement spoken by the Savior to the repentant thief, using the word *paradise*,[37] in his inspired account of the 27th Chapter of Matthew.

Christ's visit to Spirit Prison had been prophesied for hundreds of years. Isaiah records: "*And they shall be gathered together, [as] prisoners are gathered in the pit, and shall be shut up in the prison, and after many days shall they be visited*" (Isaiah 24:22). Again in Ecclesiastes: "*For out of prison he cometh to reign...*" (Ecclesiastes 4:14). The Savior himself refers to the prophecies of Isaiah in relation to himself when in the synagogue at Nazareth He proclaims that in Him Isaiah's words are fulfilled. He said, "*The Spirit of the Lord is upon me, because he hath anointed me to preach the gospel to the poor; he hath sent me to heal the brokenhearted, to preach deliverance to the captives, and recovering of sight to the blind, to set at liberty them that are bruised*" (Luke 4:18). Again in the inspired account given by Joseph Smith, these references in Isaiah, Ecclesiastes, and this second reference in Luke are not modified or corrected at all, confirming his acceptance of their accuracy.

Perhaps one of the most intriguing and illuminating testimonies to be found in the Bible of the Spirit World is a prophecy of Christ's ministry there. The Prophecy is to be found in the life story of Joseph of Egypt. The story of Joseph of Egypt is a prophetical parallel of the life of Christ and much can be learned in studying the life of this amazing man. In the details of his life, are to be found many precious truths of the Gospel and the life of Christ. Recorded over 1000 years before the first advent of the Savior, the parallels found in this story provide an almost exact blueprint of the Savior's mission on Earth. This story details the relationship of Christ to His Father and to us, His brothers and sisters. The story presents truths of the Plan of Happiness and provides many glimpses into the ministry of Christ even now not yet completed.

***The story of Joseph of Egypt is a prophetical parallel of the life of Christ and much can be learned in studying the life of this amazing man.***

The story describes the young Joseph being falsely accused and cast into the king's prison. While in prison, he is given charge over all of the other prisoners by the keeper of the prison, "*And the keeper of the prison committed to Joseph's hand all*

---

[37] It is interesting to note here that Joseph Smith uses the word *paradise* in his own additions to Matthew 27, yet he challenges its use by the KJV translators in the Luke rendering of the same account. *Indeed he challenges the validity of the word's use in this context at all.* More will be said about this later.

*the prisoners that were in the prison; and whatsoever they did there, he was the overseer of it*" (JST Genesis 39:22).

Throughout the 40th chapter of Genesis, we are told that Joseph ministers to his fellow prisoners. Anciently, dreams were thought to have been given by God, and one who could interpret them, received that gift from God as well. When Joseph interpreted dreams, he literally *spoke* for God, by which the story confirms that Joseph functioned as a prophet testifying of his Father in Heaven.

Not to be lost on us is an important detail about the divinely appointed ministry of Joseph. We are told of two particular individuals the young Joseph ministers to while in prison, the Pharaoh's own butler and baker. They are significant here in detail; the butler was eventually to be found innocent and restored to his position in Pharaoh's household (restored to the King), and the baker was to be ultimately condemned and sentenced to death (cast into outer darkness). By this telling of the story, we are to understand that Joseph (while in prison) ministered, spoke with and prophesied to both the innocent *and* the guilty.

Through the testimonial writing of prophets and apostles over a period of thousands of years, the Bible clearly foretold, and later records, Christ's entry into Spirit Prison and his ministry to the Souls there. We are taught of a Spirit World inhabited by the Just and the Unjust, of redemption and deliverance, of repentance and change. Jesus Christ Himself declared His imminent entry and that of the repentant thief into this prison, the Same Spirit World He called paradise.

## *The Spirit World of the Book of Mormon*

It is perhaps not a coincidence that without doubt the most detailed understanding of the Spirit World to be found in the Book of Mormon is another 40th chapter. In its pages, Alma provides us with the most clear and concise image of the "*state of the soul between death and the resurrection*" (Alma 40:11). Alma's sweeping discourse provides a dynamic view of this *world of Spirits*. In only four verses, Alma provides us with a glimpse of this world that we all enter immediately following our Temporal death:

> *"Now, concerning the state of the soul between death and the resurrection--Behold, it has been made known unto me by an angel, that the spirits of all men, as soon as they are departed from this mortal body, yea, the spirits of all men, whether they be good or evil, are taken home to that God who gave them life. And then shall it come to pass, that the spirits of those who are righteous are received into a state of happiness, which is called paradise, a state of rest, a state of peace, where they shall rest from all their troubles and from all care, and sorrow. And then shall it come to pass, that the spirits of the wicked, yea, who are evil--for behold, they have no part nor portion of the Spirit of the Lord; for behold, they chose evil works rather than good; therefore the spirit of the devil did enter into them, and take possession of their house--and these shall be cast out into outer darkness; there shall be weeping, and wailing, and gnashing of teeth, and this because of their own iniquity, being led captive by the will of the devil. Now this is the state of the souls of the wicked, yea, in darkness, and a state of awful, fearful looking for the fiery indignation of the*

*wrath of God upon them; thus they remain in this state, as well as the righteous in paradise, until the time of their resurrection." (Alma 40:11-14)*

At the outset, Alma reveals that his understanding was revealed to him through Heavenly ministration. Alma says that at the point of death, "*the spirits of all men, whether they be good or evil, are taken home to that God who gave them life*" (v11) (*Emphasis my own*). Alma goes on to declare that the Spirits of the righteous are received into "*a state of happiness, ...a state of rest, a state of peace, ...*" (v12) Alma calls these "*states*," paradise.

Further Alma describes the fates of the "*spirits of the wicked*." Alma says, "*these shall be cast into outer darkness*." Alma here defines the wicked in very descriptive terms; he says they are "*...evil — for behold they have no part nor portion of the Spirit of the Lord; for behold they chose evil works rather than good; therefore the spirit of the devil did enter into them, and take possession of their house...*" (v13) These wicked, as described by Alma, are the same sons of perdition described in D&C 76[38] and elsewhere.

Sweeping as this discourse of Alma may be, it is not comprehensive. Alma described the states and conditions of those Spirits on the extremes. He did not address at all the vast majority of those in the middle. No mention is made of those who died without the law or the knowledge of Jesus Christ. These passages in Alma do not address Christians who die in an unrepentant state for sins of omission, in varied states of less than complete obedience. Surely these Souls are in the Spirit World in vast numbers. Also not mentioned, are those who were disobedient in the days of Noah; they are those of whom Peter spoke. When Alma delivered this discourse to his son Corianton, these Souls were in the Spirit World yet to be visited.

Alma provides us with the most detailed scriptural look of this Spirit World. Alma tells us that, the Spirits of all men, both good and evil go to the same place, home; that the righteous obtain a "state" of peace, happiness and rest. Alma tells of the fate of the sons of perdition, of their being cast into outer darkness. In the end, however, we find that the 40th chapter of Alma details only the bookends on the shelf and leaves us looking elsewhere to read the volumes in between. Some of these volumes have been read already in the passages we have seen from the Bible. Alma here makes no attempt to provide us with a complete picture. From reading the rest of the standard works, we know that the picture is much broader. In all, it presents a panoramic view that is overwhelming in scope.

---

[38] See – Perdition, *A Special Place in Hell*, pg. 366 in this work.

## *The Spirit World of the Doctrine and Covenants*

- Section 76

Doctrine and Covenants Section 76 encompasses the Prophet Joseph's grand revelation of the Kingdom of Heaven. Under the influence of the Holy Spirit, Joseph sees the Three Degrees of Glory and of all those who dwell there. The Spirit testifies to Joseph of the states of the Souls who inhabit the three Kingdoms and of their varied accomplishments, experiences, and activities in mortality. Revelatory to our discussion here, the Prophet also describes their activities and dwelling in the intervening Spirit World; specifically, Verses 72-74 speak of those who were in *Spirit Prison* and clearly testify of them being visited by the Son, confirming the account given in 1 Peter 3.

> *"Behold these are they who died without the law; and also they who are the spirits of men kept in prison whom the Son visited, and preached the gospel to them that they may be judged according to men in the flesh; who received not the testimony of Jesus in the flesh, but afterwards received it."(D&C 76:72-74)*

- Section 138

In the vision recorded as Section 138 of the Doctrine and Covenants, the Prophet Joseph's Grand Nephew, Joseph F. Smith, saw a dynamic realm consisting of the Just and the Unjust. Emphasizing the continuity of eternal progression and the function of the Spirit World as an equalizing element in the varied histories of mankind, President Smith viewed the establishment and work of a great missionary effort carried out in the Spirit World. President Smith saw the efforts of righteous Spirit inhabitants in this cause.

As with Alma, President Smith explains that the source of this vision is the ministry of the Spirit of the LORD, and he records that he saw, "*the hosts of the dead, both small and great*" (D&C 138:11).

President Smith first details his view of the just; he begins: "*And there were gathered together in one place an innumerable company of the spirits of the just, who had been faithful in the testimony of Jesus while they lived in mortality; and who had offered sacrifice in the similitude of the great sacrifice of the Son of God, and had suffered tribulation in their Redeemer's name. All these had departed the mortal life, firm in the hope of a glorious resurrection, through the grace of God the Father and his Only Begotten Son, Jesus Christ*" (D&C 138:12-14).

As the vision progressed, President Smith saw that the Son of God appeared to this group of righteous and preached to them the "*everlasting gospel*" (v19). This is the event foretold in the Old Testament and later reported of by Peter: "*While this great multitude waited and conversed, rejoicing in the hour of their deliverance from the chains of death, the Son of God appeared, declaring liberty to the captives who had been faithful; And there he preached to them the everlasting gospel, the doctrine of the resurrection an the redemption of mankind from the fall, and from individual sins on conditions of repentance*" (D&C 138:18, 19).

President Smith also viewed within the Spirit World those who were ungodly, the wicked, the defiled and the rebellious. The vision continues: "*But unto the wicked he did not go, and among the ungodly and the unrepentant who had defiled themselves while in the flesh, his voice was not raised; Neither did the rebellious who rejected the testimonies and the warnings of the ancient prophets behold his presence, nor look upon his face. Where these were, darkness reigned, but among the righteous there was peace*" (D&C 138:20-22). [39]

It is clear the Doctrine and Covenants confirm an understanding of a very dynamic Spirit World, home to the godly and the ungodly – a place that can properly be described as both *Paradise* and *Prison*. We, however, through these passages, know still precious little about the Spirit World, of much of the inhabitants there and their interactions. Is *Paradise* really separate and set apart from *Prison*? Are they the same or different places? Is there a gulf dividing Spirits who are worthy and those who are rebellious and ungodly? Where in this Heavenly precinct can darkness reign? For these answers, we will not be able to turn to the pages of Scripture, for there they are not addressed. However, we can, and should, look at the teachings of the Latter-day founding prophets for what light they may shine on this matter.

## *Latter-Day Teaching*

Hierophantic teaching continues to express the dichotomy of terms in describing the Spirit World. As was the case with both Latter-day and ancient Scripture, the teachings of the Prophets alternately use *Paradise* and *Prison*. Furthering the confusion, as was mentioned earlier, despite his own use of the term, the Prophet

---

[39] Here we are told that the Son of God did not preach to the ungodly and unrepentant; we are specifically told that the rebellious and those that rejected the teachings of the prophets did not even see his face. At first glance, President Smith clearly contradicts the passages of 1 Peter 3:18-20 that tell us Jesus personally went and ministered to the *less than righteous* in *spirit prison*. Section 138 details the Savior preaching *only* to the righteous, but both the Biblical account and Section 76 clearly state otherwise. This *seeming* contradiction is not real; it is merely an illusion like the apparent bending of objects as they penetrate the surface of a calm pool. On closer examination, we find there is no contradiction at all.

Simply put, President Smith's vision, as it is recorded, is as incomplete a picture of the Spirit World as was Alma's discourse. He too provides us with only details of the bookends. We see the righteous, the really righteous, a grouping comprised of the very elite of God. They are not only those that are deemed faithful in their testimonies of Jesus Christ, but they are they who offered sacrifice in the similitude of the sacrifice of the Son of God. They are they who suffered tribulation in the name of the Son (refer to Matthew 10:22 for the promise made to these Souls).

Provided in contrast are the wicked; portrayed here are the same sons of perdition we saw in Alma's discourse. To these President Smith said, Jesus "*could not go personally, because of their rebellion and transgression.*" Missing in this picture are the same multitudes from the middle left out of the 40th chapter of Alma; those who died without the law or the knowledge of Jesus Christ, Christians who may not have endured the extreme tests of faith faced by those most righteous, and of course the group whom we are told were disobedient in the days of Noah.

Joseph Smith challenged the use of the word paradise in the context of the Savior's destination along with the thief on the cross on the day of their crucifixion as told in the Luke account. Of Luke's telling, the Prophet Joseph said:

> *"I will say something about the spirits in prison. There has been much said by modern divines about the words of Jesus (when on the cross) to the thief, saying, "This day shalt thou be with me in paradise." King James' translators make it out to say paradise. But what is paradise? It is a modern word: it does not answer at all to the original word that Jesus made use of. Find the original of the word paradise. You may as easily find a needle in a haymow. Here is a chance for battle, ye learned men. There is nothing in the original word in Greek from which this was taken that signifies paradise;* [40] *but it was -- This day thou shalt be with me in the world of spirits: then I will teach you all about it and answer your inquiries. And Peter says he went and preached to the world of spirits (spirits in prison, 1 Peter, 3rd chap. 19th verse), so that they who would receive it could have it answered by proxy by those who live on the earth, etc." (Teachings of the Prophet Joseph Smith, Section Six 1843-44 pg. 309)*

Additionally, the Prophet clarifies in the following: "*Hades, the Greek, or Sheol, the Hebrew, these two significations mean a world of spirits. Hades, Sheol, paradise, spirits in prison, are all one: it is a world of spirits*" (*Teachings of the Prophet Joseph Smith*, Section Six 1843-44, pg. 310).

It is interesting to note the word, paradise is only used 11 times in the standard works in total, *all but three* of those uses are in Latter-day Scripture, and all but one of those is directly attributable to the hand and mind of Joseph Smith.

It is clear from an analysis of all the Scripture and the Prophet's own words and teachings that paradise can be used to denote an un-bounded locale, wholly within the Spirit World where the righteous are comfortable and tend to gather. Alma defines it as a "*state,*" *perhaps of mind*, rather than a location or physical segregation. In fact, Alma states that at death all, good and evil, go to the same home. Paradise then can draw its definition as much from the presence and general goodwill felt among those that gather there as it can from any geometrical, geographical or supernatural physical qualifier. We will prefer this understanding here, objecting to any interpretation that presents *Paradise* as having boundaries distinct from those of the *Prison*.

A view presenting a segregated Spirit World would easily allow for an image of the Savior visiting one portion of it and shunning another, but the Latter-day prophets

[40] In the purest, technical sense, this statement by the prophet is wrong, and any "*learned*" man seeking battle on the point he makes here would be carried off the field mortally injured. The Greek word used by Luke is in fact literally, paradise, *paradeiso* (Παραδείσῳ). There really cannot be found a more literal translation of an ancient word into any modern language than is found here. Anciently, and today, this word describes a grand enclosed pleasure-ground in the upper regions of the heavens. The prophets philological inaccuracy aside, his statement here is used to cement our understanding that the Spirit World is properly understood to be accurately described as both a *Paradise* and a *Prison*.

shatter any expectation of such a divided world. Joseph Smith and Brigham Young provide detailed and revealing looks at a very interactive Spirit World.

Of the inclusiveness found in the Spirit World, the Prophet Joseph Smith said clearly, starting from where we left off before: "*Hades, the Greek, or Sheol, the Hebrew, these two significations mean a world of spirits. Hades, Sheol, paradise, spirits in prison, are all one: it is a world of spirits. The righteous and the wicked all go to the same world of spirits until the resurrection. 'I do not think so,' says one. If you will go to my house any time, I will take my lexicon and prove it to you. The great misery of departed spirits in the world of spirits, where they go after death, is to know that they come short of the glory that others enjoy and that they might have enjoyed themselves, and they are their own accusers*" (Teachings of the Prophet Joseph Smith, Section Six 1843-44 pg. 310). The Prophet details here the "*state*" of the wicked mind as opposed the "*state*" of the righteous mind as Alma described it. We can plainly see the Prophet's concept that the separation of Spirits conforms to our previous assertion that the differences between those in *Paradise* and in *Prison* in the Spirit World are based upon their "*state*" of mind.

With directness like none other, Brigham Young said, "*When you lay down this tabernacle, where are you going? Are you going into Abraham's bosom? No, not anywhere nigh there but into the spirit world. Where is the spirit world? It is right here. Do the good and evil spirits go together? Yes, they do. Do they both inhabit one kingdom? Yes, they do*" (Discourses of Brigham Young, pg. 376). President Young further goes on to emphasize:

> *"The Prophet lays down his body, he lays down his life, and his spirit goes to the world of spirits; the persecutor of the Prophet dies, and he goes to Hades; they both go to one place, and they are not to be separated yet. If we go back to our mother country, the States, we find there the righteous, and we there find the wicked; if we go to California, we there find the righteous and the wicked; all dwelling together; and when we go beyond this veil, and leave our bodies which were taken from mother earth, and which must return, our spirits will pass beyond the veil; we go where both Saints and sinners go; they all go to one place." (Discourses of Brigham Young, pg. 377)*

Saints and sinners in one place? Yes, that is exactly what these teachings detail. To further emphasize the intermingling attendant to the Spirit World, consider the following from the Discourses of Brigham Young:

> *"The spirits that dwell in these tabernacles on this earth, when they leave them go directly into this world of spirits. What! A congregated mass of inhabitants there in spirit, mingling with each other, as they do here? Yes, brethren, they are there together, and if they associate together, and collect together, in clans and in societies as they do here, it is their privilege. No doubt they yet, more or less, see, hear, converse and have to do with each other, both good and bad. If the Elders of Israel in these latter times go and preach to the spirits in prison, they associate with them, precisely as our Elders associate with the wicked in the flesh, when they go to preach to them." (Discourses of Brigham Young, pg. 378)*

The concept of a single Spirit World inhabited by righteous and unrighteous alike, freely associating with one another just as they do here, is clearly made manifest. There can be no doubt about what these passages say and mean.

## *The Spirit World – a Path to Continued Progression*

We have seen in this section that the Spirit World supports a social dynamic of Spirits not unlike what we experience in Temporality on Earth, our *states* there are a reflection of our state of mind, not the receipt of reward or punishment, *for the judgement has not yet occurred.* At the outset of this section, we said the Spirit World is a place of equalizing; it affords a second chance to many. It should not be forgotten that the *primary* role of the Spirit World is that of providing a *first* chance to *most*.

***The Spirit World is a place of equalizing; it affords a second chance to many. It should not be forgotten that the primary role of the Spirit World is that of providing a first chance to most.***

In this view, Latter-day doctrine is unique. The Spirit World allows us to answer those questions posed by the concepts of a chosen people, Calvin's *determinism* and *effective call.* Because of the Spirit World and it being a place of continued progression, we can live in a world that displays the appearance of selected election by our Father in Heaven, a world where there are a chosen people and whole masses that seem overlooked. The continuum of our existence affords many venues for each of the sons and daughters of God to achieve and move forward on differing and unique schedules. Individual as our steps may be when we look forward, toward God our Father, we see that there is just one path we all walk upon, some slower, some faster than others, but all with the same end in sight.

On the subject of our continued progression in the Spirit World, the Prophet Joseph Smith said:

> *"The situation of the Christian nations after death, is a subject that has called forth all the wisdom and talent of the philosopher and the divine, and it is an opinion which is generally received, that the destiny of man is irretrievably fixed at his death, and that he is made either eternally happy, or eternally miserable; that if a man dies without a knowledge of God, he must be eternally damned, without any mitigation of his punishment, alleviation of his pain, or the most latent hope of a deliverance while endless ages shall roll along. However orthodox this principle may be, we shall find that it is at variance with the testimony of Holy Writ, for our Savior says, that all manner of sin and blasphemy shall be forgiven men wherewith they shall blaspheme;*

*but the blasphemy against the Holy Ghost shall not be forgiven, neither in this world, nor in the world to come, evidently showing that there are sins which may be forgiven in the world to come, although the sin of blasphemy [against the Holy Ghost] cannot be forgiven. Peter, also, in speaking concerning our Savior, says that 'He went and preached unto the spirits in prison, which sometimes were disobedient, when once the long-suffering of God waited in the days of Noah.' (1 Peter 3:19, 20)*

*Here then we have an account of our Savior preaching to the spirits in prison, to spirits that had been imprisoned from the days of Noah; and what did He preach to them? That they were to stay there? Certainly not! Let His own declaration testify. 'He hath sent me to heal the brokenhearted, to preach deliverance to the captives, and recovering of sight of the blind, to set at liberty them that are bruised.' (Luke 4:18). Isaiah has it – 'To bring out the prisoners from the prison, and them that sit in darkness from the prison house.' (Isaiah 42:7). It is very evident from this that he not only went to preach to them, but to deliver, or bring them out of the prison house. Isaiah, in testifying concerning the calamities that will overtake the inhabitants of the earth, says, 'The earth shall reel to and fro like a drunkard, and shall be removed like a cottage; and the transgression thereof shall be heavy upon it; and it shall fall and not rise again. And it shall come to pass in that day, that the Lord shall punish the host of the high ones that are on high, and the kings of the earth upon the earth. And they shall be gathered together, as prisoners are gathered in the pit, and shall be shut up in the prison, and after many days shall they be visited.' Thus we find that God will deal with all the human family equally, and that as the antediluvians had their day of visitation, so will those characters referred to by Isaiah, have their time of visitation and deliverance; after having been many days in prison." (Teachings of the Prophet Joseph Smith, Section Four 1839-42 pg. 218)*

*"When speaking about the blessings pertaining to the Gospel, and the consequences connected with disobedience to the requirements, we are frequently asked the question, what has become of our fathers? Will they all be damned for not obeying the Gospel, when they never heard it? Certainly not. But they will possess the same privilege that we here enjoy, through the medium of the everlasting Priesthood, which not only administers on earth, but also in heaven, and the wise dispensations of the great Jehovah; hence those characters referred to by Isaiah will be visited by the Priesthood, and come out of their prison upon the same principle as those who were disobedient in the days of Noah were visited by our Savior [who possessed the everlasting Melchizedek Priesthood] and had the Gospel preached to them by Him in prison; and in order that they might fulfill all the requisitions of God, living friends were baptized for their dead friends, and thus fulfilled the requirement of God, which says, 'Except a man be born of water and of the Spirit, he cannot enter into the kingdom of God,' they were baptized of course, not for themselves, but for their dead." (Teachings of the Prophet Joseph Smith, Section Four 1839-42 pg. 221)*

Brigham Young spoke of the opening of the door of Salvation to those who have gone before us:

*"Jesus was the first man that ever went to preach to the spirits in prison, holding the keys of the Gospel of salvation to them. Those keys were delivered to him in the day and hour that he went into the spirit world, and with them he*

*opened the door of salvation to the spirits in prison." (Discourses of Brigham Young, pg. 378)*

## *Developing Our View of the Spirit World*

It is clear from the Scripture, both ancient and modern, and all of the Latter-day sources quoted here, that the Spirit World is a place all of us return to at the time of our death, that it is cohabitated by the righteous and the unrighteous. From the statements of Joseph Smith Jr. and Brigham Young, we see that there is open association and commingling of the righteous and unrighteous, the Spirit World is not compartmentalized or separated into pavilions of *Paradise* and halls of *Prison*.

It is the work and Glory of God to bring to pass the eternal life of ALL of mankind. "*...Jesus will bring forth, by his own redemption, every son and daughter of Adam... it is true that every person who does not sin away the day of grace, and become an angel to the Devil, will be brought forth to inherit a kingdom of glory*" (*Discourses of Brigham Young*, pg. 382). The Spirit World is designed and exists with this purpose of God and the saving grace of the redemption in mind. Our journey towards perfection does not stop with the last beat of our heart here on Earth, but continues forward into eternity. The Spirit World is but just a first step along the path forward from here.

The Bible clearly states that the Messiah was to visit Spirit Prison, gain authority by doing so and, in fact, did visit there while His body was still in the tomb. Back near the beginning of the Biblical record, we are confronted with the vivid detail of the prophetically significant life of Joseph of Egypt. In that story, we see that Joseph clearly ministers to the innocent and the guilty.

Lastly, we are left with the apparently contradictory vision of Joseph F. Smith. But a thoughtful review of his inspired vision proves that it is indeed in harmony with all else. In section 138, we are introduced to the same wicked few who are cast into outer darkness, those to whom the Son "*did not go*" for, "*Where these were, darkness reigned.*"

It may be helpful to redraw the common illustration of the divided Spirit World presenting it in a view in harmony with the teachings of Joseph Smith Jr. and Brigham Young and shown in proper context with the separate Outer darkness (*Figure 9-2*).

As we understand it here in this work, there is no division, chasm, or gulf that somehow limited the Savior's ministry. In this image, we can see that all of those who were in the Spirit World (*Paradise* or *Prison*) could receive the ministry of the Savior and receive directly of His mercy to a degree that is dependent directly on their personal worthiness and proximity to the light. In the cold and dark corners of the Spirit World, near the borders of Outer Darkness, those that sought obscurity from the presence of the Son could find a hiding place. In harmony with the vision detailed in Section 138, we can see where the Savior "*did not go*." More importantly, we can see that those to *whom* the Savior "*did not go,*" were those who *chose* to avoid His presence.

The Spirit World is indeed a realm of continued progression; within its confines many minds are changed and repentance is possible. For most, it is a chance to hear the Gospel message for the first time and act upon the change the receipt of the *Good News* brings to their Spirit. We can see the wonder, perfection and harmony of the true Gospel of Jesus Christ. We can understand the infinite mercy of the Son and God and His willingness to come unto all who will receive Him. Through this understanding, we can see that there is no need to do Him honor by entertaining false notions of the purity and sanctity of the deity of Christ as preventing Him from contacting the perceived lowest in the Kingdom. He stands only taller at the bar as our advocate with the Father by His willingness to enter the "*lowest room*," "*For whosoever exalteth himself shall be abased; and he that humbleth himself shall be exalted*" (Luke 14:10-11). Our love for the Savior grows and our respect for His prophets and the Scriptures increases.

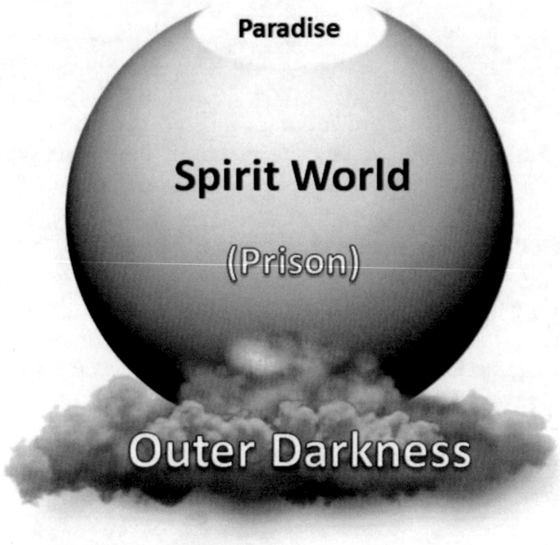

Figure 9-2 *The Spirit World.*

- Outer Darkness

Alma makes reference to "*outer darkness*" (Alma 40:13) and uses it in a context that supposes it to be remote and separate from the location of the home of "*the spirits of all men*" in the presence of "*God who gave them life*" (Alma 40:11). In that same verse, Alma describes outer darkness as a place where the, "*spirits of the wicked...shall be cast.*" Outer darkness is separate from the Spirit World; it is Perdition inhabited by those "*sons of Perdition*," we spoke of earlier.

The Endowment narrative provides us with a context which informs that Satan does not have power and or influence in the Spirit World. Our depiction of the Spirit World illustrates the shadow of Outer Darkness falling upon it, those who are wicked, seek the cover of the shadow's dark places (*See also, Figure 7-3, Page 289*).

Within the Spirit World the wicked, the rebellious, the ungodly and the unjust are there without the influence of the adversary. They are left exposed, without excuse; they find themselves there exactly who they became in this Temporal world, influenced by their past actions and hardened of heart. It is this "*state*" of mind that imprisons much more effectively than any wall, door or lock ever could.

# PART 4

# THE FAMILY OF GOD

*"...and the Word of the Lord created man in His Likeness*
*in the likeness of the presence of the Lord He created him,*
*the male and his yoke-fellow He created them."*

*Targum Jerusalem*

## CHAPTER 10

# THE LIKENESS OF THE PRESENCE OF THE LORD

The Prophet Joseph Smith said: "*It is the first principle of the Gospel to know for a certainty the Character of God.*"[1] In other words, one of the foundation stones of the restored gospel is the knowledge of what kind of being God actually is. We not only need to *understand* what kind of being God is, but we must come to *know* God in a personal manner. In the same King Follet discourse from which we just quoted, the Prophet Joseph further stated, "*If any man does not know God, ...he will realize that he has not eternal life; for there can be eternal life on no other principle.*"[2] In His great high-priestly or Intercessory Prayer, the Savior confirmed that life eternal was to "*know ...the only true God, and Jesus Christ, whom [the Father] hast sent*" (John 17:3).

Our certain knowledge, the attainment of it, is a process that cannot be completed in short order; indeed it cannot be completed in one lifetime or the totality of many life times. What Joseph spoke of as the *first principle* is a thing not attainable by any of us while yet in the Temporal world:

> *"But the natural man receiveth not things of the spirit of God: for they are foolishness unto him: neither can he know them, for they are spiritually discerned." (1 Cor. 2:14*)

What the Apostle Paul meant here is that the things of the Physical and the Spiritual spheres are incomprehensible to things that are Temporal. We have discussed concepts at length in this book that have been complicated by our need to establish an understanding of a truly *physical* nature of things most commonly understood to be *spiritual*; it has been difficult to avoid confusion. Before we progress further in this work, let us seek again to clarify the issue.

---

[1] Teachings of the Prophet Joseph Smith, Salt Lake City, Deseret Book (1938) pg. 345.

[2] Ibid. pg. 344.

*Ghost*, *Spirit*, *Intelligence* are interchanged all too freely in our language and within our minds.[3] The same is true of the many related words, *vision*, *apparition* and *visitation* among them. We are told, and speak of, "*the Holy Spirit,*" "*the Holy Ghost,*" "*the Spirit of God,*" "*the Spirit of the Lord,*" "*the Spirit of Truth,*" "*the Second Comforter*" and "*the Light of Christ,*" often in the same breath. We frequently mix and blend the separate meanings found within these entities and concepts into an incoherent amalgam that becomes inseparable in our minds, in our communications and through that, within the minds of others.

Early in the last century, the First Presidency and the Quorum of the Twelve authorized the publication of course materials for the instruction of the Seventies in the Church. This five year study returned on separate occasions to the theme of our eternal intelligences and that of our Spirits. On this subject, the training manuals present the following from the Prophet Joseph Smith:

> *"I am dwelling on the immorality of the spirit of man. Is it logical to say that the intelligence of spirits is immortal, and yet that it (i.e. the intelligence) had a beginning? The intelligence of spirits had no beginning, neither will it have an end. That is good logic. That which has a beginning may have an end. There never was a time when there were no spirits, for they are co-equal (co-eternal) with our Father in heaven...I take my ring from my finger and liken it unto the mind of man—the immortal part, because it has no beginning. Suppose you cut it in two; then it has a beginning and an end; but join it again and it continues one eternal round. So with the spirit of man. As the Lord liveth, if it had a beginning it will have an end....Intelligence is eternal and exists upon a self-existent principle. It is a spirit from age to age and there is no creation about it... The first principles of a man are self-existent with God." Joseph Smith. (Mill. Star, Vol. 23, pg. 262.)*

Even a cursory reading of this quotation will reveal that the Prophet himself freely interchanges the words and separate concepts of "*Spirit*" with "*Intelligence.*" Note also the joining of the two separate concepts where he twice speaks of the "*intelligence of spirits.*" Through the Prophet's joining of these terms, he grammatically confirms his understanding of *separate meanings* for each. For if both Intelligence and Spirit were to have *equivalent meanings*, the Prophet's joining of them would result in a redundancy of his wording, as in, intelligent intelligences, and *spiritual spirits*. Recognizing, as we have done here, that the Prophet himself interchanged the meanings of Intelligence and Spirit, and also recognizing that the same confusion of terms was common, the author of the Seventy's Course on Theology, Elder B.H. Roberts, sought to clarify by including in the approved training this important teaching:

---

[3] For another treatment of this same issue, see the quotation of Elder B.H. Roberts embedded in footnote 14 on page 306 in this book.

> *"The difference, then, between "spirits" and "intelligences," as here used, is this: Spirits are uncreated intelligences inhabiting spiritual bodies; while "intelligences," pure and simple, are intelligent entities, but unembodied in either spirit bodies or bodies of flesh and bone."* [4]

Of the "*intelligences*" themselves, the Seventy's Course on Theology, goes on to explain:

> *"They are uncreated, self-existent; but let it be observed, in passing, that nothing is here said in relation to the form of these intelligent entities, nor anything as to their mode of existence."* [5]

If we are to accept the definition provided by Elder Roberts, and by-the-way, one that was reviewed and approved by the Quorum of the Twelve and the First Presidency, a Spirit is an *embodied* Intelligence, a compilation of a distinct body comprised of immortal spiritual matter and self-existent Intelligence. A "*Spirit*" then, conforms exactly to the type of entity described in both the Preamble verses and the Endowment narrative. In the next several sections, we will keep this understanding of *Spirit* in mind as we contemplate the Physical nature of Deity. We will consider that the personages of the Father, Son and the Holy Ghost are of this form in part or in whole.

---

[4] Seventy's Course In Theology, Fourth Year, The Atonement, Lesson 3, Intelligences and Spirits, B.H. Roberts, Deseret News Press 1907, page 14.
[5] Ibid.

The understanding provided in this book of an eternal Physical sphere leaves us face-to-face with the reality of a corporeal God. Joseph Smith saw separate and distinct manifestations of the Father and the Son in his vision that has come to be known as his *First*. Though it appears his understanding of what he actually saw took years to develop,[6] eventually either the prophet's own understanding of what he saw, or direct revelation itself, caused Joseph to record the following that today is included in our canon of standard works:

> *"The Father has a body of flesh and bones as tangible as a man's; the Son also; but the Holy Ghost has not a body of flesh and bones, but is a personage of Spirit. Were it not so, the Holy Ghost could not dwell in us." (D&C 130:22)*

This verse describes *tangible* bodies for both the Father and the Son. There are many other passages throughout ancient and modern Scripture that also testify of this concept. Within the Scriptures, anthropomorphic terms are commonly used to describe both God the Father and the Son (as Jehovah) before His incarnation. We prefaced this discussion with an understanding that a *Spirit* itself is embodied; we need to understand what that embodiment entails, what its form may be.

## *The Eternal Corporeal Nature of the Personages of God*

Early in the last century, the First Presidency, consisting of Joseph F. Smith, John R. Winder and Anthon H. Lund, issued a statement detailing the form of the Spirits of the sons and daughters of God: *"'God Created man in His own image' This is just as true of the spirit as it is of the body, which is only the clothing of the spirit, its complement; the two together constituting the soul. The spirit of man is in the form of a man, and the spirits of all creatures are in the likeness of their bodies. This was plainly taught by the Prophet Joseph Smith*":

> *"...that which is spiritual being in the likeness of that which is temporal: and that which is temporal in the likeness of that which is spiritual: the spirit of man in the likeness of his person, as also the spirit of the beast and every other creature which God had created." (D&C 77:2)*

The *form of man* is to be taken literally, in every describable way:

> *"The spirit of man consists of an organization of the elements of spiritual matter in the likeness and after the pattern of the fleshly tabernacle. It possesses, in fact all the organs and parts exactly corresponding to the outward tabernacle." (Key to the Science of Theology, Parley P. Pratt, pg. 79)*

---

[6] Records made of Joseph's telling of the story to many people over the course of many years include varying details and represent a progression in the young prophet's understanding of the sacred event. For a detailed study of the various recordings, see A Pillar of Light (Matthew B. Brown, Deseret Book, Salt Lake City, 2009).

Our "*Spirits*" then, the *Personages of Spirit* that we are, what has been described as *embodied Intelligence*, is tangible, a perfunctory structure, complete with organs and bone, emulating the fleshly, Temporal form in every detail. Directly to the point of our focus here is the fact that this tangible form, we are told, was created by God to be in His own image. We are told *God Created man in His own image,* so we fully expect the details of our spiritual (Physical) bodies to be those of the Father, Son and Holy Ghost as well. Our Spirits are identical in form to the Temporal bodies (*flesh, bone and blood*) we are clothed in as we live on Earth in this Universe. We can understand then that God the Father, the Son and *even* the Holy Ghost are tangibly embodied as *Personages of Spirit*. How then does the form of God differ from the body we each hold today?

The Physical Creation, as we have described it in this work and as told in the Endowment narrative, produced an immortal physical reality that exactly mirrors the constructs of the Temporal world. In Part 1 of this thesis, we described the Temporal sphere as one that literally is a subtraction of capability from the higher Physical sphere. Nothing new was created in Temporality except uncertainty. To create uncertainty, laws were established to set boundaries to confine the operation of the substance of Temporal matter (*bohu*), the Atonement was planned for and instituted to restore perfection, and the presence of God was withdrawn (*tzimtzum*) to allow the uncertainty to function providing for the unfettered *Free Will* of man.

The *corporeal nature of the "spirit of man*" described here is the state and form of Adam and Eve in the Garden, fully embodied in complete physical detail. This Physical state, replicating that of the Temporal, was not only held by our first parents but also extended to everything in the Physical sphere. This Tangible Physical state is the one held by the Father and the Son as detailed in D&C 130; "*The Father has a body of flesh and bones as tangible as a man's; the Son also*."

Before His incarnation, Christ had a Spirit body that shared common attributes and form with that of our own Spirits and that of the Holy Ghost. Long before he walked the plains of Galilee, He declared, "*this body which ye now behold is the body of my spirit*" (Ether 3:16). "*Body of my spirit*" is another way of stating an *embodied Spirit* as per the definition of Elder Roberts, or a "*personage of Spirit*" as revealed in D&C 130:22. However, when Christ appeared to His disciples after the resurrection in the closed room, He had what He described as a body of "*Flesh and Bones.*"

The full statement made by the Savior to His disciples that day, fuels the fire of those who seek to deny that Christ has a tangible, physical body. However, as it is translated or even as it was understood to all in that room that day does not preclude the same physicality for a Spirit as we have defined it here. Christ said, "*for a spirit hath not flesh and bones, as ye see me have*" (Luke 24:39), but it must be remembered that the context by which He was using the word "*Spirit*" pertained to His disciples cowering in fear, for they feared they were seeing a ghost or an apparition. The Savior was easing the fear of his followers.

After his resurrection, when Christ made His appearance in the closed room, the disciples learned, through demonstration, that a body of "*Flesh and Bone*" can have the outward appearance and all of the expected capabilities and qualities of a ghost;

the same attributes we ascribe to a *spiritual being*, a *personage of Spirit*. For to Peter, James, John and the rest of the disciples gathered there when they saw the resurrected Christ in His full *personage of Spirit*, a ghost, a spirit, is exactly what they thought they had seen. He declared His eternal physicality within the closed room to His disciples in the same descriptive way used when He appeared in bodily form to Joseph Smith in the Sacred Grove almost 1800 years later. To the disciples in the closed room and to Joseph Smith in a grove of trees, the Savior taught that things spiritual *are* tangible; they are not ghosts, phantoms, or apparitions.

It is generally assumed that a body of "*Flesh and Bones*" as held by Christ is a residual of His incarnation; it is commonly assumed that His state then reflects also what will be our state after the resurrection. Because of our entry into the Temporal sphere, all of us, including the Son, receive a permanency of the structure we wear in Temporality. The pairing of the form of the resurrected Savior with that of our own has been codified in Scripture: "*When the Savior shall appear we shall see him as he is. We shall see that he is a man like ourselves*" (D&C 130:1). But is *Flesh and Bone* strictly a quality of that which was once mortal? D&C 130:22 states that the Father *also* has a body of *Flesh and Bones*, and in early April 1844, in front of almost twenty thousand people, Joseph Smith spoke at the funeral for his friend King Follett and said:

> *"God himself was once as we are now, and is an exalted man, and sits enthroned in yonder heavens! ...If the veil were rent today,...and the great God ...was to make himself visible, ...I say, if you were to see him today, you would see him like a man in form- like yourselves in all the person, image, and very form as a man; for Adam was created in the very fashion, image and likeness of God, and received instruction from, and walked, talked and conversed with him, as one man talks and communes with another." (Teachings of the Prophet Joseph Smith, Joseph Fielding Smith, comp., Salt Lake City: Deseret Book, 1976, pg. 345)*

In February or March of 1833, over a decade before the funeral of King Follet, Zebedee Coltrin along with John Murdock and perhaps others, witnessed very anthropomorphic manifestations of the Father and the Son while in the presence of Joseph Smith at the School of the Prophets. Brother Coltrin recorded what had been seen as follows:

> *"Joseph having given instructions, and while engaged in silent prayer, kneeling... a personage walked through the room from East to west, and Joseph asked if we saw him. I saw him and supposed the others did, and Joseph answered that this was Jesus, the son of God, our older brother. Afterward, Joseph told us to resume our former position in prayer, which we did. Another person came through; He was surrounded as with a flame of fire. [I] experienced a sensation that it might destroy the tabernacle as it was a consuming fire of great brightness. The prophet Joseph said this was the Father of our Lord Jesus Christ. I saw him...*
>
> *He was surrounded as with a flame of fire, which was so brilliant that I could not discover anything else but his person. I saw his hands, his legs, his feet, his eyes, nose mouth, head and body in the shape and form of a perfect man. He sat in a chair as a man would sit in a chair, but This appearance was so*

> *grand and overwhelming that it seemed that I should melt down in His presence, and the sensation was so powerful that it thrilled through my whole system and I felt it in the marrow of my bones. The Prophet Joseph said: 'Brethren, now you are prepared to be apostles of Jesus Christ, for you have seen both the Father and the Son and know that They exist and that They are two separate personages.'" (As provided in The Doctrine of Divine Embodiment: Restoration, Judeo-Christian, and Philosophical Perspectives, David L Paulsen, BYU studies 35, no. 4 1995-96.)*

The manifestations of the Father and the Son recorded here are consistent with the descriptions provided in the several accounts of the First Vison offered by Joseph Smith in which the young prophet referred to the Father and the Son as personages. Because of the First Vision (yet another decade earlier than the King Follet discourse), Joseph fully understood a meaning of personage as some person who was embodied, and his accounts of his initial encounter with the Father and the Son intended to communicate precisely just that. It will be noted that in stark contrast, Joseph does not use the word personage to describe Satan. Though Joseph describes Satan's actions as being very tangible and physical, he was "*seized upon*" and Satan exhibited "*influence over me as to bind my tongue.*" Note that the Prophet does not use terms that indicate an observable, bodily being. Instead he refers to the adversarial encounter as being a violent struggle with the "*enemy*" who was "*some power*" and "*some actual being from the unseen world.*"

A careful reading will reveal that Joseph emphasizes the tangibility of Satan, it was not a *feeling* that overcame him or some form of *apparition*, no, he was "*seized upon*" by an "*unseen*," "*actual being*," though of an order and class much different in category and kind than that observed and experienced with both the Father and the Son.

To these definitive statements of the corporality of the Father, we must add the following statement of the Savior to the Samaritan woman at Jacob's well: "*God is a Spirit: and they that worship him must worship him in spirit and in truth*" (John 4:24).

We understand God to be both body and Spirit just as we are. In fact, we understand Spirit itself to be an embodiment composed of eternal Physical materials. We should not be surprised that knowledge of this truth was not reserved for the Latter-day; the early church fathers taught this concept as well. Tertullian in the opening years of the third century, penned words that sound as though they were spoken by a Latter-day Prophet, he said:

> "*I assert that nothing insubstantial and void could have proceeded from God, since he himself is not insubstantial and void from whom it issued; nor can that lack substance which came forth from so great a substance and created such great substances; for he created all things which were made through him. 'Without him no thing was made;' but can he himself be nothing? Is it conceivable that an insubstantial being should have created concrete things; a void, solids; an incorporeal, bodies?...For who will deny that God is a body, although 'God is spirit?' For a spirit is a body, of its own kind, in its own form. But these 'invisible things,' whatever they are, have their own body and*

*their own form in the presence of God by means of which they are visible to God alone." (Adversus Praxean, 7)*[7]

Here John 4:24 is interpreted by one who understands, as Joseph Smith did, that God is an embodied Spirit. It must be noted, however, that although in this particular passage Tertullian is directly referencing the personage of the Son, understood in the context of his entire works is the fact that it can also refer to the Father; for, *how can an incorporeal Father beget a corporeal Son?* As is clearly shown in this passage, the early church fathers believed in a God who has a body that, in Latter-day terms, is "*as tangible as man's.*"

All the passages and statements we have examined here combine to describe an eternal Father, in possession of a tangible body of *Flesh and Bone*; the same Father whom the Savior proclaimed was Spirit. That the Father is both a tangible, immortal, physically embodied being *and* a being of Spirit, is entirely consistent with our understanding of what we in fact are, ourselves a duality of body and Spirit.

Christ was completely aware of the duality in us, and spoke of it often. Here, to the Samaritan woman at the well, He addressed it as well, when He spoke of Earthly worshipers who must do so in Spirit.

The *Flesh and Bone* of the Father's form cannot be linked directly to any transitional event in the history recorded in our Scriptures, for no such event exists and therefore must be assumed to be eternal.

***… there never was a time when Christ was not, or a time when he was not embodied; He never was logos asarkos.***

The Son is likewise embodied and has been from a time before anything our Scripture records; however, we are told that the Son was once created and His embodiment must extend from no earlier than that event. In respect to the concept of time, in the eternities there never was a time when Christ was not or a time when he was not embodied; He never was *logos asarkos* (*the Word without form or body*).

It is not made clear at what point and through which process the Savior's form became one that could be properly identified as "*Flesh and Bone*," but it is now. It has become acceptable to assume that the *Flesh and Bone* component of the Savior's current form is new to Him and residual to His incarnation, but this is not certain, nor can it be proved or disproved by any passage of Scripture. Though we cannot

---

[7] The Early Christian Fathers, Henry Bettenson (Oxford University Press, New York, 1956) pgs. 104-105.

say that the Savior's pre-incarnate body was of *Flesh and Bone*, we do know that as Jehovah, He appeared in a tangible *body of my [his] spirit*" (Ether 3:16) before His mortal ministry to the brother of Jared, and to Abraham, to Jacob and Moses. Christ and the Father both appeared in tangible form as they walked and conversed with Adam in the Garden within the Physical sphere at a time before the creation of the Temporal Universe today.

Finally, we must note here our understanding that though the Father and the Son indeed possess bodies of *Flesh and Bone*, we consider those bodies to perhaps not be continually existent. The Father and the Son may inhabit *Flesh and Bone* as a mode of their existence but are superior to any constraints that may be associated with that specific attribute. By this it is understood that the Father and the Son are *Intelligence*, *Spirit*, and *Flesh and Bone*; these attributes are separate and distinct, each one adding to the other in tiered order, each one an enhancement to the previous. I propose here that the Father and the Son, and we as well in the full realization of our future exalted state, may don *Flesh and Bone* then shed it, only to pick it up again when it is desired. In this understanding, those who mastered their independent Intelligence were added upon with the blessings and enhancements attendant to the embodiment of Spirit in the Physical sphere. The addition of immortal *Flesh and Bone* was yet still another enhancement adding to the attributes of the Spirit and in its immortal resurrected state is a unique attainment of those deified.

In every mode, and in every instance cited here, the Father and the Son were spiritual in nature in that they were embodied Intelligences, conforming to the definition of a "*Spirit*" put forth by Elder Roberts and detailed by us. Both the Father and Son (*the Son, through his incarnation for sure, and perhaps before, and at all times in the case of the Father*) also meet the defined requirements for an Eternal Soul.

## *The Holy Ghost*

> *"The Father has a body of flesh and bones as tangible as a man's; the Son also; but the Holy Ghost has not a body of flesh and bones, but is a personage of Spirit. Were it not so, the Holy Ghost could not dwell in us." (D&C 130:22)*

The Doctrine and Covenants here reveals the Holy Ghost, the third member of the Godhead to be a "*personage of Spirit*"; further stating that the Holy Ghost as a *personage* has the ability to "*dwell in us.*" We shall see that this *personage* has a body, but without *Flesh and Bone*. It is different than that possessed by the Father and the Son today; "*The Holy Spirit is the most immaterial of the divine persons, and yet even the Holy Spirit takes a personal form.*"[8]

[8] Stephen H. Webb, in his work Jesus Christ, Eternal God: Heavenly Flesh and the Metaphysics of Matter, Oxford University Press, 2012. Webb here correctly and succinctly grasps the essence of what this passage says regarding the Holy Ghost.

This verse describes *tangible* bodies for the Father and the Son. There are many other passages throughout both ancient and modern Scripture that also testify of this. Anthropomorphic terms are commonly used to describe God and as well as the Son both before and after his incarnation as Jehovah; we will not seek them out here. What is of significance for our discussion now is that this verse has been used by Latter-day Saints to support a hierophantic doctrine which states: *it is a necessary and qualifying attribute of Godhood to have a tangible, physical body.*[9] This is held to be true despite this same verse detailing that the third personage of the Godhead, of necessity fully a God, is a *personage of Spirit*, inferring at first glance that the Holy Ghost is not embodied. This presents a contradictory, or at best, confusing picture to many, but it need not be so. We have seen that a "*Spirit*" *is* an embodied Intelligence. The Holy Ghost then, being a *personage of Spirit*, would have a body as well. This is emphasized in the following quote from Joseph Fielding Smith:

> *"The Holy Ghost is the third member of the Godhead. He is a spirit in the form of a man... The Holy Ghost is a personage of Spirit, and has a spirit body only... As a Spirit Personage the Holy Ghost has size and dimensions. He does not fill the immensity of space, and cannot be everywhere present in person at the same time. He is also called the Holy Spirit, the Spirit of God, the Spirit of the Lord, The Spirit of Truth, and the comforter." (Doctrines of Salvation, Joseph Fielding Smith, vol. 1, pg. 38)*

Clearly then, the Holy Ghost is embodied. The Holy Ghost "*has a spirit body*" that "*has size and dimensions.*" The Holy Ghost is a "*spirit in the form of a man...*"[10]

---

[9] "*Three glorified, exalted, and perfected personages comprise the Godhead or supreme presidency of the universe... They are: God the Father, God the Son; God the Holy Ghost...Though each God in the Godhead is a personage, separate and distinct from each of the others...*" (Mormon Doctrine, Bruce R. McConkie, pg. 319). "*Everlasting covenant was made between three personages before the organization of this earth, and relates to their dispensation of things to men on the earth; these personages, according to Abraham's record, are called God the first, the Creator; God the second, the Redeemer; and God the third, the witness or Testator*" (Teachings of the Prophet Joseph Smith, *Joseph Fielding Smith, comp., Salt Lake City: Deseret Book, 1976*, pg. 190).

[10] The Holy Ghost is referred to as a "*Man*" here. Latter-day concepts of the Holy Ghost associate male gender with this member of the Godhead (though that concept is not to be found in modern scripture). We understand here, in this work, that gender is an essential characteristic of eternal identity and purpose. When our Intelligences were subsequently embodied to become Spirits, we were created as either male or female. It is, therefore, appropriate and true that the Holy Ghost (if a *personage of Spirit*) has gender as well. However, all is not as it seems, certainly it has not always been so.

The Holy Ghost is discussed in 43 verses within the Book of Mormon. Those verses record the Holy Ghost: is a gift given by the Father (3 Nephi 21:2), is power which both influences and is held (1 Nephi 10:17), and is to find a place in the heart of the faithful (Mor. 7:32). Alma remains gender neutral using "*it*" in referencing the Holy Ghost; he records, "*if ye deny the Holy Ghost when it once has had a place in you...*" (Alma 39:6). In fact, nowhere in the Book of Mormon is the Holy Ghost spoken of in terms of gender (and, except for when it is found in the quotes of others, we have not spoken of the Holy Ghost in terms of gender at any place in this book as well). So we see within the pages of the book of Mormon (in contrast to the understandings held by Joseph Smith that informed his later revelations recorded in the Doctrine and Covenants) the Holy Ghost is not identified as "*He.*" This is to be expected given the Hebraic background of the Nephites whose record it was.

We have seen here that the "*Breathing,*" the *Breath of Life*, the *Ruach* (רוּחַ) (the Spirit of God within us) is a Feminine word, so too is the word *Shekhinah*. This feminine understanding remained consistent in the Middle Eastern world throughout all the centuries before the development of Greek influenced

However, from what is recorded in D&C 130:22, we are to understand that the Spirit body of the Holy Ghost is not of "*Flesh and Bone*," making it distinctively different from the form held by the Father and the Son. The Holy Ghost's embodied Spirit is of the same order and form as the one possessed by each of us, the sons and daughters of God. Our Spirits are identical in form though clothed in mortal Temporal bodies (*flesh, bone and blood*) as we live today on Earth in this Universe.

D&C 130:22 implies that it is necessary for the Holy Ghost (though we know to be embodied) to be sans "*Flesh and Bone*," to fulfill the divine calling of "*dwell[ing] in us.*" We understand that the Holy Ghost can "*indwell*" while Physically embodied, as each of us has an *embodied Spirit* dwelling in our mortal frame. This same passage implies, antithetically, that the Father and Son through some restriction associated with "*Flesh and Bone*" cannot "*dwell in us.*"

It is universally accepted that the Holy Ghost has the ability to, and indeed does, dwell with all who remain worthy. We are told that through our worthiness after receipt of the gift of the Holy Ghost that we have the *right* to its continued and constant companionship. At the same time, the Holy Ghost has the ability to (through personal interaction) affect the lives of innumerable individuals simultaneously. This ability is divinely appointed, *"Because he is a Spirit Personage, he has power – according to the eternal laws ordained by the Father – to perform essential and unique functions for men*" (Mormon Doctrine, Bruce R. McConkie, pg. 359). In this context, we universally bestow omnipresence on the Holy Ghost. In fact, it can be very effectively argued that this attribute alone is what qualified the Holy Ghost as the third member of the traditional triune Godhead to begin with. *Omnipresence is an exclusive and necessary attribute of God.*

The attribute of omnipresence is specifically addressed by D&C 130:22; in fact, it is inferred through a logical declarative, *of course He could not have a body of flesh*

---

Christian orthodoxy. The Syriac word for Spirit, "*rucha.*" is also feminine. The Syriac language is derived from Aramaic, which was the language spoken by Jesus. It is clear that for the Hebrews, the Syrians and all those who spoke Aramaic, there was real meaning and intent when they called the Spirit "*She.*"

In stark contrast to the female gender of the noun used for *Spirit* in Hebrew and Aramaic, the Greek New Testament broke with the established understanding, referring instead to the *Spirit* in masculine terms, though in truth declaring masculinity may not have been the original intent. In his Gospel, John records, Jesus calling the Holy Spirit the *Comforter* (Greek, *parakletos*, Παράκλητος); when he does, John uses the grammatically necessary *autos* (a masculine form of the Greek pronoun). However, when John similarly records Jesus speaking of the "*Holy Spirit,*" he more often uses the neuter Greek *pneuma* (πνεῦμα) (John 16:13). Indeed one will struggle to find even one instance in the NT where *pneuma* is associated with any gender at all. When John (or more properly those who recorded and subsequently re-recorded the words of the Apostle) used *autos* and *ekinos* in conjunction with the *paraclete*, those words were meant to express *personhood* (a *Personage of Spirit*) and not necessarily masculinity.

The manifestations of the Spirit of God and the Holy Ghost did not assume masculine gender until the New Testament Cannon was developed in the Greek language and those grammatical constructs entered into Christian thought. That the Book of Mormon remains true to the original Hebrew constructs (in that it *never* confers masculinity upon the Holy Ghost), is evidence of its authorship in an ancient age by someone other than Joseph Smith. I make no attempt to debate the gender of the Holy Ghost here, but simply wish to divulge the complete historical record.

*and bone; for if He did, He could not dwell in us (plural us)*. It is because the Holy Ghost is, and must be, in more than one place at one time (omnipresence) that it does not have a tangible body of "*Flesh and Bone*."

However obvious this fact may seem, we are told it is an incorrect understanding. Rereading the previous quote of Joseph Fielding Smith, we find that the Holy Ghost, "*cannot be everywhere present in person at the same time.*" Can it be true there can be no omnipresence for a *personage of Spirit* in person? This constraint as detailed here encompasses not only the Holy Ghost, but also the Father and the Son. Consider the following from yet another prophet of the latter-day:

> *"The Holy Ghost as a personage of Spirit can no more be omnipresent in person than can the Father or the Son, but by his intelligence, his knowledge, his power and influence, over and through the laws of nature, he is and can be omnipresent throughout all the works of God." (Gospel Doctrine, Joseph F. Smith, pg. 61)*

Omnipresence then is *not* a fundamental characteristic of the *form* of persons of Deity but is realized through the inherent attributes of power, intelligence and knowledge of the laws of the Universe (those we have discussed here in Part 1 of this work).[11] As the Comforter, the Holy Ghost remains intimate with the confirmed faithful "*...by his influence, to be a constant companion to every baptized person...*" (*Answers to Gospel Questions*, Joseph Fielding Smith, 2:149).

What we are left with is a firm understanding of the Holy Ghost as a *Spirit* (an uncreated Intelligence inhabiting a Physical body). As an embodied Spirit, the Holy Ghost is constrained by its physical reality and construct in some way. This Spirit, though confined by size and dimension to physical existence defined in place and time, can "*dwell*" in each of us. Through its influence, its *Intelligence*, the Holy Ghost is omnipresent. We must note that the defined attributes of the Holy Ghost we have recognized here, are identical in every way to those held by the Son in His pre-incarnate state.

It is for certain that this is just a limited view of the power and form of Deity. We have much more to learn. Elder Bruce R. McConkie summed up nicely what we know of the Holy Ghost; "*In this dispensation, at least, nothing has been revealed as to his origin or destiny; expressions on these matters are both speculative and fruitless*" (*Mormon Doctrine*, pg. 359). We can, however, take what we do know about the Holy Ghost and apply what is relevant in completing our understanding of the embodiment of the Spirits of Man.

---

[11] Power, Intelligence, and knowledge of the laws of the Universe are certainly indisputable characteristics of both the Father and the Son. Because of this, the separate and distinct mission and calling of the Holy Ghost is not made clear. It must be assumed that influence of the sons and daughters of God, indeed of all the materiality of the Universe, must be equally possible for the Father and the Son whenever they may desire it. This influence may properly be attested to effects of the Spirit of God and the Spirit of the LORD or the Light of Christ.

## *The Embodied Family of God*

That the Father and the Son are embodied today, in tabernacles of *Flesh and Bone*, is universally accepted by Latter-day Saints. With this view in mind, consider the following: It does not seem possible that God, in and through His attainment of anything, through His acquiring of something that He may not have had before, would degrade or lose entirely an attribute He previously held. Nor is it possible to conclude that a creation of God, or one of His parts, has an attribute that exhibits a capability God does not have within Himself. Can it be conceived that the Holy Ghost is capable of indwelling when the Father does not have that attribute Himself?

If the Holy Ghost does possess a characteristic not held by the Father, consider what this means about the Godhead and its division of attributes and responsibilities. It is possible to understand the single God in three parts of the Trinitarian Creed as dividing His total attributes in some manner exclusive to each of His three personages. The three separate and distinct, individual personages of the Latter-day Godhead, however, cannot sustain such a view. It is not fathomable to envision any being as having an ability or fundamental attribute that our Father, the ultimate God, the source of all things, does not have. Jesus Christ, a member of the Godhead, said as much when he spoke, "*The Son can do nothing of himself, but what he seeth the Father do*" (John 5:19).

The passage D&C 130:22 dictates that the *Flesh and Bone* component of the Savior's form and that of the Father's, preclude each of them from dwelling *in* the sons and daughters of God and perhaps within anything else that has been created. What then are the attributes of the aspects of deity we refer to as the *Spirit of God* or the *Spirit* or *Light of Christ*; can these indwell another? Of course they can; what then are *their* source, if not both the Father and the Son?

We have discussed here that omnipresence is an exclusive, defining attribute of God, but it is not an attribute of all the *persons* of Deity. In fact the constraints of embodiment establish size and dimension which fix physicality to time and space. We have learned, however, that omnipresence is extended through power, intelligence and knowledge of the laws of the Universe; it can, therefore, be reasoned an embodied Father and Son can use these qualities to indwell another in a manner similar to that of the Holy Ghost.[12]

---

[12] We will explore another passage of scripture later in this work that is relevant in this discussion now. It is, "*And he [Christ] received all power, both in heaven and on earth, and the glory of the father was in him, for he dwelt in him*" (D&C 93:17). Here it is clear, the Father in His eternal, realized, being can and does have the capability of dwelling in another. This passage relates to the Father and the Son, but it is reasonable to infer that the same capacity of the Father to *indwell* extends to His sons and daughters.

Our Father is omnipresent (being everywhere influential) through His supreme power, intelligence and understanding of the laws of the Universe.[13] These attributes are His regardless of the bodily state of His being.

It is entirely consistent with all that we have learned here and with Scripture, both ancient and modern, to describe our Spirits (the immortal bodies given to us in the Physical sphere, the bodies held in the pre-existence and still held today by both our Father in Heaven and that of His Son and also the Holy Ghost, the same Heavenly form that was described by the LORD as "*the body of my spirit*") as a *personage of Spirit*, as an embodied intelligence and one possessing a tabernacle of *Flesh and Bone*.

***Our Father is omnipresent (being everywhere influential) through His supreme power, intelligence and understanding of the laws of the Universe.***

Adam was a *personage of Spirit* (clothed in a physical immortal body that was joined with his Intelligence) on the pre-temporal Earth before his entry to the Garden.

With his receipt of a *new name* and entry into the Garden, we understand here that Adam's embodied Spirit was enclosed in *Flesh and Bone*. As we understand embodiment, an embodied Adam could fulfill all of the requirements imposed on him by Scripture if the Physical body he held in the Garden was composed also of *Flesh and Bone*. There is nothing from our study here that would preclude such a possibility.

Consider the following as it relates to the embodied nature of the family of God: Within the Endowment, we are instructed that Adam receives the first token and sign of the Aaronic Priesthood (because of touch, necessarily involving a Physical, *Flesh and Bone* body) while yet in the Garden.[14] This is revealing in what it tells us about the nature of the embodiment of all involved. Adam could not receive the token without having a body, and because he receives it from either the Father or the Son also reveals to us something of the nature of the Father and the Son at this time.

---

[13] Omnipresence, its realization, may in fact be made possible by Temporality itself. This, if indeed it is factual, may be yet another reason the Temporal Sphere was created, allowing both the Father and the Son to be everywhere present here, when their forms may have precluded that capability in the Spiritual and Physical spheres. This is just presented here as food for lofty thought.

[14] It is significant to note that the name of this token is the "*New Name.*" A name we have previously identified with our receipt of a change of body type. Remember that the Garden Temple is a reflection of Temples here on Earth; we symbolically receive the New Name in Earthly Temples symbolizing our shedding of the flesh, bone and blood of our Temporal bodies and gaining the immortal *flesh and bone.* Adam in receiving the token of the Aaronic Priesthood in the Garden began his transition from immortal *flesh and bone* to the Temporal.

In further support of this conclusion (that *Flesh and Bone* is fundamental to the Physical sphere) stands the creation of things such as water, rock, plants and animals in that Physical sphere, things which we know have no correspondent equivalency to eternal intelligence as do the sons and daughters of our Father in Heaven. The immortal constructs for these inanimate, non-sentient, objects and beings nevertheless stand in physical, eternal reality. They possess form and detail in the Physical sphere without the purpose of housing agent Intelligences on their way to perfection. Their eternal existence is just one of physical structure on a similar order as *Flesh and Bone* would be for each of us.

## *The Three Grand Keys*

The 129th Section of the Doctrine and Covenants provides information that helps us understand the state of beings in the Physical sphere, and by extension the forms we hold as sons and daughters of God and even that of Satan himself. The revelation states:

> *"There are two kinds of beings in heaven, namely: Angels who are resurrected personages, having bodies of flesh and bones – Secondly: the spirits of just men made perfect, they who are not resurrected, but inherit the same glory."(D&C 129:1, 3)*

The purpose of this revelation was not to describe the form of existence in the Physical sphere, but was to provide important "*keys*" in determining the authority of messengers claiming to be on an errand from God. The "*three grand keys*" (D&C 129:9) draw upon the sense of touch and the apparent willingness of fallen angels to deceive. The "*keys,*" however, have been used to build up doctrine relating to the nature and form of each of us in the Physical sphere.

Though it is suspected that the Prophet knew of these "*keys*," or a form of them, long before this revelation was recorded, the "*keys*" were codified through revelation for the instruction and benefit of all, early in 1843. Brief and descriptive, the nine short verses of this revelation may be just as instructive by what they do not say than they are by what they do.

We have discussed before the confusion of terms that so permeates any discussion of things "*spiritual,*" and the same confusion is manifest here. The same compounding and blending of terms, this time of *embodied* and *Flesh and Bone*, leads to misunderstanding and incorrect concepts. The confusion is not to be found in the revelatory words of the Section 129 itself, for they are clear, but it rests in the many interpretations and commentaries recorded regarding this revelation and the subject it addresses.

The test detailed in the revelation is a simple one: "*When a messenger comes saying he has a message from God, offer him your hand and request him to shake hands with you.*" (v4) The messenger's response to this single gesture and the question posed are the "*three keys.*" They are provided in Verses 5 through 8 as follows:

*"If he be an Angel he will do so and you will feel his hand.* (v5)

*If he be the spirit of a just man made perfect he will come in his glory; for that is the only way he can appear – Ask him to shake hands with you, ...he will not move ...but he will still deliver his message.* (v6, 7)

*If it be the devil as an angel of light, when you ask him to shake hands he will offer you his hand, and you will not feel anything; you may therefore detect him."* (v8)

The telling and ultimately most important "*key*" is the third one, for through that one, *and that one alone*, can certainty be had that the messenger is not from God in that it forces a handshake that cannot be felt. The other two "*keys*" are inconclusive in that either response from the messenger does not represent anything out of normal experience.

The only doctrinal conclusions that can be drawn from this test are that a resurrected being with a body of *Flesh and Bone* can be sensed by touch, and Satan has a form that cannot be felt. It appears, though the deceiver *should be intelligent enough to know* you may know of the test, apparently *he still will always feel compelled to shake your hand*, thus giving himself away. Besides these three things, there is nothing new to be learned by this "key" part of the revelation.

Our understanding of a resurrected being had been well established long before 1843 and Verse two confirms that, when it references Christ's appearance to his disciples in the closed room so long ago. We have referred to that appearance many times as well in this work as an example of tangible attributes of immortal, spiritual (Physical) beings.

What is of *new* doctrinal importance in Section 129, is the difference between the feel of a resurrected being of *Flesh and Bone* and the lack of physicality of the devil and perhaps that of his followers as well.

Left to itself, *right at this point*, we could walk away with conviction that Satan and his followers are without embodiment – but that is an incomplete picture.

At a general conference of the church held at Nauvoo, October 2 and 3 of 1841, the Prophet spoke on a subject related to this. It must be noted that we have minutes from this meeting and notes taken of the discourse; importantly, the *terms* used in the record are from the understanding of whoever penned it:[15]

*"He explained the difference between an angel and a ministering spirit; the one a resurrected or translated body, with its spirit ministering to embodied spirits – the other a disembodied spirit, visiting and ministering to disembodied spirits. Jesus Christ became a ministering spirit (while his body was lying in the sepulcher) to the spirits in prison, to fulfill an important part of his mission, without which He could not have perfected His work, or*

[15] Either Elias Smith or Gustavus Hills, both were appointed clerks for this conference.

> *entered into His rest. After His resurrection He appeared as an angel to His disciples. Translated bodies cannot enter into rest until they have undergone a change equivalent to death. Translated bodies are designed for future missions. The angel that appeared to John on the Isle of Patmos was a translated or resurrected body [i.e. personage], Jesus Christ went in body after His resurrection, to minister to resurrected bodies." (History of the Church Volume 4:425)*

Immediately it becomes apparent that this teaching, as it is recorded, conflates the *terms* of embodiment through compounding; it may be assumed the conflation came from the mind of the recorder.

As written, the first statement says that an Angel is a resurrected or translated body, housing a Spirit, who ministers to embodied Spirits.[16] Through the context, we must conclude that the "*embodied spirit*" is a Temporal man with a body of *Flesh and Bone and Blood*, in the fallen state. By this statement, we are to assume that only Spirits tabernacled in the body of Temporality (translated) or in the perfected Temporal body (resurrected) minister to the sons and daughters of God within this Universe. What is ignored here at great peril to a clear understanding is that (as we have shown) a Spirit is of itself an embodiment of an independent Intelligence. So we have a Spirit in a body that is in another body as well.

The "*ministering spirit*" of the first statement we are told is *disembodied*, ministering to *disembodied spirits*. This statement, however, is offered with a descriptive example, that of Christ ministering to the *Spirits* in *Prison*. Further it is emphasized that this occurred while the Temporal body of Christ's incarnation was yet lying in His tomb. So it is clear here, that by *disembodied*, the author and by extension the speaker (if the speaker is properly understood by the recorder) defines embodied and disembodied as a Spirit (itself an embodied Intelligence) that is either with a *temporal* (Flesh and Bone) tabernacle or without one.

We are told that after His resurrection, Christ appeared (as what is defined as an Angel) to his disciples in the closed room. This is so because (as we have previously seen), though He was endowed after His resurrection with a body of supernatural power, the Savior still disclosed it was tangible and of *Flesh and Bone*.

These three things are clear and in accord with all that we have discussed here. Additionally, we have an understanding of translated beings that is properly supported by the teaching here. Translated beings are worthy, just men, who lived here in Temporality and reserved from death until some future point after they have accomplished their mission on Earth. Their mission required their mortal lives to be extended through the intervention of Deity.

---

[16] In the Endowment, Peter (who had not yet received a Temporal body) could not be translated for he had not yet *lived* in temporality and, therefore, could not be *resurrected*, for he could not have tasted of death, is depicted as physically touching Adam while yet in his pre-mortal, embodied state. This paradoxical challenge will be discussed later in this work; see The Spirits of Just Men Made Perfect, pg. 508.

Examples of translation are Moses and Elijah, both of whom the Scriptures detail translation events and their subsequent, centuries later, appearance with the Savior on the Mount of Transfiguration. Translation was required for both of these prophets because the end of their natural sojourn on Earth and their much delayed subsequent return both *preceded* the resurrection of Christ, who was of necessity and by decree the *First Born of the Resurrection*.

After Christ's resurrection, translation is no longer necessary, for Just Men can go straight to the resurrection, enabling them to minister as Angels with a resurrected body, as Moroni and Peter and James did to Joseph Smith.

The last statement as recorded encompasses yet another conflation. It strongly infers that a translated and/or resurrected being, a Spirit in a fleshly tabernacle, through translation or by resurrection, is the definition of a *personage*. This is not exclusive, and it would be incorrect to narrow the definition of personage in this way. We have seen that the Holy Ghost, clearly not translated or resurrected and without a fleshly tabernacle, is called a *personage* as well; the Holy Ghost is called a *personage of Spirit*.[17]

---

[17] Finally, the very last part of the recording of this teaching appears to be an error in inscription. The last statement declares that "*Jesus Christ went in body after His resurrection, to minister to resurrected bodies.*" There simply is no mention of such an event in Scripture. There are two possibilities here: *First*, what is meant here is the Savior will minister eternally to all of us in our resurrection as a resurrected being himself. Or *second*, the record should read that Christ in resurrected form ministered to the disciples after his rising, to the Nephites and so ministers today. I submit it is clear here, that what the Prophet said was dramatically altered in transcription through interpretation by preconceived or incomplete understandings of the recorder.

## Jesus Christ, the Eternal Son of God

*"In the beginning was the Word, and the Word was with God, and the Word was God." (John 1:1)*

It is fitting that John would begin his Gospel with these words, identifying the Savior as both eternal and fully God. We accept both of these terms without reservation here. God is the Creator, and is eternal; John assigns both of these deific attributes to Jesus Christ. To identify the Savior as the Creator, John purposefully uses the Greek, *Logos* (λόγος) (translated in the KJV as *Word*) as his preferred appellation for Christ. John was aware he was not the first to use this term for the Creator; he was not the first to use *Logos* to identify a separate and distinct character of God.[18] John's usage, however, added dimensions to the previous understanding and changed its recognizable character forever.

John carefully selected the Greek *Logos* because he wanted to identify Christ as the Creator of the world. He chose *Logos* because it was an equivalent of the Aramaic word *Memra* that had, by John's time, long been used to identify the Word and Power of God.[19]

To emphasize the eternal nature of Jesus Christ, John begins his telling of the Gospel of the New Testament, exactly where the Old Testament did, at the very beginning; at *archē* (ἀρχῇ), literally, *the most ancient of times*.

John's Gospel used phrasing (calling the Son the Word) which resonated with those who were familiar with the Genesis 1 account. Moses used similar wording to describe the creation of the Temporal sphere; "*and this I did by the word of my power, and it was done as I spake*" (Moses 2:5). In yet another Biblical account, the eternal nature of the one called Wisdom, *chokmah* (חָכְמָה) of the Hebrew, and *Sophia* (σοφία) of the Greek, the one whom Philo first called the *Logos*, is detailed. Speaking of the *chokma* (Wisdom), Solomon said:

> *"The Lord possessed me in the beginning of his way, before his works of old. I was set up from everlasting, from the beginning, or ever the earth was. When there were no depths, I was brought forth; when there were no fountains abounding with water. Before the mountains were settled, before the hills was I brought forth: While as yet he had not made the earth, nor the fields, nor the highest part of the dust of the world. When he prepared the heavens, I was there: when he set a compass upon the face of the depth: When he established the clouds above: when he strengthened the*

---

[18] See our previous discussion of this subject, - Wisdom is God Himself, in Chapter 7, pg. 274.

[19] A fascinating treatment of the concept that John carefully and intentionally identified the Christ as the Word of God, because Jewish thought and beliefs already understood the Power of God in creation and in matters of God's dealing with man on Earth were attributable to the Word (Memra) of God. The Targum Jerusalem reads: *"and the Word of the Lord created man in His Likeness, in the likeness of the presence of the Lord He created him, the male and his yoke-fellow He created them"* (The Word, William S Eldridge, 2010).

> *fountains of the deep: When he gave to the sea his decree, that the waters should not pass his commandment: when he appointed the foundations of the earth: Then I was by him, as one brought up with him: and I was daily his delight, rejoicing always before him." (Proverbs 8:22-30)*

This passage of Scripture describes the Son as with the Father at the beginning, yet created in that same beginning age. We attest to the accuracy of the many statements of our LORD which have Him professing His distinctness and subservience to the Father. We believe Christ was honest and straightforward, that He spoke plainly, not in riddles, in that gentle, very intimate moment with Mary near the open tomb when He said, "*go to my brethren, and say unto them, I ascend unto my Father, and your Father: and to my God, and your God*" (John 20:17). We attest that we are His brethren, He is our brother and we share a common Father with Him, *the God in Heaven we both look up to*. We believe Paul when he says that Christ was but the "*image of the invisible God,*" that the Savior of the Universe was and is "*the firstborn of every creature*" (Col. 1:15).

We understand things that were *once made* to *also be eternal*, and Jehovah, the Firstborn, Jesus Christ, is of that order of being. He has a special and revered association with the Father. Not only *Firstborn*, He was formed with attributes of wisdom, mercy and with creative powers above all others.

John picks up where Solomon left off. Solomon describes Wisdom as being possessed by God in the beginning, brought forth before all the rest of creation, and with God when all was formed and instructed. John, however, tells us He was not only present in the beginning, beside God, but the *Logos* was the active source of creation and life as we know it; "*All things were made by him; and without him was not any thing made that was made. In him was life; and the life was the light of men*" (John 1:3, 4).

The *Intelligenc*e given to Jehovah by the Father *far exceeds* that given to any other creature. We entertain the notion here that the Intelligence parted from the Father to form the independent Son, encompassed much of, indeed included the best parts of the Father's capacity to love and create; the Son is the creative power behind the several spheres of God. By that we can understand that He is the embodiment and personification of that creative force within the Father, perhaps most of this complete part of the Father was given to Him when He was created. Jehovah, the Christ, is the creator of ALL that has been created, save perhaps the Holy Ghost and our Mother in Heaven.

The Son is subordinate to the Father exceeding us in Glory by orders of magnitude that match those magnitudes of Glory by which the Father exceeds His: *"We say that the Son and the Holy Spirit excel all created beings to a degree which admits of no comparison, and are themselves excelled by the Father to the same or even greater degree" (Origen, Comm. in Ionnem, xiii. 25).*[20]

---

[20] The Early Christian Fathers, Henry Bettenson (Oxford University Press, New York, 1956) pg. 233.

We ascribe nothing more to the being of Satan than the Scriptures do. Much is said about Latter-day Saints equating Satan to the Son. This is not true and emanates from those who do not understand and seek to defame. They employ false logic, or at the very least, distorted logic to press their errant accusations. They establish an equation with variables that are not our own. The math they employ on the surface looks simple but is ultimately deceiving.

***Christ is our elder brother because we share the same source for our independent Intelligence, but because of His unique eternal nature with the Father, the Son is also our Creator and He created Lucifer as well.***

Consider this: The Christ of traditional Christianity is in fact the eternal unchanging God, the same today yesterday and tomorrow. As God, fully so, He was incarnated and became as one of us. He died and was resurrected, His tomb is empty, and there is no body there, for He took it up again. In that same body, He appeared to His disciples. He became like us and remains like us today. John says, when we die "*we shall be like him*" (1 John 3:2). Traditional Christianity agrees, God became like us and remains like us. We simply assert that a changeless God must then always *have been* like us. That He created us just like Him.

We assert that Christ was at one time created, to that we can quote Paul, without need of amendment or modification: "*the firstborn of every creature*" (Col. 1:15); Christ is the firstborn, He is our *elder brother*. Those who do the simple math and place the Son on one side of the equation, while placing Satan on the other side, do no injustice to the truth. Both were created by the same God, the Father, of the same order; we will not deny that math here, for it is proclaimed in Scripture. That such a conclusion seems degenerate is dependent on the observer's ability to understand the truth and should not take anything away from the truth that it is.

Christ is created, just like we and Lucifer, but we are not equal with the Son, neither is the Devil. Jehovah is the Firstborn; from the moment of inception, His intelligence was embodied, both with a Physical body and with one of *Flesh and Bone*.[21] All things thereafter were made by Him and through Him. Jehovah already stood at the Father's right hand, fully a member of the Godhead, when our

---

[21] See our subsequent discussion of <u>Eternal Procreation</u> in the next chapter of this book, pg. 521.

Intelligences (including that of Lucifer) were parted from the Father; John attests, "*The same was in the beginning with God*" (John 1:2). The Scriptures attest to the Son's "*unique*" position with the Father. He is called (in Greek) *monogenés* (μονογενής) most often translated as the only-begotten, but more properly rendered into English as *unique*. Christ is our *elder brother* because we share the same source for our independent Intelligence, but because of His unique eternal nature with the Father, we and the *Satan* were created through Him. The Son then, along with the Father, is our Creator – *and He created Lucifer as well.*

## *Christ the Creator of the Physical Sphere*

The Endowment narrative informs us that the creation of the Physical sphere was a collaborative effort; at the direction of the Father, Jehovah was joined by Michael in the work. We are not informed, however, of what creative powers might have been wielded by Michael in the process. We are left to wonder if Michael was truly a co-creator, or merely an aid, providing assistance and input to Jehovah throughout. Perhaps in the Endowment description of both Jehovah and Michael creating the Physical sphere, we can understand that the two together fulfill a priesthood requirement of two-by-two.

John tells us that nothing was made that was made without him; "*All things were made by him; and without him was not any thing made that was made*" (John 1:3). The precise detail offered by the Endowment narrative is the most definitive instruction we have regarding the creative attributes of Jehovah. The Biblical telling of the Genesis 1 account ascribe creation to Elohiym, which is a plurality in its original Hebrew understanding; Moses describes the Creator merely as God, and Abraham returns to the Hebrew plurality by describing the creation as being carried out by *the Gods*. These accounts are of the Temporal Creation and not of the Physical sphere, but without the Temple instruction, they are the only records of the creative powers of God and are incomplete in their identification of Jehovah, Christ as a creator.

The Endowment narrative is clear in its description of the creation of the Physical sphere through the efforts and powers of Jehovah, the Son of God. He created, from the unorganized material of physicality, *all* that there is. We understand the Physical creation to be complete and all encompassing. We also understand that the Temporal sphere is not a new structure but exists as the realization of the things of the Physical sphere after the divine presence of God is withdrawn in part. The Temporal results from a subtraction of what was created by Jehovah in the Physical sphere.

The Father delegated the creation of all things Physical to the Son, except the creation of the embodiment of the Intelligences making up the sons and daughters of God. It should not go unnoticed that in the creation of our Spirits (the embodiment of our Intelligences), the Father joins the Son in the process. The union of the creative powers of the Father and the Son together were to be repeated again when our Temporal bodies were formed.

## *Christ the Creator of the Temporal Sphere*

Christ proclaimed Himself to be the *Great I Am*. To the Jews who believed on Him, He said, "*Before Abraham was I am*" (John 8:58), and to another believer, the Samaritan woman at Jacob's well, He declared, "*I...am he*" (John 4:26). With these proclamations, Jesus Christ unambiguously identified Himself as the only God the Hebrews had ever known. He identified Himself as the God of the Bible, the Creator of all things. To His statements we add the testimonies of His disciples, the Apostle John and Paul:[22] "*He was in the world, and the world was made by him, and the world knew him not.*" (John 1:10); "*[God] Hath in these last days spoken unto us by his Son, whom he hath appointed heir of all things, by whom also he made the worlds*" (Heb. 1:2).

As we have seen, the Genesis 1 accounts speak of the creation of the Temporal sphere by a plurality (*Elohiym*) in the Bible. We are told the *Gods* created in Abraham, and a singular God describes Himself as being the Creator in Moses. This, however, should not be interpreted as one alone; all of these readings are correct, without contradiction. We have detailed the creation of the Spiritual, the Physical and the Temporal as being a combined effort of the Father and the Son and even Michael in various and specific ways.

The multiple renditions of the Genesis 1 account can be understood as follows: The most direct rendition is in the book of Moses, and here we have the Father recorded as He directly spoke to Moses. The creation of the Temporal was in fact created by and through the withdrawal of the Father's presence. The presence of Jehovah, remained as the Light of Christ, and it informs all things Temporal of the Law by which they are governed. In a metaphysical sense, these laws are those of morality, choice and consequence, fall and redemption, faith, hope, charity and the Atonement. Practically speaking, the Law is also the laws of physics which enable the existence and control the actions of all non-sentient and inanimate things.

Abraham's account and the account as it is rendered in Genesis, properly identifies the creation as being accomplished by the Father and the Son. We must remember the difficulty in describing the distinction of the Father and the Son. None of us, including the Son, operate as truly independent of the Father. The Prophet Joseph understood this truth to be true; "*And he [Christ] received all power, both in heaven and on earth, and the glory of the father was in him, for he dwelt in him*" (D&C 93:17).

Let us be clear here: The Temporal is not a new creation, it is a Fallen state of the Physical Creation, the creation of which was accomplished by Jehovah (and Michael) at the direction of the Father. Christ then *created* all that there is, for "*without him was not any thing made that was made*" (John 1:3). The Temporal is only possible because the Father actively withdraws, leaving the Son alone, His *light* in the vacuum. We understand that the Temporal was created through two additive

---

[22] See also: John 1:3; Eph. 3:9; Heb. 1:2, 10; Heb. 11:3; Rev. 4:11.

components, *the withdrawal of the Father* and the *active creative, instructive accomplishments of the Son*.

The creation of the Temporal sphere was a process that continues to this day. It began with the creation of a vacuum (the *tohu*) in the emptiness left when the presence of God was withdrawn. The elements (the *bohu*) were instructed and informed of the law that would empower them to fulfill the measure of their creation when Jehovah infused them with His knowledge in Day 1; Jehovah is the Divine *Logos*, He is *Wisdom*. It was Jehovah who performed the work (providing the *bohu* with information) as the Father oversaw, anxiously brooding along with His Son over the nascent Universe.

- The Very Voice of God

The work performed on the first day was sufficient to put in place all that was needed for the fledgling Universe to develop on its own, through all of the processes known by science. This it did do but not beyond the careful oversight of its Creator. We have detailed here several occasions along the path of the reflexive and autonomic development of the Universe we live in, wherein God intervened to guide and direct. We understand here that the personage of the Godhead who conducted these interventions was the Son. It is His intervention that is marked in the Biblical account when it is recorded that God *spoke* or *spake*. John identified the Son as the Word; it is the Word of God that the Scriptures identify as speaking:

> *"For behold, by the power of his word man came upon the face of the earth, which earth was created by the power of his word. Wherefore, if God being able to speak and the world was, and to speak and man was created, O then, why not able to command the earth, or the workmanship of his hands upon the face of it, according to his will and pleasure?" (Jacob 4:9)*

Comparing Genesis 1:1 with John 1:1, the phrase, "*In the beginning*" identifies a specific point in time. In the former, Moses starts from this point, "*the beginning*" and works towards our present age, coming forward in time detailing the evolutionary steps of the informed *bohu* as the laws of physics and the elements formed the Universe we live in today. St. John, however, starts at the same point and works backwards; he takes us upwards into the eternities before time itself existed. In Genesis 1:1, we are told that God created the Temporal sphere, an act done in time, as we have seen it was the *very act* that created time itself. John declares, "*in the beginning the Word was*." The Apostle here very antithetically separates out the Word from the coming-into-being of the things of the Universe spoken of in the Genesis 1 account. John implies the absolute existence of the Son (*the Word*) before the *initial* creative moment,[23] for he said, *in the beginning* the Son "*was.*"

---

[23] D&C 93:6-10.

## A Truly Eternal Father

The three creations described in this work dictate that God is Eternal, the word being understood in its infinite form, *without beginning or end*. We have described the Father as the source of both eternal intelligence and the principles of matter, both of those eternal and infinite only in that they find their source in God; *they* have always been because *He* has always been.

God the Father is the *eternal cause*, the original cause of causes. We have shown "*The glory of God*" to be "*intelligence, or in other words, light and truth*" (D&C 93:36). It is also God's work, the *prime mover* of the *eternal cause* to bring to pass eternal life for His children; "*For behold, this is my work and my glory—to bring to pass the immortality and eternal life of man*" (Moses 1:39).

There is an unnamed law, one that is not detailed that has been foreordained in heaven, upon which all blessings flow. That greatest of all blessings, *the whole of all others combined* is eternal life. That blessing and all others flow from the Plan of Happiness put in motion by God's *eternal cause*. We are here to understand that the unnamed law, the primal law, frames the precepts in which we progress to be like our Father, precepts that find their foundation in His everlasting *eternal cause*.

In the next few paragraphs, we will explore the eternal attributes of the Father, considering them against the flawed concepts of tradition. We will properly draw the correct picture of Him as told in Scripture and revealed in His three creations.

In His eternal state, God is perfect and described as unchangeable. The Bible is replete with passages which describe him as such: "*For I am the LORD, I change not*" (Malachi 3:6). Solomon, singing out, spoke of God as follows, "*But thou art the same and thy years shall have no end*" (Psalm 102:27). The Apostle James added, "*Every good gift and every perfect gift is from above, and commeth down from the Father of lights, with whom is no variableness, neither shadow of turning*" (James 1:17).

Not only old world, but new world prophets proclaim this invariability: "*For he is the same yesterday, today, and forever*" (1 Nephi 10:18). Nephite prophets were consistent in this message, and the Book of Mormon testifies of this truth from its first pages, to its last:

> *"For do we not read that God is the same yesterday, today, and forever, and in him there is no variableness neither shadow of changing? " (Mormon 9:9)*

> *"For I know that God is not a partial God, neither a changeable being; but he is unchangeable from all eternity to all eternity." (Moroni 8:18)*

Though we have no continuum of record in the New World which would express how these words of the Nephite prophets were interpreted, we have nearly two thousand years of Old World Church history to delve into for an understanding of what the Biblical passages meant to the Christian Church that came to be.

Intellectual thought in the Old World (in the opening centuries of the Christian era) was entrenched in Greek philosophy; it permeated every word that was written by man. Anyone who could read and write had been schooled in Greek thought, logic and concepts. The world, everything in it and about it, was viewed as if it were truly through *Corinthian* (Greek) glass, tainting *all* things known in the process.[24]

As Christianity developed and its thinkers expounded on the Scriptures, their theology took on an ever more Greek flavor. It is no wonder the *traditional* concepts of the being and nature of God mirrored precisely those that had been taught by Greek scholars for centuries: following definitions, fulfilling expectations, completing the images held and taught by Greek philosophers that had died hundreds of years before Christ's Earthly ministry. Their teachings and notions were framed before the Savior provided the very example of what He, His Father, His Gospel and our relationship to them in fact are.

## *God of the Empire*

- The God of traditional Christianity is immutable; He "*is unchanging in his being, perfections, purposes, and promises.*"[25] Traditionally, this is understood to mean that God does not, is incapable of, change in any manner or extent, either extrinsic or intrinsic. God is static, for theologians reasoned, change of any sort must be either for the better or for the worse. Their rationality expressed, if God was capable of change for the better, He wasn't already perfect; if He could change for the worse, God then would lose His indispensable quality of perfection.

- The God of tradition is immovable. To the ancient Greek monistic philosophers, God was envisioned as a universal oneness. This philosophical concept of Deity was called the "*Unmoved Mover*" by Aristotle. He and the rest of the *monists* understood God to be the cause of all motion but unable to move. Motion is something that is recorded when one thing moves relative to another. Since this singular being is eternal, infinite, and perfect, it encompasses all, and therefore is the ultimate frame of reference and cannot be imagined to move relative to itself.

We understand within this work, however, that what is described in the Scriptures does not support these traditional theological concepts. God, as described here in this work, is not an unmoving mover, and perfection is not static and unchanging.

We understand Eternity to be a very long time, and through it all our Father continues to grow and progress:[26]

---

[24] This truth is so strongly evidenced, it must be concluded that Paul's iconic phrase was intended to be understood *literally* and not just metaphorically. See 1 Cor. 13:12.

[25] Bible Doctrine, Wayne Grudem (Zondervan, Grand Rapids, Michigan, 1999).

[26] We understand here the eternal progression of the Father to be self-evident, though some may disagree. As we have before in this work, we can find President Joseph Fielding Smith in opposition to what this work affirms to be true; "*It should be realized that God is not progressing in knowledge, truth, virtue, wisdom, or any of the attributes of godliness. He has already gained these things in their fullness*" (emphasis my own). Mormon Doctrine (Joseph Fielding Smith – cited by Bruce R. McConkie, 1958, pg.

> *"God himself is increasing and progressing in knowledge, power, and dominion, and will do so, worlds without end." (Wilford Woodruff – Journal of Discourses 6:120)*

> *"...there is progress for our father and for our Lord Jesus. There is no such thing as standing still in the eternal work of our God. It is endless progress, progressing from one degree of knowledge to another degree." (George Q. Cannon – Millennial Star 61:117)*

The immutability and immovability of God in His being is not described in the Scriptures. God is steadfast in purpose and unwavering in His goal, but that does not speak to the immutability of His form, His actions or His desires. We see His singlemindedness in Samuel, "*The Strength of Israel will not lie nor repent: for he is not a man, that he should repent*" (1 Samuel 15:29). Here we must add that there is hierarchy to the singlemindedness of the LORD, for though we understand Him to be resolute and unwavering in His determination to "*bring to pass the immortality and eternal life of man,*" we have seen, at times, the means He had employed to effect such a task (as in the case of the *ha-a-dam*, הָאָדָם) had, "*repented the LORD*" (Gen. 6:6).

***The immutability and immovability of God in his being is not described in the scriptures. God is steadfast in purpose and unwavering in His goal, but that does not speak to the immutability of His form, His actions, or His desires.***

Because Scripture cannot be found to support an immutable or unmovable God (that He, doesn't change in any way), many Christian teachers today reject the radical concepts developed by the Greek philosophers and the mid period Christian theologians influenced by them. They emphasize that "*God does act and feel emotions, and He acts and feels differently in response to different situations.*"[27] They recognize that God is said to be both wrathful, "*Now therefore let me alone, that my wrath may wax hot against them, and that I might consume them*" (Exodus 32:10), and merciful with longsuffering "*And the LORD passed by before him, and proclaimed, The LORD, The LORD God, merciful and gracious, longsuffering, and abundant in goodness and truth, Keeping mercy for thousands, forgiving iniquity and transgression and sin*" (Exodus 34:6, 7). They

---

221). In this belief, President Smith not only conflicts with what we present here he also conflicts with many other Prophets of the Latter-day. How can it be imaginable for each of us to be capable of eternal progression, when the Father himself, is not.

[27] Bible Doctrine, Wayne Grudem (Zondervan, Grand Rapids, Michigan, 1999).

recognize, as we do, that God is capable of changing His mind: "*And God saw their works, that they turned from their evil way; and God repented of the evil, that he said that he would do unto them, and he did it not*" (Jonah 3:10).

We recognize here that though God is unchanging in His eternal nature and He is perfect forever in His purpose and His promises, He can and does change, not for the better or for the worse, not in a vertical fashion, but horizontally. Our Father, who righteously destroyed Sodom and Gomorrah and mercifully saved Nineveh, the same God who kept His promises to Abraham, Isaac and Jacob, and the God who sent His Son to save us all two thousand years ago, is still the same God today, remaining perfect but recognizing change around Him;[28] our God progresses eternally.

We understand our Eternal Father to be a thinking, living, emotional being with an essential set of principles that never vary; among those are love, holiness and justice. He is approachable and can be appealed to. The Scriptures tell us that if a person comes to God with a genuine heart, God will react; "*If any of you lack wisdom, let him ask of God ...and it shall be given him ...But let him ask in faith, nothing wavering*" (James 1:5, 6).

We know him to emotionally change direction as the circumstances of our agent actions dictate. (*Because he had repented of making man in the Earth, He brought on the great flood, Genesis 6:5-7; God changed His mind after Moses plead for the people, Exodus 32:14; and He regretted He had made Saul King, 1 Samuel 15:11.*) Though we attest that these Scriptures testify of the true character of our Eternal Father, we recognize that the modern Christian apologist will be quick to point out that in these things the omniscient God did not *really* change his mind. They will insist that the wording of the Bible (it being delivered to man in man's perspective) only makes it appear that God had second thoughts. Such thinking can only be considered a challenge to the inerrancy of Scripture and its ability to perfectly communicate the message of God they also so boldly profess.

- Why Do We Pray

One of the greatest flaws in the logic that insists in a perfectly divine, immutable God is revealed in the scriptural understanding of the efficacy of prayer.

Omniscience (as it is most commonly understood) dictates that God knows the future; in fact, He dictates it. Because He is immutable, He cannot change the future, and neither can He change the past, for it was perfectly accomplished according to

---

[28] An example of this type of change has been described by one modern Christian commentator in the following manner, *paraphrased here*: '*God may change from knowing that it's 2:30 PM right now to knowing that it's 2:31 PM in the next moment. This is a change, but it's not a change for the worse. If God is a perfect being who knows all truths, we would expect Him to always know the right time!*' (However, it can be argued that time itself is a Temporal construct and its *changing* (progression) is evidence of the corruption of this sphere. God would not experience the change of time, because he is changeless and incorruptible making this argument for His *horizontal* changeability void.)

His perfect plan. An unchanging being cannot know changing truths. How then could God truly answer prayer or intervene on behalf of the supplicatory? If He did respond by altering expected circumstance, He would be changing His predetermined plan. [29]

Though we adamantly affirm God's *complete sovereignty* throughout the entire schema of the multiple creations we have described here, our understanding of God's interaction with the Temporal sphere most closely resembles what is often called "*Open Theism.*"[30] We understand God to be unchanging (immutable) in His natural essence, in His mission, purpose and in His trustworthiness over time, but God is open to change depending on the reality of the situation or experience.

Because of the divine plan and its dependence on our growth and experiences in Temporality, God does not, more properly, *chooses not*, to know all possible future outcomes. This is often described as the *loose* view of God and is not accepted by the traditionalist and fundamentalist Protestant Christian community, because they feel it suggests a God that is not all powerful, all knowing, unchanging and perfect.

Joseph Smith once said, "*By proving contraries, truth is made manifest.*" This issue, the true nature and being of our Eternal Father, is full of complexity and it may be seen to be teneted by Scripture full of contraries, and as such the truths they hold must be of enormous worth. I believe this to be immensely so.

God has provided us with a formula to discern the truth in difficult situations. This formula never fails. It begins and ends with prayer, and in the middle we are to search the Scriptures and ponder the question out in our own minds; then and only then, will the Spirit reveal the answer to us.

"*Even God cannot change the past.*"

(Agathon c. 448-400 BC)

---

[29] To the reformed Christian, this conundrum is alleviated by their insistence that man is not free to act on his own, that all interaction with God on the part of man is in response to election and God's effective call. In their logical construct, a man prays for God's intervention as a result of God making him do so. Effectively then, the man is told to ask God to do what God already planned on doing. If that sounds absolutely unintelligible – It is.

[30] Through the multiple created spheres and their dissimilar attributes and purposes we can avoid such logical absurdities as those presented in Molinism. Because we understand the Temporal construct to be initially designed by the Father to provide human free will and that free will is made possible by His intentional withdrawal, there is no need to discuss the existence of, or the point in which God attained, *Scientia Media* (Middle Knowledge). The purpose of Temporality *is* the attainment of experiences unknown, for both man, and through man, by God. God's foreknowledge was an awareness that the unknown *would* occur, and to counter its detrimental effects, He set in place the Atonement in advance.

All of Genesis Chapter 1, and Verses 2 and 3 of Chapter 2, speak of the creation of the Temporal sphere by *Elohiym* (a plurality).[31] Abraham correctly identifies the creation of the Temporal sphere as the work of Gods. We discussed what the particular wording of Abraham meant previously in this book. The Physical sphere was created by Jehovah with the assistance of Michael at the direction of the Father. Both of these creative accounts culminate, however, with specific language that details both the Father and the Son creating the separate embodiments of man; this partnered process is repeated in both the Physical *and* the Temporal spheres. This detail is significant. We will explore here briefly what the interpretation of the three creations described in this work draws out of these truths.

Our understanding of the Endowment narrative, as it describes the creation of the Physical sphere, informs us that the Son along with Michael together formed the Physical sphere from unorganized materials. The materials originally unorganized are eternal, yet they themselves were created at some time more ancient by the Son. The Father directed the creation of the Physical sphere but allowed the actual creation of it to the powers and forces attendant with the Son. We previously described that the Son guided the materials of Temporality (infusing them with information through His *Light*, and at times guiding them through direct intervention) as the Universe we live in developed from its inception, when the presence of the Father withdrew. The Son then created the Physical and Temporal spheres themselves, but we are told that when (in both creations) it came time to craft the attributes of the sons and daughters of God, He was joined by the Father, and the Son took a subordinate role in the specific creative processes that were to follow.

In the Endowment, on the sixth day, the Father changed the established creative rhythm of the narrative when He announced, for the first time, that He Himself would join with Jehovah and Michael[32] in the creation of man. The Father asked, "*...is man found upon the Earth?*" When told that man was not on the Earth, our Father replied, "*...then let us go down and form man in our own likeness and in our own image, male and female, and put into them their Spirits...*" The Genesis 1 accounts of the creation of the Temporal sphere echo this same detail: "*And I, God, said unto mine Only Begotten, which was with me from the beginning: Let us make man in our image, after our likeness*" (Moses 2:26). Why did the Father do this? What can be learned from this scriptural detail?

First, the Father clearly shows His affinity towards the sons and daughters of God, and an unfaltering desire for intimacy with each of us that is beyond comprehension. We have described in this work a novel understanding that the Intelligence within each of us is eternal, uncreated, because it is in fact a piece of

---

[31] See our previous discussion of this subject, Who Are The Gods Of Creation?, in Chapter 6, pg. 184.

[32] We have already discussed the significance of Michael not responding to this call of the Father as he had done previously on days 1 through 5 of the Physical Creation.

the Father, literally from Him and of Him. In that respect, it is understandable why His desire of intimacy with each of us is so strong, so driven. The Father's investment in each of us is of extreme value to Him. His personal interest in the development of the tools that are necessary for each of us to become like Him is clear.

We can only assume there may be certain elements of the creation of our Physical bodies that require the exercise of the powers of Deity that are held by Him alone, not shared with the Son. In His mortal ministry, Christ declared, "*my father is greater than I*" (John 14:28); perhaps in the creation of the sons and daughters of God, specifically, we can find evidence that this is so. Whatever the case, the Father, co-creates man in both the Physical sphere and the Temporal sphere.

The processes of the Physical Creation of man resulted in the embodiment of our independent Intelligences, within immortal Physical bodies that exactly mirror those we inhabit in mortality. This embodied Intelligence we have defined here as "*Spirit.*" The verses of the Genesis 1 accounts and the Endowment narrative inform us that this embodiment is in the *image* and the *likeness* of the Father and the Son. We will speculate later in the next chapter of this work that after his entry into the Garden, concurrent to his receipt of his new name and his becoming "*as a little child,*" Michael (now to be known as Adam) received the overlay of *Flesh and Bone* to his Spirit through some wonderful process involving both Our Heavenly Father and Mother. Within the Physical sphere, both of these processes (the Physical embodiment and the overlay of *Flesh and Bone*) intimately involved the Father.

***The processes of the Physical Creation of man resulted in the embodiment of our independent Intelligences, within immortal Physical bodies that exactly mirror those we inhabit in mortality.***

We have described the Temporal as necessary, providing for the uncertainty required for a truly "*free*" agency. The Father's involvement in the creation of the Temporal is largely confined to the withdrawal of His Divine presence. We spoke at length in Chapter 3 of the *Ruach* and the *Neshama*.[33] The *Ruach* being the breathing of the Spirit of the Father into our Temporal bodies and the binding it provides to firmly attach His Spirit (the *Neshama*) to that of our own. Within the Temporal, through the insertion of the Father's Spirit, the *Neshama*, and with the

---

[33] See IN THE IMAGE OF GOD, in Chapter 3 of this work, pg. 84.

acquisition of a body of *Flesh and Bone*, we all take on the *image* and *likeness* of both the *Father* and the *Son*.

The Son is the co-creator of our Physical Bodies. He established the organization necessary for that form to take place and accompanied the Father at the moment our individual Intelligences were placed into that form. The Son shepherded the elements of the Temporal sphere as well, as the Father withdrew, creating the conditions necessary for the uncertainty that reigns here. That uncertainty required information and caretaking to ensure the conditions developed that would eventually provide the perfect conditions for the bodies of Temporal man to thrive.

The Son remains in place to watch, guide and intercede when required to safeguard the sons and daughters of God, empowering us to return again to the precincts of our Father in Heaven. Through His development of (His brooding over) the very elements of Temporality, and by His direction of the organization of spiritual matter in the Physical sphere, the Son, Jehovah, is in a very real sense our Creator and can rightfully speak of Himself as so.

## God is Spirit

We believe that God the Father and the Son are Personages of Spirit Exalted in their *True Form* with an immortal body of *Flesh and Bone*. We understand that a *True Form* body has an attribute that fixes it in place in some way within the Physical sphere, though we admit to not knowing fully how that placement is sensed and realized in the Temporal sphere.

We do know there are attributes of both the Father and the Son which are comprehended, and influence all places, all at once. These attributes have been given various names: Glory, Shekinah, Spirit of God and Light of Christ. By extension, these attributes are also realized and spoken of in other ways, among them are: the Logos, Wisdom, the reason within each of us, the Word of God and the Spirit of Truth.

We have discussed that Personages of Spirit have the power to indwell, but have surmised that this power does not extend to beings who possess the *true form*, that of Spirit tabernacled in *Flesh and Bone*. We must, therefore, address the omnipresent "*spiritual*" attributes (or perhaps *essences*) of the Father and the Son, being careful not to confuse them with those of the Holy Ghost.

### *The Neshama*

The Savior said, "*God is a Spirit...*" (John 4:24). We have discussed previously that the Greek word translated in the KJV as "*Spirit*" is *pneuma* (πνεῦμα), and we understood this to mean that God is substantial, not at all ethereal. Spirit in the original context, *the Greek context*, was Air, by extension breath, and was one of the four basic elements making up the material world. *Pneuma* (Spirit) then is material, tangible and measurable. To the Greek world, it was fundamental and was substance. We understand by this that God is tangible, embodied, but we also know that His compound nature includes an element that is ubiquitous, permeating all of Creation, fundamental to all that there is. We attest here that, *His Glory* (intelligence) is something intangible, not physically sensible, yet ultimately most persuasive.

When we entered the Temporal sphere, God added to our Spirit His own – *the Neshama*. It is bound to ours through the *Ruach* or the Breath of Life. The Neshama is the Spirit of God within us that is spoken of in Scriptures; "*Know ye not that ye are the temple of God and that the Spirit of God dwelleth in you?*" (1 Cor. 3:16).

*The Shekinah (Sh'khinah)*

The Shekinah is a localized representation of the Spirit, it appears confined to place and time. The Shekinah was within the Temple of Solomon; "*Now when Solomon had made an end of praying, the fire came down from heaven and consumed the burnt offering and the sacrifices; and the glory of the LORD filled the house*" (2 Chr. 7:1). It was seen proceeding the children of Israel in the wilderness, "*And the LORD went before them by day in a pillar of a cloud, to lead them the way; and by night in a pillar of fire, to give them light; to go by day and night*" (Ex. 13:21). It is the divine presence represented as light or fire.

Shekinah is represented by *Memra* in Aramaic. The word *Shekinah* or its Aramaic equivalent does not occur in the Scriptures. There is also no occurrence of the word in pre-rabbinic literature; its first appearance is to be found in the targums. The Semitic root of the Hebrew word *Sh'khinah* means *to settle, inhabit or dwell*; it can also mean *neighbor*, and it is often used to refer to birds nesting and nests. The Shekinah is referred to as the manifestation, *presence*, of the Spirit of God dwelling in Temporality with man.

In Rabbinic literature, the Shekinah is directly associated with the transforming Spirit of God understood to be the source of prophecy as described by Samuel:

> *"...when thou art come thither to the city, that thou shalt meet a company of prophets coming down from the high place with a psaltery, and a tabret,, and a pipe, and a harp, before them; and they shalt prophesy: And the spirit of the LORD will come upon thee, and thou shalt prophesy with them, and shalt be turned into another man." (1 Samuel 10:5, 6)*

Prophets of the Old Testament make frequent reference to visions of the presence of God (the Shekinah), particularly in the context of the Tabernacle or Temple. Ezekiel speaks of the "*presence*" of God within the Temple: "*And behold, the glory of the God of Israel was there, according to the vision that I saw in the plain*" (Ezekiel 8:4).

In passages that suggest a presence, later described in the Targums as the *Shekinah*, the prophets often spoke of visions of grandeur, of thrones and of a royal train. Isaiah wrote, "*I saw also the Lord sitting upon a throne, high and lifted up, and his train filled the temple*" (Isaiah 6:1).

Kabbalism advanced an understanding of the Shekinah as feminine. This essential innovation became so important and has remained so, that it has been said no other element of Kabbalism has won such a degree of popular approval. It is this framing of the Shekinah as a *feminine* divine presence, which perhaps more than anything else, distinguishes Kabbalistic literature from earlier Jewish literature.

## *The Light of Christ*

There is no doubt that our understanding of things spiritual is incomplete and often confused. Previously, we addressed this confusion of terms relating to things spiritual.[34] We recognize our ability to distinguish is diminished when we really don't understand. Such is the case with the "*Light of Christ.*" No doubt there are times when the "*Light of Christ*" is conflated with the *Neshama*, which emanates from the Father; we here in this work have come close to making this conflation ourselves at times. There may be some room found in our void of information for us to consider that the influence of the Holy Ghost, as we often understand it, is misappropriated and is in fact the actual workings of the "*Light of Christ.*"

The "*Light of Christ,*" we will discover, is fundamentally different than the two manifestations (realizations of the Spirit of God) we just discussed: the *Neshama* and the *Shekinah*. Here we will make a case for the "*Light of Christ*" to be the "*substance*" of law in the Universe, the application of reason, Wisdom itself. In this we will attempt to do nothing more than ascribe literal interpretations to the Scriptures themselves. In the end, we will take liberty to extrapolate and extend the understanding to suggest that the reach of *our* spiritual influence is unfathomable as well. In fact, we will find it is something we share all of the time, unknowingly for most. But for those who do understand, it is a real component of our eternal selves and the power behind the binding of Souls, one to another.

The Mosaic account of the creation details: "*the Spirit of God moved upon the face of the waters*" (Genesis 1:2). Abraham chronicled the same event: "*the Spirit of the Gods was brooding upon the face of the waters*" (Abraham 4:2). In both instances, the revealed Word of God is speaking of the infusion of the "*Light of Christ*" into the elemental substance (*the bohu*) of the nascent Temporal sphere.

We have described here that the "*brooding*" identified in these verses was the process God used to put into the elements of Temporality information they would need to fulfill their purpose.[35] We understand that information to be the totality of the laws of Physics, Biology and every other principle that governs and controls the operation of everything in the Universe. It was the Spirit of God(s) which was doing the moving (brooding), but what was deposited, put in place and infused into the substance of Temporality, is what is known as the "*Light of Christ.*" It should not go unnoticed that this brooding period literally (physically) transformed the Universe, separating out what was darkness and introducing the light (the *Light of Christ*):

> *"And the earth was without form, and void; and I caused darkness to come up upon the face of the deep; and my Spirit moved upon the face of the water; for I am God. And I, God, said: Let there be light; and there was light. And I, God, saw the light; and that light <u>was good</u>. And I, God, divided the light from the darkness." (Moses 2:2-4)*

---

[34] See <u>Describing an Intelligence</u> within this book, particularly footnote 14, pg. 306.

[35] See the discussion detailed in <u>Day 1</u> within this book, pg. 39.

We understand the "*Light of Christ*" to be intrinsic to everything, it is the guiding principle of the law; "*the light which is in all things, which giveth life to all things, which is the law by which all things are governed*" (D&C 88:13). As such, the "*Light of Christ*" is fundamental to all that there is in Temporality. The "*Light of Christ*" is the truth that powers the Universe. It is the force of law which governs the Heavens, the Sun, Moon, and stars and *all* the matter from which we and the Earth itself are made. It is the light that "*proceedeth forth from the presence of God to fill the immensity of space*" (D&C 88:12). Through the law, "*he comprehendeth all things,*" and "*all things are before him, and all things are round about him.*" By the law, "*he is above all things, and in all things, and is through all things, and is round about all things.*" Because of the law, "*all things are by him, and of him, even God, forever and ever*" (D&C 88:41). We attest here that "*law*" *is* the "*Light of Christ.*"

The Law, infused into the very substance of all that became the Temporal Universe, is the information that guides everything in their courses. By it, gluons are gluons, protons are protons, electrons find their orbits, and atomic nuclei find their balance. Particles form out of energy and are attracted by gravity against electrical resistance; stars are born in the gathering, and Hydrogen is transformed into Helium in the immense heat of their stellar cores. The result of these cosmic furnaces is the forging of the entire periodic table of elements. The Law fashions life by stringing base pairs together to form DNA, which governs the formation of an enormous spectrum of proteins all repeatedly fashioned in exacting detail. Energy is produced by the inanimate, absorbed by the living and converted into the processes of life. All of these things are for the "*benefit and the use of man,*"[36] that our Spirits may have a Temporal home, fit for Agency, which leads to our eternal progression.

The "*Light of Christ*" is not a *personage* or an entity or even an *essence* of Deity, but it is Divine Knowledge shared. Knowledge is the power of God; His Glory is His intelligence. In this context, we can understand where the "*Light of Christ*" was sourced, why we are to follow it, listen to it, why it is universally applied to all things, and how it is used by the Holy Ghost which is distinct from it.

Confusing our understanding, the "*Light of Christ*" has been given many names in the Scriptures by many different prophets. Joseph Smith himself has identified this fundamental aspect of creation as, "*the Spirit of Jesus Christ*" (D&C 84:45), "*the Spirit of the Lord*" (Mosiah 25:24), "*the Spirit of truth*" (D&C 93:26), "*the light of truth*" (D&C 88:6), "*the Spirit of God*" (D&C 46:17) and "*the Holy Spirit*" (D&C 45:57). We understand the Holy Ghost to be "*a personage of Spirit*" (D&C 130:22), and as such distinct from the "*Light of Christ,*"[37] but they are often conflated into a

---

[36] Doctrine and Covenants 59:18.

[37] Brother Bruce R. McConkie describes this distinction perfectly: "*The light of Christ is neither the Holy Ghost nor the gift of the Holy Ghost; but that member of the Godhead, because He along with the Father and the Son is God, uses the light of Christ for His purposes. Thus spiritual gifts, the gifts of God meaning faith, miracles, prophecy, and all the rest -- come from God by the power of the Holy Ghost. Men prophesy, for instance, when moved upon by the Holy Ghost. And yet Moroni says: 'All these gifts come by the Spirit of Christ' (Moroni 10:17), meaning that the Holy Ghost uses the light of Christ to*

single thing, confusing law with a being or entity, even among those who recognize the distinction between the two:

> *"The Holy Ghost and the Light of Christ are different from each other. While they are sometimes described in the scriptures with the same words, they are two different and distinct entities." (The Light of Christ, Boyd K. Packer, June 22, 2004, Ensign April, 2005)*

We understand, here in this work, three distinctions:

- The "*Spirit of God*" within each of us, is intelligence that is parted directly from our Father, and it is His *Ruach* which binds that intelligence to our Intelligence within our Temporal form.

- The "*Light of Christ*" and the "*Spirit of the Lord*" are the laws by which Temporality functions. They are what provide us with the conditions necessary for our progression. The Son is their source and the reason He is the Deity we are bound to in Temporality; here the Son represents the Father. He is Jehovah, the Creator of the Temporal sphere and the ultimate redeemer of the whole of that creation.

- The Holy Ghost is independent of the laws of Temporality and our individual Intelligences but works with them to remind us of the truth of the light within us. Of this, the Savior (the source of the light) said: "*the Comforter, which is the Holy Ghost, whom the Father will send in my name, he shall teach you all things, and bring all things to your remembrance, whatsoever I have said unto you*" (John 14:26). By His lack of a body composed of Temporal elements (even perfected, resurrected, perfect Temporal element), the Holy Ghost is not confined by the laws of Temporality or any restrictions they may enforce.

The laws of the Universe are, of course, universal, equally applying to all things. As such, they are fixed and everlasting; they cannot perish. President Harold B. Lee explained:

> *"That light never entirely goes out ...Its glow may be so dim that we can hardly perceive it, but it is there for us to fan into a flame that shall burn brighter with understanding and with knowledge..." (The Teachings of Harold B. Lee, Clyde Williams 1996, pg. 101)*

The "*Light of Christ*" is everywhere, falling upon all, just as sunlight and the rain do. The presence and sense of direction we receive from the "*Light of Christ*" is perceived in the same way as a blind man can feel the warmth of the day on his face and is always able to turn towards the Sun even though he cannot see: "*the light of truth; which truth shineth. This is the light of Christ. ... He is in the sun, and the light of the sun, and the power thereof by which it was made*" (D&C 88:6, 7).

---

*transmit his gifts. But the Spirit of Christ, by which the Holy Ghost operates, is no more the Holy Ghost himself than the light and heat of the sun are the sun itself*" (A New Witness for the Articles of Faith, Bruce R. McConkie, 1985, pgs. 257-258).

For the inanimate, the non-sentient, and other lower forms of life, the law which is the "*Light of Christ*" is narrow in scope, defining function without choice. To the freely agent, the sons and daughters of God, that law gains dimensionality, described now also by an inner sense of right and wrong, morality, a conscience; *"[all] are instructed sufficiently that they know good from evil*" (2 Nephi 2:5). This part of the "*Light of Christ*" is often referred to as *truth* itself.

Truth always has been, and it was part of us before we were born; "*Ye were also in the beginning with the Father; ...even the spirit of truth; ...Man was also in the beginning with God. Intelligence, or the light of truth, was not created or made, neither can be.*" (D&C 93:23, 29) Truth will remain at our side every moment that we live. It is this extended set of laws, the added dimensionality of truth, the conscience within, that the Holy Ghost most often works with when He effects our actions; we are admonished to, "*quench not the Spirit*" (1 Thes. 5:19) when He seeks to influence us.

When we enhance our understanding of the laws of the Universe, we can accelerate the acquisition of more knowledge of the things of God; "*the light which shineth, which giveth you light [and] enlighteneth your eyes, which is the same light that quickeneth your understandings*" (D&C 88:11). Early in the last Century, the First Presidency, using terms a bit different than the ones we have developed here, wrote:

> "*There is a universally diffused essence which is the light and the life of the world, 'which lighteth every man that cometh into the world,' which proceedeth forth from the presence of God throughout the immensity of space, the light and power of which God bestows in different degrees to 'them that ask him,' according to their faith and obedience.*" *(Receiving the Holy Ghost, Improvement Era, March 1916, pg. 460)*

We have presented in this work the concept of the Physical sphere providing the unseen substrate of the Temporal; we identified that substrate as the *Zero Point Field*. The "*Light of Christ*" permeates this field as well. We understand that this field informs all in a physical manner that imitates the fashion by which all is influenced by the "*Light of Christ*."

The "*Light of Christ*" defies description and is beyond Temporal comprehension, so too the *Zero Point Field*. The *Zero Point Field* is (as is the "*Light of Christ*" as well) to be found everywhere. *They* are within us and within all things. *They* surround us and surround all things. *They* fill the voids within the things that make up the Earth and the Heavens, indeed the entirety of the Universe. In all the immensity, without exception, the "*Light of Christ*" indwells; it is forever immanent, never absent.

The "*Light of Christ*" is knowledge, information, and the means by which that knowledge and information is received, exchanged and transmitted; in this it again fully emulates the *Zero Point Field*. We propose here the "*Light of Christ*" uses that field as the means of its propagation – perhaps itself being the *substance* of the *Zero Point Field* – the lasting reverberation (the energy) of the very creative words *spoken* by God so long ago.

The "*Light of Christ*" is not independent; it does not exist to act, but to be acted upon; it is not agent to itself, but it perfectly expresses the will of its source. It is not an entity nor a person nor a personage; it has neither form, nor personality. The "*Light of Christ*" and the *Zero Point Field* is not substantial, is not material, at least as we may be allowed to measure things.

Each of us is integral to this light. We leave our mark on the *Zero Point Field* by our every action and with each thought we entertain. Everything we do, every machination of our mind, sends out vibrations in this substrate, influencing, touching and impressing others who likewise interact with the *Field* and through it with us.

**_The "Light of Christ" is knowledge, information, and the means by which that knowledge and information is received, exchanged and transmitted._**

These motions in the *Field* make up the symphony of sound that joyfully, *at times distressingly*, sings the song of the combined chorus of all creation. Because of this *Field* and because of the "*Light of Christ*" from whence it came, none of us are ever alone; we each are, in a very real sense, sharing our journey in Temporality with *everyone* else.

I submit that had this understanding of the "*Light of Christ*" been fully realized in the opening centuries of the Christian era, there may have been no development of the concept of the Holy Ghost. Obscure New Testament readings that led to a realization of the Holy Ghost as an individual entity might never have been interpreted the way they were had the understanding of the "*Light of Christ*" we have put forward here been known. The eventual Arian controversy would not have led to the development of a Trinitarian doctrine, for there would not have been a concept of three, but only two. This would have been significant, in that without an understanding of the Holy Ghost as a *person*, yet ethereal, the embodied nature of the incarnate Christ might have represented a different context by which to understand His relationship to the Father.

In many respects, what the Trinitarian creed did was to strike a balance between the tangible and intangible elements of God, the primary intangible of that *triune* being the Holy Ghost. If all of the occurrences within Scripture of the Holy Ghost were attributed to the "*Light of Christ*" at work within the Temporal sphere, the divine Christ would have been understood in a fundamentally different way.

From its inception, the Church of Jesus Christ of Latter-day Saints accepted the triune arrangement of the Godhead that had long been accepted by Christianity of Tradition. It did so despite Joseph's strong testimony of direct association with both the Father and the Son. Today Latter-day Saints espouse a firm belief in the Holy

Ghost through experience with *its* promptings and the comfort they receive from *its* presence in their lives. At confirmation, the newly baptized are given the Holy Ghost as a *gift*, as if *it* can be held. In fact, the faithful are told they have a *right* to *its* companionship, as long as they remain worthy – admittedly a strange concept of right and privilege, even ownership and demand, given an understanding that the Holy Ghost is an independent, agent, fully realized member of the Godhead.

## Hearing the Softest of Voices, Our Heavenly Mother

Though a belief in a Mother in Heaven has become ingrained into Latter-day Saint thought today, Her actual existence (absolute confirmation of it) remains outside of our official theology. We recognize here that we fashion out of the substance of reason alone the existence of our Mother in Heaven. Though some may search for hints of Heavenly Mother within the Scriptures, definitive statements of Her actuality do not exist anywhere within the pages of the Bible, the Book of Mormon, the Doctrine and Covenants and the Pearl of Great Price, nor is there to be found reference to Her in the Temple narrative.

It should not go unnoticed that these multiple settings span millennia of recorded revelation and encompass a wide variety of contexts addressed by hundreds of prophets without a single word spoken about a Deity (for certain) of supreme importance. Despite this silence, *some call it a sacred silence*, we still feel compelled to believe in Her, as if we are drawn instinctively in our belief towards something we all know and feel deep inside; we follow an intuition born in the substance of Faith which seeps out of our Soul.

Within this work we have rejected the traditional Christian concept of creation *ex nihilo.*[38] We believe that all of the sons and daughters of God are the actual spiritual offspring of God, our Eternal Father; "*...we have had fathers of our flesh which corrected us, and we gave them reverence: shall we not much rather be in subjection unto the Father of our spirits...*" (Heb. 12:9). We accept literally the teaching of Paul: "*The Spirit itself beareth witness with our spirit, that we are the children of God*" (Rom. 8:16).

Our theology teaches us that through the power and by the eternal nature of the sealing ordinance, families and procreation are eternal. The LORD declared that those who keep the most sacred covenants of eternal marriage will inherit "*thrones, kingdoms, principalities, and powers, dominions, all heights and depths—*," for "*Then shall they be gods, because they have no end;*" and they will receive "*a continuation of the seeds forever and ever*" (D&C 132:19-20). Because we understand these blessings promised to us are attributes that make us like our Father, we understand those attributes to be His as well. God "*has no end,*" and that means He has a "*continuation of the seeds forever.*" It is upon the backdrop of this perspective, *that we understand parenthood to require both a father and a mother*, wherein the concept of Heavenly Mother comes into full view. We reason that if both a Father and a Mother are required for the creation of our Temporal bodies, and we are sealed eternally in pairs with the promise and expectation of eternal increase and procreation in mind, then both a father and mother must be required for the creation of immortal Physical bodies as well.

This reasoning was applied early in the church, and though there is no record of a formal revelation to Joseph Smith on this doctrine, some claimed he had on

---

[38] See <u>Creation out of Nothing</u> in Chapter 8 of this book, pg. 308.

occasion spoken of a Mother in Heaven.[39] The first formal recognition of the doctrine of Heavenly Mother was penned in 1909, by the First Presidency of Joseph F. Smith: "*man, as a spirit, was begotten and born of heavenly parents, and reared to maturity in the eternal mansions of the Father,'* as an "*offspring of celestial parentage;*" it further declared that "*all men and women are in the similitude of the universal Father and Mother, and are literally the sons and daughters of Deity.*"[40]

Also early in the 20th century, the Apostle John A. Widtsoe wrote that the eternal sphere "*is given radiant warmth by the thought that* ... [we have] *a mother who possesses the attributes of Godhood.*"[41] Finally in 1995, as that same century was about to close, the Church issued an official statement, "*The Family: A Proclamation to the World*;" In it, the Church codified that each person is a "*spirit son or daughter of heavenly parents.*"

How then are we to understand the concept of our Heavenly Mother in the vacuum left by the dearth of Scripture referring to Her? Let us consider two possible understandings here, *emphasizing the speculative nature of both.*

- Equivalency with the Holy Ghost

Some early Church teachings suggest the concept that the Holy Ghost is Heavenly Mother. Though never addressed before in this manner, I propose here that this concept can be understood in the following way: The Father (God) divided himself, creating the Son and the Holy Ghost, the Son created all afterwards, becoming truly a Father to all, but the Holy Ghost further subdivided forming each of our individual Spirits.

We have already discussed the concept that the embodied Spirit within each of our Temporal bodies is bound to the Temporal and back to God. Through the *Ruach* and the receiving of the *Neshama*, we all share something of the Father. However, if we are formed out of, being truly of the "*stuff*" of the Holy Ghost, we, as is true of the intimate physical relationship we have with our Earthly mothers, contain within us the very essence of the Holy Ghost. In this sense, if true, the Holy Ghost may rightly be considered as the Mother of us all.

---

[39] One such occurrence was recalled by Zina Diantha Huntington Young. When her mother died in 1839, she remembered the young prophet Joseph Smith telling her "*not only would she know her mother again on the other side, but 'more than that, you will meet and become acquainted with your eternal Mother, the wife of your Father in Heaven*'" (*History of the Young Ladies' Mutual Improvement Association of the Church of Jesus Christ of Latter-Day Saints,* Susa Young Gates, [Salt Lake City: Deseret News, 1911], pgs. 15–16).

[40] *Teachings of Presidents of the Church: Joseph F. Smith*, The Church of Jesus Christ of Latter-Day Saints, 1998, pg. 335

[41] John A. Widtsoe, "*Everlasting Motherhood,*" Millennial Star 90, (May 10, 1928), pg. 298.

Scripturally then, denying this possible truth, denying our heritage, through the pride of self-esteem over all else, could be understood to be *the sin that is unpardonable*. This understanding presents the Holy Ghost as perhaps something other than an independent third personage of the Godhead, something akin to the collective consciousness of God through which we all are intimately tied to our Father in Heaven.

***The endowment account identifies Elohiym as the name of our Father, depicting him as a separate and distinct singular personage of Deity. But that representation may only be symbolic.***

The analogy is structured as follows: In Temporality our material bodies contain elements of both our father and mother within every cell, combining together to structure and code our individual DNA. Our mothers, however, have a more intimate bond with each of us for we organically formed within her; the materials that comprise us are distinctly originally part of her.

A mother's love is strong; it is unfailing and ever hopeful. No matter what one may do to push it away, a mother's love always draws her back to us. Though our Father in Heaven has a perfect love beyond comprehension, the Holy Ghost remains intimate with us here in Temporality in ways that cannot be imagined. The Holy Ghost seeks us out when we stray; it loves us when we don't love ourselves. Through its fidelity, we can come back to the Father no matter how far we may wander. In these aspects, those that describe the fidelity of the Holy Ghost to each of us, it can certainly be understood that the Holy Ghost expresses itself – as only a Mother can.

- The Plurality of Elohiym

We have discussed here that the Hebrew word *Elohiym* suggests a plurality; it is the plural form of the Caananite *El* or the Hebrew *Eloah*. We have understood this plurality to reference the combined works of both the Father and the Son; "*So God* [Elohiym] *created man in his own image, in the image of God* [Elohiym] *created he him, male and female created he them*" (Genesis 1:27). This interpretation of the plurality suggested by *Elohiym* is universally accepted by Christianity, and we have also adopted that construal here in this work as it applies to both the Father and the Son accomplishing the tasks of creation, but perhaps *Elohiym* suggests something more.

The Endowment account identifies *Elohiym* as the name of our Father, depicting Him as a separate and distinct singular personage of Deity. But that representation may only be symbolic, and in fact as understood by the plurality suggested by the name, the one called *Elohiym* is in fact two, a male and a female perfectly bonded into one (*echad*), neither of them *ever* to be considered without the other,

"*Nevertheless neither is the man without the woman, neither the woman without the man, in the Lord*" (1 Cor. 11:11). This understanding is revelatory in that it identifies the *one* we call Elohiym as *two*, our parents, both Father and Mother. [42]

By this we can understand that Adam and Eve were visited, taught, and accompanied by *both* a Father and a Mother in the Garden when God walked with them "*in the garden in the cool of the day*" (Moses 4:14). When *Elohiym* announces in the Endowment narrative and again in similar fashion in the Genesis 1 accounts, "*let us go down and form man in our own likeness and image, male and female*," it was both a Father and a Mother who would do it. We can extend this understanding to the Spiritual sphere to the creation of the Son and our independent intelligences. The Creator there was really two inseparably bound into one, our Father and Mother in Heaven.

This understanding provides a pattern for us, the sons and daughters of God. It explains the importance of *celestial marriage* in the eternal progression of each of us and is entirely consistent with all that the Scriptures and the Endowment have revealed. Heavenly Father and Mother are a bonded pair, we understand that we too were formed in bonded pairs, "*both male and female created he them*." This bond is eternal and extends back to our creation and forward into our eternal future. Our individual realities are meant to be joined with another's in a way that is not to be recognized as separate, "*Nevertheless neither is the man without the woman, neither is the woman without the man, in the Lord*" (1 Cor. 11:11). What does this important context teach us? Gender is eternal; it was fixed simultaneous to our initial creation long ago as immortal perfect pairs, both male and female. Our Spirits are not gender neutral, gender has eternal purpose.

As is the case with many of the most valuable truths of the Gospel, our knowledge as it stands today about Heavenly Mother is incomplete. Perhaps the most enigmatic component of the Gospel is the concept of a Mother in Heaven. We cannot speak for certain here of anything regarding Heavenly Mother. Nonetheless, because of the Spirit, we have received enough knowledge to appreciate the sacredness of this concept and to comprehend the divine pattern it establishes for us as children of heavenly parents. Though we cannot read of Her in the Scriptures, we can all sense that She is there.

---

[42] Consider this understanding as it relates to and draws some confirmation from Genesis 5:2: "*Male and female created he them; and blessed them, and called their name Adam, in the day when they were created.*" Here, two separate and distinct individuals, a male and a female, a mother and a father, are considered by God to be one and given a single name which they both share.

## Post Hoc Ergo Propter Hoc

Temporality is a degradation of something more complete and more perfect. It is what fills the vacuum when God withdraws. It is created through subtracting from the whole. More importantly, Temporality is a catalytic state, a period representing not more than a few moments in a grand, eternal process of enhancement, change and progression. It is not a primal state, *nothing finds its beginning* or original identity here or anywhere like unto it. Understanding this is critical in avoiding the false cause fallacies that follow attempts made to equate God with man in times most distant past.

### *Finding the Beginning in a Perfect Past*

The Universe is not the perfect state it once was thought to be; Joseph Smith spoke in terms that suggested he had an understanding that it was. Because of revelation, and not by any intellectual process, Joseph learned of the eternal nature of matter and intelligence. Joseph *understood* that matter and intelligence to be the substrate and effect of the everlasting.[43] They are the underpinnings of eternity, the building blocks of the Universe we inhabit, the home of the family of God. Joseph and his associates understood God to exist inside of, or beside, this timeless reference frame; though He coexisted with these eternal things, He was not their source.[44] To them though, because of what their combined intellects had come to know, the eternal sphere began at the boundaries of space, just a few miles above their heads, or perhaps the boundaries of the Solar System but not much farther. We have presented here a different view. Joseph and his associates knew nothing of the science we have incorporated into the cosmological theology detailed in this work. We now look beyond the Universe to find the eternal.

For the Prophet Joseph Smith and much of those who followed after him, the Universe was static and unchanging, infinite and eternal in nature. Such would be the thoughts of Religionists and Scientists alike until 1923, just over 100 years after Joseph's first encounter with God. To all then, the Universe was a perfect eternal

---

[43] "*Intelligence, or the light of truth, was not created or made, neither indeed can be*" (D&C 93:29). "*There is no such thing as immaterial matter. All spirit is matter, but is more fine and pure...*" (D&C 131:7). Joseph Smith taught, "*Element had an existence from the time God had. The pure principles of element are principles which can never be destroyed: they may be organized and reorganized, but not destroyed. They had no beginning and can have no end*" (History of the Church, Volume 6, pg. 309).

[44] Parley P. Pratt wrote, "*Matter and spirit are the two great principles of all existence. Everything animate and inanimate is composed of one or the other, or both of these eternal principles... Matter and spirit are of equal duration: both are self-existent, they never began to exist, and they never can be annihilated... Matter as well as spirit is eternal, uncreated, self-existing.*" (History of the Church, Volume 4, pg. 55). Elder Pratt's comments here are riddled with conflated terms and mixed definitions. He speaks of "*spirit*" when he means to describe "*intelligence*" as we have defined it here. He describes matter and spirit as separate eternal things, when we know that "*spirit*" is a combination of organized and formed eternal matter and uncreated eternal intelligence. He improperly states that things within this sphere can be composed of spirit and/or matter; matter as he describes it here is the perfunctory component of the Temporal World, for things without "Spirit" can be composed of it. Yet this perfunctory matter is eternal in nature suggesting that to Elder Pratt the things of this Universe are everlasting.

thing. In their time, galaxies had not yet been discovered, let alone understood to each house perhaps hundreds of millions of dying stars. These, as of then yet to be discovered galaxies, themselves fantastically distant, are the primary building blocks of the structures in the vast expanse, not the stars as had been supposed. To Joseph, it must be remembered, the stars, because of their apparent faintness of light, their relative dimness compared to the Sun and the Moon, were of the lowest order of the heavenly things. He had no idea that our own Sun, representing the grandest of all in his schema, the *celestial*, is in fact just average in size and brightness of those lights he called stars, the *telestial*.

Much has changed about our understanding of the Universe over the millennia, most of it in the last century; our developed theological concepts must change with them. We must find the truths within the teachings of Joseph and others, separating those truths from the assumptions they drew when they applied their knowledge to them. Then we must apply those disclosed truths to the broader understandings we have today.

Joseph's "*principles*" of matter and intelligence are eternal; because this Universe is Temporal they are not intrinsic to our sphere as Joseph had supposed them to be. We have described matter and intelligence to be eternal because they emanate from (are partitioned attributes of) our Heavenly Father; they are fundamental to His being, and as such, primary to the Spiritual sphere. They are uncreated and indestructible because they are literally the substance of the Eternal God. With the conceptualization of the Big Bang model, the once static, perfect, eternal Universe became a created, changing thing, bounded in time. We know today that for *intelligence* and *matter* to be everlasting, they both must have their source from beyond the vastness of space stretched out above our heads.

No longer shackled to the Universe around us, because we have associated them to the Spiritual sphere, intelligence and matter have a more ancient foundation. Joseph said, "*Element had an existence from the time God had;*" he understood the principles of element to have "*had no beginning and can have no end.*" Taken together these statements have an important reciprocal understanding; they define matter and intelligence as eternal, but they also disclose that the image of the Father held by Joseph was an eternal one, one which always was.

Eternity is found in two parts: the uncreated self-existent, and created things having a confined past and defined beginning but having no end of existence extending into an infinite future. God, intelligence, and the principles of element are not of this latter created eternal order, but having "*no beginning,*" are infinite and unbounded by time.

In the next few pages we will discuss the desire of some to construct a chain of cause extending deeply into the past. In constructing the endless chain, the builders must eventually come to a First Cause. A First Cause has unique attributes that distinguish it from all other causes. No matter how many steps we choose to place between us and the *true* beginning, eventually we will get there, coming face to face with the God Joseph knew, the God of Isaiah 43, the God of the Bible, the Book of Mormon, the Pearl of Great Price and the Doctrine and Covenants. Indeed all volumes of Scripture speak of this ultimate God, the first and Universal Cause, the *first* Creator, our Father, He who sustains <u>all</u> things:

> *"He comprehendeth all things, and all things are before him, and all things are round about him; and he is above all things, and in all things, and is through all things, and is round about all things: and all things are by him, and of him, even God, forever and ever." (D&C 88:41)*

God, the source of intelligence, and the principles of element, is the substrate, the essence of which "*all things*" are "*before*" and "*consist*" (Col. 1:17). He becomes the focus of the eternal nature of all things. In His eternal state, God is perfect and described as unchangeable:

> *"For I know that God is not a partial God, neither a changeable being; but he is unchangeable from all eternity to all eternity." (Moroni 8:18)*

> *"For do we not read that God is the same yesterday, today, and forever, and in him there is no variableness neither shadow of changing?"*[45] *(Mormon 9:9)*

Both old world and new world prophets proclaim this invariability: "*For he is the same yesterday, today, and forever...*" (1 Nephi 10:18); "*For I am the LORD, I change not...*" (Malachi 3:6). Solomon singing out spoke of God, "*But thou art the same and thy years shall have no end*" (Psalm 102: 27); and the Apostle James added, "*Every good gift and every perfect gift is from above, and commeth down from the Father of lights, with whom is no variableness, neither shadow of turning*" (James 1:17). We must understand, however, that what is described here is not the traditional theological concept of immovability; God is not an unmoving mover, and perfection is not static and unchanging.

Eternity is a long time, and through it all our Father continues to grow and progress:

> *"God himself is increasing and progressing in knowledge, power, and dominion, and will do so, worlds without end." (Wilford Woodruff – Journal of Discourses 6:120)*

> *"...there is progress for our father and for our Lord Jesus. There is no such thing as standing still in the eternal work of our God. It is endless progress, progressing from one degree of knowledge to another degree." (George Q. Cannon – Millennial Star 61:117)*

The immovability of God in his being is not described in the Scriptures. God is steadfast in purpose and unwavering in His goal, but that does not speak to the immutability of His form. We see His singlemindedness in Samuel; "*The Strength of Israel will not lie nor repent: for he is not a man, that he should repent*" (1 Samuel 15:29). Here we must add that there is hierarchy to the singlemindedness of the LORD, for though we understand Him to be resolute and unwavering in his determination to "*bring to pass the immortality and eternal life of man*" (Moses 1:39), we have seen that the means He had employed to effect such a task in the case

---

[45] Where did Moroni read this text?

of the *ha-a-dam* (הָאָדָם) had "*repented the LORD ...*" (Gen. 6:6). For those still not convinced that God changes, moves, and interacts, consider the following statement of invariability from the Apostle James, "*Jesus Christ the same yesterday and to day, and for ever*" (Hebrews 13:8). James understood that the "*same*" Christ was Pre-existent God, Creator of the Universe and the Physical sphere, yet he left that behind to become incarnate in Temporality. He was born a babe, "*grew, and waxed strong in spirit,*"[46] doing all that He was commanded of the Father. He wept for the loss of a friend; He cried out. He plead for mercy asking that if it were possible "*take away this cup from me*"[47] at Gethsemane. He was beaten and crucified; He died. He preached to the Spirits in Prison and was resurrected on the third day. He hid Himself from His disciples on the road to Emmaus and later appeared in recognizable form to those gathered in the closed room. He ascended to Heaven in glory then appeared to Paul on the road to Damascus. This represents a lot of changing (waxing) for Deity who we are told is *forever the same*. Change then is an attribute of God, an expectation of it must be part of any correct concept of the Father.

As this, the *Dispensation of the Fullness of Times* opened, Joseph saw God. The First Vision established for Joseph what had been known by many prophets before him, that God the Father has an embodied form. This began a progressing understanding within the boy prophet that led to the realization that man was of the same family as God and eventually led in the last years of Joseph's life to the truths of the Atonement and our destined place as exalted beings, heirs to all that the Father has. The *deification* of Man was not a new concept, though its full understanding had been lost millennia ago. Through Joseph, the Doctrine of Theosis was to be given new birth.

Deification, as we described it in this work, does not bring God down to man, but restores man to God from whence we came. We understand that we are created sons and daughters of God; because we are of the family of God; the traits of deity are intrinsic to us. We concur with the Apostle Paul in condemning those who lessen the station of our Father as He is now, and as He ever has been. We declare that those who do, profess "*themselves to be wise*," but "*they became fools*" because they have changed "*the glory of the incorruptible God into an image made like to corruptible man*" (Romans 1:22, 23).

Whatever Glory we hope to attain in an eternal future, we *always* remain subordinate to the Father, becoming gods, not God. We are creatures destined by God's Grace to be *like* Him, not equal to Him. Though He may desire that we attain such a lofty station, it is not possible because every step we take toward Him, moves Him forward as well. Our growth is His growth, adding the total sum of all that is gained by each of us to His already superior presence — this is truly Infinite and Eternal increase.

---

[46] Luke 2:40.

[47] Mark 14:36.

We have covered a lot of ground here. We have detailed that God is the primary eternal, having no end, but more importantly having no beginning. He is the First Cause; this is as Joseph understood Him and as all Scripture, ancient and modern, attest.

Intelligence and the Principles of Matter are eternal in this same respect, neither created nor capable of destruction. We have affirmed that this is so, because both emanate from God who is their original source. Our God continues to progress and change through the acquisition of experience (through and by us) and adds continually to His knowledge, wisdom and power. God is perfect, though not static; there is no degradation from that state possible, either in the future or at any time past. These are the truths revealed to Joseph through revelation. His understanding of what these truths meant, and how he applied what he understood to the world around him, however, was flawed, for it was not complete. Joseph lacked only in knowledge, not in wisdom. With this in mind let us press forward and examine how limited understanding can distort the truths of God fully and completely revealed.

## *Failing Logic, the Reason Within*

Before the century that saw the birth of Joseph Smith was over, the massive amount of information received and dispensed by the young prophet had begun to be interpreted by others. Among those interpretations, one stands out: "*As man now is God once was; As God now is, man may become.*" This oft repeated utterance of the fifth president of the Church, Lorenzo Snow, is a couplet; as such, it is composed of two separate phrases. Though never officially canonized, it finds its way into Mormon conversation all too often, if however infrequent. It may sometimes be chewed whole, but most often it is swallowed only in part. The second phrase of the couplet is all but universally accepted and doctrinal, but not without modification, for no Latter-day Saint accepts that any man who ever was will equal God, in *any metric* at *any time*. The first phrase of the couplet is almost always discarded as undecipherable, unimaginable, non-canonical and certainly not core to our Faith.

***Whatever Glory we hope to attain in an eternal future, we always remain subordinate to the Father, becoming gods, not God.***

When asked if God once was a man, Gordon B. Hinckley (while he was the fifteenth prophet and President of the Church in the modern era), said, "*that gets into some pretty deep theology that we don't know very much about. I don't know that we teach it. I don't know that we emphasize it. I haven't heard it discussed for a long time in public discourse. ...I understand the philosophical background behind it. But I don't know a lot about it and I don't know that others know a lot about it.*"[48] To put President Hinckley's parsing of words in perspective, consider another exchange that followed the first by just moments. When a follow up question was asked about the Latter-day belief in the divine potential of human beings, "*Do Latter-day Saints believe that men may become gods?*" President Hinckley responded, "*Well, as God is, man may become. We believe in eternal progression. Very strongly.*"[49]

How did a belief that God was created a mere mortal and progressed to Godhood from that base-state get started? How has it become to be associated with the doctrines of the Restoration? Is it even correct?

First, let us answer the last question. No, a belief that God was mortal before He became God is not correct; Joseph Smith did not teach that. Such a notion was debunked earlier in this work when we discussed the concepts of First Cause and Eternal Cause. Our Universe, the three creative spheres, the existence of all things, indeed existence itself, demands there to be a singular universal cause that is a true First Cause for all things. Though it is true that our Creator could have had a creator before him, logic demands that chain to come eventually to an end. Which end would have to be the First and Universal Cause, an infinite uncreated, Creator, truly God. It is not possible to extend a chain of Fathers (or causes) indefinitely into an eternal past.

Infinity is possible; it is a concept we have embraced in this work, but it has two distinct forms. Though science understands our Universe to have had a beginning point, it expresses both Space and Time in infinite terms; that, however, is a resultant of our internal frame of reference, a construct of the fabric of the Universe itself. It is a false image of infinity, properly expressed as, *finite but unbounded*[50] –it *is not* infinity at all. The logical affront caused by this is mitigated by some by suggesting that this Universe is just one of an *infinite* chain of such places, *a multiverse*[51] that pushes both space and time into *something* beyond both. This

---

[48] Interviews with San Francisco Chronical (Don Lattin, "Musings of the Main Mormon," Apr. 13, 1997) and Time Magazine (David Van Biema, "Kingdom Come," Aug. 4, 1997, pg. 56).

[49] Ibid.

[50] Space and time are both constrained and made infinite by the very confines of the Universe itself. Before the Big Bang, there was neither space nor time to measure with. Space itself is what the "*stuff*" of the Universe is indexed against; without it, in any place beyond the Universe, there is neither any way of measuring nor anything to measure. It is not possible to step outside of the confines of the Universe (though we may imagine such a threshold); it cannot be reached or crossed. Space then is finite, with no boundary. The same construct occurs with time. It did not exist before the initial instant of the Big Bang, for there was no way of measuring it, theoretically then there exists the condition of no time. Without time, there is no before, so we have a beginning without a construct in which to place it. Time once started can never end, it is measured against something, and as the expanding Universe moves toward its distant totally empty end, time will forever slow extending it in theory into an infinite future though it is understood there will be no way of determining its running.

[51] The concept of a multiverse is completely unsupportable by any discipline of science. It is a fallacy and can only exist in the minds of those who stubbornly seek a creation by another form, simply to avoid God. See the Appendix, Boltzmann's Paradox, pg. 537; Entropy the Great Equalizer, pg. 538, Do You Feel Lucky, pg. 539 and The Impossible Universe, pg. 540.

logical fallacy employed by atheistic modern scientists is broken; it runs headlong at full speed into the stops of First Cause. The second form of infinity is truly unbounded, for it is without beginning or end. Any construct that does not have a beginning or end, no matter how big, falls inside of this form of the infinite.

President Snow's couplet follows the formula, *Post hoc ergo propter hoc*. This phrase is Latin, meaning, "*after this, therefore because of this*." The logical structure can be stated as follows: "*Since event Y followed event X, event Y must have been caused by event X*." In the case of "*As man now is God once was; As God now is, man may become,*" the logical argument is turned around. President Snow did not express his thoughts as a revelation from God, so we can be certain he did not express them as facts; as such, *is* and *may* can be taken as *can be* and *could*. Properly formed and phrased the argument should be, "*Because man can become God, God could once have been man*." The couplet receives its force because it assumes a relationship between the variables. President Snow (who wrote it) and the many that have formed their opinions of the eternal nature of God around the couplet, assume that our Heavenly Father became God in the same manner we are to become like Him.

A correlation between two variables, however, no matter how closely related they may be, does not always imply that one causes (or follows) the other. Consider this: *When rock salt sales increase, traffic deaths also seem to increase. Hence, buying rock salt leads to traffic accident deaths*. In this case, a *confounding variable* is a factor that impacts both the cause and effect. Generally, during the extreme cold of the winter months, rock salt sales increase to be used on slick roads fouled by ice. Independent of that, yet related, the very ice the rock salt is purchased to deter is a cause of dangerous conditions—that would, of course, contribute to increased traffic accidents leading to an increase in auto accident fatalities. The correlation between the two variables, however, does not give one license to conclude that increased rock salt sales are a *cause* of traffic related deaths. In this case, the logical argument is flawed.

Forming an argument about the being of God around this logic is not new; it was recognized and took form early in Christianity. Late in the second century, using almost the same cadence as President Snow, Irenaeus wrote, "*How then will any be a god, if he has not first been made a man? How can any be perfect when he has only lately been made man? How immortal, if he has not in his mortal nature obeyed his maker? For ones duty is first to observe the discipline of man and thereafter to share in the glory of God*." (Adversus *Haereses, IV. xxxix. 1-2*) [52]

Irenaeus needed no modern day revelation on which to base his argument for the progression of God; he found ample source for his logic within the nascent pages of the New Testament and the Gospel he heard preached by those who were taught by the Apostles themselves. President Snow, however, found the basis for his belief in the hierophantic teachings of the Prophet Joseph Smith.

---

[52] The Early Christian Fathers, Henry Bettenson (Oxford University Press, New York, 1956) pg. 69.

Lorenzo Snow was not alone in the beliefs expressed in his couplet. Indeed, he did not even originate the concept that God was once a man. Throughout the extended history of the Church, many leaders have taught the basics of President Snow's couplet, though perhaps less famously.

An overarching theme found in this work, is one which suggests that revelation given is not always received; it is often not understood, nor perhaps is it intended to be initially.[53] We have seen that it was most often impossible for the recipients to fully understand, for they not only lacked the ability to fully comprehend spiritual things, but they also lacked knowledge of the workings of things Temporal, in this Universe.[54]

Joseph attempted to understand what was revealed to him, often not fully aware of just what it was he had received.[55] (Such was true of the First Vison. It took him over a decade to fully absorb what he came to know that initial contact had revealed. This is apparent in his varied tellings of the same scene and in its belated official recording altogether.) Joseph, over time, began to comprehend what had been revealed (understandably, imperfectly), and related what he learned to others; his ability being hindered by both the human condition, which prevents us from grasping fully the things of God and the limitations of language, itself not capable of accurately describing the supernal.

Whatever Joseph said or recorded was subsequently received and processed into the minds of others, developing yet another layer of realization, further compounding

---

[53] Orson Pratt spoke of the revelatory process as one of iteration, incomplete at first and incremental: "*But some may inquire, are not the Almighty's revelations perfect when they are first given? And if so, where was the propriety of the Lord's adding any thing to them, when they were already perfect? We reply that every word of God is Perfect; but he does not reveal all things at once, but adds 'line upon line, precept upon precept, here a little, and there a little,' revealing as the people are able to bear, or its circumstances require.... The Lord, therefore, adds to His own revelations whenever he thinks proper.*" (Explanation of Substituted Names in the Covenants, Orson Pratt, The Seer 2.3 (March, 1854); pg. 229.)

[54] Brigham Young (himself the recipient of the revelation published as Section 136 of the Doctrine and Covenants) said of this limited ability of man to receive and understand: "*I do not even believe that there is a single revelation, among the many God has given to the Church, that is perfect in its fullness. The revelations of God contain correct doctrine and principle, so far as they go: but it is impossible for the poor, weak, low, groveling (sic) sinful inhabitants of the earth to receive a revelation from the Almighty in all its perfections. He has to speak to us in a manner to meet the extent of our capacities... The laws that the Lord has given are not fully perfect, because the people could not receive them in their perfect fulness; but they can receive them a little here and a little there, a little today and a little to-morrow, a little more next week, and a little more in advance of that next year, if they make a wise improvement upon every little they receive...*" (The Kingdom of God, Brigham Young, Journal of Discourses (8 July 1855) v2, pg. 314.)

[55] Elder John A. Widtsoe would explain: "*The language [of the Doctrine and Covenants], with exception of the words actually spoken by heavenly beings, is the language of the Prophet. The ideas were given to Joseph Smith. He wrote them in the best language at his command. He was inspired at times by the loftiness of the ideals so that his language or words are far above that ordinarily used by a backwoods boy of that day.*" (Message of the Doctrine and Covenants, John A. Widtsoe, edited and arranged with foreword by G. Homer Durham (Salt Lake City, Utah, Bookcraft, 1969), pgs. 4-9.)

the error inherent in the *Natural* man. In time, simple truth assumed complexity of thought, and details were added that never were in the first iteration. What began as a desire of our God to demonstrate the divine nature of man, was turned around and transformed into heresy, resulting in an image of God in some minds that had Him assuming attributes which became ever more human.

As the theology developed, the god it described became mortal, born as a child, progressing not only in stature but elemental knowledge:

> *"Remember that God, our heavenly Father, was perhaps once a child, and mortal like we ourselves, and rose step by step in the scale of progress, in the school of advancement; has moved forward and overcome, until He has arrived at the point where He now is."(Orson Hyde – Journal of Discourses 1:123)*

> *"He is our Father-the Father of our spirts, and was once a man in mortal flesh as we are, and is now an exalted Being." (Brigham Young – Journal of Discourses 7:333)*

The fatal flaw in this logic extends the concept further, pressing the error into previous supposed generations. We are left to wonder: *If God was once mortal and was born a child in Temporality like us,* would God then have had a father as well, perhaps even an endless chain of fathers before Him? These questions have been considered before:

> *"How many Gods there are, I do not know. But there never was a time when there were not Gods and worlds, and when men were not passing through the same ordeals that we are now passing through." (Brigham Young – Journal of Discourses 7:333)*

> *"God is a natural man... Where did he get his knowledge from? From his Father, just as we get knowledge from our earthly parents." (Heber C. Kimball – Journal of Discourses 8:21)*

> *"But if God the Father was not always God, but came to his present exalted position by degrees of progress as indicated in the teachings of prophet, how has there been a God from all eternity? The answer is that there has been and there now exists an endless line of Gods, stretching back into the eternities." (B.H. Roberts – New Witness for God 1:476)*

> *"God is an exalted Man... The Prophet taught that our Father had a Father and so on."(Joseph Fielding Smith – Doctrines of Salvation 1:10, 12)*[56]

---

[56] The proliferation of the views expressed in these quotes was so widespread among the brethren that it has been said: *"Mormon prophets have continuously taught the sublime truth that God the Eternal Father was once a mortal man who passed through a school of earth life similar to that through which we are now passing. He became God-an exalted being." "Yet, if we accept the great law of eternal progression, we must accept the fact that there was a time when Deity was much less powerful than He is today." "Thus He grew in experience and continued to grow until He attained the status of Godhood."* (The Gospel Throughout The Ages, Milton R. Hunger – pgs. 104, 114-15.)

Joseph Fielding Smith, as late as the 1970's, carried within him the concept of an eternal Universe consisting of the eternal matter and intelligence Joseph had described.[57] He would, as all the other men we have quoted here, as well as many scientists of the time, reject the premises of the nascent Big Bang model.

On theological grounds, these men and Joseph Fielding Smith would have understood the Big Bang to have been a challenge to their cosmology, denying, in their view, the Glory of God and presenting a secular explanation for creation. Though he may likely have been very much aware of the Big Bang theory, it was beyond Joseph Fielding Smith's ability to understand that, for the first time in scientific history, here was a theory that demanded a Creator and the existence of an external, eternal construct. The Big Bang placed scientific theory face-to-face with a beginning and all that beginning would encompass. Eventually the *hard stop* of a beginning would lead to a logical need for a Creator, a secular conundrum that purportedly was resolved by a purely imaginary construct, an endless chain of beginnings, *a multiverse*. Such a multiverse cannot be so; our understanding of First Cause and entropy says otherwise.

The lack of understanding that the entire Universe is Temporal, and as such it is a creation within creations, led to gross misunderstanding of the beginning state of God. Contrary to the *man god* that came to be understood, the God Joseph described was much different, not very much unlike that described by the orthodox Christian Trinitarian concept. That God is a single deity in three defined distinct personages, one of which was at one time mortal, lived on Earth as we do now and is now exalted both man and God; Christ *became* what his Father is. Joseph, however, had seen the Father, so his developed theology had to incorporate that truth. Like the ancients of old, the young prophet dealt with Scripture that told two stories. The bishops of the fourth century resolved the issue by conflating two, and eventually three, into one; Joseph laid the three side-by-side and end-to-end.

What is clear is Joseph Smith did not imagine the Father having had a father; that was an error made later. For Joseph, and for us in this work, God had always been God, just not always what He is today, for Joseph's God and ours, eternally progresses.

---

[57] See our previous discussion Finding the Beginning in a Perfect Past, pg. 453.

## *Returning to the Source*

The birthplace of the misunderstanding about an endless chain of fathers stretching back into the past can be found among the sayings of Joseph Smith. Perhaps the most direct teaching we have (certainly the best documented and most repeated) was delivered by the Prophet at the funeral of his friend, King Follett. Joseph Smith was recorded as having said:

> *"God himself was once as we are now, and is an exalted man, and sits in yonder heavens! That is the great secret. If the veil were rent today, and the great God who holds this world in its orbit, and who upholds all the worlds and all things by his power, was to make himself visible, —I say, if you were to see him today, you would see him like a man in form —like yourselves in all the person, image and very form as a man; for Adam was created in the very fashion, image and likeness of God, and received instruction from, and walked, talked and conversed with him, as one man talks and communes with another.*
>
> *In order to understand the subject of the dead, for the consolation of those who mourn for the loss of their friends, it is necessary we should understand the character and being of God and how he came to be so; for I am going to tell you how God came to be God. We have imagined and supposed that God was God from all eternity. I will refute that idea, and take away the veil, so that you may see.*
>
> *These are incomprehensible ideas to some, but they are simple. It is the first principle of the Gospel to know for a certainty the Character of God, and to know that we may converse with him as one man converses with another, and that he once was a man like us; yea, that God himself, the Father of us all, dwelt on an earth, the same as Jesus Christ did; and I will show it to you from the Bible." (King Follett Discourse, Apr. 7, 1844, available at josephsmithpapers.org)*

Here we find the answer to the question: How did the Mormon belief that God was created a mere mortal and progressed to Godhood get started? The source of the doctrine found in the Lorenzo Snow couplet was uttered by Joseph Smith; "*God himself was once as we are now*." Discovering what the Prophet Joseph Smith meant the day King Follett was memorialized, is critical to our understanding as it relates to the subject of this book. Was it the intent of the Prophet to communicate that God was once merely mortal? A thorough examination of his complete statement, within the contexts of this work, will show us that his intent was otherwise.

God, Joseph says, "*holds this world in its orbit*." That may be a limited God, but Joseph goes on to say, He is the one "*who upholds all worlds, and all things by his power*." The God Joseph described encompassed *all* things, sustained *all* things, and upheld *all* things by His power alone. This is the God who speaks to us in Isaiah 43. Simply put, this is NOT a God who was created and has a God greater than He or even an equal. Everything else Joseph said that day must be understood in this context. The "*great secret*" Joseph was to disclose wasn't that God *once was* a man, but that in His glorified, eternal deity, He *remains* a man to this day.

The intent of Joseph's remarks were not to describe a God who progressed from a lesser to a greater state, but the young prophet desired to underscore the personal nature of our Father in Heaven. Joseph sought to emphasize that our God is a "*person*;" He is an "*Exalted man*," in the "*image and very form as a man.*" Joseph paraphrased the Genesis 1 account to elaborate: "*for Adam was created in the very form and fashion, image and likeness of God, and received instruction from, and walked, talked and conversed with him, as one man talks and communes with another.*"

The context of the Prophet's statement that day was "*consolation of those who mourn for the loss of their friends.*" Joseph said to do so, "*it is necessary we should understand the character and being of God.*" For Joseph, and as we have detailed in this work, God was not and is not immutable, immovable or impassable. To express such belief, Joseph said he would refute the idea that, "*God had been God from all eternity.*" Joseph sought to express here that the Eternal God had not always been in the station He holds now. God is capable of change, evolution is a function of His true being; progress is who He is.

Joseph envisioned a God who had come down from the lofty heights of Heaven, for he saw Him in the Grove. Joseph's God does move, experience passion and has transformed. The God Joseph described that day to those who had gathered for the memorial, was accessible and active in our lives because He had experiences that can directly relate to ours, His children.

These evolutions are not to be understood as progression of a *creature* from a lowly state to one of grandeur, but as growth inherent to a dynamic eternal divine. God is not static perfection; He becomes *more* each day. It is this accumulation of knowledge and experience, acquired through motion and feeling, that has made our Father the God He is today, something other than the God He once was; it is literally the process by which "*God came to be [the] God*" He is today.

When Joseph spoke, "*God is an exalted man*," he did not intend to convey that God was ever anything other than an exalted being, for such was not possible in the Prophet's eyes. So what did the Prophet mean when he said: "*he was once a man like us*"?

Christ, in His mission to minister to the sons and daughters of God at the meridian of time said, "*The Son can do nothing of himself, but what he seeth the Father do*" (John 5:19). Joseph drew upon this Scripture as a proof text to substantiate what he was about to declare: "*God himself, the Father of us all, dwelt on an earth, the same as Jesus Christ did.*" Joseph understood the Savior's words literally.

An incarnate God, the very Son of the Father, had laid down His life to take it up again, because He said, "*this commandment have I received of my Father*" (John 10:18). In this context, Joseph, and we as well, can understand that the Father had also sacrificed (at some point in the past) His perfected *True Form* to assume an imperfect, Temporal incarnation for some purpose as yet unknown to us.

***...Joseph, and we as well, can understand that the Father had also sacrificed (at some point in the past) His perfected true form to assume an imperfect, Temporal incarnation...***

We must emphasize here the importance of the context – Joseph did not state or imply that the Father was once a *creature* of a God before him. This truth is made clear; Joseph said, God "*dwelt on an earth, the same as Jesus Christ did.*" God did not dwell on an Earth as Adam, or you and I do, but as an incarnate perfect divine, as a Creator taken in Temporality in a manner only experienced on Earth by the Savior. This reality is what was meant by Joseph Smith when he uttered what he did in the King Follett discourse.

We understand here then the basic understanding behind the famous couplet, *As man now is God once was; As God now is, man may become,* to be flawed. *Post hoc ergo propter hoc* represents, in concept, a logical fallacy stemming from a questionable cause. The correlation between the two variables does not imply that one is caused by the other. Though we are His children, and like Him, of the family of God, our path to exaltation does not infer God followed a like one, having a similar beginning.

Our God was not created by a God before Him and evolved from a carnal state to one of exaltation through processes of Grace and redemption performed on His behalf by another. God was not at any time *created* out of uncreated eternal intelligence but is the source of that intelligence, because He is Himself eternal and uncreated.

## *An Alternate View —The LDS Trinity?*

There is perhaps some room for us to consider yet another interpretation of the sayings of Joseph Smith and the revelations he received, another way to sound out the depths of what we can know about God. We have discussed here the facts that the Trinitarian God of Christian tradition was Himself once both fully mortal *and* a son of His Father. These truths conveniently align the words spoken by Joseph in the King Follett discourse with Christian Dogma without modification. Latter-day Saints reject wholeheartedly the Trinitarian Creed as a viable conceptualization of God and proudly denounce it as corruption of divine truth. However, in fact, such a compounded concept cannot be so easily distanced from our Scriptures. Section 49 of the Doctrine and Covenants is one place where this point is made (perhaps even painfully) clear.

In it we read: (Verses 5-7)

> *"Thus saith the Lord; for* ***I am God, and have sent mine Only Begotten Son into the world*** *for the redemption of the world, and have decreed that he that receiveth him shall be saved, and he that receiveth him not shall be damned— And they have done unto the Son of Man even as they listed; and he has taken his power on the right hand of his glory, and now reigneth in the heavens, and will reign till he descends on the earth to put all enemies under his feet, which time is nigh at hand— I, the Lord God, have spoken it; but the hour and the day no man knoweth, neither the angels in heaven, nor shall they know until he comes."*

And: (Verse 28)

> *"****Behold, I am Jesus Christ****, and I come quickly. Even so. Amen."*

This section of the Doctrine and Covenants dutifully records a single entity who declares Himself to be *both* God the Father and the Son, Jesus Christ. Here there are no errors of translation, no corruption of text, no dogmatic preferences altering interpretation. In one single contiguous revelation, just 826 words in length, received in one single sitting, to one modern prophet, in modern English, we have one Deity speaking as if He were two.

By this we can understand that the lines between the Father and the Son are not as clearly defined as we would perhaps like them to be. The separation between the Father and the Son (though distinct in actuality as witnessed by Joseph in the First Vision) on paper, in the revealed Word of revelation, is sometimes hopelessly entwined. This entanglement of terms and phrases leaves us less than capable of boldly declaring our theology above and beyond the blatant contradictions and heretical conclusions associated with Trinitarian doctrine. We are (as it seems Joseph Smith was) in the end, at times left confused ourselves, incapable of expressing the divine completely in terms understandable to mortal man.

Joseph, saw both God the Father and His Son Jesus Christ as separate and distinct personages in the Sacred Grove. The concept of their individual beings was core to his faith and all that he taught. Yet here, in this revelation received late in the spring of 1831, Joseph would hear the Son, "*Behold, I am Jesus Christ,*" say that He was the Father who had sent Him, "*I am God, and have sent mine Only Begotten Son into the world.*" If Joseph, knowing what he did, could record such a message, it should not be hard to imagine the difficulties which faced those much less informed and aware, when they tried to understand things of God only available to them in ancient languages recorded by those from lands and cultures far distant from their own.

It is possible in the context of the passages of section 49 that Joseph's words describe a God not unlike the Trinitarian God of tradition, a God who was mortal on an earth, one who had a father yet somehow also a God who was uncreated and forever unchanging and immortal.

- Sorting it Out

Though we "*very strongly*" believe in the eternal progression of man, it does not mean we must believe that God was once a created being like we are, nor should our belief imply that.

There are truths to be learned from the parallels developed by such a concept though. Any understanding that God is an exalted man serves the purpose of illustrating the potential that is within each of us, helping us in conceptualizing the magnitude of the gift we have received from our Father in Heaven. It is descriptive in the way it closes the ontological gap between God and man, leading us to recognize that we are of the same *nature* as God our Father.

The ontology is not arrogant or self-aggrandizing. As we have detailed the relationship here, we understand that we are literally the children of our eternal unbegotten Father. He has shared in a very literal way of His own substance (intelligence) to create each and every one of us. We are of God, because God is within us; the essence of the Father is our reason in every sense of the word; it is the source of individuality, self-awareness, and it was and is our purpose for being.

This work is replete with information that reframes revealed Word– captured and understood in ancient contexts, into modern understandings. What we know today of necessity presents us with interpretations of revelation that was not possible to the ancients no matter how superior their position may have been to the original source. This is true if the time frames involved are millennia or just the few generations encompassing a century or two. In the next chapter, we will consider the nature of man in the context of how it has been transformed by the things we have most recently come to know in this work.

# CHAPTER 11

# WHAT IS MAN?

Much more important to our eternal progression than knowing who and what God is, is an understanding of what Man is. God is who and what He is. He was the same before we came into being and forever will be. The knowledge of the being and form of God, our comprehension of Him, is peripheral to the core and foundational truths that declare who we are, His literal sons and daughters.

In the previous chapters, we have detailed three separate creative eras, their related but disparate spheres and ages, and the courses of God (begun, completed and ongoing) which have led to the space we inhabit today and made us what we are. We have described an eternal past, and within it, an ancient beginning for each of us, intimate with our Father. Of these truths we are left in no doubt. The future state of our being is a direct result of what has happened before and is assured by our Father's singular focus on what He declares is His work and His glory, "*to bring to pass the immortality of man*" (Moses 1:39).

Contention with traditional Christianity begins here, with our understanding of a near infinite pre-existence of man. To that is added our Latter-day concept of an eternal god-like inheritance for man, one that insists we are children of our Heavenly Father, not just in deed, when we live worthily, but by birth; because of this, we can become like God, heir to all that He has. In this conflation of ideas, the whole of which describes a divinely appointed progression for man, Latter-day teachings stand out today as unique, at odds with that described as mainstream Christian doctrine. Within this chapter, we will explore that what canonical Latter-day teaching expounds is consistent not only with Biblical revelation, but it is consistent with the earliest interpretations of the inspired passages by those who were taught the very precepts by the original apostles of the LORD themselves.

Within this work, we have developed an understanding that the primary being of man is eternal in nature, an embodied *personage of Spirit*. In this chapter, we will delve in to what a *personage of Spirit* is. We will revisit, this time in depth, the state of each of our separate existences into the distant past, to the point of our primal creation. But first we will visit what is disclosed about who we are to become.

# *Deus factus est homo, ut homo fieret Deus*

*"God Became Man that Man Might Become God"*

This sometimes inflammatory statement is one of the best-known quotes from Athanasius (ca 298–373 AD), whom we have already met; it comes from his influential On the Incarnation, first published in Latin in the fourth century. What is expressed in this phrase is the concept of Theosis or Deification of man (*Figure 11-1*). It is an understanding that mortal men can have a real and integral union with God, that we can become like God to such a degree that we participate with the divine nature.

This precept, however, was not exclusive to Athanasius but was understood and taught by virtually all of the early church fathers, Irenaeus, Clement, and Justin Martyr among them. It continued to be taught as indicated by its explication by Athanasius in the 300's and even later by the likes of Augustine and others; it remains an Orthodox teaching and wholly Catholic to this day, highlighted prominently in the first few pages of the Roman Catholic catechism[1] and espoused emphatically by all of the Eastern churches.

Deification can be considered as similar to the Wesleyan teaching of holiness or perfection with the significant addition of a somewhat *mysterious* union of man with God through Christ. This union empowers both the desire to seek and the method for gaining perfection. Theosis, as it is understood today in an entirely modern theological context, has been identified as "*the gradual process by which a person is renewed and unified so completely with God that he becomes by grace what God is by nature.*"[2] Protestants today may refer to this process as *sanctification;* perhaps a term more familiar to Western Christianity would be *mortification*; the putting to death of sin in a believer's life. We shall see here the modern theological context denies something that was fundamental to Deification as it was originally understood, *that we as mankind are in fact destined for godhood because we are gods by nature.*

---

[1] Theosis (divinization) remains an important Catholic teaching today. The following is a set of quotes from various Catholic Fathers included within a quote from today's Catechism of the Catholic Church: "*The Word became flesh to make us "partakers of the divine nature" (2 Pt 1:4.) "For this is why the Word became man, and the Son of God became the Son of man: so that man, by entering into communion with the Word and thus receiving divine sonship, might become a son of God." (St. Irenaeus, Adv. haeres. 3, 19, 1: pg. 7/1, 939.) "For the Son of God became man so that we might become God." (St. Athanasius, De inc. 54, 3: pg. 25, 192B.) "The only-begotten Son of God, wanting to make us sharers in his divinity, assumed our nature, so that he, made man, might make men gods." (St. Thomas Aquinas, Opusc. 57, 1-4.)*

[2] Fr. David Hester, in his booklet, The Jesus Prayer.

The fullness of the Gospel as understood and accepted by Latter-day Saints terms the process of deification as Exaltation and entirely encloses the concept within the progression of man detailed in the Plan of Happiness. Exaltation as revealed in the last days to modern prophets does not represent some new doctrine, but is truly a reiteration of what was already known.

## *Deification in the Biblical Account*

Deification as a concept can be found throughout the Bible; passages that more than suggest that man is and may become like our Father in Heaven can be found in both the Old and the New Testaments. After emphasizing that man was created in the very image of God, Genesis records the Words of the Father: "*man is become as one of us*" (Gen. 3:22).

Note that this statement, though associated with the Fall, is recognition of an accomplishment, a progression, a change of state, a step in the direction of God.

Figure 11-1 *Icon of the Ladder of Divine Ascent. (the steps toward theosis as described by St. John Climacus) showing monks ascending (and falling from) the ladder to Jesus. (Saint Catherine's Monastery, Sinai, Egypt.)*

In his songs of praise to the Father, David quotes the voice of the Creator: "*I have said, Ye are gods; and all of you are children of the most High*" (Psalm 82:6). That this is true can find no greater testimony than that offered by the Savior himself when he recounted these very words again to the Jews who accused Him of blasphemy and were ready to stone Him for equating Himself with the Father: "*Is it not written in your law, I said, Ye are gods?*"[3] (John 10:34). Despite what some pundits proclaim, this comment by the Savior at that moment was not satirical,[4] it was not said mockingly, but it was

[3] Note here Christ associates this phrase with the "*law*" (Torah) not the "*prophets*" (Nevi'im) as it would properly be considered to be in the Psalms. The significance of this is uncertain. Does it mean the LORD was referring to a passage that He attributed to Moses and is unknown today, or did He include the sayings of the Prophets to be "*law*" as well?

precise and demonstrative of an eternal truth. The Savior's statement encompassed a double entendre as well, for here Christ speaks knowing that it was indeed He Himself that David quoted. Jesus Christ, Jehovah, was the one who said, "*Ye are Gods*" to the Israelite king before, and the Son was doing so again here to His accusers at Jerusalem.

There are a set of passages in the Gospels that are often overlooked by many Christians; they include a promise made by the Savior to his disciples because of their faithfulness. They speak directly to participation in the very Glory of God. Because they had forsaken all and followed Jesus, continued with Him in His "*temptations*," the Savior said to the twelve apostles, "*...I appoint unto you a kingdom as my Father hath appointed unto me; That ye may eat and drink at my table in my kingdom, and sit on thrones judging the twelve tribes of Israel*"[5] (Luke 22:29, 30). Kingdoms, thrones, and the authority to judge for the twelve? These three are attributes of God and deity; those who would attain them could be called, even seen, as gods. Some insist that these promises were extended to the Apostles only, that theirs was a special case (it must be noted at this time, the *Twelve*, still included Judas Iscariot), but regardless of the case, they were still just mortal human beings, being raised (because of *their* efforts in following Christ) to a status of Glory associated with Deity. What can be done for twelve, can be done for all.

The New Testament includes an abundancy of passages pointing to the doctrine of Theosis. In the Sermon on the Mount, the Savior spoke of progression to those who gathered around him, realizing their ultimate potential, He uttered, *perfect* (Greek, *teleioi*, τέλειοι), as He commanded His disciples to become "*perfect, even as your Father which is in heaven is perfect*" (Matt. 5:48). *Teleioi* can also be rendered in English as *complete*, thereby implying a process leading to a result, an objective, perhaps distant, a divine goal.

Having heard the Savior on the mount, the Apostle Peter later referred to the Savior's "*exceeding great and precious promises*" given there. Peter understood that we might become "*partakers of the divine nature*"[6] (2 Peter 1:4). Here the commandment to be perfect has become a promise associated with the "*partaking*" of divinity; this is what Paul refers to when he admonished us to "*work out [our] own salvation with fear and trembling*"[7] (Philippians 2:12).

As Peter did, Paul spoke often of the divine destiny of the sons and daughters of God. When Paul addressed the Greeks at Mars Hill, he said, we are "*the offspring of*

---

[4] I must note here that any interpretation of this statement by the Savior as satirical or ironic would degrade the efficacy of scripture by belittling its message, something that the Savior would not do. Additionally, such an understanding if expressed and understood by the hearers, would have left His defense of His actions without force and impotent.

[5] See also Mathew 19:28.

[6] The Greek word translated in the KJV as "*partakers*" is κοινωνοὶ, *koinonoi*. *Koinonoi* is also rendered as *fellowship*, *participants*, *sharers in*, etc. *Koinonoi* in these contexts is congruent with the Johannine and Pauline language describing deification as the concept that declares, "*man might become God.*"

[7] A decidedly Arminian and Mormon thing to do.

*God*" (Acts 17:29). Later, while in Corinth during the winter of 57-58 AD, he composed a letter to the church at Rome. In it, the prolific Apostle underscored what he previously taught in Athens. He wrote, "*...we are the children of God: and if children, then heirs; heirs of God, and joint-heirs with Christ*" (Romans 8:16, 17). The doctrine could not be more precise.

The Gospel of John and his letters stand out for they speak in a voice much different than the synoptic Gospels of Mark, Matthew and Luke. Within them John describes the Son, His relationship to the Father and to us, in terms that allude to an *oneness* like no other. Ofttimes referred to as "*Johannine Mysticism*," the doctrine of deification can be seen most clearly in the words of John. Speaking of *oneness* (Greek, *hen*, ἓν), the Gospel of John reveals the Savior's words at Gethsemane:

> *"Neither pray I for these alone, but for them also which shall believe on me through their word; That they all may be one; as thou, Father, art in me, and I in thee, that they also may be one in us: that the world may believe that thou hast sent me. And the glory which thou gavest me I have given them; that they may be one, even as we are one: I in them, and thou in me, that they may be made perfect in one; and that the world may know that thou hast sent me, and hast loved them, as thou hast loved me." (John 17: 20-23)*

The Savior's entreaty was that we are to be *one* with each other, with the Savior and through Him with the Father This is an oneness not to be described in the dysfunctional term of the Trinity, but in the very real way Adam and his equal Eve became one (*echad*) in the Garden. This *drawing up* towards God described by the LORD on the Mount of Olives is deification in its essence.

To John this elevation was literal and real; it was much more than metaphor. Though he could not in his natural state *describe* it, he could express his understanding that in the end we all are of the "*stuff*" of God; "*Beloved, now we are the sons of God, and it doth not yet appear what we shall be: but we know that when he shall appear, we shall be* <u>*like*</u> *him, for we shall see him as he is*" (1 John 3:2). Hearkening back to our previous discussion of the substance of God,[8] we understand the Greek word *homoioi (*ὅμοιοι), translated as *like* in this passage when used to relate the Son to the Father, means that the Son (though separate and distinct from the Father) was no less Deity. In this same context, John understands that man, being "*like*" (*ὅμοιοι*) the Savior, will be deity as well.

The Apostle Paul also spoke of Deification; he described it in terms of receiving the fullness of God, the Greek *pleroma* (πληρωμα). In his letters to the saints at Ephesus and at Colossae, Paul uses *fullness* in an instructive way. First, he describes Christ using the term, "*For it pleased the Father that in him should all fullness dwell*" (Col. 1:19). This verse is the capstone of a chain of verses that proclaim the Christ to be the deliverer, the Creator and sustainer of all things, the firstborn. In the next chapter, Paul adds to this understanding of *pleroma*; he writes of the Savior,

---

[8] See <u>The Great Arian Controversy</u>, pg. 326, in this work.

"*For in him dwelleth all the fullness of the Godhead bodily*" (Col. 2:9). These two verses are perhaps the strongest testimony of the Deity of Christ to be found anywhere in the Greek Testament.

After having associated *fullness* (*pleroma*) with the Deity of Christ, the Apostle Paul describes our attaining the same fullness (*pleroma*), which is the essence of deity; to him such a prospect was undeniable. Paul said:

> *"And to know the love of Christ, which passeth knowledge, that you may be filled to all the fullness of God." (Ephesians 3:19)*

> *"Till we all come to the unity of the faith, and in the knowledge of the Son of God, unto a perfect man, the measure of the stature of the fullness of Christ." (Ephesians 4:13)*

Within this work, we have described Temporality as a creation by subtraction (*tzimtzum*), that what we experience about us today is something less than what it once and actually is. If we apply this concept to that of Paul's "*fullness*," his *pleroma* (πληρωμα), understanding him to be referring to a restoration of what was once removed, then Paul was envisioning a return of things to the perfection of the Physical and Spiritual states from whence they came.

The Bible today ends with an apocalyptic vision of the Apostle John.[9] It is not surprising that the Book of Revelation contains the same themes of Deification found in John's other inspired writings, for they all were received from the same source. In the Book of Revelation, John records the words of Jesus Christ saying, "*to him that overcometh will I grant to sit with me in my throne, even as I also overcame, and am set down with my Father in his throne*" (Revelation 3:21). There can be no ambiguity here, the enduring Christian Faithful are to join with Christ in Deity, in the same way that Christ Himself shares Deity with the Father. As the vision of John approached the final events of the world at the beginning of the *eschaton*, John records Christ saying of each of us, "*He that overcometh shall inherit all things; and I will be his God and he shall be My son*" (Revelation 21:7).

---

[9] The current position of the Book of Revelation as the last and final book of the New Testament is a result of Martin Luther moving it from a more central position in the collection to an appendix of texts he formed because he deemed them not to be part of the cannon and perceived them to contradict nascent Protestant doctrines. The books Luther rejected include: Hebrews, James, Jude and Revelation; these books are ordered last today because of this. It must be noted here that the Book of Revelation (perhaps written as early as the decade following AD 69 and no later than 96 AD) is not even the last of the canonical works of John to have been completed by him; his Gospel is the last to be penned (AD 90-110). This is of serious import to those who hold that Rev. 22:18 officially closes the canon, for if it did, John in participating in the development of his own Gospel, added to the Bible and should have suffered the plagues spoken of.

## *Deification as Understood by the Early Church*

We have seen the concept of Theosis is found in the first pages of the Old Testament, in the last pages of the New and scattered everywhere in between. Though these passages are taken in many different ways among the Protestant sects today, Latter-day Saints, Catholic faithful and all of the Eastern Churches see these same passages as *direct revelatory confirmation of the divine potential of man*, if not in fact actual testimony of man's divine nature. The earliest generations of Christians read deification into these texts as well. In the first centuries of the Church, the early fathers spoke ubiquitously of the concept of Theosis.

It would be impossible in the limited space we have here to include all of the statements made by the early church fathers in support of the concept of deification of man. It is important, however, that we recognize just how literal their understanding was. We will examine here just a few of the many declarations we still have access to today.

Irenaeus, Bishop of Lyons (c. 130–202), addressed deification precursory to Athanasius using almost the exact same words. He posited: " *[T]he Word of God, our Lord Jesus Christ, who did, through His transcendent love, become what we are, that He might bring us to be even what He is Himself.*" (Adversus Haereses, V. Preface). Irenaeus also wrote, "*Do we cast blame on him [God] because we were not made gods from the beginning, but were at first created merely as men, and then later as gods? Although God has adopted this course out of his pure benevolence, that no one may charge him with discrimination or stinginess, he declares, 'I have said, Ye are gods; and all of you are sons of the Most High.' ... For it was necessary at first that nature be exhibited, then after that what was mortal would be conquered and swallowed up in immortality*" (Adversus Haereses, IV. XXXVIII).

It must be noted here that the modern convention of using a capital "G" to distinguish God the Father, the God of the Universe and monotheism, from mere *gods* is a modern construct of language not at all present in the original. Any attempt to differentiate is an overlay of modern thinking and does not in fact exist in Irenaeus' writings. If there is any doubt that Irenaeus spoke of the eventuality of true divinity for man, consider the following passage on the subject, a passage we will return to later in this chapter. On Man's *created* nature Irenaeus said:

> *"Men, being created by God, are by nature sons of God; but they are not sons in their deeds. For as among men disobedient sons, disowned by their fathers, are by nature their sons but are alienated in law, since they are no longer heirs of their natural parents: in the same way, with God, those who disobey him are disowned and have ceased to be sons."* (Adversus Haereses, *IV. xli. 2-3)*[10]

---

[10] The Early Christian Fathers, Henry Bettenson (Oxford University Press, New York, 1956) pg. 73.

A contemporary of Irenaeus and mentor of Origen, was Clement of Alexandria (c. 150–215 AD). Clement like Irenaeus anticipated the later words of Athanasius that are to this day quoted in the Catechism when he wrote: "*the Word of God became man, that thou mayest learn from man how man may become God*" (Protrepticus, Chap. 1). Like Irenaeus, for Clement, Deity was not only an eventual state but something integral to who we are as well. He stated, "*[i]f one knows himself, he will know God, and knowing God will become like God. ...His is beauty, true beauty, for it is God, and that man becomes a god, since God wills it*" (Pædagogus, III, 1).

We should not be surprised to find others echoing both Clement and Irenaeus in the belief of divinity within man, and indeed we do. In the generation before Irenaeus and Clement, Justin Martyr (c. 100–165) (who was one step closer to the apostles themselves) also spoke plainly of Deification. In his esteemed dialogue with the fictional Jewish sage Trypho, he wrote:

> *"And when I saw that they were perturbed because I said that we are the sons of God, I anticipated their questioning, and said, Listen, sirs, how the Holy Ghost speaks of this people, saying that they are all sons of the Highest; and how this very Christ will be present in their assembly, rendering judgment to all men. ... the Holy Ghost reproaches men because they were made like God, free from suffering and death, provided that they kept His commandments, and were deemed deserving of the name of His sons, ... all men are deemed worthy of becoming 'gods,' and of having power to become sons of the Highest; and shall be each by himself judged and condemned like Adam and Eve."*[11]

Here yet another early Christian father spoke of Theosis, of the divine nature of man. To his imaginary foe Trypho, Justin insisted that men were in the beginning, "*made like God, free from suffering and death,*" and that because of their initial state, all men are accordingly "*deemed worthy of becoming gods, and of having power to become sons of the highest.*" Justin, like Irenaeus who would follow him, understood that men were made sons of God when they were created; divinity was not just a future expectation for man but intrinsic to our being and is our true nature. In the mid-second century, Justin Martyr spoke in terms familiar and in complete accord with Latter-day theology that was restored independently by inspiration to a young, uneducated Joseph Smith.

That his intent could be made clear, so it could be (without question) understood that he believed in the literal Deification of man, Athanasius expanded upon the quote with which we started this discussion. He added, "*The Word was made flesh in order that we might be made gods.*" Athanasius went on to clarify, "*... Just as the Lord, putting on the body, became a man, so also we men are both deified through his flesh, and henceforth inherit everlasting life*" (Orationes contra Arianos, Ora.1, 39; Ora. 3, 34, respectively).

---

[11] Justin Martyr. *"Dialogue with Trypho," Ante-Nicene Fathers*, (Ed. Alexander Roberts. Vol. 1. New York: C. Scribner's Sons, 1905, pgs. 261-262).

These statements make it clear that the early Church fathers understood Deification in terms that are expressed by Latter-day Saints today. To them and we there is no ambiguity about such things as: the family of God, our kinship with the Son, mankind's divinity by *nature* (*not through Abrahamic adoption, but because we are of the Father, itself a reality that assures our potential for eventual realization of His fullness*), heirship to the throne, sanctification, and perfection in Christ.

Despite this, the doctrine of Deification began to be challenged as the Roman Empire declined and the dark ages began to settle in. References to the concept became infrequent in medieval times. Within three hundred years, teachings about man "*becoming God*" assumed a tone that was much different than what they once did.

The capacity of man for Godliness was diminished and lessened in scope. In the sixth century, Pseudo-Dionysius the Areopagite defined deification as: "*the attaining of likeness to God and union with him so far as is possible.*" This of course had always been true. Deification never described an attainment of Godliness that surpassed, let alone equaled that of the Father, but here the pure power of the promises made by the Son to his disciples and spoken of by John, Peter and Paul, was not to be found. The *likeness* that could be attained, as envisioned by the Areopagite, was not much removed from the state of the man Adam in the Garden before the Fall.[12]

With such amendments, the concept of Deification remained foundational Christian thought in the East, if however obscured it seems to have become in the West. As late as the 13th century, Catholic theologians of such prominence as Thomas Aquinas still, however, could be found espousing the concept as follows: *"The only-begotten Son of God, wanting to make us sharers in his divinity, assumed our nature, so that he, made man, might make men gods"* (St. Thomas Aquinas, Opusc. 57, 1-4).

Deification, as it has come to be understood in the Christian church, is a mystical, even philosophical notion involving two distinct humanly hypothesized qualities of God – His essence and His energies. Entirely non-Biblical, the concepts of essence and energies are fundamental to modern orthodox understandings of Theosis. In short, *essence* is what something is; the *energies* of this *something* are manifested in what that something does.

What God is, His *essence*, what we have termed His "substantive *nature*," is unknowable. Both sectarian Christian doctrines and this work agree on this point. Though we have made an attempt to do so here, we know that our definitions and

---

[12] This image of Adam in the Garden is far removed from the one we have described in this work. The *Adam*, Pseudo-Dionysius and the rest of the Christian Church imagined in the sixth century, was strictly a Temporal creature, native to the sphere we inhabit today, without knowledge or power, immortal but not godly or of the substance of Deity. He was not the offspring of God; he did not share any of God's eternal attributes nor could he.

understanding of the nature of the divine are incomplete because it is impossible to define God entirely using Temporal terms and language.

We comprehend our Father, are aware of Him, most completely through His actions and His stated purpose; these are understood in the Eastern Church to be His (*energies*). These actions and His purpose are God, what He is. Like the Father, we too are all definable not just by what we are, but also by what we do. The actions of God (His energies) are how we understand Him, experience Him, and it is through emulation of these actions that we seek to become like him.

Modern Christianity does not accept that we are of the same essence/nature of our Father, but it espouses that we are deified through participation in His actions, which are both expressed in the mission of, and in fact personified entirely by, the incarnate Christ. Their belief is that through our emulation of these actions, an interactive relationship is established in which the *energies* of God are shared with us until we attain perfection in Christ.

It is expressed that Christ revealed this completed state destined for man on the Mount of Transfiguration, as He assumed, before Peter, James, and John, the full realization of what mortal man is to become when perfected in Him. "*And after six days Jesus taketh Peter James and John his brother, and bringeth them up into an high mountain apart, And was transfigured before them: and his face did shine as the sun, and his raiment was white as the light*" (Matt. 17:1, 2).

The transfigured Christ was glorious to behold and inspired the desire to worship from the three amazed apostles with a force they had not experienced before. That this is understood, even by the most orthodox of Christians today to be representative of the eventual state of man, is telling of a very literal understanding of Deification.

As we have seen here, Deification is scripturally based, historically attested, and though modified almost beyond its original understanding; it remains foundational to most of Christianity today. Despite this, it is (in its truest expression) considered heresy by much of the Low-Church Evangelical denominations of America; the nearest they approach the thought is the Wesleyan conception of Christian Perfection. It has been surmised by more than just a few Christian commentators that the effective refusal of American Protestant churches to teach Deification is that it is so strongly associated with Mormon doctrine of Exaltation.

*Exaltation*

*It is a serious thing to live in a society of possible Gods and Goddesses. To remember that the dullest, and most uninteresting person you can talk to may one day be a creature which, if you saw it now, you would be strongly tempted to worship."*

*(C.S. Lewis)*[13]

Gods and Goddesses... This phrase seems mythical and from a far distant age to most modern western ears, but it is familiar and credible to the Latter-day Saint. For all its outward appearances, the doctrine of Theosis or Deification is *not* the same as what we understand the doctrine of Exaltation to be. In this variance, Theosis is not lacking in its final outcome, for it delivers man to the same place as Exaltation (to a fullness of Deity, co-heirs with the Son of God, glorified on His throne, perfect as the Father is). The difference between Latter-day Exaltation and Deification is rooted at the beginning and not at the end.

Deification, as it came to be understood (if however it was not originally understood by the early fathers)[14] involves making gods and goddesses out of something that was not of God to begin with. In contrast, Exaltation expresses an understanding of improvement upon, the progression of, divine beings from one state of experience and influence to another more empowered and capable. Deification is the making of something that *is not of God* into something that *is like God*. Exaltation is the attainment of enhancement for the already godly, *returning what is of God to God* with experience, wisdom and knowledge that makes it (them) *more like God*.

Our conceptions of the Physical and Spiritual realms are (of necessity) limited in the Temporal sphere, so we struggle to sufficiently articulate eternal glory. Of this struggle, the Apostle Paul wrote, "*Eye hath not seen, nor ear heard, neither have entered into the heart of man, the things which God hath prepared for them that love him*" (1 Cor. 2:9).

We understand Exaltation to be a delivery of sorts, to a perfected state, back to where we began, back to the presence of the Father. We have been directed towards home; "*be ye therefore perfect, even as your Father which is in heaven is perfect*" (Matt. 5:48). We can attain perfection because we were first made *perfect* (*τέλειοι*). Our journey to exaltation has been one of constant progression, though on the

---

13 The Weight of Glory, (Harper Collins 1949, New York, NY).

14 See our previous quote of Irenaeus at the bottom of page 475. There Irenaeus details in allegory his belief that man is fallen from a once held deified state, through disobedience we have "*ceased*" to be what we once were, *of the same nature as God.* This is the context Deification must be fitted into when discussing it as it was understood by the early, original, Christians. This context and understanding is what gave root to the appellation (Kingdom of Arrogance), which was fixed on the nascent Christian church by its Jewish antagonists.

surface it appears less than so; clothed in fallen mortal bodies today, in a vast Temporal Universe, we are visitors from a world much different than the one we inhabit today.

Perfection is not just a goal but our native state; it most closely describes our nature. When the Savior bade His listeners to be perfect, it was not a demand or charge but a gentle entreaty that they stop to ponder and remember who they really were. Just a few moments earlier He said: "*...Love your enemies, bless them that curse you, do good to them that hate you, and pray for them which despitefully use you...*" (Matt. 5:44). This He said they should do that they "*may be the children of your Father which is in heaven...*" (Matt. 5:45). All present were already the children of Heavenly Father, so what Christ was making an appeal for was their return to the state of perfection from which they came. It was not a figurative or symbolic request; it was spoken by one who knew most certainly with full anticipation and literal expectation. C. S. Lewis understood this concept and expressed it compellingly:

> *"The command "Be ye perfect" is not idealistic gas. Nor is it a command to do the impossible. He is going to make us into creatures that can obey that command. He said (in the Bible) that we were "gods" and He is going to make good His words. If we let Him—for we can prevent Him, if we choose—He will make the feeblest and filthiest of us into a god or goddess, dazzling, radiant, immortal creatures, pulsating all through with such energy and joy and wisdom and love as we cannot now imagine, a bright stainless mirror which reflects back to Him perfectly (though, of course, on a smaller scale) His own boundless power and delight and goodness. The process will be long and in parts very painful; but that is what we are in for. Nothing less. He meant what he said." (Mere Christianity Macmillan, 1952)*

It would be an egregious error to characterize the attainment of Exaltation as some sort of end state, the capstone or pinnacle of achievement – it is far from that. It is just another step we take along our path of eternal progression. We should have no expectation that the road (or our journey along it) *beyond the milestone* will become flatter or easier, for it will not. Lewis stated it correctly, "*The process will be long and in parts very painful; but that is what we are in for. Nothing less.*"

Exalted beings continue to progress and remain mutable. As we have described the Father in this work, He, Himself is capable of progression and change; He is not immutable. Our God weeps and repents; He is limitless and without bounds. These last two attributes insist that He continues to grow in both knowledge and power through experiences that are new and somehow unknown to Him. This we have addressed is one of the purposes of Temporality itself, of our experiences in it, perhaps even a reason for our creation in the first place.

Many Church fathers saw something in Theosis that went far beyond the simple restoration of the Children of God to their common state before the Fall. They saw Deification as a process that *unites* the natures of the human and the divine. This unification was first accomplished in the incarnation of Christ and finalized for *all* through His death and resurrection; "*For as in Adam all die, even so in Christ shall all be made alive*" (1 Cor. 15:22).

The Atonement makes it possible for man to experience intimacy with the Father to an order of magnitude that was not possible within the Garden before the Fall. Because of this enhanced fellowship, man can be more like God than it was possible for Adam and Eve to ever have been. Some Eastern Orthodox theologians expanded this thought to such a degree that they conveyed a belief that even if Adam and Eve had not fallen, Jesus would have taken on the incarnation and the trials and death that accompanied it for nothing more than this reason alone.

As we have shown in this work, much of what was surmised by these men is indeed true. Exaltation, through its returning of us to precincts that were once our home, includes our acquisition of knowledge, experience and procreative powers that are far more than a restoration of what we once had.

***The Atonement makes it possible for man to experience intimacy with the Father to an order of magnitude that was not possible within the Garden before the Fall.***

Through entry into the Temporal sphere, we gained access to the materials of uncertainty, allowing for a truly "*free*" agency. It is Temporality that makes us human, the *fallen* matter of this Universe is joined to the pre-existent divine, and finally sealed into immortality by the redemptive powers of an infinite Atonement effected by the sacrifice of the Son of God.

The Plan of Happiness, as originally expressed in the Grand Council and fully realized through the creations of God, does make it possible for us to be more like the Father than we could have been without it; *that is the purpose for our Temporal existence.* The great Plan of Progression was not a resultant of sin, but sin is a catalyst for the experiences the Plan requires. The incarnation of Christ was not reactionary for the purpose of correcting unintended sin and evil in the world but for the benefit and advancement of the sons and daughters of God on a road to perfection and exaltation.

These understandings, fundamental to our Latter-day beliefs, provide us with a much different view of the Atonement than that defined by modern sectarian Christianity. Today much of Christian theology venerates the enormousness of the salvific grace of the Atonement by imposing and widening a divide between God and Man, degrading and mischaracterizing God's greatest creation. The Scriptures do not describe such a rift. The Savior does not speak of *total depravity*, to the contrary, His mission testifies of the infinite worth of sons and daughters of God.

The greatness of the Savior's atoning sacrifice is conveyed in the reality of the Exaltation (*deification*) of man and the infinite potential it makes possible. The Atonement of Christ does much more than provide forgiveness from the terrible consequences of sin and a triumph over the chains of Temporal death, but it restores and redeems; it heals our aggrieved Spirits removing obstructions to our eternal progression, enabling each of us to develop the attributes of our Father in Heaven.

We, *all of us*, are literal children of our loving Father in Heaven. Within us is to be found the source of divinity, intelligence, the principle of the Glory of God. We do not proclaim ontological *equality* with God, for He is our Father. We do understand that we are as much *of Him* as we are *from Him*. We are of His family, each of us intimately associated with the Father, the Son and the rest of the Children of God.

We do not profess that we in some way merge or co-mingle with the *being* of God, for we understand Him to have His own essence. By His grace and because of unbounded love, God encourages us, His children, to realize our eternal potential, a potential only made possible by the Son's atoning sacrifice, not through any work we are capable of doing on our own.

As we understand it, Exaltation is total participation in Jesus Christ, not just in a theological sense but in a very literal way. Exaltation is an unimaginable, limitless gift; through it we realize the destiny spoken of by Maximus the Confessor (c. 580-662 AD), who said, "*All that God is, except for an identity in being, one becomes when one is deified by grace.*" All that God is, is God by definition.

The doctrine of Exaltation, which teaches of the potential of humans to progress eternally to become like their Father in Heaven, is fundamental to the Gospel taught by the Son. It inspires faith, hope and charity and instills gratitude in the hearts of the sons and daughters of God. By lighting our way forward, it illuminates our past, closing the loop of our journey, identifying the place where we are headed with the place from which we came – *home*.

**Divine by Nature**

The doctrine of Exaltation describes a return to the precincts of God for those who truly belong there by nature. Our understanding of the continuity presented in the three creations of God presents a picture of movement along a line, a progression in parts, of that which comes *from* God back *to* Him, enhanced by experience, improved through struggle and empowering independence. For Latter-day Saints, divinity is not a future potential state of Man but is truly a fundamental aspect of our nature. We were created divine and existed in an immortal state before we took on a Temporal form in the Universe we inhabit today. We remain eternal at our core, literally of the family of God, *from the outset* children of our Father in Heaven.

It is this belief that sets our doctrine apart from what has become "*traditional*" though not *original* Christianity. Theologically, and in the contexts we have developed in this treatise, the sons and daughters of God are those who existed in the eons past as independent Intelligences with our Heavenly Father and His first born Son. They are those who received their immortal physical bodies at the same time as Adam and are those who through participation in the effects of the Fall received Temporal (mortal) bodies. We understand that it was necessary for God to participate in mortality in order to bind Deity to the Temporal sphere with a bond so strong that it could overcome death for all and restore us to the perfection required in our eternal home. It must be understood that the, *we*, *those*, *they*, and *us* in the preceding sentences, are the sons and daughters of God, all of us, wherever we may find ourselves within the Temporal Universe. We all became incarnate to come here, and so did the Son. We all have only one trip here to this Fallen, subtracted sphere and so did the Son; once and for all, He was the Sacrifice for sin, restoring all to the perfection of the Physical sphere:

> *"For Christ is not entered into the holy places made with hands, which are the figures of the true; but into heaven itself, now to appear in the presence of God for us: Nor yet that he should offer himself often, as the high priest entereth into the holy place every year with blood of others; For then must he often have suffered since the foundation of the world: but now once in the end of the world hath he appeared to put away sin by the sacrifice of himself. And as it is appointed unto men once to die, but after this the judgment: So Christ was once offered to bear the sins of many; and unto them that look for him shall he appear the second time without sin unto salvation." (Hebrews 9:24-28)*

> *"That by him, and through him, and of him, the worlds are and were created, and the inhabitants thereof are begotten sons and daughters unto God."*
>
> *(Doctrine and Covenants 76:24)*

As this gospel dispensation opened, Protestantism had dominated the English speaking world for more than 250 years, and it had been a full three centuries since Martin Luther had posted his 95 thesis on the door of All Saints' Church in Wittenberg. The false doctrine and teachings of the Trinity and creation *ex nihilo*, the limited Atonement and other fallacies of the Counter Remonstrance of 1611 and belief in a God who existed "*without body, parts, or passions,*"[15] had become standards of orthodoxy. When Joseph Smith began to receive revelations that would fundamentally change our understanding of the family of God, the truths that would again be delivered had long since been forgotten by man. At the outset, the young Joseph was completely unaware that what he would come to understand, as much as that was possible, was but a restoration of things once known. Joseph was told that we are fundamentally intelligence, that intelligence is eternal in nature and uncreated:

> "*I am dwelling on the immortality of the spirit of man. Is it logical to say that the intelligence of spirits is immortal, and yet that it (i.e. the intelligence) had a beginning. The intelligence of spirits had no beginning, neither will it have an end. That is good logic. That which has a beginning may have an end. There never was a time when there were no spirits,*[16] *for they are co-equal (co-eternal) with our Father in heaven...I take my ring from my finger and liken it unto the mind of man—the immortal part, because it has no beginning. Suppose you cut it in two; then it has a beginning and an end; but join it again and it continues one eternal round. So with the spirit of man. As the Lord liveth, if it had a beginning it will have an end....Intelligence is eternal and exists upon a self-existent principle. It is a spirit from age to age and there is no creation about it... The first principles of a man are self-existent with God.*" *Joseph Smith. (Mill. Star, Vol. 23, pg. 262.)*

We have explored here in depth what these words of Joseph mean and concluded that the *Intelligence* he spoke of, that which comprises each of us, is from God. Since intelligence is uncreated, self-existent and indestructible, we have proposed that its source is God Himself; the Intelligence within each of us is a part of our Father, both us *and Him* at the same time. That realization bespeaks of an intimacy

---

[15] This phrase finds its origin in the Westminster Confession of Faith. Drafted by the Westminster Assembly of 1646, it represents the standard doctrine and defines the mode of worship and the government of the Church of England. The Westminster Confession of Faith has been adopted as authoritative by many other Protestant churches in the centuries since it was first composed.

[16] We have observed the conflation of terms spirit and intelligence in this statement before.

with God that is real. We literally are of the "*stuff*" of God; at the most basic level, we are of His Nature, His essence.

We (our Intelligences) were agent unto ourselves from a very early period, self-aware and fully sentient. Our Intelligences exercised will and gained knowledge and capability beyond that which was original to them.[17] Joseph Smith saw the results of agency at work within the Intelligences revealed in his inspired interpretation of the vision of another. Through the eyes of Abraham, Joseph, and we, have come to understand the dynamic nature of our primordial state; "*Now the Lord had shown unto me, Abraham, the intelligences that were organized before the world was; and among all these there were many of the noble and great ones; And God saw these souls that they were good, and he stood in the midst of them, and he said: These I will make my rulers; for he stood among those that were spirits, and he saw that they were good*"[18] (Abraham 3:22, 23). Abraham clearly described our pre-existent state and spoke of those who had already accomplished nobility and greatness to the extent that God called them good and proclaimed their divinely appointed destiny.

Joseph taught us that intelligence is our first principle component, our fundamental essence. He said this "*first principle*"[19] is eternal, that it is, was and forever will be,

---

[17] Elder B.H. Roberts detailed the nature of our divine intelligence; "*The Nature of Intelligencies: There is in that complex thing we call man, an intelligent entity, uncreated, self-existent, indestructible, He—for that entity is a person; because, as we shall see, he is possessed of powers that go with personality only, hence that entity is 'he,' not 'it,'—he is eternal as God is; co-existent, in fact, with God; of the same kind of substance or essence with deity, though confessedly inferior in degree of intelligence and power to God. One must needs think that the name of this eternal entity—what God calls him—conveys to the mind some idea of his nature. He is called an 'intelligence;' and this I believe is descriptive of him. That is, intelligence is the entity's chief characteristic. If this be a true deduction, then the entity must be self-conscious, and 'others—conscious,' that is, he must have the power to distinguish himself from other things—the "me" from the 'not me.' He must have the power of deliberation, by which he sets over one thing against another; with power also to form a judgment that this or that is a better thing or state than this or that. Also there goes with this idea of intelligence a power of choosing one thing instead of another, one state rather than another. These powers are inseparably connected with any idea that may be formed of an intelligence. One cannot conceive of intelligence existing without these qualities any more than he can conceive of an object existing in space without dimensions. The phrase 'the light of truth' [Doc. & Cov., Sec. xciii.] is given in one of the revelations as the equivalent for an 'intelligence' here discussed; by which is meant to be understood, as I think, that intelligent entities perceive the truth, are conscious of the truth, they know that which is, hence 'the light of truth,' 'intelligence.' Let it be observed that I say nothing as to the mode of the existence of these intelligences, beyond the fact of their eternity. But of their form, or the manner of their subsistence nothing, so far as I know, has been revealed, and hence we are without means of knowing anything about the modes of their existence beyond the fact of it, and the essential qualities they possess, which already have been pointed out.*" (B. H. Roberts, Seventy's Course in Theology, Second Year [Salt Lake City: Deseret News, 1907-1912], part 1, lesson 1, pgs. 8-9).

[18] Note here again a conflation of the terms *Intelligences*, *Souls* and *Spirits*. We have correctly shown elsewhere in this work the separate and distinct meanings of these words and have described them in relationship to three distinct states of our individual beings in the context of the three created spheres. It is important to note here this lack of specificity and exactness within the recorded word of God. Where then is the blame for this obvious error to be placed? *Squarely on man's inability to fully comprehend and express that which has been revealed.*

[19] King Follett Sermon *(Teachings of the Prophet Joseph Smith, Joseph Fielding Smith, comp., Salt Lake City: Deseret Book, 1976, pg. 353).*

"*self-existent with God.*"[20] This salient view of the Prophet Joseph was reinforced by direct revelation from God; Intelligence, the core of human existence, central to each of our Souls, "*was not created or made, neither indeed can be*" (D&C 93:29).

Our Intelligences, though of the same element of the Father, of His *essence*, did not possess the fullness of His substance; we lacked form and the physical attributes associated with it; we were yet incomplete. If our beings were to progress further, we would have to become more like our Father, indeed this is His will; "*For behold, this is my work and my glory—to bring to pass the immortality and eternal life of man...*" (Moses 1:39). God next directed the creative events of the Physical sphere.

***Physicality was a step forward in our eternal progression; it brought us closer to our eventual goal of becoming not just like the Father but attaining all that He is.***

To our first principle component, *Intelligence*, was added an eternal physicality, tangible and substantive, organized from eternal matter, out of a substrate that continues to envelope all that there is today, the *Zero Point Field*. It is not serendipitous that science most closely identifies this field as the manifestation of energy from a dimension beyond our observation, for that is what it is. Drawing upon terminology developed in the sixteenth century to describe God as essence and energies, we see that we too can be described in such a manner. Our Intelligence is from the essence of the Father, for it is His glory, and our eternal physicality comes from His Energies, the *Zero Point Field* that subtends, infuses, empowers and sustains all that there is.

Physicality was a step forward in our eternal progression; it brought us closer to our eventual goal of becoming not just *like* the Father, but attaining all that He is. With our Physical bodies, we could touch and feel, hear, see and smell; our reality now became experiential like it could not have been before. God the Father recognized the risks associated with our embodiment and explained them to each of us, but He emphasized that the rewards would be great; "*spirit and element, inseparably connected, receive a fulness of joy*" (D&C 93:33).

Along with our embodiment, an entire Physical sphere was created so that it could be experienced through our new found senses. This new "*world*" provided each of us, and the Father through us, joy and wonder and knowledge. Our bodies were

---

[20] Ibid, pg. 354.

crafted by the Father and the Son with purpose; they emulated the bodies they each possessed. Christ testified of this fact to the Brother of Jared, "*Seest thou that ye are created after mine own image? Yea, even all men were created in the beginning after mine own image. Behold this body which ye now behold, is the body of my spirit; and man have I created after the body of my spirit; and even as I appear unto thee to be in the spirit, will I appear unto my people in the flesh*" (Ether 3:15-16).

The pre-existence of man as *Spirit* (physically embodied Intelligence) is a pillar of the Restoration though it is not a concept of modern origination. Its foundation is found in the truths of the fullness of the eternal Gospel. As such it has been known and taught by those who communed with God in all ages; it was *restored* for it was known before, "*...when I laid the foundations of the earth... When the morning stars sang together, and all the sons of God shouted for joy...*" (Job 38:4, 7).

Though the specific details regarding eternal intelligence and the tangibility of the Physical world were Latter-day additions, *much spoken of for the first time in context within this work*, the general concept of the pre-existence of mankind was understood and taught by the Early Christian Fathers (again those are they who were taught by the apostles directly or those who were taught by those whom the apostles taught). Origen, already familiar to us in this work, taught early in the third century that the human "*Soul*"[21] was pre-existent. His beliefs on this subject have proven to be enigmatic to Christian leaders who have taken exception to it throughout the intervening centuries. Origen's beliefs, however, were firm and far from an enigma to him, he reasoned as follows:

> *"God cannot be called 'omnipotent' without the existence of subjects over which he may exercise his power; and therefore in order that God may be displayed as 'omnipotent' it is essential that everything should subsist . . . Since 'all-sovereign' entails 'subjects' it follows of necessity that God created these subjects in the beginning, and that there was no time when they did not exist. If there were such a time . . . it would follow that the unchangeable and unalterable God would be altered and changed. For if he made everything later on it is clear that he would change from not-making to making." (De Principiis, I. ii. 10)*[22]

A careful reading of Origen's comment will reveal several essential elements. First, it is clear that Origen based his beliefs on three universally accepted principles of God's nature: His omnipotence, His eternal character, and His consistency or unchangeableness (what is known as *His immutability*). Second, beginning with these attributes of God, Origen reasons that an eternal, unchanging and sovereign God must have always had subjects. Last of all, Origen presents the following

---

[21] We must forgive Origen here, or more properly, the translation of Origen's work for its use of "*Soul*" to describe the pre-existent nature of man in a general sense. We have shown in this work that the Garden of Eden was within the Physical sphere, itself pre-existent to the Temporal sphere and that Adam, while within the Garden, was properly a *Living Soul*. It is proper then to speak of pre-existent man, at least in this special case as a *Soul*. We understand here that our pre-existent states were, for most of the eons past, most properly defined as unembodied Intelligence and embodied *personages of Spirit*.

[22] <u>*The Early Christian Fathers*</u>, Henry Bettenson (Oxford University Press, New York, 1956) pg. 189.

thought: Since *we* are now, if ever there was a time when *we* were not, that in-and-of-itself would demand that God had changed. To carry this last thought a little further, one could surmise that the very essence of God the Father is somehow defined by the very existence of mankind. This truth is testified of again in a Latter-day rendition of a very ancient vision, "*For behold, this is my work and my glory—to bring to pass the immortality and eternal life of man*" (Moses 1:39).

To some, the concept of pre-existent man, eternal in nature, seems absurd and may even be considered blasphemous, but it can only seem so if Man is counted as some kind of an afterthought, as an almost inconsequential element in the broader landscape of creation. This latter view of Man and the world around him has come (in most recent years) to be the one taken by an ever growing secular majority and unfortunately so too within Christianity of tradition. It is, however (despite its popular appeal) incorrect and destructive to the very fiber of a proper understanding of God, Man and the relationship between the two.

When the existence of mankind is understood in its proper context (that which is provided by the Scriptures), it becomes apparent that our relationship to the Father is exactly as Origen had thought it to be. From the six days of creation to the crucifixion and resurrection of Christ, one thing stands clear – mankind, through the Grace of God, is eternally important; our state and salvation are of utmost concern to the Father. His actions are consistent with those of a caring father, one who has already invested an eternity of love and kindness in each and every one of us. Our Father in Heaven, it would seem, is measured in much the same way Earthly parents are – by the impact we have on our sons and daughters, by the truths we instill in them, by the values they come to hold on to, by the love our children return to us, and last of all, by the love we unselfishly give.

From the creation of our Intelligences through this physical embodiment resulting in our existence as spiritual personages,[23] we all were considered *sons and daughters of God*, destined to follow the same progressive course through time and the several created spheres. However, agency is not without its pitfalls, and at some point, either within the Spiritual sphere or later within the Physical, dissention broke out around a Plan for our exaltation presented at a great Grand Council. When Lucifer presented his pretentious alternative plan, many chose to rebel against the Father and follow the devil, becoming then, "*the children of the devil*" (1 John 3:10). From this inauspicious divide forward, there would be a parting, contention, and separate courses for the sons and daughters of God and the devil and his children.

---

[23] "*The intelligent entity inhabiting a spirit-body make up the spiritual personage. It is this spirit life we have so often thought about, and sang about. In this state of existence occurred the spirit's 'primeval childhood;' here spirits were 'nurtured' near the side of the heavenly Father, in his 'high and glorious place;' thence spirits were sent to earth to unite spirit-elements with earth-elements—in some way essential to a fulness of glory and happiness (Doc. & Cov. Sec. xciii: 32-35)—and to learn the lessons earth-life had to teach. The half awakened recollections of the human mind may be chiefly engaged with scenes, incidents and impressions of that spirit life; but that does not argue the non-existence of the uncreated intelligences who preceded the begotten spiritual personage as so plainly set forth in the revelations of God.*" (B. H. Roberts, Seventy's Course in Theology, Second Year [Salt Lake City: Deseret News, 1907-1912], Part 1, Lesson 1, pg. 11.)

The sons and daughters of God chose to accept the progress afforded by the Plan presented by the Father and agreed to face the challenges and dangers of a Temporal sphere. In our view here, the Universe we inhabit today was not yet extant and the first hints of the Temporal sphere were presaged in a Garden Temple formed in a corner of the Physical sphere not too far distant eastward in Eden. There the first of the sons and daughters of God to be clothed in bodies made of a new form of matter were removed one further step away from the perfection of the Father when they were placed beyond a veil of forgetfulness. They would be nurtured in the Garden to adulthood, exposed to the confliction attendant to Temporality and its weighing of competing values and face a struggle to find a path to good. Adam and Eve would choose, and that choice made them more like God; He said, because they had chosen they had, "*become as one of us, to know good and evil*" (Gen. 3:22).

The Fall precipitated the creation of the Temporal sphere, not as a form of punishment but as a means for each of us to experience just as Adam and Eve had, the power of weighing the costs, making a choice, and fully experiencing the consequences of willful agency, being for the first time fully "*free.*"

The cost of Temporality is death, both spiritually, for God is withdrawn from us, and physically for the Temporal is, in this iteration, mortal. God understood these resultant facts and put in place safeguards for our return to Him. First and foremost is the protection offered by Temporality itself. We see this fact attested to (beginning) in the same verse of Genesis we just discussed. God spoke:

> *"lest he put forth his hand, and take also of the tree of life, and eat and live for ever: Therefore the Lord God sent him forth from the garden of Eden, to till the ground from whence he was taken. So he drove out the man; and he placed at the east of the garden of Eden Cherubims, and a flaming sword which turned every way, to keep the way of the tree of life" (Gen. 3:22-24)*

Temporality then is a protective measure as well, in that it allows for the experiences necessary for our progression in an environment that shields us from its negative effects in eternity.

We have seen in this work that the Temporal Universe was created in accordance with the many scientifically identified truths we understand today and that that creation is in agreement with the several Genesis 1 accounts recorded in Scripture. We understand that despite our various Pre-existent states which predate the event we call the Big Bang by many, many eons, the sons and daughters of God are newcomers to the Earth we inhabit today. On our planet, on this third rock from the star we call the Sun, some few, perhaps tens of thousands of years ago, the Fallen, fully matured and schooled Adam and Eve from the Garden, were placed as strangers in this land to populate this world with their children, the *Be-ne- ha-e-lo-him* (בְּנֵי־הָֽאֱלֹהִים), *the sons* (the children) *of God.*[24]

---

[24] See our previous discussion in this work, Is There Anyone Out There?, pg. 127. We have affirmed within this work that there indeed may be other worlds within THIS universe that are inhabited by the sons and daughters of God. We further affirmed that the Atoning sacrifice of the Son was performed but

We understand here that our mortal (Earthly) birth is not the beginning of our individual histories, is not the beginning of an individual's life, for: "*Man was also in the beginning with God*" (D&C 93:29). It is an entry from a more perfect sphere into one that was created by subtraction (*tzimtzum*). Our appearance here in this Universe is a Fallen state of our being, something less than it once was. It by now should not be surprising to learn that this truth was known to those who understood and taught the Gospel as it was originally written. On Man's descent from God, Origen wrote:

> *". . . there was a descent from a higher to a lower state not only on the part of those souls which had deserved it by the various motions [of their will] but also on the part of those who were brought down from the higher and invisible to this lower and visible realm, for the service of the whole world, although they did not wish it . . . on account of those for whom this was needful, the visible world was constituted . . . The world must be supposed to have been made of a nature and a size to contain all the souls which were appointed to be trained therein . . ." (De Principiis, III. v. 4-5)*[25]

Our Temporal existence is a necessary step; it is advancement, not a reflexive action of God imposed punitively. Though our Temporal beings are less than the perfect they once were, we have been added upon as part of our eternal progression. In mortality we gained a truly "*free*" agency, procreative powers and an ability to exercise principles of faith. To these positive attributes are added the structure of matter, here somehow different than that of the matter bound to us in the Physical sphere, that which comprises the Spirits that we are. This Temporal matter though now mortal, decaying, suffering the effects of entropy, emanates from the energies of God and is destined to clothe each of us in a perfected immortal form throughout eternity.

The "*stuff*" of this Temporal world was not created *ex nihilo* in the Big Bang, but as we have shown here, was transformed from the *Zero Point Field* through the input of negative entropy. Joseph Smith spoke of this reality in 1839 well before it had been surmised by modern science when he said: "*Earth, water &c — all these had their existence in an elementary State from Eternity.*"[26]

Some would say that the Atonement of Christ is limited to only those who God calls to salvation; their theology is broken and has been corrupted by the philosophical

---

once for all at Jerusalem on this planet at the meridian of time as reckoned by us. We can surmise that because Christ conducted his mortal ministry here, the original human pair from the Garden also accomplished their mortal probation on this sphere; this, however, may not be so. The habitation of another world may have begun with that pair, or the story of the Garden may be one that is repeated for each world that is to be home to the sons and daughters of God, making there Adams, many, and Eves, many. It appears to me to be inconsistent with the Plan of Salvation to support a theory that the original pair lived out a life on more than one planet, for we understand that, "*...it is appointed unto men once to die...*" (Hebrews 9:27).

[25] The Early Christian Fathers, Henry Bettenson (Oxford University Press, New York, 1956, pg. 191).

[26] The Words of Joseph Smith: The Contemporary Accounts of the Nauvoo Discourses of the Prophet Joseph (Joseph Smith, remarks, delivered before Aug. 8, 1839, in Andrew F. Ehat and Lyndon W. Cook, eds. (1980), pg. 9).

musings of man. A proper understanding, as expressed in this work, identifies all of the children of Adam and Eve as the sons and daughters of God, all sent to mortality with a purpose, that of divinely appointed progression.

We understand here, however, that all are not to follow the same path in mortality. The vast majority of the sons and daughters of God who will *ever* live will not be exposed to the Gospel; they will not be acquainted with the name of Jesus Christ, nor will they receive the gospel message: they will not live a "*Christian life*," firm in the faith of a loving Father in Heaven and the mercy extended to them by the Son (*Figure 11-2*). We do not here attribute this disparity to God having only extended His call to a handful, for we affirm He has called all, intends to save all, will redeem all, and extend exaltation to as many as will avail themselves of it. Any theology that does not address these billions of Souls, provide an explanation for their purpose here in mortality, accommodate their receipt of the gospel message, and cover them with the love of their Father in Heaven, is immoral and without any understanding of who God is. [27]

- That All Men are Created Equal?

The effects and benefits of Temporality are afforded to all regardless of their circumstances. We all, the sons and daughters of God, experience death and the glory of birth; our Spirits are associated with Temporal bodies that will one day clothe us in eternity after the resurrection. Each of us see in our lives the effects of wrong choices made, whether or not those effects are due to our own actions or the actions of others. *These are the primary purposes of life in mortality, the reasons we come here.* However, not all of us here are equally exposed to the *secondary* purposes of mortality. The free exercise of agency, the joys and responsibilities attendant to procreation, and the profound growth to be gained through faith are not available to all who live; these of course, must also be addressed by God's Plan.

At the beginning, when our Intelligences were made, it may be assumed they were created equally, but we know from the glimpses we have seen of our existences througout the intervening eons, progression and growth was accomplished for certain by all, but for many, much more was attained – some of that is shown in the

---

[27] Elder Neal A. Maxwell spoke of the numbers of Souls who have passed away as follows: "*On the other side of the veil, there are perhaps seventy billion people. They need the same gospel, and releases occur here to aid the Lord's work there. Each release of a righteous individual from this life is also a call to new labors. Those who have true hope understand this. Therefore, though we miss the departed righteous so much here, hundreds may feel their touch there. One day, those hundreds will thank the bereaved for gracefully forgoing the extended association with choice individuals here, in order that they could help hundreds there. In God's ecology, talent and love are never wasted. The hopeful understand this, too*" (Notwithstanding My Weakness, General Conference, October 1976.)

Elder Maxwell's count of 70 billion did not include the almost 10 billion more who have been added to the departed and that are alive on Earth today, so his count would be closer to 80 billion Souls, still short of the 108 billion counted by science as shown in (*Figure 11-2*). The discrepancy can be accounted to science counting all homo sapiens that have ever lived and Elder Maxwell considering only the sons and daughters of God in his figure. Either way the number of Souls that have ever lived, by far outnumber those that have been exposed to the Gospel and the commandments of God in mortality.

vision of the "*nobel and great ones*" of Abraham. Greatness and ability beyond comprehension is shown in what has been revealed of the pre-existent Michael; his progression in the Spiritual sphere was so marked that he was chosen to join with Jehovah in the creation of the Physical sphere.

We understand that by the time the Temporal sphere was created, our individual progress along a metaphoric ladder to Exaltation positioned each of us at many different points, with many different abilities, needing many different challenges to effect the maximum allowable advancement towards our ultimate goal. That this is indeed so, is evident in the many different experiences of the many different peoples who have inhabited this world alone. It is unfathomable to think of the variety of challenges facing Souls who may have entered Temporality on any number of distant worlds alien to ours.

Our progressive course to become like our Father in Heaven did not start with our Earthly birth, it will not end with our death. It began a near infinite time long ago and will continue for an eternity to come. The Plan calls for all to come unto Christ, hear His words and accept His loving grace. Our understanding of our return to the Physical sphere after our death allows for such an eventuality.

In the end, the chasm between the Temporal and the Physical spheres is drawn up and closed, as perfection is added back to where it was taken away by the *subtraction* (*tzimtzum*) that created the Temporal sphere. The Atonement of Jesus Christ perfects us; it redeems and restores the sons and daughters of God to what they once were. The experience we obtain here and the attributes associated with the addition of perfected, now immortal, matter to our being, will cause us to be more than we once were. This is the progression attainable only by entry into Temporality and participation in Christ. These are the blessings afforded to the sons and daughters of God, the same that were denied to Lucifer and the children of the devil.

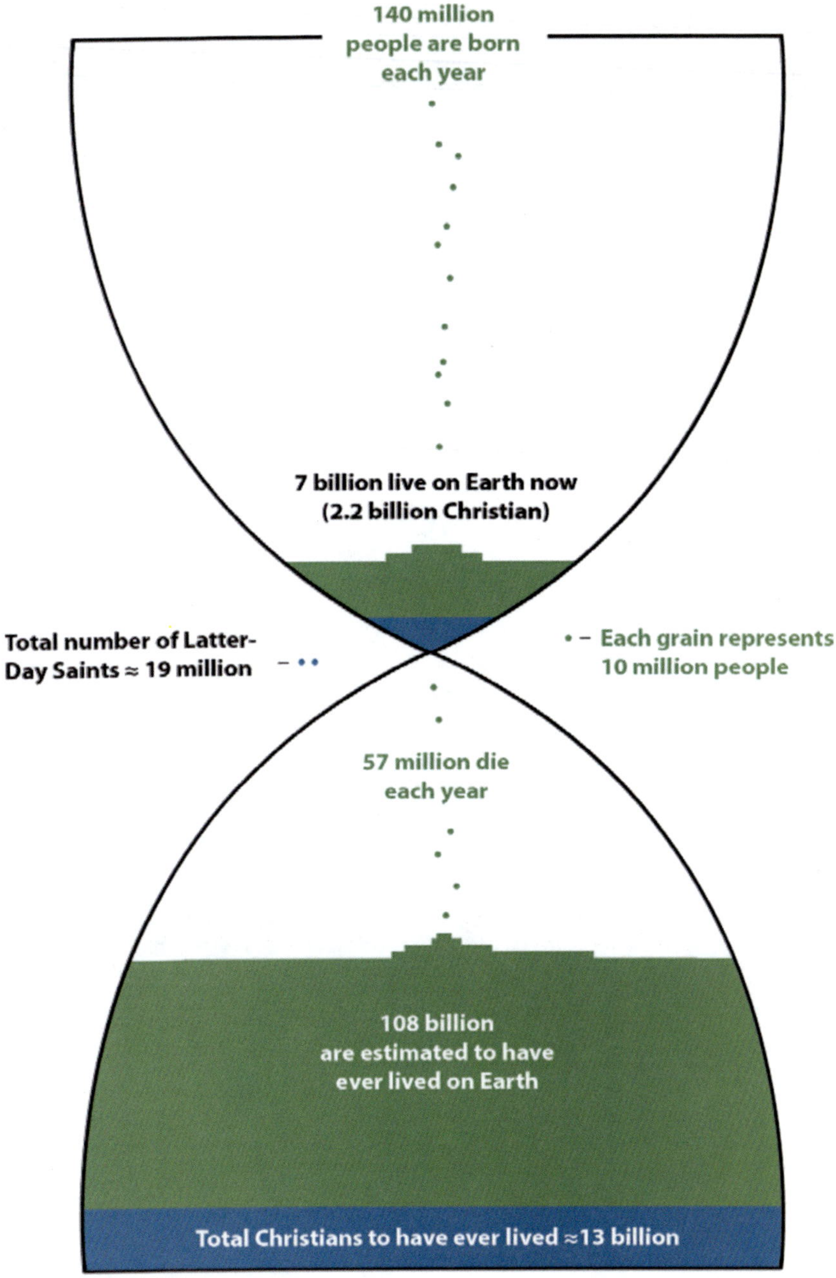


Figure 11-2 *An Estimation of the Earthly Population of the Sons and Daughters of God. Though there have been billions of us to have lived on Earth, God knows each of us intimately, He said, "I have called thee by thy name; thou art mine" (Isaiah. 43:1).*

"*What kind of a being is God?*" Joseph Smith once asked.[28] Willard Richards recorded the prophet's reason for asking this question: Human beings needed to know, he contended, because "*if men do not comprehend the character of God they do not comprehend themselves.*" In that simple expression, just 14 words, the Prophet closed the chasm that almost two millennia of apostasy and misperception had created between God the Father and humanity. Human nature at its core is divine.

This work has detailed this divine nature of humanity as consisting of two parts: first, eternal Intelligence sourced and parted directly from the Father, and second, immortal physical and tangible bodies. Together these comprise a Spirit that is divine both in appearance and in fact. Our Temporal state does not reflect our true nature. Mortal man with his weaknesses, imperfections and faults is not who we are, nor does mortality reflect our initial created state.

Joseph is famously quoted as having said, God *"was once as one of us;"*[29] we affirm here that this means: eternal, embodied, intelligent, agent and capable of progressing.[30] God remains so in all these aspects today. Joseph inferred that there was a time, most distant, when God was once *less* than He is today and by that the young prophet was equating our present state as something less than we are to become. The context of the Prophet's teaching was not centered on the being of God but upon the divine potential of the sons and daughters of that God. In that same discourse, Joseph Smith expanded on these thoughts by expressing that, *"all*

---

[28] *King Follett Discourse, Apr. 7, 1844, recorded by William Clayton, available at josephsmithpapers.org.* While this discourse, given at the funeral for Joseph's good friend King Follett, is often considered to be Joseph Smith's most detailed account of his understanding of divine nature and exaltation, its context, remarks made at a funeral concerning the resurrected status of man, must be remembered. Joseph began speaking that day noting the high winds that would make his remarks to the large crowd difficult to hear. The most quoted sermon text is a compilation of four partial accounts transcribed that day by Willard Richards, Wilford Woodruff, William Clayton, Thomas Bullock and a newspaper report of the address printed August 15, 1844, in the Times and Seasons. Because of the conditions that day and the limitations imposed on those who transcribed the discourse, it remains unclear what exact words Joseph Smith used. Though what is recorded gives us insight into what Joseph thought and provides understanding that enhances the meaning of many passages of scripture, it does so imperfectly. The King Follett discourse was never canonized, and though it was contributed to other un-canonized teachings by subsequent church leaders, the sermon and the teachings that draw upon it are not to be understood as representing standard church doctrine.

[29] Ibid.

[30] It is not entirely clear what the Prophet meant when he spoke at the funeral for King Follett, but the first sentence quoted here coupled with a belief that "*flesh and bone*" are residual to mortality has spawned many an adventuresome expedition into the wilderness of speculation, bearing confusing fruit about the possibility that God once was a *mortal* man, Himself just a creation of another God before Him. These forays into the heavy woods of conjecture need not be taken; they are fraught with pitfalls and quicksand and pose problems with a firm understanding of eternity and the critical concepts of first and universal cause we have discussed in this thesis. See our previous discussion in this work, Post Hoc Ergo Propter Hoc, pg. 453.

*the spirits that God ever sent into the world"* are on a path of progression and by that path, "*susceptible of enlargement."*[31]

The *relation of being* spoken of by Joseph Smith, that *God was once as one of us* can be turned around to explain, *we were once as God*. In this form there is original Christian teaching for support. On Man's *created* divine nature Irenaeus said:

> *"Men, being created by God, are by nature sons of God; but they are not sons in their deeds. (3) For as among men disobedient sons, disowned by their fathers, are by nature their sons but are alienated in law, since they are no longer heirs of their natural parents: in the same way, with God, those who disobey him are disowned and have ceased to be sons." (Adversus Haereses, IV. xli. 2-3)*[32]

We do not here claim that Irenaeus was making a statement of equivalency; he did not intend to suggest that the created sons and daughters of God are in any way equal with God in capability, intelligence, knowledge and power; that would not be possible, for He is greater than all and precedes our existence. Irenaeus, however, just as Joseph Smith later would do, expressed his belief that we are of the same family of God, because we are by nature of the same substance, essence, of our Eternal Father.

Joseph Smith taught that divinity and the attainment of all of its attributes was inherent to our being from the beginning. William Clayton's transcription of the King Follett discourse records the Prophet as having said that long before the world was formed, God found *"himself in the midst"* of these beings, His sons and daughters and *"saw proper to institute laws whereby the rest could have a privilege to advance like himself."*[33] Wilford Woodruff recorded that same day that Joseph had said, we were designed to be *"exalted"* with Him.

In this work, as Joseph Smith did then, we understand man to be of the family of the divine, God then, in embryo. We fall short of this eventuality today, not even appearing to ourselves to be the same species. This is understandable, without a knowledge of biological growth, someone who views an infant at birth would find it impossible to fathom that one day (through natural processes) that newborn could become a strong man perhaps over six feet tall and 200 plus pounds. We know this is true, however, and think nothing about accepting it. We understand and anticipate the *character* of the child by *comprehending* that of the father. In this context, paraphrasing what Joseph said, we *comprehend* ourselves by understanding the *character* (nature) of our Father, God.

---

[31] *King Follett Discourse, Apr. 7, 1844, recorded by William Clayton, available at josephsmithpapers.org.*

[32] The Early Christian Fathers, Henry Bettenson (Oxford University Press, New York, 1956) pg. 73.

[33] *King Follett Discourse, Apr. 7, 1844, recorded by William Clayton, available at josephsmithpapers.org.*

Because we are the same species (understood in this context, composed of the same intelligence and of the same spiritual matter organized in a way that is the *likeness* of our Father), we are capable of becoming closely comparable to God. This understanding is explicit in our theology and what has been presented here.

Because we are divine by nature, our eternal existence is intrinsic. This immortality is what was spoken of by the Savior when He said, "*Marvel not at this for the hour is coming, in which all that are in the graves shall hear his voice, And they shall come forth; they that have done good, unto the resurrection of life; and they that have done evil, unto the resurrection of damnation*" (John 5:28, 29). The eternal progression attendant with exaltation, however, is conditional.

In a vision he received in February of 1832, Joseph Smith saw and heard the "*testimony of the gospel of Christ*" (D&C 76:50). The Prophet understood that only those, "*who overcome by faith, and are sealed by the Holy Spirit of promise which the Father sheds forth upon all those who are just and true*" are, "*priests and kings*" and are "*of the church of the Firstborn.*" Those sealed receive of the fullness and Glory of God. Of them, "*it is written, they are gods, even the sons of God*" (D&C 76:53-58). Within a year, in December of 1832 and early January of 1833, another revelation confirmed what it meant to be "*gods, even the sons of God*;" it said, "*the saints shall be filled with his glory, and receive their inheritance and be made equal with him*" (D&C 88:107).

Equal with God? We do not imply an *equivalency* with our Father in power, Glory, knowledge or other identifiable attributes, but our understanding speaks to the attainment of those qualities to a measure that allows us to progress in the manner He has. It is the full realization of our becoming *like* Him, the process that was started in the Garden (a Temple of Temporality) with the Fall when God declared, "*I, the Lord God, said unto mine Only Begotten: Behold, the man is become as one of us to know good and evil*" (Moses 4:28). Our progression is infinitely behind His, and we will never approach His state, though we can head in His direction. Exaltation is not understood then to be an *end state* but the means to get there.

Joseph Smith told those in attendance at the funeral for King Follet, "*You have got to learn how to be a god yourself.*"[34] That learning process requires that we seek godliness in our lives, desire to be more like God and allow Him to affect change in our lives.

The process is eternal, it is never ending. It requires first, a reliance on the atoning sacrifice of the Savior and continues with the exercise of faith, a willingness to keep the commandments and remaining in a repentant frame of mind; it will require great patience.[35] We do not learn all at once, nor will we come to a full knowledge of

---

[34] *King Follett Discourse, Apr. 7, 1844,* recorded by William Clayton, *available at josephsmithpapers.org.*

[35] Joseph Smith taught that for Salvation, actually Exaltation, to be effective in our lives, *work* on our part was required. For that proclamation, he and the church have received much condemnation by Christian philosophers of this day. We fully understand that the working out of one's own salvation is not

godliness here in the Temporal sphere. We need to understand the "*first prin[ciples] of the Gospel,*" and when our time comes, continue to acquire beyond Temporality until we "*learn the last prin[ciples] of the Gospel.*"[36] "*It is not all to be comprehended in this world,*"[37] the young prophet said, "*It will take a long time after the grave to understand the whole.*"[38]

Within Temporality, we are handicapped by an inability to fully comprehend the broad scope of the creations of God. Paul wrote: "*...the natural man receiveth not the things of the Spirit of God: for they are foolishness unto him: neither can he know them, because they are spiritually discerned*" (1 Cor. 2:14). With the help of the Spirit, guided by the written word, we have been able to glimpse here things beyond our Temporal realm, but our conceptions of this eternal reality are necessarily limited. For millennia, religious thinkers and the faiths they have worked to establish have struggled to effectively express their visions of eternal Glory.[39]

The effects of such a restricted view have constrained man's view of himself and his relationship within the family of God. The Apostle Paul lamented of this reality more than once when he wrote to the Corinthians. He said, "*Eye hath not seen, nor ear heard, neither have entered into the heart of man, the things which God hath prepared for them that love him*" (1 Cor. 2:9). Later, with hope of a yet future restoration he said, "*For now we see through a glass, darkly: but then face to face: now I know in part: but then shall I know even also as I am known*" (1 Cor. 13:12).

The dark view spoken of by Paul led most theological constructs to see man as debase, thoroughly corrupt and vacant of anything worthy of praise. This understanding may in some way be an attempt to express humility, but most often (as in the reformed religions for example) this concept is espoused by those that are

---

possible; the Grace of God through the atonement of Christ is both the only effective way of achieving that and it alone, without any other action, is fully sufficient. However, "*work,*" a *participation* of the individual with Christ in the salvific effort in their own life and in the lives of others, is a resultant evidence of divine transformation in the life of the saved. This truth was taught by the original Christian fathers as well. "*How then will any be a god, if he has not first been made a man? How can any be perfect when he has only lately been made man? How immortal, if he has not in his mortal nature obeyed his maker? For ones duty is first to observe the discipline of man and thereafter to share in the glory of God.*" (Adversus *Haereses, IV. xxxix. I-2)* The Early Christian Fathers, Henry Bettenson (Oxford University Press, New York, 1956) pg. 69.

[36] *King Follett Discourse, Apr. 7, 1844,* recorded by Thomas Bullock, *available at josephsmithpapers.org.*

[37] Ibid, as recorded by William Clayton.

[38] Ibid, as recorded by Willford Woodruff.

[39] These "*natural*" limitations lead to vocabulary frequently twisted into simplistic terms that can be satirically expressed in cartoonish caricatures when represented in popular culture. The deep peace and overwhelming joy of Salvation as expressed in the scriptures, for example, are often depicted by images of resurrected men and women as winged angels, each sitting on their own cloud, playing harps throughout eternity. This is a thoroughly inadequate and near derogatory representation of the truly faithful's understanding of eternal joy. Latter-day doctrines of exaltation, which detail enormous eternal creative potential for the sons and daughters of God, are similarly represented in derogatory terms through satirical caricatures of the exalted receiving their own planets to govern as a god.

foremost vain and arrogant, proclaiming themselves to be *the elect* while condemning most of the sons and daughters of God, their literal brothers and sisters to predetermined damnation through no fault of their own.

Most theologies never see the good that is present in most of us, certainly at birth. We understand here that we are literally of the "*stuff*" of our Father. He is in fact *in* us, in a very actual way, not at all unlike our mortal father's DNA is in us and comprises our genetic make-up, making us who we are.

In this Universe we are Temporal (mortal), but within each of us dwells a Spirit that is immortal, physical and tangible and centered on an Intelligence. Fundamentally we are this Intelligence, itself the very "*stuff*" of God and His Glory; it is eternal and without a beginning. This understanding empowers our belief that we are divine, capable of attaining all that the Father has. We are not here in Temporality to be punished because of an unplanned and disobedient Fall, but we are here to be empowered by experience and the exercise of a truly "*Free*" *Agency*.

Born of our Father so long ago, our theology expresses a firm belief that we, the sons and daughters of God, will forever remain His children; He will always remain our God and worthy of praise, worship and adoration. Whatever heights we attain through the progression He affords us, we will forever remain less than what He has reached.

Our understanding of God is not static or immutable. Our Father continues to achieve, grow and progress. How this is so is unknowable to us, but it is a mystery we can happily live with. We understand here unity in the family of God. Though we remain separate and distinct from the Father and the Son, a reality that cannot be questioned, through participation with Christ, *keeping the commandments and living as He did*, we ultimately will, because of the Atonement, attain the oneness expressed by the Savior, "*That they all may be one; as thou, Father, art in me, and I in thee, that they also may be one in us"* (John 17:21).

*We declare these truths to be self-evident.*

## The Nature of the Family of God

### *A Personage of Spirit*

Understanding and defining what each of us are today as Temporal sons and daughters of God, is difficult enough. Describing our true eternal nature in terms we can discuss and comprehend is near impossible. Nonetheless it is necessary for us to know the form of our eternal being, for our Temporality is but just a moment in time, a flash of light along the eternal continuum of our actual existence. What we truly are, forever will be, and ever have been, is something much different than the mortal shells we inhabit today.

We in fact are something that has been defined as *Spirit*, what we have described here as truly tangibly Physical, yet immortal. That we are created in the image of God, means we are like Him; *"...now are we the sons of God, and it doth not yet appear what we shall be: but we know that, when he shall appear, we shall be like him; for we shall see him as he is*" (1 John 3:2). In the previous chapter, we described the nature of God, the Father, Son, and Holy Ghost. We detailed the attributes of their nature. Here we will apply what we learned of them to develop an understanding of us.

We each began at some point in the distant past with the creation of our independent Intelligences; we have described that the *Intelligence* within us is eternal in that it always existed within God, who is without a beginning. At the point of our first creation, our Father, God, parted Himself, divided His eternal Intelligence to form individual, independent, eternal, sentient beings, agent unto themselves and capable of choice, experience, and intellectual growth. This was the state of our being throughout the unknown eons that encompass the Spiritual sphere before the creation of the Physical.

The 6 days of the Physical Creation were sequential, they began with the forming of the immortal *earth* and other like spheres and ended with the creation of immortal Physical bodies to enclose and empower our independent eternal Intelligences. The planting of the Garden, and the eventual placement of our first parents within it, surely followed these six creative periods.

The point of Michael's embodiment to become a Spirit perhaps immediately followed his co-creation of the Physical sphere, but for certain it followed his leading the defense of the Great Plan against the forces of Lucifer. Subsequently, he was enclosed in the materials of Temporality as he was placed within the Garden, immortal, embodied, *Flesh and Bone*, yet innocent because a veil of forgetfulness had been placed about him. He is declared a "*Living Soul*" as he became a reflection of the being we are to be in the resurrection. This point of *Temporal embodiment* is marked in the creative story by Michael's receipt of a new name, Adam, as he entered the *Garden Temple* for the first time.

Unfortunately, much too often our interpretations of Scripture do not clearly identify between what is purely *Intelligence* and what is *Spirit*. If these interpretations and Scriptures ever were clear on this subject, they certainly are not now. Great confusion circumscribes this matter. The lack of clarity stems from many factors, not the least of which is the limited capability of natural man to understand fully the things of God and the spheres beyond our Temporal existence. The limitations of properly transferring meaning of things barely understood in the original language to other languages in translation aside, our individual conceptions of things unseen are no doubt at great variance with the conceptions of others. Even internal to our own vocabulary each of us exchanges certain words and concepts freely without regard at times for any nuance that may prove to be critical.

We find that *Spirit* and *Intelligence* are interchanged without regard in our Latter-day theological language and within our minds, often mixing and blending the separate meanings found within these concepts into a single understanding that confuses and misrepresents truths that are ultimately important. Perhaps in no case can these multiple factors be viewed together more than the one in question here.

## *The Spirit of Man*

Much of what we learned from our study of the personage of the Holy Ghost can be applied to our understanding of the personage of the Spirit of Man. We are the same as the Holy Ghost in the form of our "*Spirit.*" We each are a Spirit comprised of the blending of eternal Intelligence and spiritual (immortal) matter into the form of an eternal Physical body. We have confirmed here that an embodied Spirit can indeed indwell a Temporal body; in fact, that is the case with each of us and *every* living thing in this sphere within this Universe.

Early in the last century, the First Presidency, consisting of Joseph F. Smith, John R. Winder and Anthon H. Lund, issued a statement (The Origin of Man, Improvement Era, November 1909, pgs. 75-81) detailing the form of the Spirits of the sons and daughters of God: "*'God Created man in His own image' This is just as true of the spirit as it is of the body, which is only the clothing of the spirit, its complement; the two together constituting the soul. The spirit of man is in the form of a man, and the spirits of all creatures are in the likeness of their bodies. This was plainly taught by the Prophet Joseph Smith*:"

> *"...that which is spiritual being in the likeness of that which is temporal: and that which is temporal in the likeness of that which is spiritual: the spirit of man in the likeness of his person, as also the spirit of the beast and every other creature which God has created." (D&C 77:2)*

The "*form of a man*" spoken of by the First Presidency is to be taken *literally*, in every describable way:

> *"The spirit of man consists of an organization of the elements of spiritual matter in the likeness and after the pattern of the fleshly tabernacle. It possesses, in fact all the organs and parts exactly corresponding to the outward tabernacle." (*Key to the Science of Theology*, Parley P. Pratt, pg. 79)*

The Physical Creation, as we have detailed it in this work, and as told in the Endowment narrative, produced an *immortal physical reality* which is a perfected realization that precisely duplicates the imperfect constructs of the Temporal world. We pronounced the Physical to be the *template* for the Temporal. In Part 1 of this thesis, we described the Temporal sphere as one that literally is a *subtraction* of capability from the higher Physical sphere. Nothing new was created in Temporality except uncertainty.

To create and control uncertainty, laws were established to set boundaries for the substance of Temporal matter (*bohu*). The Atonement was planned for and was to be instituted to restore the subtracted perfection at the conclusion of the Temporal period, and the presence of God was withdrawn (*tzimtzum*) to allow the uncertainty to function providing for the unfettered "*Free*" Will of man. This subtraction "*created*" Temporal matter and form in the vacuum it left behind.

The "*spirit of man*" as described by Elder Pratt, reflects a parallelization of the Temporal and the Physical; the Physical emulates that which is Temporal. He says: "*...spiritual matter [is] in the likeness and after the pattern of the fleshly.*" Elder Pratt's description is reversed in fact from what is obviously true, for we know the Physical sphere preceded the Temporal; it is the "*fleshly*" *matter* of the Temporal that is in the likeness of the more ancient and senior Physical form. Of significant interest is that the exactness of this "*likeness*" extends to the fine details of structure as well; "*It possesses, in fact all the organs and parts.*" What is the function of all of these organs in immortality? An answer to this question strains the imagination.

***The Creation described by the Preamble verses and the Endowment narrative resulted in rocks of granite, trees of wood, water that is wet, flowers that are fragrant, fruit that tastes sweet…***

This physical state, matching the perfect and ideal form of things Temporal, extends to everything in the Physical sphere. We are told "*God Created man in His own image,*" so we fully expect the details of our tangible Physical (spiritual) bodies to be those of the Father, Son and Holy Ghost as well. Our Spirits are identical in form to the physicality we have today, though in Temporality we are clothed in (*flesh, bone and blood*) as we live on Earth in this Universe.

With a straightforward reading of the Scriptures and an understanding of the Physical sphere that we have provided here, it is easy to comprehend that objects and beings beyond the veil have a Physicality that is actual, making them capable of being perceived in the same sense that tangible things are beheld here in this Universe. The creation described by the Preamble verses and the Endowment narrative resulted in rocks of granite, trees of wood, water that is wet, flowers that are fragrant, fruit that tastes sweet, and animals of all kinds. Some of these forms are unthinking, some inanimate, yet we understand them to have been given an immortal, literal, tangible physicality. It is clear the Physical Creation included immortal perfunctory constructs. Could not these non-sentient structures be termed "*Flesh and Bone*" using the broadest application of the definition we have developed for Resurrected Temporal things?

## *We Shall be Like Him*

Let us return briefly to the Garden story narrative, "*And the LORD formed man of the dust of the ground, and breathed into his nostrils the breath of life; and man became a living soul*" (Gen. 2:7). Of key interest here is the final phrase, "*...and man became a living soul.*" We have established here a firm definition of what is considered a "*soul.*" It is, we said, in three parts, Intelligence, bonded to a Physical body that together comprise an embodied Spirit, and that two part Spirit is then bonded to the "*stuff*" of Temporality through the *Ruach* of God to become a *Living Soul,* either understood to be one mortal (in Temporality) or immortal (after the resurrection) (*Figure 11-3*).

The "*stuff*" of Temporality of which we speak is termed "*Flesh and Bone.*" As a "*living Soul*" then, the *personages* of Adam and Eve are fully embodied in complete physical form and detail. It can be derived that the bodies of both Adam and Eve within the Garden Temple[40] were of what has been called "*Flesh and Bone.*" This understanding remains consistent with the scriptural record, for we are told, "*I, the Lord God, created all things, of which I have spoken, spiritually, before they were naturally upon the face of the earth*" (Moses 3:5). *Flesh and Bone* certainly are included under the umbrella of *all* things that were created.

Further we are told, "*God Created man in His own image.*" The Father, we have determined, has a body of *Flesh and Bone* and has had one from at least an ancient, distant time that stretches out far beyond the Physical Creation.[41] So it is not beyond conjecture that by creating us physically in His own image, beginning with Adam and Eve in the Garden, our Father intended us to be clothed in *Flesh and Bone.*[42] In this construct of our beings, it is the addition of blood (if only symbolically) that sets mortals apart from the realms of pre-existent and resurrected immortality. That concept aligns as a matter of fact with doctrines long held by many Christian traditions.

---

[40] The Garden Temple (it will be remembered) served as a bridge with Temporality and as such encompassed unique conditions allowing for Temporal elements in the Spiritual sphere.

[41] We understand here, God to be "immutable" in purpose, mind, and desire; that, as is the case with His Son, He is, "*the same yesterday, and to day, and for ever*" (Heb.13:8). We, however, cannot state that an eternal, all-powerful God is limited in a way that would preclude Him from accomplishing anything He would desire. With that understanding, we can only say that He possesses a body of "*flesh and bone;*" we cannot speak in any certain way that such has always been the case.

[42] We have previously surmised that *flesh and bone* are eternal components of our Eternal Fathers being, that our attainment of them in immortality (after the resurrection) is part of our continued progression to be "*like*" the Father.

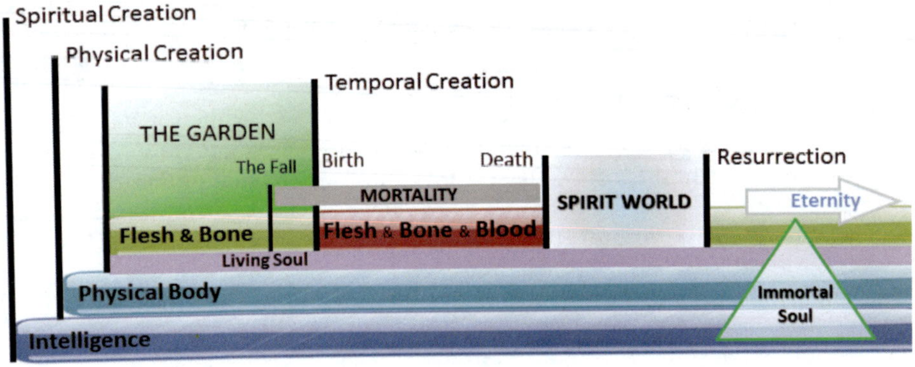


Figure 11-3 *The Three Part Nature of the Children of God. Each of us is an eternal being composed of three parts. We are Eternal Intelligence parted from the Father long ago in the Spiritual sphere. We are embodied with an Eternal Physical Body that combines together with our Intelligence to form our Spirit. Added to these two is our Body comprised of the substance of Temporality both mortal and later immortal in nature. The third part of our nature is Flesh and Bone; during our mortal existence, it is accompanied by Blood (however symbolic that accompaniment may be). Upon our death, we return to the "Spirit World" (appropriately named) where we lose the third part of our being and return again to the "embodied Spirits" we once were, there awaiting the restoration that comes through the Resurrection.*

God is what we hope to someday be; as our Father, He is and always has been, what we are capable of eventually becoming. In the last chapter, we described God, our Father, as composed of intelligence, immortal matter (collectively His Spirit) and an eternal body of *Flesh and Bone*. We have shown here that His form, an embodied Intelligence clothed in immortal *Flesh and Bone* was also held by Adam and Eve in the Garden; this form, held precursory to the bodies they would receive in mortality, was indicative of their "*true character*" as children of God. After the Fall, the immortal bodies of our first parents were *transfigured* (at this time *downward* from perfection) and their *Flesh and Bone* was merged with *blood*, so symbolic of the Savior's sacrifice and His Atonement. This transfiguration is represented for us today in the Endowment and the Genesis 1 accounts through their being clothed by God in the *coats of skins* of animals in the Garden and our *like* receipt of Priesthood Robes (on the left shoulder) and Garments, which we are told represent those skins, within the confines of a Temple as well.

It should not be lost on the observer that the laws of obedience and sacrifice were given to Adam and Eve in conjunction with their receipt of these clothing skins. Neither should we miss that those skins themselves required the shedding of blood and that none of us can return to God and our *true character* but through the processes of justification, sanctification and the atoning *sacrifice* of the Savior, all of which require shedding (*separating from*) blood and *obedience* to the requirements of the laws given. Blood then, both symbolically and factually, is the corrupting component in our being.

Within the Garden Temple, Adam and Eve were specially empowered in the Physical sphere to choose and exercise their agency freely. This was accomplished by enclosing them within (placing them beyond) a veil of forgetfulness. There, the *chosen* first couple was free to experience their agency in a way that Temporality is designed to deliver to us all.

Temporality allows for the uncertainty necessary for free will; that uncertainty is made possible through the withdrawal of the presence of God, resulting in imperfection. The form of all things in this Universe (including that of the stars, planets, and the life upon them) are deficient copies of the perfect, immortal objects and beings within the Physical sphere of which they are a reflection.

Temporality includes limitation in the processes of life and physical laws. Those imperfections are to be blamed for deformity and disease that make some of us even less than the flawed images of our Physical (spiritual) selves we all are.

It is often this disconnect (*sometimes in the extreme*) between what we truly are in the Physical sphere and what our bodies become in Temporality, that leads to many conditions associated with depression, identity confusion and some mental illness. As we age and when severe illness overtakes our body and our form, the Temporal becomes less and less an image of who we are; these too are processes that disconnect our mortal bodies from harmony with our eternal Physical self.

We understand here that it is not the bad choices we make which separate us from God but the *conditions that afford our ability to choose* that does. For our *Free Agency* here in Temporality to be possible, our beings (by necessity) are disconnected from the perfection of God. This disconnect eventually results in our Physical death and the separation of not just blood from our being but *Flesh and Bone* as well.

Despite the binding power of the *Ruach* and the *Neshama* within us, Temporality and the effects of sin, our choosing courses contrary to the will of God (or the effects we experience through those contrary actions made by others) separate us from God bringing spiritual death upon us. Spiritual Death is the acquisition of attributes within our character that prevent us from receiving the full Glory of God (His energies) in our being. The destructive effects of sin need not accumulate or be measurable to kill; *any sin at all in our lives is fatal*, anything short of perfection, is to God, "*an abomination in his sight*" (JSH 1:19).

For each of us, Temporality results in Physical death (separation of our embodied Spirit from *Flesh and Bone and Blood*) and Spiritual Death (the inability of our Spirits, because of imperfection, to return to the presence of our Heavenly Father). We return to the Physical sphere in habitation within the *Spirit World*, where as Spirits (embodied Intelligences) we await the resurrection and its restoration of our "*true character*" that was lost.

## *Our Resurrected Bodies*

In the last chapter, we discussed that we cannot say for certain that the Savior's pre-incarnate body was of *Flesh and Bone*, but we did say we know that as Jehovah, He appeared in a tangible *body* before His mortal ministry to many ancient prophets, the brother of Jared, Abraham, Jacob and Moses among them. This tangible body shared common attributes and form with that of our own Spirits and that of the Holy Ghost.

Long before He walked the plains of Galilee, He declared, "*this body which ye now behold is the body of my spirit*" (Ether 3:16). "*Body of my spirit*" is another way of stating, an embodied Spirit as per the definition of Elder Roberts or a "*personage of Spirit*" as revealed in D&C 130:22. Before He created the Temporal sphere, the Son and His Father walked with Adam in the cool of the day within the Garden. These pre-incarnation appearances aside, the most telling and direct evidence we have of the "*true character*" of the Son, was provided when Christ appeared to His disciples after the resurrection in the closed room. There He had what He described as a body of "*Flesh and Bone.*"

The full declaration made by the Savior to his disciples that day has fueled the fire of those who seek to deny that God, in His natural state, that of a *Spirit* (John 4:24), has a tangible, physical body. The statement made by the Savior, however, as it is translated (or even as it was understood by all in that room that day), does not preclude the same physicality for a Spirit as we have defined it here.

Christ said, "*for a spirit hath not flesh and bones, as ye see me have*" (Luke 24:39); indeed this is factually accurate. A "*Spirit*" (an embodied Intelligence) by itself, though both Physical and tangible, does not have *Flesh and Bone*. However, it must be remembered that the context with which the Savior used the word "*spirit*" in that place, that day, pertained to His disciples cowering in fear for they thought they were seeing a ghost or an apparition.

The Savior sought to ease the fear of his followers, proclaiming to those that were afraid that He was not a ghost, an ethereal essence. In His comforting utterance, the very Son of God was not endeavoring here to define His eternal physicality in word; *that understanding was intended to be discovered by the disciples for themselves as He sat, dwelt and dined with them that day, as they touched and felt His real and substantial form.*[43]

---

[43] To the non-Latter-day Saint of today, this idea likely seems foreign and unnatural. Although to some, the Nicene God without "*body, parts and passions*" seems somewhat "*puritan*" and unfathomable. The image of an embodied God is even more untenable, after all, they say, "*God is a Spirit.*" This seems to be a major point of divergence between the Latter-day doctrine and that of the traditional Christian faith. The gap between these views is a result of, as often is the case, constriction and limitation. Traditional Christianity, in its efforts to "*define the faith,*" has a tendency to frame the attributes of God into small restrictive packages, easily understood by Man but woefully insufficient in providing a true picture of what God is. These limitations take on many forms and are primarily due to a lack of vision, a failure to see the whole structure of the Gospel. We have previously addressed in this work the ancient understanding that "*spirit*" (the Greek *pneuma*) was elemental and tangible though ever so pure. As a Latter-day Saint, others have often read to me John 4:24, "*God is a spirit.*" They then further go on to

After His resurrection, when Christ made His appearance in the closed room, the disciples learned, through *direct demonstration*, that a body of "*Flesh and Bone*" can have the outward appearance and all of the expected supernal capabilities and qualities of a ghost; the same attributes we traditionally, if however loosely, ascribe to a spiritual being, a *personage of Spirit.* For to Peter, James, John and the rest of the disciples gathered there, when they saw the Christ in His full resurrected form, in His "*true character,*" that of a *personage of Spirit enclosed in immortal Flesh and Bone*, a ghost, a spirit, or an ethereal apparition, is exactly what they thought they had seen.[44]

***Because of our entry into the Temporal sphere and the redemptive power of the Atonement, all of us, including the Son, through the resurrection, receive a permanency of the structure we wore in Temporality.***

The resurrected Son of God declared His eternal physicality within the closed room to His disciples not by word alone (because words could not be expressive enough), but He did so in the same descriptive way used when He appeared in bodily form to Joseph Smith in the Sacred Grove almost 1800 years later. To the disciples in the closed room and to Joseph Smith in a secluded grove of trees, the Savior taught that things spiritual are tangible; they are not ghosts, apparitions or ethereal.

It is traditionally understood that the body of "*Flesh and Bone*" held by Christ was residual of His incarnation. It is commonly assumed that His state then after His resurrection reflects also what our state will be after our resurrection. We understand here "*Flesh and Bone*" to be structural. It is nothing more than the perfunctory component of an immortal sentient being; all thoughts, emotions and senses are experienced by the *Intelligence* within.

---

explain, quoting Luke 24:39 in the Saviors own words, "*...a spirit hath not flesh and bones, as ye see me have.*" I am often bemused and somewhat amused by those that do, for they errantly impress a modern understanding onto the original Greek meanings. This uninformed teaching also comes from someone who either had said, or was just about to say, that Jesus Christ is the one and only God in the flesh and that there is none other than He. Now, if God is a Spirit and he does not have a body of *flesh and bones*, and Jesus Christ is the one and only God, and he does, what *exactly* is the point?

[44] It must be noted that Peter, James and John were included among those who were frightened and confused by the appearance of the resurrected Christ who stood before them in the closed room. Whatever His form there was, it was unfamiliar, displayed differently and had a significantly dissimilar effect than what had been witnessed before when those same three had been present as Christ was transformed on the Mount of Transfiguration.

Because of our entry into the Temporal sphere and the redemptive power of the Atonement, all of us, including the Son, through the resurrection, receive a permanency of the structure we wore in Temporality. This pairing of form, the *perfected, immortal* Temporal to the *eternal* Physical (that which was and still is possessed by the resurrected Savior) is to be our future state as well, for it has been codified in Scripture: "*When the Savior shall appear we shall see him as he is. We shall see that he is a man like ourselves*" (D&C 130:1).

However uncertain we have had to be to this point about what relationship *Flesh and Bone* had to our eternal being, without a doubt the term *Flesh and Bone* connotes the immortal state of the Physical bodies we will possess after the resurrection and the boundless eternities which are to follow. The truth so clearly taught through demonstration by the resurrected Christ is also boldly declared in Scripture:

> *"The spirit and the body shall be reunited again in its perfect form; both limb and joint shall be restored to its proper frame, even as we now are at this time; and we shall be brought to stand before God, knowing even as we know now, and have a bright recollection of all our guilt... Now, this restoration shall come to all... and even there shall not so much as a hair of their heads be lost; but every thing shall be restored to its perfect frame..." (Alma 11:43, 44)*

We understand here the resurrection, through its permanent connection of Temporal flesh with immortal Spirit, assures eternal progression through continuation of the enhanced free agency we were afforded here. Perhaps *only* through this melding of the *Eternal* and the *Temporal* can we have the privilege of parenting Eternal son and daughter Spirits of our own. Of these two potentialities we are not certain, but we may most probably conclude.

This work has described a detailed, progressive and evolutionary course for the being and form of man; this coincides with detailed progression and evolution of the three spheres of creation we have described here.

Of necessity, due to our limited capacity to understand things eternal in Temporality, we have perhaps forced definitions where they don't belong. We have perhaps unintentionally shuttered windows and put up walls blocking the views of infinite vistas and preventing us from going where we are meant to go. We have clear indication that the partitions we have described here are often broken and the spheres spill over on to each other at times for the benefit of man and in accordance with the Will of our Father. In consideration of these things, let us turn to a unique revelation from God.

In the winter of 1843, Joseph Smith received the following revelation:

> *"There are two kinds of beings in heaven, namely: Angels, who are resurrected personages, having bodies of flesh and bones— For instance, Jesus said: 'Handle me and see, for a spirit hath not flesh and bones, as ye see me have.' Secondly: the spirits of just men made perfect, they who are not resurrected, but inherit the same glory. When a messenger comes saying he has a message from God, offer him your hand and request him to shake hands with you. If he be an angel he will do so, and you will feel his hand. If he be the spirit of a just man made perfect he will come in his glory; for that is the only way he can appear— Ask him to shake hands with you, but he will not move, because it is contrary to the order of heaven for a just man to deceive; but he will still deliver his message. If it be the devil as an angel of light, when you ask him to shake hands he will offer you his hand, and you will not feel anything; you may therefore detect him. These are three grand keys whereby you may know whether any administration is from God." (D&C 129:1-9)*

To properly place this in the context of the *Three Creations of God* we have addressed in this work, we must first understand that *heaven*, as used here, describes the *Spirit World*, a specific locale or area fully encompassed within the broader Physical sphere both in space and time.[45] It may not and most probably is not bounded in any way we understand boundaries, but those that inhabit the *Spirit World* are associated with the sons and daughters of God, who inhabited this Earthly sphere. As we understand there to be the possibility of other habitations within the Temporal, each called Earth and populated with sons and daughters of God, we equally understand there to be a possibility of multiple heavens each with a Spirit World associated with its own planetary orb or similar habitation. To this point consider the following:

> *"And the Lord God spake unto Moses, saying: The heavens, they are many, and they cannot be numbered unto man; but they are numbered unto me, for they are mine. And as one earth shall pass away, and the heavens thereof even so shall another come; and there is no end to my works..." (Moses 1:37, 38)*

The verses of D&C 129 identify two conditions or states of beings in Heaven and by extension the Physical sphere; we are told there are "*Angels*" and "*the spirits of just men made perfect*."

The first classification of being disclosed by the revealed word is that of the *Angels*. We are told they are those who are resurrected and as such they are *personages of Spirit* who are joined to an immortal body of *Flesh and Bones*. Of the second type of being spoken of, we know *something*, if, perhaps, not all. We know that a Spirit is a physically embodied Intelligence; so we are to understand that "*the spirits of just men made perfect*" are embodied physically for comprehension of those in the Temporal sphere and glorified (*made perfect*) through the actions of God.

---

[45] See Figure 7-3 <u>The Physical Sphere and its Environs</u>, pg. 289, and footnote 70 on that same page.

This passage continues by providing us with information regarding the sensible nature of these two types of beings. First we are told that the Angels can be felt, that their physicality is tangible and sensible to the touch. This, it is inferred, is due to their *resurrected* state and its attendant bodily attributes of *Flesh and Bone*.

The revelation draws a direct connection with the state of these Angels to the state of the Savior when He appeared after His resurrection to the disciples in the closed room. Jesus said: "*Handle me and see, for a spirit hath not Flesh and Bones, as ye see me have*" (Luke 24:39). Touch is key here; Christ said "*handle me*" to His disciples. He expected to be sensed by touch, and the revelation details that an Angel's hand can be felt in the hand of a Temporal man when a handshake is exchanged.

The attributes possessed by "*the spirits of just men made perfect*" are less well defined in this revelation. We are told that if we ask such a being to "*shake hands,*" that being will not do so, for "*it is contrary to the order of heaven for a just man to deceive.*" Why would it be a deception for the Spirit of a just man made perfect to shake hands? The answer is not entirely clear, but it can reasonably be inferred that it is because the personage of Spirit in its usual state has a physicality that is not sensible by touch, that such an attempt to touch the Temporal is not possible.[46] We are left then with the conclusion that the embodiment we attained in the Physical sphere, a *personage of Spirit*, an *embodied Intelligence*, is not sensible by touch.

Further this revelation also confirms that the Devil's being cannot be felt (sensed) by the hand of a Temporal man.

- The Just and the Unjust

What are we to understand about a *just man made perfect*? The phrase is applied to what is termed *Spirits*, not to the Angels who have already received their resurrection. All are to be resurrected, "*And have hope toward God, which they themselves also allow, that there shall be a resurrection of the dead, both of the just and unjust*" (Acts 24:15). We understand the resurrected state to be one of perfection, for in it, "*...all things are restored to their proper and perfect frame*" (Alma 40:23).

We can infer here that though a *personage of Spirit* is immortal and properly a being of the Physical sphere, it is somehow considered less than perfect, needing to be *made* so, to interface with that of the Temporal. This revelation addresses those

---

[46] There should be no relationship to this set of circumstance and that of Mary meeting with and clinging to the Savior outside of his tomb on Resurrection Morning. There is a mistranslation in John 20:17 that has long given the impression that Jesus prevented Mary from touching him, something that would have bearing on the subject here. However, the Greek used by John gives a different sense than what is rendered in the King James text. John uses a present imperative coupled with a negative which means to cease doing (i.e., to stop an action that is already taking place). The Greek word this grammatical convention is applied to is "*hapto*" (ἅπτω), which means to *embrace*. Christ's words to Mary that morning are most properly translated as, *Mary, you are going to have to stop clinging to me, you are going to have to let me go, I have got to go and do what my father wants me to do*.

special messengers of God made perfect for entry into the Temporal sphere. They include those who are *personages of Spirit*, not yet resurrected but considered *just*, either through transference of that attribute from God when they were *made* perfect by Him or considered *just* because of their acceptance of Salvation and their subsequent actions in a previous mortal[47] existence.

The touch of an Angel is sensible then because they are resurrected, being *just* is not a consideration, though being a messenger of God strongly implies such is the case. The "*spirits of just men made perfect*," however, begs consideration of two separate attributes involved, perhaps distinct from one another in source. We are left with the conclusions that *embodied Spirits* are not yet perfect having to be made so, and that a *just* quality is required to make that perfection available. In parting from this discussion, consider Lucifer, an *embodied Spirit*, denied the qualities of *Flesh and Bone* associated with the sense of touch and in fact, by that, the ability to perform the ordinances of the priesthood.

- An Apostles Touch

As we have seen, the Endowment liturgy provides another perspective to the story told in the Scriptures that is more than enlightening; in the present discussion, such is also true.

The Endowment is in the form of a play with four acts. First we are shown the creation, followed by an act which takes place in the Garden; we witness Adam and Eve in their initial stages of Temporality, trailed by instruction focused on significant priesthood symbolism. The third act is of keen interest regarding the revelation detailed in section 129. The Endowment account portrays Peter and his fellow Apostles before their incarnation; they are sent "*in their true character*" to minister to Adam and Eve. In this task, Peter touches Adam freely, resting his hand on Adam's shoulder and exchanging the tokens of the Priesthood after Adam is mortal within the Temporal sphere. This Peter does to authenticate his mission as a messenger of God in the ***very*** manner described in Section 129.

Peter, James and John are depicted in the Temple as *pre-existent* beings. They are ministering messengers sent by the Son in direct response to a request made by the Father, of this there can be no doubt. These three future, and it seems then present, Apostles were *personages of Spirit*, in possession of immortal physical bodies having received them, along with Adam, when all the rest of the sons and daughters of God did, before (or perhaps in conjunction with) Adam's entry into the Garden.

Peter, James and John were *personages of Spirit* before their birth into the Temporal sphere. Yet in their pre-existent form (as *embodied Spirits*), one at least, Peter, touches Adam, and it is certainly unambiguously implied that Adam can feel his hand as they exchange the token of the Priesthood. It is not clear if the physical attribute of sensibility by touch was already inherent to the being of Peter before his

---

[47] We must also consider here the possibility of those pre-mortally existent.

entry into the Temporal sphere or was endowed upon him by God and/or the Son for this specific mission, but it is clear that Adam could *feel* Peter's physical form. According to the keys given in Section 129, this should not be so; at least Peter should refrain from physical contact.

Perhaps a clue to this conundrum can be found in the specific phrase, "*in their true character*" spoken by the Father in His command to the Son to dispatch the Apostles on their mission to Adam. What then does this phrase mean? What is the *true character* of any son or daughter of God? For those who *keep their first estate*, eternity is to be spent as an immortal Soul, a three part being comprised of Intelligence, a physical body and immortal *Flesh and Bone*. Is this somehow representative of the state of these three apostles when they ministered to Adam in the Telestial world?

Though the Endowment account of the interaction of Peter and Adam leaves us uncertain, we can with effort develop an understanding which suggests that an embodied *personage of Spirit* is not fully sensible to the Temporal; it exists beyond the plane of our ordinary experience. It cannot be touched. We are told here that a tangible body, one that can be sensed by touch in the Temporal sphere, is like that of the Savior's after His resurrection, that this state of being is not exclusive to the Son, but is achieved by all those who are resurrected. Additionally, we are to understand that there exists a midpoint between our death and resurrection, a state held by those declared *just* who can be made perfect in Christ before their resurrection.

Section 129 makes no statement at all regarding the physicality, tangibility or possible association of *Flesh and Bone* with any Spirit in its pre-existent form. What Section 129 *does* say is that a resurrected being has a physicality that relates to the Temporal sphere in such a way that it cannot only be seen and heard, but also felt by Temporal man. Section 129 also states that the devil (by extension any fallen angel) attempting to deceive (as an angel of light) does not have a form that can be detected through the sense of touch by a mortal man.

The premortal mission of Peter, James and John to Adam was not the only supernal priesthood action of this revered trio. They would be called upon again to instruct and inform and do so in another manner that would prove to be most revealing. At the dawn of the last dispensation, Peter and James would return to the Earth they had left behind, but for John the story was to be much different.

- What About John

In the sixteenth chapter of Matthew, the words of the Savior to His disciples are recorded, and in the twenty-eighth verse, one of His most profound statements is given; the Savior said:

> *"Verily I say unto you, there be some standing here, which shall not taste of death, till they see the Son of man coming in his kingdom." (Matt. 16:28)*

What did the LORD mean when He said this? The LORD's coming in His kingdom, an event millions of Christians have anticipated and prayed for nearly two thousand years, is yet to happen. How could someone standing there, in Galilee, in the presence of the Savior, be alive today? That the Savior referred to Himself, as some may suggest, is untenable, for by the Scriptures we know that Jesus Christ truly died and later was resurrected; if the Savior then, was not referring to Himself, then who? The answer to this question is provided in the Scriptures and rich traditional stories that originated in the early Christian era.

In the closing verses of his inspired Gospel, the Apostle John added his comments to the record of a subsequent discussion of the promise as recorded by Matthew. He says:

> *"Then Peter, turning about, seeth the disciple whom Jesus loved following; which also leaned on his breast at supper, and said, Lord, which is he that betrayeth thee? Peter seeing him saith to Jesus, Lord, and what shall this man do? Jesus saith unto him, If I will that he tarry till I come, what is that to thee? follow thou me. Then went this saying abroad among the brethren, that that disciple should not die: yet Jesus said not unto him, He shall not die; but, If I will that he tarry till I come, what is that to thee? This is the disciple which testifieth of these things, and wrote these things: and we know that his testimony is true." (John 21:20-24)*

In this statement, John acted to quell a rumor that had falsely gone abroad about him that he, John, would not die. John very quickly and correctly stated that the Savior never told Peter that His beloved would never die, but that he would merely tarry, in service, until the second coming. But just as important and perhaps more so than what John said is what he did not say. John made no attempt to deny that he was the one the Savior spoke of, and as incredible as it must be, John never attempted here to modify or recall in any way the commonly held belief that the LORD had actually promised the miraculous extension of his own life until the Son of Man returned in His Glory. John's statement and his silence take on greater meaning when all the contexts of his life and experiences are applied. On the basis of all that had happened to him since that day in Galilee, John understood just what the Savior's promise really meant.

Throughout much of the history we have described in this work, and beside many the men involved, stood John. In the first century, John actively evangelized and taught the gospel; he organized the Church and directed its leaders. Like all of the Apostles, John labored under stiff persecution and was often in mortal danger. With faith in Jesus Christ he was undaunted. Perhaps he was encouraged to continue on against all odds because he had received this special promise from the LORD. Whatever the source of his faith was, just like the other of the LORD's Apostles, he never stopped carrying out his mission.

As the first hundred years of the Christian era passed by, John had heard of the death of the other Apostles one by one. As the century came to a close, he alone stood as a special witness to the ministry of the Son of God. Those that were close to him and cherished his special testimony pleaded with him to compose his own very special account of the Savior's Earthly sojourn. John at first was hesitant to do so, believing that the other three Gospels presented the story completely and clearly. However, in time, his friends prevailed and John agreed to present his own Gospel but only if after praying on the matter he was directed to do so by the LORD. John prayed and received instructions to prepare the work.

The Gospel of John was from its inception to be a different type of account than the three synoptic Gospels were. John dictated every word of it under inspiration to his scribe Papias (Bishop of Hierapolis). The influence of the Spirit on the text of this Gospel is readily apparent. The Gospel of John fulfills the hopes and expectations of all of those who petitioned so hard to get it; it is truly unique and provides insight into the Savior's mission on Earth that could only have come from John, the LORD's "*Son of Thunder*."

With his Gospel complete and what to others appeared to be the closing years of his mortal existence upon him, John began a transition that few have ever experienced. During the following years, the record of John began to slip from the pages of history. At the dawn of the second century, what is known of the Apostle John starts to blend with that of another, one called John the Elder. The blending, modest at first, in time becomes so complete, that in the end, it is impossible to differentiate between the two. John the Elder is thought to have been a disciple of the Apostles. His authority comes directly from them. It is understood that he was taught by them; what he knew of the teachings and life of the Savior came to him by way of the witness of others. History records that John the Elder was universally accepted as a leader of the church in Asia Minor.

By the second century, however, the record of this great man becomes hopelessly entangled with that of the Apostle John. Statements and teachings that can only have been made by the Apostle John (by nature of his personal witness of the Savior) are now routinely attributed to John the Elder. Often statements that clearly show that the speaker personally knew Jesus are made in a historical context that places the time of their utterance well into the second century, well beyond any reasonable life expectancy for the Apostle John. The confusion of the record is enhanced by those who would have known both of these *Johns*. Often these individuals will use the designations Elder and Apostle interchangeably and without consistency.

Finally, in the end, the confusion extends to the grave. Traditional burial places exist for both Johns along the coast near Ephesus – two burial sites, both said to entomb the remains of John the Apostle or John the Elder. It is unclear who was buried where or if anyone was buried at either site at all. John the Elder undoubtedly passed away; from the Scripture record, we know John the Apostle did not. With purposeful intent, perhaps under the cloud of this confusion, the Apostle John moved on.

Wherever John went, we can be assured that his mission continued; the Gospel was in the best of hands. Although the Church of the Firstborn was in years ahead to fall into the great apostasy, the promise of the Savior to John was to be kept, and through it, the LORD could keep another. Through John and the priesthood which he held, the Church would remain intact, however small.

John would be the lone voice in the wilderness fulfilling the promise made by Jesus to Peter when he said, "*thou art Peter, and upon this rock I will build my church; and the gates of hell shall not prevail against it*" (Matt. 16:18). In the long ages that would follow, nothing is known of John until the middle of the nineteenth century, when under the direction of the Spirit, John was to use his priesthood to transfer keys and authority to others chosen by God to receive them.

Late in the spring of 1829, in a quiet grove of trees on the banks of the Susquehanna River near the New York/Pennsylvania border, the same John who walked the dusty trails of Palestine so long ago, stood next to Peter and James and conferred the Priesthood of our LORD to a kneeling and humbled Joseph Smith. Perhaps it was just moments before this sacred and long awaited event took place that he walked into the small clearing, holding back the tender branches as the leaves gently touched his face. Or he may have made his way to the neighboring town days before, a traveler, a stranger in the land. However, and whenever he came, his arrival in Harmony was much different from that of Peter and James. John was mortal; he made his way to Joseph's side on the soles of ancient feet, feet that had been bathed by the Savior's own hands, long centuries ago.

John, the one whom Jesus said would tarry, at long last was about to perform an ordinance that was pivotal to the ushering in of the last dispensation of time. How long he may have stayed with Joseph is unspoken. We are not told of anything this wise and experienced servant of the LORD may have said. Joseph must have cherished his time with John. Surely, he would not have wanted it to end. There was so much to be said, so much to be learned. Evenings in the home around the fire must have been enchanting and wonderful. And then, like so many times before, John moved on. Today, somewhere, this most blessed of the LORD's Apostles, "*The Beloved*" of the Savior, continues his mission, serving the LORD in all that he does, living in anticipation of the Holy day when he can go home. For him the change as "*the twinkling of an eye*" (1 Cor. 15: 52) will be sweet and joyful. The Savior, I am sure, looks forward to the return of this dedicated and great disciple.

— *God's Speed, John.*

The Apostles mission in both of these appearances, one to Adam at the beginning of the sojourn of the sons and daughters of God on Earth and the second to Joseph and Oliver at the opening of this, the last dispensation, punctuate a necessary component of priesthood power – that of *Physical Hands*. We described how through the process of a "*just man made perfect*" such physicality was possible for these three when they appeared to Adam. With Joseph and Oliver, Peter and James returned from the precincts of Heaven embodied in the perfection of the resurrection. John stood there beside them with ancient (transfigured), mortal hands, to ordain. In the Garden, Satan wears an apron signifying his priestly power; it seems the true *garment* of the Priesthood is made of "*Flesh and Bone.*" Make note of this extremely important fact as you reflect back to the words we discussed that are associated with our Garments and Robes in the Temple – *the circle of connection closes completing a perfect ring.*

## *Keeping Our First Estate?*

> *"...And they who keep their first estate shall be added upon: and they who keep not their first estate shall not have glory in the same kingdom with those who keep their first estate" (Abr. 3:26)*

With some confidence, we have placed the War in Heaven before the 6 creative periods of the Physical sphere,[48] but of this we cannot be certain. What remains undecided is where the War in Heaven fits in and how it relates to the timing of the attainment of our immortal Physical bodies. This is significant as it has bearing on the state of our being and the attributes held by each of us at the Grand Council, which led up to the great confrontation.

***...the decision made by each of us at the Grand Council is arguably the most important choice we have or will ever have made.***

If Day 6 preceded the Grand Council, then each of us would have certainly been at that event as *embodied Intelligences*. Placing the Grand Council after the creation of our Physical bodies late in Day 6, is not a consideration to be taken lightly, as it would embody not only the sons and daughters destined for mortality in this Temporal sphere but also Satan and our brothers and sisters who chose to follow him. What are the consequences of this possible chronology? Is it possible that Satan and his followers are *embodied Spirits*? Or does even the possibility of such an actuality preclude the Grand Council occurring during or after the closing hours of Day 6?

Of the Grand Council we are informed by Scripture, that we all were participants, and we all were empowered with the agency and the associated knowledge and intelligence to make considered decisions. In fact, the decision made by each of us at the Grand Council is arguably the most important choice we have or will ever have made, attendant to perhaps the most significant consequence we will ever face.

Certainly, any understanding that would indicate that our knowledge of things eternal and our agency (our ability to make thoughtful choices) was somehow not fully developed when we stood in that Great Council, must be logically unsupportable. It is clear those who stood in opposition to us in the Grand Council had full agency when they did so. What the timing of the Grand Council places in question here is: were Satan and his eventual followers physically embodied at the time they chose to oppose God? Before we can answer that question, we must first establish a few definitions and examine some important details.

---

[48] See The War in Heaven in Chapter 6, pg. 195.

Our first estate could not have been *kept* unless there was an achievement of something; choosing correctly in the Council could be that accomplishment or at least part of it. What then was the reward? Receiving an immortal physical body, becoming a *personage of Spirit*?

-If yes, then all of us in Temporality have accomplished this first task and are on to our next challenge (adjusting to and mastering *Flesh and Bone* and the challenges of free agency in Temporality) with the potential of another reward ahead.

-If no, if the Physical body was attained as a matter of course, not in association with an accomplishment of any sort, then keeping the first estate could literally mean we were allowed to *keep* the bodies we had been given, and perhaps those who did not keep the first estate *lost* what they had been given.

Does God take gifts back; does He *regress* His children's progress? If the answer to this question is no, then most likely keeping the first estate means we are rewarded through the giving and our receipt of something we did not have before, but that reward need not have been Physical embodiment, the making of our Spirits.

If the *Grand Council* followed our having been embodied, then our reward most probably is further progression and increased intimacy with God through mortality (as detailed earlier in this work in Send Me).[49] Those then who kept *not* their first estate are cursed; they may not have lost the Physical body they had, but most certainly they are denied progression into the Temporal sphere, *the attainment of Flesh and Bone*, the priesthood that seems to be associated with it, and the increase of connection with the Intelligence of our Divine Father we can experience here.

- A Most Critical Conclusion

What are the consequences of the creation of our physical, immortal Spirits occurring before the conflict in Heaven? Is it possible that Satan and his followers are embodied Spirits? Has anything we have learned here precluded this possibility all together?

We can answer the last of these questions first. Nothing we have discussed here prohibits a conclusion that Satan and his followers are embodied *personages of Spirit*. We have shown that, while embodied, they can fulfill all of the activities ascribed to them in the Scriptures. We in fact are each a physically embodied intelligence that indwells a Temporal shell. We have shown in this work that the Holy Spirit (the Holy Ghost) is indeed physically embodied as well, yet the Holy Ghost can both indwell others and influence all of the billions of God's children *directly, individually*, at the same time. Satan, the Devil and his minions, can be embodied in these respects as well and influence, possess and perform all of the maleficence ascribed to them in the Scripture.

---

[49] See Send Me on pg. 204 of this work.

The primary answer to the first question posed above is obvious. If the embodiment of Spirits occurred before the rebellion that occurred, at or soon after the Grand Council, then *all* of the Intelligences, including those that rebelled, have *bodies* as we have defined a body here. This is a significant issue because it has been commonly understood among Latter-day Saints that receiving a *body* was the *exclusive reward* of those who kept their "*First Estate*." This is the interpretation most often ascribed to this utterance of the Prophet Joseph:

> *"The spirits in the eternal world are like the spirits in this world. When those have come into this world and received tabernacles, then died and again have risen and received glorified bodies, they will have an ascendency over the spirits who received no bodies, or kept not their first estate, like the devil. The punishment of the devil was that he should not have a habitation like men."*[50]

Joseph here, in using the term *first estate*, is referring to the words recorded by Abraham, "*...they who keep their first estate shall be added upon*" (Abr. 3:26). Joseph's words put in the proper context, as discussed in this work, can be amplified when restated as follows:

> *The physically embodied Intelligences in the Physical sphere are identical in form to the physically embodied Intelligences in this Temporal sphere, who remain embodied Intelligences though they are enclosed in Temporal bodies.*
>
> *When those embodied Intelligences who have come into this Temporal sphere and received Temporal bodies, then died and again have risen and received glorified, resurrected bodies, that are perfected Temporal physicality enclosing their immortal Spirits, they will have an ascendency over the embodied Intelligences who received no Temporal existence and its benefits, or kept not their first estate, like the devil.*
>
> *The punishment of the devil was that he should not have the empowering attributes associated with a resurrected, perfected, immortal body, the experiences of mortality, and increased association with the Father like the sons and daughters of God.*

Joseph clearly understood the differences between the *immortal Physical Body* and the *Temporal body,* though he did not have the language we have developed in this work to express it. In Joseph's mind and in the theology he expressed, there was no equivalency made between the Physical embodiment that comprises our Spirit and the Temporal embodiment of the Spirit which comprises our *living Soul*. The repeated compounding of the multiple contexts of embodiment and the unfortunate (yet common) misuse and misunderstanding of what is meant by the word Spirit, has led to the improper interpretations that have emerged.

---

[50] Discourse given by Joseph Smith in Nauvoo, Illinois, May 21, 1843; recorded by Willard Richards.

Within this work, we have established a progressive path for the state of the sons and daughters of God. This path maps out an evolution of our individual being from the first moment of our creation, becoming an independent entity, through to an attainment of an exalted, glorified state, in intimate contact with our Heavenly Father. We can summarize this path as follows:

1) We each were parted from the Intelligence of our Father and created as independent, sentient, Intelligent Entities –

2) Next we received a physical, tangible, embodiment of our Intelligence forming each of us into an Immortal Spirit –

3) In preparation for entry into the Temporal .sphere, Adam and Eve's embodied immortal Spirits were enclosed for the first time in a body of *Flesh and Bone* (the new name) within the Garden Temple, before the Fall (*for the Remainder of the sons and daughters of God there is no preparatory state but the embodiment in Flesh and Bone and Blood occurs at our Earthly, Temporal conception*). Through the *Ruach,* God adds His *Neshama* to each of us making each a tripartite being –

4) It is the uncertain Temporal sphere (in effect the *presence* of choice), not wrong choice or even "*sin*" that separates the Spirit from God the Father when Man Falls (Adam and Eve chose this path to eternal progression). The immortal becomes mortal and blood is added to our once perfect *Flesh and Bone*. (*Note that it is the shedding [discarding] of this blood that is the focus of the Law of Sacrifice and the Atonement of Christ*) –

5) Finally death separates the Spirit from the Temporal body –

6) After Temporal death, we each depart the Temporal body of *Flesh and Bone and Blood* and return as embodied Spirits to the *Spirit World* (a habitation within the Physical sphere) –

7) Because of Christ, restoration through the resurrection returns the immortal *Flesh and Bone* to the embodied Spirit becoming an *Immortal* Soul; "*Therefore, as the soul could never die… it was expedient that mankind should be reclaimed from this spiritual death.*" (Alma 42:9) –

8) Our active acceptance of the redeeming gift of the Atonement restores the Spirit of man to God in His Kingdom. (Through sealing ordinances completed in the Temples of the Garden, Temporal Earth and in Heaven) –

We understand here then, the bodies denied those that failed to keep their "*first estate*" are the Temporal bodies we indwell in the Fallen, mortal Universe. We, therefore, must understand that there are qualities of value associated with the acquisition of a Temporal body. We have explored many of these qualities in this work already, among them are:

- A *causal* association between a Temporal body and the power of procreation.
- The attainment of a truly "*Free*" Agency associated with the laws of uncertain matter.

- Experiences individually acquired because of agency and the sharing of those experiences with the Father through the integration of our Intelligence with His.
- Personal intimacy with the Father through His sharing of His Spirit (the *Neshama*).

The mortal and eventual immortal resurrected body we each acquire because of our entry into the Temporal sphere is bound to our Spirit making us a *triune* Soul (Intelligence, Physical embodiment and Temporal embodiment), described when appropriate as either *living* or *immortal*. This Soul makes us whole in a way that can never be appreciated by those who do not experience it. A Soul has powers beyond that of *Spirit* alone. Because they "*kept not their first estate,*" it is clear Satan and his followers are denied a Soul in both of its forms, mortal and immortal.

So, though we understand Lucifer and those that chose to follow him in rebellion, "*they who keep not their first estate*" (Abr. 3:26), are denied a *Temporal* body (the benefit of fully realizing their potential as a son or daughter of God with a Soul), we are still left to wonder if they are *embodied* (spiritually) in the Physical sphere.

In recognition of this unknown, we asked at the outset, *is it possible that Satan and his followers are embodied Spirits?* Answering this would be made very simple if we could place the time of the Grand Council precisely in the order of the creative events we have detailed here, but we cannot, at least not directly.[51] Determining then if it is possible that Satan and his followers are embodied will rely on a more circuitous approach.

The Biblical account of the Garden story tells of Satan appearing as a serpent. As the serpent of Christian tradition, it can be inferred that Lucifer appeared in that way as a means of interfacing with the Physical Eve. Lucifer's use of the form of a physically embodied creature (the snake) to communicate alludes to the deceiver not having a body of his own; this is in fact the universally accepted interpretation in "*traditional*" Christianity. The Endowment account, however, relates a much different story that provides a strong indication that Satan is indeed embodied.

The Endowment account depicts Lucifer in the Garden in human form,[52] speaking and directly interacting with both Adam and Eve. We have shown conclusively that Adam and Eve were in the Garden in a physically embodied form, yet our first parents interact with Satan as if he too were embodied as well, as if he were of the same order of creation as they, the Garden and all of its contents around them. Eve

---

[51] We have established a strong circumstantial case that the War in Heaven was fought not in the Spiritual but in the Physical Sphere because we have inferred that such rebellion and discord could not find a place in the perfect precincts of the Father. See our previous discussion, <u>- Michael and His Angels</u>, pg. 353 and footnote 70 on pg. 354.

[52] This is in fact also the case for the scriptural accounts as they are properly understood in the original languages and contexts we have shown in this work.

fully accepts Lucifer's claim to be her brother, a fact not tenable had he been in the form of a reptilian snake.

There simply is no way of harmonizing the exceptionally clear doctrine of the real physicality of the Physical sphere we have detailed here with any understanding of *Spirits* that are not embodied. We firmly understand here that *all* of us possess and *are* an embodied (tangible, physical) Intelligence called a *Spirit*. We were all embodied at the same time within the Physical sphere, before the Temporal sphere was created. We reason that embodiment to be *irrevocable*, we know it does not regress at death, and it is eternal.

We have detailed here that Adam and Eve were embodied Physically within the Garden and have determined that they were also embodied there in the same immortal state (*Flesh and Bone*) we will all acquire in the resurrection. With this being the case, Satan's appearance to them then, as unremarkable as it seemed to our first parents, speaks strongly to an understanding that the deceiver was embodied, *a personage of Spirit*, within the Garden as well.

Within this work then, based on all that we have learned here regarding the three creations of God and applying that knowledge to the teachings of the Prophets Joseph Smith and Abraham, we will conclude that Satan and his followers both received and retain their Physical bodies to this very hour.

When they kept not their "*first estate,*" then they were denied the progression afforded by entry into the Temporal sphere as mortals, and therefore they cannot experience agency freely, participate in the blessings of procreation and receive the power associated with an immortal Soul.

Any discussion regarding the embodiment of our Intelligences to form Spirits and the subsequent enclosure of those immortal, physical, tangible Spirits within Temporal bodies to become Souls, must always remember that the Temporal is in fact a *fallen* state, a subtraction from a more perfect, pre-existing sphere, a fall from a higher to a lower order and realm. There is *nothing* to be taken away from this realm through the resurrection that we did not already have in a more perfect form in *the time* before, *nothing* that is, but: experience and the knowledge we can only gain through truly unfettered free agency, the power of procreation and the Priesthood associated with the attainment of *Flesh and Bone*, and our ability to love in a truly selfless way. These are the greatest fulfillments to be had from living in mortality.

Our path to progression points only upwards toward God, but it is not without significant detours and pitfalls. It is probably not entirely ironic (indeed it is perhaps fundamentally intentional) that the greatest detour of all, along our road to perfection, Temporality itself, is specifically designed to demonstrate and warn of the crookedness of our eternal path and is most excellent for teaching us how to make our course straight in the face of great evil.

## Eternal Procreation

As his tenure as Prophet of the Restoration lingered, Joseph Smith continued to receive revelation. One of the last he received before his martyrdom, connected exaltation to eternal marriage – the sealing of a man and wife within the confines of the Temple. What Joseph came to last understand then, will be the last thing we discuss in this work as well. As we conclude, always remember what was taught by the Savior, "*the last shall be first, and the first last*" (Matt. 20:16). In this case, the Savior's admonition is most appropriate.

In 1843, the LORD declared to Joseph that those who keep the sacred covenants entered into within the Temple, including the covenant of eternal marriage, will inherit "*thrones, kingdoms, principalities, and powers, dominions, all heights and depths.*" The revelation goes on to say, "*Then shall they be gods, because they have no end;*" for they will receive "*a continuation of the seeds forever and ever*" (D&C 132:19–20).

Eternal procreation is a concept that is *expressly* understood within this work. We declare here that indeed a foremost purpose of the multiple creation schema is the development of the sons and daughters of God. It is the intent of our Father God that we will one day achieve *exaltation,* and like He, have "*no end*" and "*a continuation of seeds forever*;" we are to exist for "*everlasting to everlasting*" (D&C 132:20).

We have spoken of the pairing of male and female Spirits in an intimate way from a time before the Garden, describing that intimacy between a man and a woman is profound and deep beyond comprehension.

Gender we have seen is a primary attribute of our eternal being. When we were first embodied, we were created in divinely appointed pairs; Adam and Eve are the symbolic representatives of this truth. This is of significance and bespeaks a verity about the chain of life and the genetic heritage of all of mankind, one that extends beyond the veil into the Physical sphere and (it seems) before. The LORD has declared: "*Nevertheless neither is the man without the woman, neither the woman without the man, in the Lord*" (1 Cor. 11:11). The eternal nature of this principle forward into eternity is certain; *by inference,* it can also be extended into the distant past.

God ordained marriage; such union is not to be for time only but extends into eternity. The first marriage was solemnized in the Garden Temple by God himself. Though Adam and Eve, and all men and women who have since been joined in this sacred and eternal manner, were separate and independent beings, they were to be one in the eyes of God, one in heart and soul. Adam understood this fully; "*Therefore shall a man leave his father and his mother and shall cleave unto his wife and they shall be one flesh*" (Moses 3:24). We already discussed the descriptive use of the Hebrew words *Wedevak* (וְדָבַק) and *Echad* (אֶחָד) in the Garden account. We understand that the LORD, Adam, and Moses used those words to communicate a closeness of two Souls that is indescribably powerful and permanent. To Adam, the bonding between a man and a woman is to be inseparable; they are to "*cling*" or to "*stick*" to each other in a tightness that emulates the binding of flesh to bone. Our first father Adam understood his relationship to Eve was to be closer than any other, even closer than his relationship with God.

In all of the creative events we have detailed here in the creation of Intelligences and of our Spirits, the Scriptures and our interpretations of them, within this work, make no mention of a maternal/paternal process. In fact, consider the creation of the Spiritual sphere, completed by Jehovah and Michael at the direction of *Elohiym* and the creation of the Physical form of man by the Father and the Son together. These creative processes did not involve any known aspect of male/female procreation – they were not sexual in their nature. We described the creation of our Intelligence, which generated the most fundamental component of our eternal being as a division of the Intelligence of the Father alone. Where is a Mother in Heaven to fit in?

Why then this eternal gender and intimate pairing if there is no indication of an eternal function for it? What does it mean to have eternal increase, a continuation of seeds forever? We know that the contexts of these phrases are that of eternal procreation, but what does that mean? How are we to understand it?

Those who are parents, in the proper traditional Christian sense of the word, those who bare, deliver, nurture and raise children as mothers and fathers to adulthood and beyond, eventually watching and participating with those children in raising children and grandchildren of their own, know that the joys and blessings of parenting encompass so much more than the simple sexual act of conception. It is the nucleus of a family and its extension in which the *heights and depths* of joy are experienced, it is the *continuation* of those family ties into the future, even generations beyond our death where *forever and ever, everlastingly* lies; this is eternal procreation.[53]

We understand then, that eternal increase envelopes, for God and for His faithful sons and daughters, the relationships and experiences of eternal families and the growth that comes through them generation-upon-generation. This in fact can be seen in the promise of the Covenant of Abraham. Abraham it will be remembered was promised in his old age that his seed would be "*as the stars of the heaven, and as the sand which is upon the sea shore*" (Gen. 22:17). Abraham, at the time was ninety and nine years old; his marriage to date had been entirely unfruitful. Though Abraham had a single son Ishmael, and Sarah was to bare Isaac, God's promise, "*I*

[53] Public discussion of these principles is all too often controlled and driven by those who seek to maliciously distort the meanings of this sacred understanding. Contexts of sexual intimacy in the eternities, intercourse, and polygamous innuendo abound – all spoken of in disdain as if the calumny were true, it would be worthy of scorn. Not all understandings of physical intimacy in the eternities between husbands and wives find their generation outside of the Latter-Day faithful; many sources of such context can be found from within the church especially from the early days in Utah. Nothing we have spoken of or discovered here in this work supports such suppositions at all.

*will make thee exceeding fruitful*" (Gen. 17:6), did not (and could not) relate to Abraham's personal and direct *physical* endeavors.[54]

What is the process of procreation in Heaven? We do not know; it has not been revealed to us, but perhaps some understanding can be found by consideration of a specific set of events at a critical period in the progression of the sons and daughters of God who share our story.

Within Temporality we all have both a mother and a father; [55] this truth extends in an unbreakable chain back to the most ancient of times on Earth. For the sons and daughters of God, our pedigree traces back to Adam and Eve when they first conceived after expulsion from the Garden and their entry into the Temporal sphere. Procreation in Temporality is understood to us – *it is physical.*

We have developed an understanding in this work that Adam was placed in the Garden as an infant child, and that he was raised and taught there by loving parents. Of this time in the life of Adam we have no record, which it seems is in accord with the same dearth of revelation we have regarding our Heavenly Mother; this perhaps is not coincidental.

Let us propose here that Adam was raised to adulthood by both a Mother and a Father in the Garden. That his initial *Flesh and Bone* conception there, in that special precinct, involved both a father and a mother as does all *Flesh and Bone and Blood* conceptions do in Temporality. Of kind, we make no claim; we propose no theory as to how the conception proceeded, other than we understand that it involved gender both male and female, our Father in Heaven, *Elohiym*, and our Heavenly Mother.

It will be noted here that we have left Eve out of this discussion, for we are told (if symbolically) that she was a derivative of Adam. We have discussed why that may have been so. We will refrain from speculation here other than to say, perhaps Eve was conceived in a manner similar to Adam while somehow preserving the significance of her generation from the "*stuff*" of Adam.

There has been no conversation of male/female intercourse or plural wives in our discussion here. Our understandings, those of which we are certain, do not include

---

[54] Abraham later was to take another wife (concubine) Keturah, who would bare him six more sons, all the founding patriarchs of Arabic nations: Zimran, Jokshan, Medan, Median, Ishbak and Shuah. Even though this would bring the total count of Abraham's sons to eight, including Ishmael, the promise of God to Abraham would ultimately be fulfilled not by his reproduction, but by the extension of his family through the blessings of God in the generations that were to follow.

[55] The Genesis 1 account records that Adam uses the Hebrew *wedevak* (וְדָבַק)(cleave) to describe his relationship with Eve. *Wedevak* (cleave) comes from the root *davak* (Hebrew, *debek*, בַדָּק), which means to *cling* or to *stick* like flesh does to bone. Perhaps there is a double entendre at play here; it may be revelatory that Adam associates his relationship with Eve, to the *flesh and bone* we understand is vital in the procreative attributes of mankind.

these things as a requirement, nor do they even suggest them.[56] We understand that we fashion out of the substance of reason alone the existence of our Mother in Heaven, sprinkling in some of the intuition that seeps out of our Soul. The reasoning and feelings are so strong at times that it has been said: "*Our theology begins with heavenly parents. Our highest aspiration is to be like them.*"[57]

***Within this work, we can certainly affirm that our theology begins with the creation of both sons and daughters of God, in the eternities past and extends as bonded pairs – men and women, husbands and wives, into an eternal future.***

Within this work, we can certainly affirm that our theology begins with the creation of both sons and daughters of God, in the eternities past and extends as bonded pairs – men and women, husbands and wives, into an eternal future. We declare that our Father and perceived Mother in Heaven continue in increase, for they "*have no end*" and possess fruitful "*seeds forever and ever*" (D&C 132:19, 20). Of the Father this is certain; for a hoped for Mother, it is inferred. Because of Joseph Smith, we know that the eternal procreative state held by our Father will be gifted to the faithful who receive *exaltation*.

It is important that we remember we have no information in the Scriptures detailing the specifics of eternal procreation; it is not described at all. Nothing that we have discovered here about the creation of the three spheres of existence, the creation of our Intelligences and our eternal Physical bodies, speaks of male/female conception anywhere but the Temporal sphere. Any teaching that extends beyond these borders, that attempts to detail the processes, or certifies as fact the things we have defined as mere proposal here, *does so ever so tenuously*, stepping out where the informed dare not go, drawing, at best, conclusions of fact based on false cause relationships.

---

[56] Similarly, some have chosen to speculate on the manner of the conception of the Savior at the meridian of time. We can declare nothing more than what we know. Mary was a virgin *at the birth* of the Savior, and the Father of Christ was God, *the Father himself.* A virgin birth is just what it professes to be, *virgin* in *every sense* of the word. Anything that claims otherwise is FALSE.

[57] "*Apostasy and Restoration*," Elder Dallin H. Oaks, (*Ensign,* May 1995), pg. 84.

# AFTERWORD

## A Superior Theology

Christian religion is fractured between various sects that espouse alternate views, sometimes hopelessly at odds with one another. Today, an, almost infinite, array of denominations contend that they are the purveyors of truth. Low Church and High Church alike, fall short of meeting the needs of the Christian faithful in so many aspects of theology. A spectrum of belief systems can be found, analogous to the range of grade level education familiar to us all. Some advocate basic theologies void of specifics and representing a rudimentary, elementary belief system. Other churches take a higher view, disseminating detailed doctrine steeped in particulars, however lacking they may be in fundamental salvific power; these can be viewed as secondary level Christian education. Still others rooted in millennia of tradition and scholarly enterprises deliver a graduate level theology to those that are willing to receive it. Latter-day Saint theology, especially as it has been utilized here, represents a post-graduate education (in its highest form) of the things of God. Where others read within the pages of Scripture of talking snakes and a creator who formed elephants and giraffes whole, for nothing more than the sake of whim, our theology begs a fuller application of the facts and accepts all that can be accurately known to man as a single continuum of truth, providing answers to all of the *why*s an inquiring mind may present.

Our understanding here does not develop a theology that presents an image of the Son as a created being who is insufficient in producing an infinite Atonement. Christ is Jehovah, the *Creator* of both the Physical and the Temporal spheres; He stood side-by-side with the Father when our eternal Spirits and our Temporal bodies (Adam & Eve on Earth) were created. He was there as Wisdom, Himself of the very intelligence of God, when our Intelligences were made. If the Father Himself accepts His sacrifice as payment in full, we feel no need to justify its power and efficacy by means of the application of false attributes stemming from reason or logic that conform to the precepts of Greek philosophy, no matter how lofty that logic may appear to be.

We believe the Universe itself, and all that is within it, testify daily of the Creator, revealing great truth when it is looked for. A true understanding of God can only be had after we *read* all that He has revealed in every place it was left to be found. Truth can be uncovered everywhere when we let the Spirit guide our vision; this process of attaining knowledge is the one ordained by God. The Word of God is not confined to images of ink on sacred pages of Scripture; the canon is not closed. It was the hand of God that dropped the apple on Newton; the Spirit of the LORD inspired him on to *feel* the force of Gravity when no one else had. The voice of God can be heard in the words of Lincoln and Jefferson; He walked with and whispered in the wind through the trees to Emerson and Thoreau at Walden's pond.

Those who espouse the concept of "*Sola Scriptura*" must be prepared to answer the following questions: Who wrote the Book of Hebrews? When was it penned? Do *any* of the books of the New Testament *declare* themselves to be the Word of God? Such a proclamation is lacking even in the Vision disclosed in the Book of Revelation. Who penned them all? Where are the original autographs to be found? If Scripture is "*sufficient alone*," why translate at all; why do you expound? What translator: Tyndale, Wycliffe, Erasmus, or Jerome claimed their work to be divine and perfect? What *is* the original language and the original text?

For Sola Scriptura to stand (its included concept of inerrancy, and to a lesser extent sufficiency), it must do so on a firm foundation of substantial answers to all of these questions posed, none of them lacking. Sola Scriptura must entail an unbroken chain of controlling divine influence over every process of transmission of the sacred texts to every eye and ear that receives them from any source. There can be *no* error, for the concept demands perfection in totality; after all, it is claimed to be God Himself who provides the assurance.

What we have presented here is the full, broad spectrum of truth displayed in the Universe around us, truly the depository of the mysteries of God. Because our picture includes all, it is as complete as our understandings will allow today; it supports a theology that is superior to any that seeks to limit the view of man to any one sector of the knowable.

We have described the creation without falling into the absurdities espoused by "Creationists." Our view is one of the expansive power of God to create and command. We do not, and cannot, sustain theories that support a young Earth and the Universe, or those that espouse instantaneous generation of galaxies and animals from nothingness at the command of God. Adam's origin did not coincide with the appearance of other life on Earth created in the same 168 hour[1] week. Death was present on the Earth long before the arrival of man. The Grand Canyon and majestic Mount Everest were not formed in the last 6000 years, neither were the planets that circle our Sun. Light does not approach us from the farthest reaches of the Universe through anisotropic time frames.[2] Noah did not embark dinosaurs with him on the Ark. We have shown such false suppositions are not supported by Scripture, though many defend such views because they think they are.

The Word of God speaks to us daily from all quarters of our experience; it is not limited to out-of-date interpretations of a few words recorded on even fewer pages, no matter how revered those pages may be.

*The heavens declare the glory of God;*
*and the firmament sheweth his handywork.*

(Solomon, Psalm 19:1)

---

1 As measured in the Earthly time reference we use today.

2 Anisotropic Synchrony Convention.

## Standing in Awe at the Edge of Everything

Writing this book has been an adventure, or more properly an odyssey. It took me places I never intended to go. I started out wanting to simply describe a novel understanding of a singular thing, the Creation of God, in three separate parts. However, I was soon overtaken and overcome by the fact that what I had actually set out to do included everything there is, quite literally, not only everything that is but everything that has been or ever will be. In this process, I discovered that to detail the things created, we had need of understanding the Creator, why He made the things that were made, and what their purpose was. Doing so revealed deeper meaning and broader vistas at every turn. The journey was always upwards toward God. At times it was daunting, but in the end, so very worth the climb.

***...to detail the things created, we had need of understanding the Creator.***

What I am left with is a feeling that I know so little, sensing that I have but examined a master's painting, *in whole*, just one brushstroke at a time. I realize that each stroke I have followed and detailed, I do not fully understand; no one does. We follow its contour, define its hue and attempt to describe its composition and the way it relates to the strokes around it, but without stepping back, away from the detail to view the whole canvas, what we can know of the single stroke is never enough to appreciate the full splendor of the Masterpiece.

All that I have discussed here is like that single stroke of the brush. I have described, in great detail at times, each subtle turn of the master's hand, knowing full well that I do so from a viewpoint of examination of just the remnant effect He has left on the cloth. I fully appreciate that we are without any possibility of wholly understanding His thoughts or intentions, the way He held the brush, moved His hand, or even clearly knowing the expanse of the scene He is completing.

I do know that the strokes we have discovered and blended together are truly His; that they were put down with extreme care and for divine purpose.

Finally, I must observe that I believe revelation to be a fuzzy thing. The message and images as received are never really clear or fully understood, but I believe that this result is the intent of the sender. The Word of God, His truth, is too large to be contained on paper, in ink, within a book; even the Universe is not large enough to contain it. It is far too much to be comprehended, even perhaps by the whole of mankind. The things of God, certainly in their wholeness, and even just one small piece at a time, cannot be really, completely understood by one lone man, no matter how great a Prophet he may be.

True revelation is timeless, it transcends the limitations of language, culture and the limitations of knowledge the receiver may have. Its intent is not only to reveal but to inspire, testify and encourage growth, curiosity, wisdom and hope. It is limited, not by God, but by our ability to believe and comprehend. Despite our weakness, it is always sufficient, releasing more truth each day to the observant who seeks it. The true value of any worthwhile prize rests not so much in the realization of it, but is found within the glory and blessings attained as we put forth effort to achieve it.

Revelation is given variously through vision and observation afforded to the blessed or through direct dictation of words and messages received. Prophetic power and ability is not always based on the receiver's capacity to understand and comprehend what is being delivered, in fact they rarely can, but it is found in their willingness to be true and accurate messengers. It is my belief that *most often* revelation is never fully understood by the Prophet receiving it, and in most instances, it is never intended to be. Such is the case with the creation accounts, the Garden accounts and even the Endowment narrative discussed here.

The Pearl of Great Price informs us that both Abraham and Moses received revelation through vision and as it was dictated to them. Of what they saw, they communicated what they could comprehend, limited by their knowledge and their ability to communicate Heavenly things with Earthly language and understandings. When these two great prophets wrote what God told them about creation, they no doubt were in awe with what was revealed, assuredly they were not capable of processing fully in their minds the words and concepts they received. We have seen here that what was described to them then, thousands of years ago, have only recently begun to be understood by science. Still truth was revealed and continues to be revealed by the Word of God as recorded.

The Endowment liturgy similarly contains much that was not understood, realized or comprehended by those who received and fashioned it almost two centuries ago. They may not have been aware of the truths hidden in the detail. They may not have known why the changing of the clothing was completed or the reason behind the different order of the creative periods and the meaning of the two stages for each Priesthood. But we have seen great truth is revealed in these things. In many cases, they are *only* revealed through juxtaposition with other revealed truths that themselves have only recently been understood in the light of yet tertiary discoveries.

The stories and narratives discussed here are replete with symbolism; they are built upon it. But that must not diminish the expectation that intricate detail is to be found in the metaphor. We do a great disservice to ourselves and to our God by dismissing hard things as archaic elegy, whimsical in form, somehow unimportant to the true picture.

Though Albert Einstein admitted late in his life that a "*cosmic religious feeling*" underlay his scientific work, he once famously wrote:

> *"...the Bible a collection of honorable, but still purely primitive, legends which are nevertheless pretty childish. No interpretation, no matter how subtle, can change this for me. For me the Jewish religion like all other religions is an incarnation of the most childish superstition." (Letter to Eric Gutkind, Princeton, March 1, 1954)*

The fact that the creation stories and narratives discussed here were received, recorded, revered and diligently repeated by those who did not, and could not, fully understand the truth they contain, testifies of the divine nature of the message and the power of God to reveal truth to us despite our limitations and weakness. A pretender, liar or charlatan would not, could not, derive intricate detail that only has value when understood beyond his own level of comprehension. No, it is the divine that is at play when ancient prophets record sophisticated detail in antique writings that can only now be understood in the most modern of times through advanced scientific findings.

> *"Faith, to be faith, must center around something that is not known. Faith, to be faith, must go beyond that for which there is confirming evidence. Faith, to be faith, must go into the unknown. Faith, to be faith, must walk to the edge of the light, and then a few steps into the darkness. If everything has to be known, if everything has to be explained, if everything has to be certified, then there is no need for faith. Indeed, there is no room for it...In a world filled with skepticism and doubt, the expression 'seeing is believing' promotes the attitude, 'You show me, and I will believe.' We want all of the proof and all of the evidence first. It seems hard to take things on faith. When will we learn that in spiritual things it works the other way about--that believing is seeing? Spiritual belief precedes spiritual knowledge. When we believe in things that are not seen but are nevertheless true, then we have faith." ("What Is Faith?" in Faith, Boyd K. Packer, 1983, pgs. 42–43)*

The Book of Mormon, the Pearl of Great Price and the Revelations of the Doctrine and Covenants are as correct as they could be, as they were penned, when they were penned. They were not and are not superior to ancient Biblical Scripture but more relevant for their time. Received most recently, they were more synchronized with the common knowledge, understandings and popular discourse of their day. I have presented here a further update to our Gospel understanding, reflective of knowledge and information yet another 20 decades more recent.

This work is not complete, nor do I expect it to be. After over 550 pages and many years of research, study and writing, I have found my understandings of many of the concepts here to have evolved, often resulting in dramatic change as I came upon new truths, discovering them line-upon-line, precept-upon-precept. More than once, I have had to return to what I had written before to make changes. Such is the process of the attainment of the knowledge of God. I fully expect to be able to return to these pages again in the future, finding them in need of revision; that expectation will not cease to be true. However, truth does not change. What has truly been revealed by God does not change; it is our ability to comprehend that forever improves, when we turn our eyes in the direction of the LORD.

*I will instruct thee and teach thee in the way which thou shalt go: I will guide thee with mine eye.*

(Psalm 32:8)

# APPENDIX

## A Zero Point Field

Within the space occupied by the quantum electron, exists the fabric of space time itself, the *Zero Point Field*, or Quantum Foam. Fluctuations in this field are on an order of magnitude of $10^{-32}$ cm, one hundred million billion times smaller **(0.0000000000000000000000000000000001)** than the electron, which is itself tiny compared to another particle of the atom, a proton, some 1,836 times larger.

The energy generated by the miniscule vibrations (negative entropy) in this field account for the Dark Energy of the Universe, driving the expansion of space itself and accounting for 78% of everything there is.

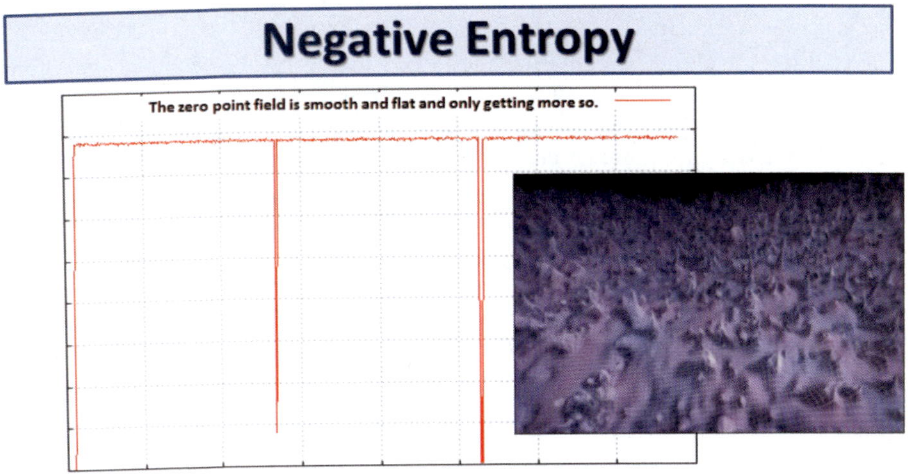


The Zero Point Field, or Quantum Foam, fluctuates due to remnant energy from the Big Bang. Today these fluctuations are on an order of magnitude of $10^{-32}$ cm. As this energy diminishes, space itself expands.

## Just Six Numbers

The Universe is governed and exists because parameters are found within it producing the laws of physics. All of these parameters are precise in value, many of them extremely fine-tuned. In this section, we will discuss six of those numbers, a few others and derivatives, that get their value and significance from them.

| | |
|---|---|
| Gravitational Constant | Cosmic Expansion |
| Cosmological Constant | Cosmological Inhomogeneity |
| Mass Lost to Fusion (4H – 1He) | Dimensionality |

**N** *Gravitational Constant*

The cosmos is so vast because of this crucially important number, the first of the critical six. The number associated with **N** is huge (10 Undecillion or 1 trillion, trillion, trillion), **1,000,000,000,000,000,000,000,000,000,000,000,000**.

> **N** *is the ratio of the strength of gravity to that of electromagnetism; this number measures the strength of the electrical forces that hold atoms together, divided by the force of gravity between them. It is so large because Gravity is so feeble. If N had a few less zeros, only a short lived miniature universe could exist. No creatures could grow larger than microbes, and there would be no time for biological evolution.*

**λ** *Cosmological Constant (Dark Energy)*

The number lambda is the newest addition to the list of critical numbers and it is the second of the six. **λ** (Lambda) is a "new force," an "antigravity" This Cosmological Constant was originally formulated by Einstein to prevent his dynamic Relativistic Universe from evolving. **λ** represents the force which controls the expansion of our Universe, a kind of pressure that acts against the force of gravity to push the Universe apart. **λ** is now called "dark energy" and it makes up almost 75% of the entire Universe. Even though it has no noticeable effect on scales less than a billion light years, dark energy (**λ**) is destined to become the dominate force in nature.

> *Fortunately for us, and very surprisingly to science,* ***λ*** *is very, very small, on the order of* $10^{-122}$ (0.138 X $10^{-122}$) *smaller than the incredibly small Planck's constant. If it were not, it would have stopped galaxies and stars from forming, and the development of the Universe would have been halted even before it began.*[1]

---

[1] It is straightforward to show that if this constant were any larger (in the slightest), matter could not coalesce into stars and galaxies, and life as we know it could not form. It can be argued that if it were less, a more natural Zero, or a negative number, matter would clump together more effectively and life *could* (?) form more easily, suggesting that this number is not optimally tuned for life. However, such a conclusion is narrow-minded and discounts the optimization of universal resources: The need of time for life, and the elements it requires, to develop; the necessary scales and size of stars and galaxies and their

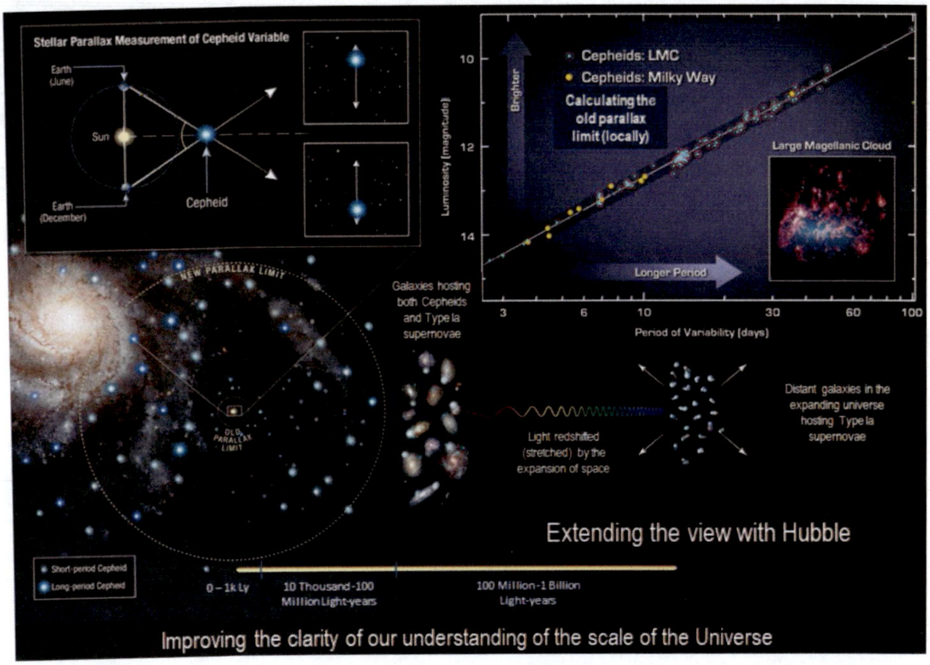


*The upper right corner of this chart illustrates the Cepheid period-luminosity relationship, which establishes that if the period (or timing) of a Cepheid star's pulse is known, its intrinsic brightness can be determined. A star's distance can then be calculated by comparing intrinsic brightness to observed brightness, because it appears dimmer as it moves farther away.*

*Building on the work completed by the Spitzer telescope, observations using the Hubble Space Telescope as directed by the SHOES (Supernova Ho for the Equation of State) team of the Space Telescope Science Institute (STScI) connected data collected by Spitzer with Type 1a Supernovas in distant galaxies to push out, and therefore improve, the distance scale dramatically. Results of the SHOES team research were published in the April 1, 2018, issue of the Astrophysical Journal (just five months after the first edition of this book was printed). Hubble measured the expansion rate of the universe more precisely than ever before, narrowing the uncertainty in the measurement to within just 3.3 percent while fixing the observed expansion rate within the range previously detailed.*

Results from combining data derived from Cepheid and Type 1a Supernova images attained by the Spitzer and Hubble Space Telescopes reveal that **Space itself is pulling apart at the seams, expanding at a rate of 45 plus or minus 1.855 miles per second per mega parsec** *(a mega parsec is roughly 3 million light-years).*

---

productive lifetimes; and the requirements of expanse between stars to secure habitable zones for life to flourish. That this number is so small, but positive and non-zero, is not only "*the biggest mystery of the Universe,*" and its "*worst fine-tuning problem*" (Dr. Lawrence Krause); but it is evidence of an intelligent creator at work producing conditions necessary for precisely the type of life that we are – the sons and daughters of God.

***Doing the Math of adding these critical numbers together…***

The numbers are incredibly fine-tuned, but they do not work alone to define our Universe, they must be combined together. When you combine their degree of precision, one with another, the resultant precision is itself even more remarkable. Consider this effect for just the first two of the numbers we have discussed so far.

## Gravitational Constant

The value of the force of gravity is exact to a degree of precision, similar to that of a ruler stretched out as long as the radius of the universe with increments of one inch defined on it. The value of the force of gravity can be represented by just a single three inch long segment. If the value of the gravitational force is moved to the right or left just a few inches, the effect on life would be catastrophic; this represents precision to within just 6 inches in forty six billion light years.

## Cosmological Constant

The small cosmological constant has to be precise to the degree of ***One part in a hundred million billion billion billion billion billion*** 1:100,000,000,000,000,000,000,000,000, 000,000,000,000,000,000,000,000,000.

*This precision can be likened to traveling hundreds of miles into space, then throwing a dart back at the Earth and hitting a target one trillionth of a trillionth of an inch in diameter, an area smaller than the width of a single atom.*

*- Considering just these two parameters alone, and combining their necessary accuracy of fine-tuning, renders a level of precision of just one part in a* ***hundred million trillion trillion trillion trillion trillion trillion.*** *That preciseness is comparable to choosing just one atom from the total number of atoms in the entire known Universe.*

**1:100,000,000,000,000,000,000,000,000,000,000,000,000,000,000,000, 000,000,000,000,000,000,000,000,000,000,000.**

**ε** *Ratio of Mass lost to Energy in the Fusion of Hydrogen to Helium.*
The third critical number is derived from another fundamental Physical constant, The Strong Nuclear Force. This number defines how firmly atomic nuclei bind together and how all the atoms in the Universe were made. Its value is precisely 0.0071. This number is critical because it first determined the ratio of Hydrogen to Helium in our Universe and subsequently the development of the other elements in the Periodic Table as well.

*This value controls the power of stars, how they make all the other elements on the periodic table from hydrogen. Because of this number, Carbon and Oxygen are common, Gold and Uranium are rare.*

**E** *Strong Nuclear Force*

Critical to the value of **ε** is the value of the Strong Nuclear force, (E). E is itself fine-tuned to a remarkable degree.
**1:10,000,000,000,000,000,000,000,000,000,000.**

*If the demanding value of the* <u>*strong nuclear force*</u> *were to be decreased by just one part in ten thousand billion billion billion billion, the only element in the Universe would be hydrogen, and chemical life would be impossible.*

**Ω** *Cosmic Expansion*
The fourth of our critical numbers is **Ω** (Omega). Omega measures the amount of material in our Universe – galaxies, diffuse gas, and the mysterious "dark matter." **Ω** is a derivative number and tells us the relative importance of gravity and the expansion energy of the Universe. It is the ratio between the actual density and the critical density required to collapse the Universe.

*If this ratio were too large relative to a particular "Critical" value, the Universe would have collapsed long ago; had it been too small, the expansion would have been so fast no galaxies or stars could have formed. The initial expansion speed seems to have been precisely and finely tuned.*

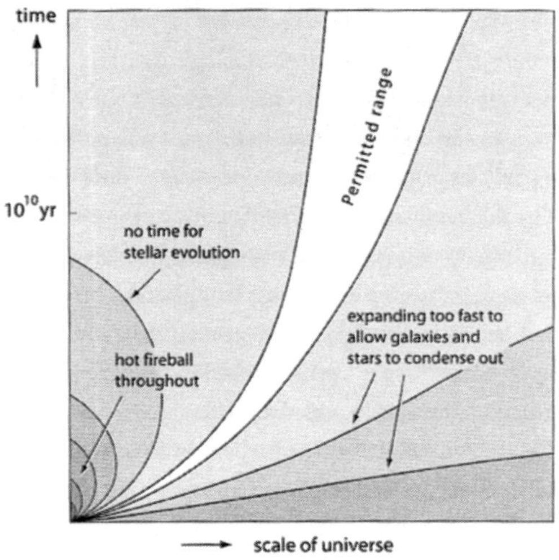


***This drawing illustrates the various trajectories of possible universes. Despite the uncertainty of the present value of W, the initial conditions must have been tuned with remarkable precision in order for our universe to end up in the permitted range.***

**Q** *Cosmological Inhomogeneity*

The seeds for all cosmic structure – stars, galaxies, and clusters of galaxies – were all imprinted in the uniformity of the Big Bang. The fabric of our Universe depends on one number, Q, which represents a ratio of two fundamental energies. Q is about 1/100,000 in value and is the fifth of the six critical numbers we will discuss here.

> *If Q were smaller, the Universe would be inert and without structure; if it were much larger, it would be a violent, destructive place in which no stars or solar systems could survive. The Universe would be made up almost exclusively of vast black holes.*

***D*** *Dimensionality*

The sixth crucial number has been known for thousands of years; it is intuitive to us all, though it is now viewed in a new perspective. It is the number of spatial dimensions in our Universe. ***D*** equals three.

> *Life could not exist if **D** were two or four. Time is a fourth dimension but distinctively different from the others in that it has a built-in arrow; we "move" only towards the future, never in the direction of the past.*

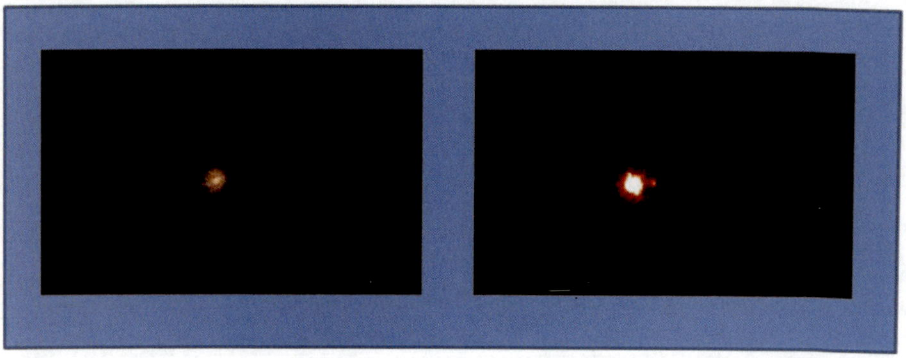

Scientists led by Mikkel F. Andersen at the University of Otago in New Zealand used lasers to slow down the frenetic movement of a group of rubidium-85 atoms and then capture them inside optical tweezers. The method could catch and isolate atoms 83 percent of the time. Using two lasers acting as a kind of tractor beam, Andersen's team was able to hold a single rubidium atom in front of a camera designed for use in space and snap its picture (left image). The Hubble Space telescope captured this look at Pluto and its companion Charon in the image on the right.

## Conditions Necessary for Life on a Planet are Rare and Exceptional

In addition to the numbers we have described that point to a finely tuned Universe, there have been twenty (or more) parameters identified for advanced life to form on a planet supporting the concept of a Rare Earth. Assign a conservative 1 in 10 probability rate to each of these 20 parameters, and the odds of finding a habitable planet are 1:1,000,000,000,000,000 that is $10^{-15}$, or one, one thousandth of one, one trillionth. It should be noted that there are only approximately one hundred billion (100,000,000,000) stars in our galaxy. That would mean you would need 10,000 galaxies like the Milky Way before you could expect to find one planet like our own. *There are many probabilistic resources in the galaxy, but the other side of the coin contains all these factors making a planet fit for advanced life.*

## Boltzmann's Paradox

Boltzmann brains are often referred to in the context of the "*Boltzmann brain paradox*" or "*Boltzmann's problem*;" they have also been referred to as "*Boltzmann babies*."

The concept arises from the need to explain why we observe such a large degree of organization in the Universe. The second law of thermodynamics states that the total entropy in a closed universe will never decrease. We may think of the most likely state of the Universe as one of high entropy, closer to uniform and without order. So why is the observed entropy so low?

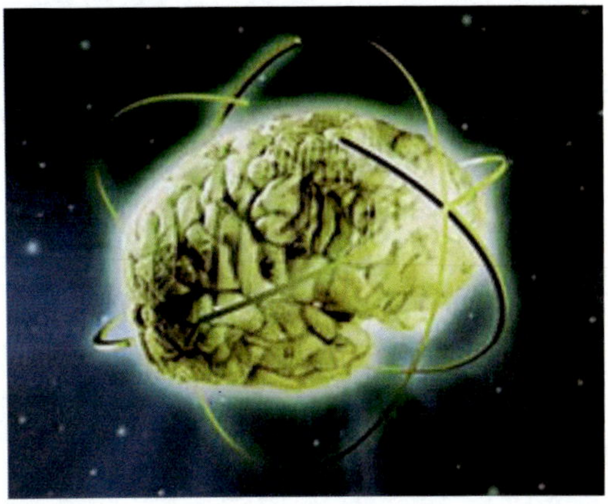

Boltzmann proposed that we and our observed low-entropy world are a random fluctuation in a higher-entropy universe. Even in a near-equilibrium state, there will be stochastic fluctuations in the level of entropy. The most common fluctuations will be relatively small, resulting in only small amounts of organization, while larger fluctuations and their resulting greater levels of organization will be comparatively much rarer. Large fluctuations would be almost inconceivably rare but are made possible by the enormous size of the universe and by the idea that if we are the results of a fluctuation, there is a "*selection bias.*" We observe this very unlikely Universe because the unlikely conditions are necessary for us to be here, an expression of the ***anthropic*** principle (which states that the Universe must be suited for those who occupy it).

If our current level of organization, having many self-aware entities, is a result of a random fluctuation, it is much less likely than a level of organization which only creates stand-alone, self-aware entities. For every universe with the level of organization we see, there should be an enormous number of lone Boltzmann brains floating around in unorganized environments. In an infinite universe, the number of self-aware brains that spontaneously randomly form out of the chaos, complete with false memories of a life like ours, should vastly outnumber the real brains evolved from an inconceivably rare local fluctuation the size of the observable Universe.

The Boltzmann brain paradox is that any observers (self-aware brains with memories like we have, which includes *our* brains) are therefore far more likely to be Boltzmann brains than real evolved brains, thereby at the same time also refuting the selection-bias argument.

# Entropy

## THIS IS THE LAW

Entropy of an isolated system ***never decreases***, because isolated systems ***spontaneously*** evolve towards ***thermodynamic equilibrium***

— the state of ***maximum entropy***.

Despite this law, which has been empirically verified in *every* circumstance, there are those that insist that given an infinite amount of time, in an infinite amount of space, the Zero Point Field can somehow spontaneously generate on its own, <u>without any external impetus</u>, quantum fluctuations, so deep and so large that all of the Universe, with all of its complexity, could come into being entirely on its own.

# Negative Entropy

**0.00000000000000000000000000000001** **$10^{-32}$**

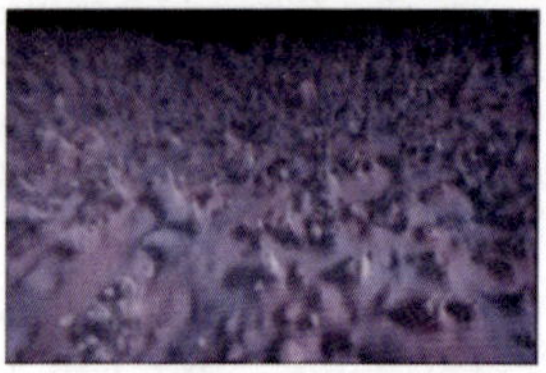

**An electron is *one hundred million billion* times larger than fluctuations in the zero point field.**

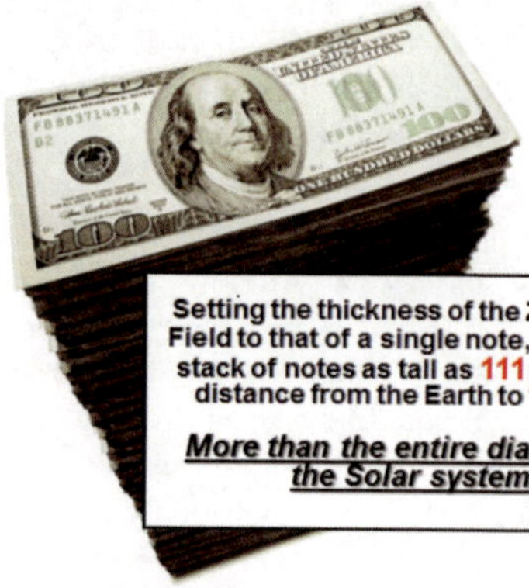

**Setting the thickness of the Zero Point Field to that of a single note, requires a stack of notes as tall as 111 times the distance from the Earth to the Sun.**

***<u>More than the entire diameter of the Solar system.</u>***

**0.000000000000001** **$10^{-15}$**

**Copernican principle**, named after Nicolaus Copernicus, states that the Earth is not in a central, specially favored position. Today, the principle has been generalized to the relativistic concept that humans are not privileged observers of the Universe. In this sense, it is equivalent to the mediocrity principle, with important implications for the philosophy of science.

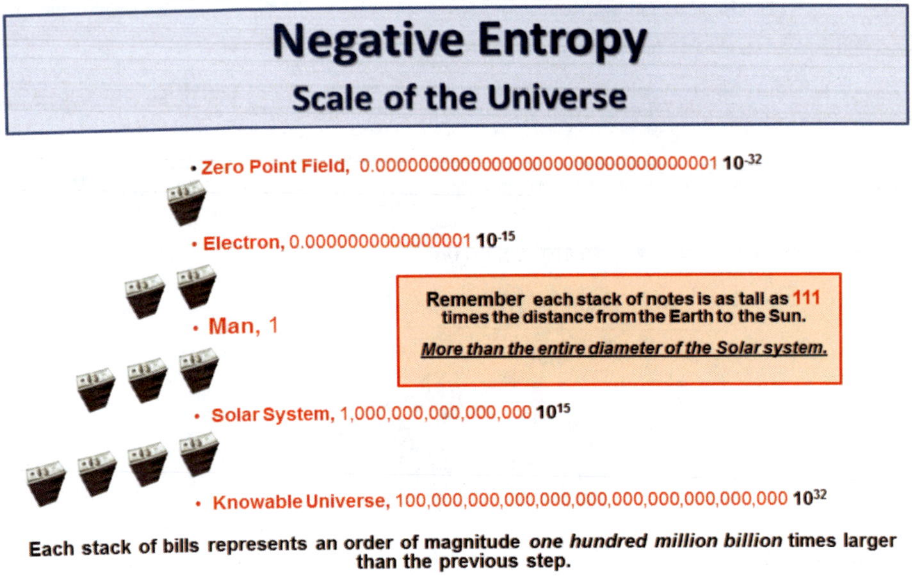


## Do You Feel Lucky?

The latest "Multiverse" theory put forth by Stephen Hawking and others, postulates that the Big Bang was brought about by a spontaneous, random fluctuation that occurred in the *Zero Point Field*. The "*fluctuation,*" if it can be *called* that, for it was HUGE, needed to be on an order of magnitude *ten billion billion billion billion billion billion billion* times larger than the field itself.

**(10,000,000,000,000,000,000,000,000,000,000,000,000,000,000,000,000,000,000,000,0 00,000,000)**

Such an enormous "fluctuation," it is told, formed in an entropic field that has not generated a single measureable fluctuation on the order of the size of a single subatomic particle. If over time it can spawn a complex, dynamic universe, why haven't we observed it creating particles, planets, galaxies and even Boltzmann's Brains from nowhere at all?

When we account for the six numbers we identified earlier in this Appendix (Just Six Numbers) and apply statistical analysis to determine the probability distribution of those conditions occurring in any one universe, we conclude that the needed attempts a random Multiverse would need to make to get a single universe like ours, is almost infinite.

**In our discussion we have seen a lot of zeros, representing numbers very large and very small; dimensions immense and very tiny; numbers defining incredible precision.**

- **1:100,000,000,000,000,000,000,000,000,000,000,000,000,000,000,000,000,000,000,000,000,000,000, 000,000,000,000,000** *one part in a hundred million trillion trillion trillion trillion trillion trillion*
*(Precision required in the combined Gravitational Force and Cosmological Constant).*

- **100,000,000,000,000,000,000,000,000,000,000** *one hundred million trilllion trillion.*
*(Size of the Universe in centimeters).*

- **10,000,000,000,000,000,000,000,000,000,000,000,000,000,000,000,000,000,000,000,000,000** *ten billion billion billion billion billion billion billion.*
*(Fluctuation needed in the Zero Point field to create a Universe).*

*The Law of Entropy demands that given an infinite amount of time, in an infinite amount of space, the Zero Point Field will go completely flat and inert. Quantum fluctuations within the field will remain, though as we have seen, these are on a very, very small scale and getting even smaller with time. Still it is conceivable, where all things are possible, that a fluctuation on an order of magnitude large enough to spawn a universe could occur, though extremely, extremely rare.*

**If we were to throw the six numbers we have discussed here up in the air like dice, how many throws would it take for them to come down just the way they have in this Universe?**

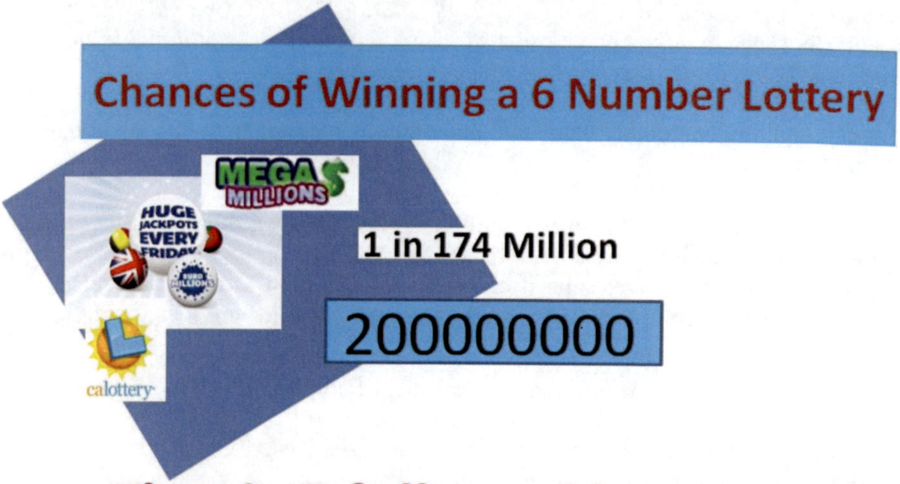


# That is 2 followed by 8 Zeros

***Chances of Winning a 6 Number Lottery are 1 in 174 Million, approximately 200000000 $2.0x10^8$. A Lottery uses whole numbers only, from a usual pool of 52 numbers: 1,2,3,4,5,6… The real world uses real numbers and is not limited by a pool to pick from.***

***1.36547, 8.000002564367529, 132,687,454, 2.090463***

## Minimum Iterations Required for a Life Friendly Universe to Form

**Minimum Iterations Required for a Life Friendly Universe to Form**

*100,000,000,000,000,000,000,000,000,000,000,000,000,000,000,000,000,000,000,*
*000,000,000,000,000,000,000,000,000,000,000,000,000,000,000,000,000,000,000,*
*000,000,000,000,000,000,000,000,000,000,000,000,000,000,000,000,000,000,000,*
*000,000,000,000,000,000,000,000,000,000,000,000,000,000,000,000,000,000,000,*
*000,000,000,000,000,000,000,000,000,000,000,000,000,000,000,000,000,000,000,*
*000,000,000,000,000,000,000,000,000,000,000,000,000,000,000,000,000,000,000,*
*000,000,000,000,000,000,000,000,000,000,000,000,000,000,000,000,000,000,000,*
*000,000,000,000,000,000,000,000,000,000,000,000,000,000,000,000,000,000,000,*
*000,000,000,000,000,000,000,000,000,000,000,000,000,000,000*

**That is 1 followed by 500 zeros....**

*This is the **MINIMUM** number of throws needed; the real number is closer to infinity. This is the largest number we have seen by far.*

*The number representing the total number of atoms in the Universe contains only 80 zeros. There have only been approximately **473,000,000,000,000,000** seconds since the Big Bang.*

## The *Instantaneous* Development of Advanced Life

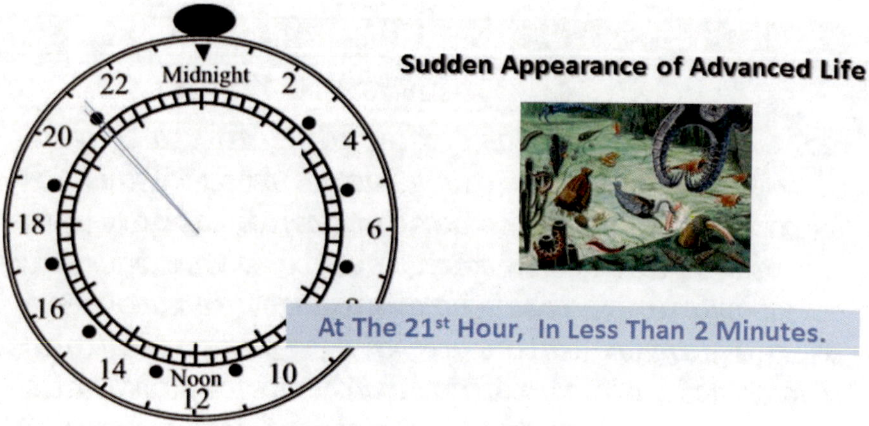


It has been universally accepted that the first forms of life, bacteria and photosynthetic algae appeared 3.8 billion years ago, almost immediately, in geologic terms, after the Earth is formed. These simple forms of life remain unchanged for billions of years.

If we set the entire history of life on Earth, from its first appearance until the present day, to a scale represented by just 24 hours on a clock, we can illustrate the significance of the Cambrian Explosion and the rapid appearance of its varied life forms, phylum, and biological advancement.

Start the 24 hour clock ticking 3.8 billion years ago when the first single cell life forms appear; 6 hours pass and life remains unchanged; 12 hours pass, the same thing; 18 hours pass and still the simple single cell life forms are the only life on the Earth. Then when the clock reaches 21 hours, in the span of just 2 minutes, advanced life appears representing almost all of the phyla that ever existed. Many of these original forms remain unchanged and in existence today (without evolution).

## The Introduction of the Sons and Daughters of God

The latest paleogenetic studies (Evidence mounts for interbreeding bonanza in ancient human species, Nature, 17 February 2016[2]) confirm early modern humans and other Homo Genera (Denisovan, Neanderthal and another unkown Hominin) interbred with each other on multiple occasions in the past 100,000 years (marked as episode/event 1 through 5 on the chart below). Many of these events 1, 2, and 3, are eluded to in the Scripture as precipitating the repentence of God leading up to the Great Flood.

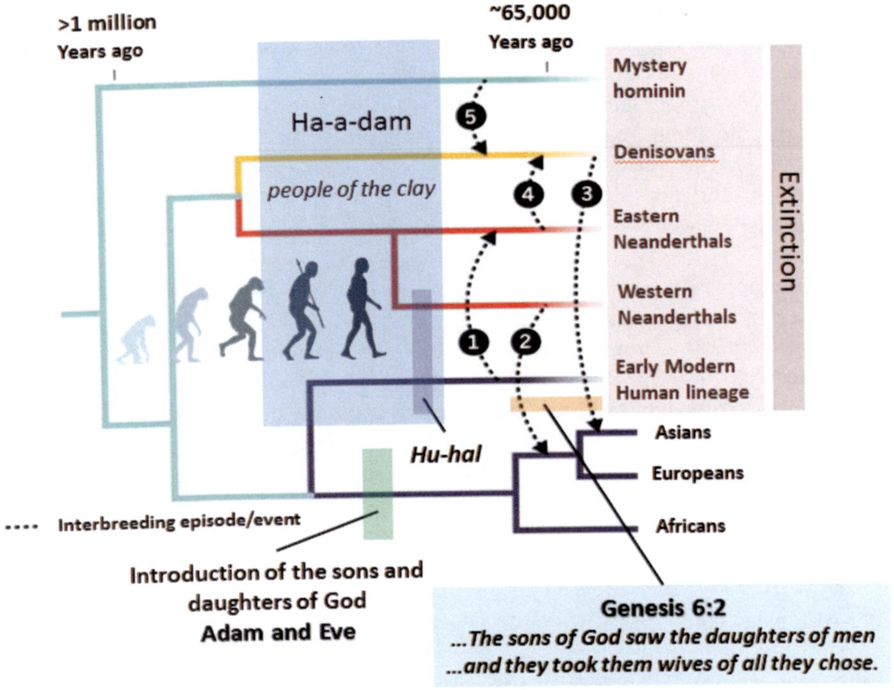


Within this work, we have identified all of the populations marked with the large blue block as the *ha-a-dam* (הָאָדָם), the *people of the clay*, and both the scriptural and paleontological records identify them as having gone extinct (pink block).

---

[2] Ancient gene flow from early modern humans into Eastern Neanderthals (Kuhlwilm, Martin. et.al., Nature, Vol. 530, 25 February 2016, pg. 429). "*It has been shown that Neanderthals contributed genetically to modern humans outside of Africa 47,000-65,000 years ago. Here we analyse the genomes of a Neanderthal and a Denisovan from the Altai Mountains in Siberia together with the sequences of chromosome 21 of two Neanderthals from Spain and Croatia. We find that a population that diverged early from other modern humans in Africa contributed genetically to the ancestors of Neanderthals from the Altai Mountains roughly 100,000 years ago. By contrast, we do not detect such a genetic contribution in the Denisovan or the two European Neanderthals. We conclude that in addition to later interbreeding events, the ancestors of the Neanderthals from the Altai Mountains and early modern humans met and interbred, possibly in the Near East, many thousands of years earlier than previously thought.*"

We have further identified a subgroup of these populations within this work as the *hu-hal* (הוּהַל), who through their association with Cain and his offspring (in the land of Nod) began to "*call upon the name of the Lord*." The *hu-hal* are the most closely associated of these extinct groups to the sons and daughters of God, the *Be-ne- ha-e-lo-him* (בְנֵי־הָאֱלֹהִים).

Here we show the late introduction of Adam and Eve (green block) to the pre-extant human population on Earth. That introduction, of much advanced genetic material and the *Neshama*, drastically changed the population of man in very short order.

Today the karyotype of modern man dispays a change that, if it was in place from the beginning, would have curtailed, if not entirley prevented such interbreeding (certainly it would have made it extremely difficult). Modern Man has 23 pairs of chromosomes, and all of the other extant hominids have 24. (Some most recent research suggests that Denisovan man and perhaps even Neanderthal may also have had 23 chromosomal pairs, but because there is no living tissue to display, choromosomes can only be surmised by other means and as such will forever remain just speculative.)

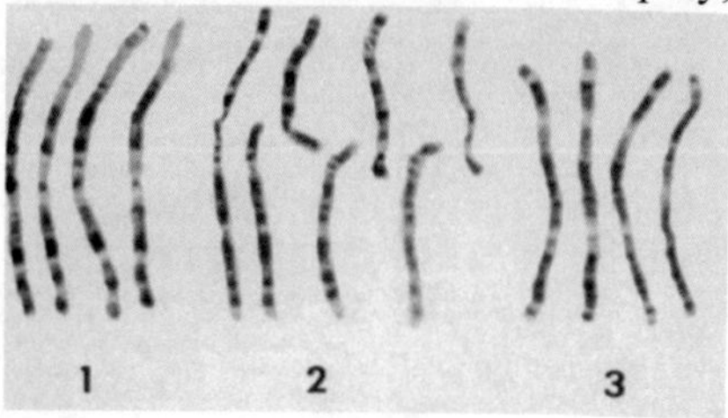


The picture here shows a side-by-side (left to right) comparison of chromosomes 1, 2, and 3, of humans, chimpanzees, gorillas, and orangutans. For chromosome 2 it will be noted that chimpanzees, gorillas, and orangutans each have two smaller shorter segments, identified as 2p and 2q, where humans only have 1. This is significant, for though the human chromosome 2 has been shown to be a fusion of two smaller shorter chromosomes very similar to Chimpanzee 2p and 2q, it is not exact, and the fusion makes it impossible (in all practicality) for any viable offspring between man and ape to develop.

The evidence here most probably shows: the fusing of two into one was accomplished late in the overall history of development, that it was a game changing event in the history of man, and the fusing prevented the interbreeding that is understood to have occurred in the past. Perhaps this change is a trace of the events that transpired when it *"...repented the Lord that he had made man on the earth."*

- Genes are Not Everything

It must be noted here that the number of chromosomes is secondary to the information that they contain. It is not the number of chromosomes that we have that distinguish us as separate from the great apes, but it is the information contained in our chromosomes *and* the *Neshama* that distinguishes Modern Man (the sons and daughters of God) from the rest of created life.

In recent years, research projects have finalized the complete mapping of Human and Chimpanzee Genomes. These mappings, complete to the most minute detail, demonstrate that Modern Humans and Chimpanzees display indentical sequencing throughout 98.8 percent of the code (98.8 percent of the approximately 3 billion base pairs[3] are identical), a match that is nothing short of astonishing considering how different the two species are. This almost *exact* match begs a question: *Are all of the differences between species really just because of sequencing in the genetic code?* This question is not easily resolvable by what is known to science. Richard Dawkins is quoted as saying: "*A tiny number of differences account for all the really quite large difference that we see between humans and chimpanzees.*"[4] Dawkins does so without offering any substantive explanation for the sequence differences as they relate directly to differentiation in the species. He seems to infer that man, most definitely, evolved from apes *just because* the genome is so similar; that similarity he projects is irrefutable proof of human evolution.

But his very "*evidence*" signals a much *different* conclusion; the sameness of the coded genome tells a tale of resistance to random change through mutation of the DNA code. Consider the statistical relevance of two enormous sequences and the small number of differences between them. Given usual expectations, how many of these differences are in fact there in the code or are simply a result of "noise" or error in the process or in fact are differences associated with the individuals tested and not the species as a whole. No two humans are genetically identical. On average, their DNA sequences vary by as much as .5 percent; this *average* variation accounts for as much as 1/3 of the number associated with differences between humans and chimpanzees (1.2). In many cases, the number of differences in the genome from human to human approach and exceed the number of differences averaged between man and ape. Yet, we never see the gulf established by the remaining .7 percent variation crossed; Humans do not spontaneously mutate into Chimpanzees during mitosis and vice versa.

Changes to the complex, very large pattern found in DNA on scales we have discussed here (at or less than .7, or 1.8 million pairs) do not result in change, though gradualism *insists* that *change must occur* for life to have evolved randomly on Earth. It is clear there is something much more than DNA sequencing at work in speciation and the remarkable consistency throughout the history of life on Earth.

What we espouse here is not evolution as it is understood by armchair scientists and espoused by ardent athiests, but we believe that *informed* matter developed teleologically. The genetic code within us all is comprised of the *bohu* we have discussed in this work, the same *bohu* that was infused with the knowledge of God in Day 1. That *bohu* has worked with that knowledge, at times by the direction of God (the *bara* events of the the Genesis 1 accounts), to produce life in all its variety and with all of its adaptablity to survive in an uncertain Universe. That the elements

---

[3] Note that as the outward expression of the complexity of life increases (the more evolved it appears), the numbers or base pairs actually *decreases*. Humans and the great apes have approximately 3 billion pairs in their DNA, an amphibian (Newt) some 15 billion, and a simple Lily genome has 100 billion pairs.

[4] See the Richard Dawkins Foundation, For Reason and Science, Comparing the Human And Chimpanzee Genomes – Nebraska Vignettes #3, available on YouTube.

of life can adapt and transform yet remain consistent in their purpose, bespeaks of intelligent design and of an intelligent and purposed designer.[5]

We understand here that our intended forms were established in a previous Physical reality which overlays the Temporal reality. That overlayment adds to the epigenetic landscape, contributing (perhaps most significantly) to the replication of those preexistent images from Temporal matter in accordance with the laws of physics, chemistry and biology. It is the primary purpose of the DNA code to govern and direct that overlayment and record the information necessary to make that possible (Life assumes the primary forms it does in the Temporal, because it imitates the forms it holds in the Physical; DNA then is more like a construction materials *build list* and less like a complete set of *architect's blueprints*). The remarkable consistency in the form of life in the face of uncertainty which works to counter it, stems from an affinity life holds for the Physical form that is inherent within it.

There is yet another element of our being at play which directs our development within the Temporal sphere. It is the addition of the *Nefesh* and *Neshama*.

We attest here that we understand and accept the good science of evolution; we agree that life developed and increased in complexity from its beginning on Earth so long ago. Within this work, we acknowledge the shared structures of all life and recognize that God has arranged the *bohu* in such a way that life itself is intimately related in all of its forms.

---

[5] For a very interesting study of how an atheist evolutionary scientist seeks to avoid the issue of an intelligent creator, God, see Ben Stein interview Richard Dawkins in Ben Stein's documentary Expelled. Mr. Stein asks Dawkins: "*What do you think is the possibility that intelligent design might turn out to be the answer to some issues in genetics or evolution?*" In what turns out to be a very uncomfortable response, Dawkins struggles for an answer then states: "*It could be that at some earlier time, somewhere in the Universe a civilization evolved, by probably some kind of Darwinian means, to a very, very high level of technology and designed a form of life they seeded onto perhaps this planet. Now that is a possibility, and an intriguing possibility, and I suppose it is possible that you might find evidence of that, if you look at the details of biochemistry and molecular biology you might find a signature of some sort of* ***designer****... And that* ***designer*** *could well be a higher intelligence from elsewhere in the Universe.*" Dawkins still can't concede his a-teleological bias as he continues, almost catching the words he let slip from his mouth: "*but the higher intelligence would itself have had to have come about by some explicable, or ultimately explicable process, it couldn't have just jumped into existence spontaneously, that's the point.*" The exchange is amazing and reveals to what strained reason an atheist will turn to in avoiding the obvious, GOD. Dawkins' explanation runs out of time though, life appeared on Earth 3.8 billion years ago, and his "*higher intelligences*" would have needed at least that much time (perhaps much more), to have developed themselves a "*very, very high level of technology*" before striking out like *Johnny Appleseed* to spread their wealth throughout the Universe - that is at 8 billion perhaps 10 billion years. That puts our benevolent gardeners beginnings dangerously close to the Big Bang, certainly at a time when the very first generation of stars were beginning to form, at a time before the Galaxies appear. Without the presence of elements found higher up on the periodic table (elements that can only be formed in the furnaces of those first stars and created through the death of those stars in novae and supernovae), where did our "*creators*" get the carbon they needed to exist and the carbon they needed to make the life they introduced to this planet? Dawkins' tale is a telling explanation. By it, he reveals that he is aware of the corner modern science has backed deniers like himself into, as it has revealed the most ancient occurrence of life on Earth, in its earliest age, and its modern disciplines of biochemistry and molecular biology have begun to show the signature of intelligent design. I must point out here that at one point Dawkins said, "*you might find a signature of some sort of designer.*" In truth, it is clear; he knows that signature has already been found.

# Missing Generations in the Genealogy of Christ

| Kings | Matthew 1 | Luke 3 | Old Testament |
|---|---|---|---|
| | | Adam | |
| | | Seth | |
| | | Enos | |
| | | Cainan | |
| | | Malaleel | |
| | | Jared | |
| | | Enoch | |
| | | Mathusala | Mathuselah |
| | | Lamech | |
| | | Noe | Noah |
| | | Sem | Shem |
| | | Arphaxad | |
| | | Cainan | |
| | | Sala | Salah |
| | | Heber | Eber |
| | | Phalec | Peleg |
| | | Ragau | Reu |
| | | Saruch | Serug |
| | | Nachor | Nahor |
| | | Thara | Terah |
| | Abraham | Abraham | |
| | Isaac | Isaac | |
| | Jacob | Jacob | |
| | Judas | Juda | |
| | Phares of Thamar | Phares | |
| | Esrom | Esrom | |
| | Aram | Aram | |
| | Aminadab | Aminadab | |
| | Naasson | Naasson | |
| | Salmon | Salmon | |
| | Booz of Rachab | Booz | Boaz |
| | Obed of Ruth | Obed | |
| | Jesse | Jesse | |
| | David | David | King David |
| | Solomon | Nathan | |
| Rehoboam | 1Roboam | Mattatha | |
| Abijam | 1Abia | Menan | |
| Asa | 1Asa | Melea | |
| Jehosaphat | 1Josaphat | Eliakim | |
| Joram | 1Joram | Jonan | |
| 1Ahaziah | *Ahaziah?* | Joseph | |
| Athaliah | | Juda | |
| 1Joash | *Joash?* | Simeon | |
| 1Amaziah | *Amaziah?* | Levi | |
| 1Uzziah/Azariah | 1Ozias | Matthat | |
| Jotham | 1Joatham | Jorim | |
| Ahaz | 1Achaz | Eliezar | |
| Hezekiah | 1Ezekias | Jose | |
| Manasseh | 1Manasses | Er | |
| Amon | 1Amon | Elmodam | |
| Josiah | 1Josias | Cosam | |
| Jehoahaz | | Addi | |
| 2Jehoiakim | *Jehoiakim?* | Melchi | |
| 1Jehoiachin | 1Jechonias | Neri | |
| 2Zedekiah | *Pedaiah?* | | |
| | 2Salthiel | Salathiel | |
| | Zorobabel | Zorobabel | Zerubabel |
| | Abiud | Rhesa | |
| | Eliakim | Joanna | |
| | Azor | Juda | |
| | Sadoc | Joseph | |
| | Achim | Semei | |
| | Eliud | Mattathius | |
| | Eleazar | Maath | |
| | | Nagge | |
| | | Esli | |
| | | Naum | |
| | | Amos | |
| | | Mattathias | |
| | | Joseph | |
| | | Janna | |
| | | Melchi | |
| | | Levi | |
| | Matthan | Matthat | |
| | Jacob | Heli | |
| | Joseph* | | |
| | Mary? | Joseph | |
| | Jesus | | |
| | 27 Generations | 42 Generations | |

Matthew gives the account to the Jews
making the case that Jesus is King of the Jews.
Matthew gives the lineage through Mary.....
Joseph here is Mary's Father.

Luke gives the lineage through Joseph.
Luke writes to the Gentiles pleading the case that
Jesus is the Savior of all mankind and details his humaness.

**2 Samuel Chapter 3**
2 And unto David were sons born in Hebron: and his firstborn was Amnon,
of Ahinoam the Jezreelitess;
3 And his second, Chileab, of Abigail the wife of Nabal the Carmelite;
and the third, Absalom the son of Maacah the daughter of Talmai king of Geshur;
4 And the fourth, Adonijah the son of Haggith;
and the fifth, Shephatiah the son of Abital;
5 And the sixth, Ithream, by Eglah David's wife.
These were born to David in Hebron.
**2 Samuel Chapter 5**
13 And David took him more concubines and wives out of Jerusalem, after
he was come from Hebron: and there were yet sons and daughters born to David.
14 And these be the names of those that were born unto him in Jerusalem;
Shammua, and Shobab, and Nathan, and Solomon,
15 Ibhar also, and Elishua, and Nepheg, and Japhia,
16 And Elishama, and Eliada, and Eliphalet.
**1 Chronicles 3**
9 These were all the sons of David, beside the sons of the concubines,
and Tamar their sister.
10 And Solomon's son was Rehoboam, Abia his son, Asa his son,
Jehoshaphat his son,
11 Joram his son, Ahaziah his son, Joash his son,
12 Amaziah his son, Azariah his son, Jotham his son,
13 Ahaz his son, Hezekiah his son, Manasseh his son,
14 Amon his son, Josiah his son.
15 And the sons of Josiah were, the firstborn Johanan, the second
Jehoiakim, the third Zedekiah, the fourth Shallum.
16 And the sons of Jehoiakim: Jeconiah his son, Zedekiah his son.
17 And the sons of Jeconiah; Assir, Salathiel his son,
18 Malchiram also, and Pedaiah, and Shenazar, Jecamiah, Hoshama,
and Nedabiah.
19 And the sons of Pedaiah were, Zerubbabel, and Shimei: and the sons
of Zerubbabel; Meshullam, and Hananiah, and Shelomith their sister:
20 And Hashubah, and Ohel, and Berechiah, and Hasadiah, Jushab-hesed, five.
21 And the sons of Hananiah; Pelatiah, and Jesaiah: the sons of Rephaiah,
the sons of Arnan, the sons of Obadiah, the sons of Shechaniah.
22 And the sons of Shechaniah; Shemaiah: and the sons of Shemaiah;
Hattush, and Igeal, and Bariah, and Neariah, and Shaphat, six.

*Greek, *ander* from root, *aner*. Should read father, not husband.
This makes 14 generations listed from the return. (verse 17)

## Letter from Eusebius Pamphilus to the Church of Cesarea

It is probable, beloved, that you have already learnt from another source, what has been done respecting the ecclesiastical faith in the great council convened at Nice, as common fame usually outruns an accurate report of facts. But as rumor of this kind may have represented things differently from what they actually were, we have thought it necessary to send you, first, the form of faith proposed by us, and afterwards that which was set forth by the bishops who made some addition to ours. Our own form, then, which was read in the presence of the emperor, and appeared to be right and proper, is expressed in these terms. As we have received from the bishops who preceded us,—as we have been taught in the rudimental instructions of our childhood, and when we were subjects of the baptismal rite, and as we have learnt from the divine scriptures; as we have believed and taught, both in order of presbyter, and the episcopal dignity itself, and as we now believe, we present to you our profession of faith. And it is this. We believe in one God, the Father Almighty, Maker of all things, visible and invisible; and in one Lord Jesus Christ, the Word of God, God of God, light of light, life of life, the only begotten Son, the first born of every creature, begotten of the Father before all ages, by whom all things were made; who for our salvation was made flesh and conversed among men; who suffered, and rose again the third day, and ascended to the Father, and will come again with glory to judge the living and the dead. We also believe in one Holy Spirit; believing every one of these to be and to subsist, the Father truly the Father, the Son truly the Son, and the Holy Spirit truly the Holy Spirit; as our Lord, when he sent his disciples to preach, said, "*Go, and teach all nations, baptizing them in the name of the Father, and of the Son, and of the Holy Ghost.*" We solemnly affirm that we thus hold and thus think, and have so held formerly, and will hold even unto death, and will always continue in this faith, anathematizing every impious heresy. We testify before Almighty God and Our Lord Jesus Christ, that we have believed this sincerely, and from the heart, from the time when we were capable of knowing ourselves, and now also truly think and speak, being prepared to show by sufficient proofs, and to convince your minds, that we have so believed in times past, and have preached accordingly.

Having made this representation of our faith, there was no pretence for contradiction. But our pious emperor himself was the first to declare, that it was extremely well conceived, and that it expressed his own sentiments, exhorting all to assent to, and sign it, that they might unite in its doctrines, with the addition only of the single word consubstantial; which he himself explained by asserting that he did not use the term with reference to corporeal affections, and that the Son did not subsist from the Father, either by division or abscission, since it was impossible that an immaterial, intellectual and incorporeal nature could admit of any bodily affection; but that it must be understood in a divine and mysterious manner. It was thus that our most wise and religious emperor argued on this subject. But the bishops, taking occasion from the word consubstantial, committed to writing the following form:—

> We believe in one God, the Father Almighty, Maker of all things, visible and invisible; and in one Lord Jesus Christ, the Son of God, the only begotten of the Father, that is of the substance of the Father; God of God, light of light, true God of true God,

> begotten not made, consubstantial with the Father, by whom all things were made, both in heaven and in earth; who for us men, and for our salvation, descended, was incarnate, and was made man, and suffered, and rose again the third day; he ascended into heaven, and shall come to judge the living and the dead: And in the Holy Spirit. But the holy catholic and apostolic Church of God anathematizes those who affirm that there was a time when the Son was not, or that he was not before he was begotten, or that he was made of things not existing; or who say, that the Son of God was of any other substance or essence, or created, or liable to change or conversion.

When this form was dictated by the prelates, their expressions "*of the substance of the Father,*" and "*consubstantial with the Father,*" were not suffered to pass without examination. Hence, therefore, several questions arose, and answers were made, and the sense of these terms was carefully considered. They admitted that the words "*of the substance*" signified that the son was of the Father, but not as a part of the Father. We thought it well to assent to this explanation, as conveying the pious doctrine, that the son was of the Father; but not, however, a part of the Father. We therefore agreed to this opinion; nor did we reject the word consubstantial, having in view the promotion of peace, and being anxious to avoid a departure from the right belief. For the same reason we approved also of the words "*begotten, not made,*" since the word made, they said, was common to the other creatures which were made by the Son, and to which he has nothing similar; and that therefore he is not made like those who were created by himself, but is of a more excellent substance than any created being. The divine oracles inform us, that he was of the Father, by a mode of generation, which can neither be conceived nor expressed by any created intelligence.

The question whether the Son is consubstantial with the Father being thus examined, it was agreed that this was not to be understood according to the manner of material things, nor that of mortal beings, since it could be neither by division, nor abscission, nor by a change of the paternal essence and power, since the unbegotten nature of the Father is foreign from all these things. But by the expression "*consubstantial with the Father*" nothing else is intended, than that the Son of God has no similitude with created beings, but resembles in all things the Father only, by whom he was begotten, and that he is of no other substance or essence than that of the Father. The proposition being thus explained, we thought that we might justly accede to it; since we knew that some of the most learned and distinguished of the ancient bishops and writers had made use of the term consubstantial, in treating of the divinity of the Father and the Son.[††]

This is what I intended to say concerning the faith which was declared, and to which we all gave our consent; not, however, without inquiry and examination, but according to the senses adduced, which were discussed before our most

[††] Eusebius, must here be referring to some documents of which he had access to that are now lost. The lack of any such documents today have led most all to conclude that the term consubstantial originated at Nicaea.

religious emperor, and for the reasons already mentioned, unanimously approved. We also agreed without difficulty to the anathema put forth by the prelates, and subjoined to the form of faith, because it prohibits the use of unscriptural expressions, from which nearly all the confusion and disturbances of the Church have arisen. Since, therefore no divinely inspired writing has made use of the phrases, "*of things not existing,*" and "*there was a time when he was not,*" and others which are added to them, it did not seem proper that they should be spoken or taught. We therefore consented also to this salutary decree, not having been accustomed, in times past, to the use of such terms.

We have sent you this information, beloved, that we may clearly show you, with what care and deliberation we conducted our inquiries and examination, and gave our assent, and with how much reason we resisted at first, and continued our opposition to the last hour; so long indeed, as any thing being written otherwise than correctly, afforded occasion of offence. We finally embraced, without further contention, those expressions which were found to be unexceptionable, when, on a candid examination of the sense of the words, it appeared that they entirely agreed with those admitted by ourselves, in the exposition of faith which we at first proposed.

## Adam in the New World

Why is Adam-Ondi-Ahman on the North American Continent?

When our first parents were introduced to this planet, God (no doubt) carefully selected a location best suited for their success. That spot we understand was central North America.[6] Why this spot? The answer can be found in examining the conditions on the North American continent 60,000 – 6,000 years ago. We have discussed that the evolutionary processes which developed early man and its ancestors began some 200,000 years ago, yet despite all of this time, those advanced animals had not spread beyond Africa, Southern Asia and Europe, and Australia when Adam was placed on the Earth.

At that time, the Americas, both North and South, were entirely devoid of human population. The earliest human remains (Clovis) found in the Americas date far more recent than they do anyplace else in the world (not before some 13,000 years ago, and when they do appear, they are entirely *Modern* in anatomy indiscernible from the sons and daughters of God). It no doubt was God's intention to colonize man here, away from those populations of early and modern man (who were simply animals possessed with the *Nefesh* only). God's choice of North America as the original home of the sons and daughters of God was wise, for it insured their civilization would develop and blossom free of the external forces that eventually brought about the need for the flood. So, how did we get from the City of Zion to the Ark adrift on the waters and the complete destruction of early Man, Denisovan and Neanderthal?

- The Migration of Cain

> *"And Cain went out from the presence of the LORD, and dwelt in the land of Nod, on the east of Eden." (Gen. 4:16)*

This verse, though short, is full of issues that need to be unpacked. It is clear, Cain, born of Adam and Eve in Temporality, never dwelt in the "*presence of the LORD*" as we have described it here. Cain, on Earth, was a Temporal being, not an inhabitant of the Physical or Spiritual spheres when he committed the murder of his brother; neither (while in the flesh) was he living in any place from which he could *relocate*

---

[6] There are no scriptural references to Adam's Temporal habitation having been located on the North American continent (D&C 116 does not speak of Adam inhabiting the region round about Spring Hill Mo., but it describes his future coming to those environs at the end of time). The evidence that Joseph spoke of Adam dwelling in the Missouri area is found only in secondhand accounts found in journals and personal letters written by others, e.g., "*The Prophet Joseph called upon Brother Brigham, myself and others, saying, 'Brethren, come, go along with me, and I will show you something,' He led us a short distance to a place where were the ruins of three altars built of stone, one above the other, and one standing a little back of the other, like unto the pulpits in the Kirtland Temple, representing the order of three grades of Priesthood; 'there,' said Joseph, 'is the place where Adam offered up sacrifice after he was cast out of the garden.' The altar stood at the highest point of the bluff. I went and examined the place several times while I remained there.*" (Life of Heber C. Kimball, pgs. 209-210)

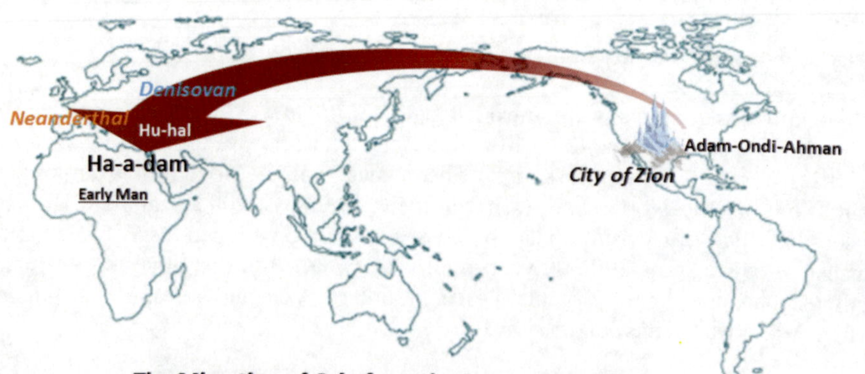


**The Migration of Cain from the *New* to the *Old* World**

to any place *East* of Eden. What then are we to understand by this strange interpretation of words so anciently written?

Hebrew was never intended to be a "*written*" language; it was to be transmitted verbally. What we see today, of the ancient texts, is a form of shorthand, notes meant only to bring to mind the words to be spoken by priests and elders (primarily) to the lay class, most of whom could not read. Because the written text was a form of "*shorthand*," the words and phrases were incomplete, intended only to bring to mind things remembered through repetition of things spoken.

The Hebrew alphabet only included consonants, in fact the very concept of verb characters did not occur until the Greek period fully eight centuries years after some of the earliest Hebrew texts were penned (and were not added to the Hebrew texts until the sixth century AD).[7] Without verbs, the words written were incomplete, needing context to fill in the gaps left in understanding. Often, as in the case of many of the words in the above verse, the same image (string of consonants) was used to express multiple words with disparate meanings.

---

[7] The "*origin of the vowel-points is to be ascribed to the effort which the Jewish learned men made to preserve the pronunciation of their sacred language at a time when its extinction as a living tongue endangered the loss of the traditional memory of its sound. Therefore, the vowel points were for the purpose of pronunciation. This pronunciation was understood and kept in collective Jewish memory as long as Hebrew was a living, spoken language; however, as its spoken use began to disappear, Jewish scholars determined to preserve this pronunciation by use of vowel points*" (McClintock and Strong Bible Encyclopedia, Volume IV, pg. 137). Jerome (345- 320 AD) and the Talmud (completed around 500 AD) knew nothing of the vowel points. Neither are they to be found in *any* Hebrew religious, commercial and legal documents dating from periods before that. Crude vowel points began to be used around the sixth century after Christ. The vowel points laboriously developed until the current system was congealed around the tenth century when it began to be used by the Masoretes (and students of biblical Hebrew today). This begs a question about the concept of *Sola Scriptura* and (as we have shown here) the accuracy of interpretation of the Word of God without a written language fully capable of communicating meaning, without the application of reason of the influence of the Spirit.

The Hebrew *qdm* (קדם) (shown here as originally written, without the vowel points [called *niqqud*]) is the word translated as *East* in the KJV verse cited above. These same three characters, in this same sequence, have several meanings, among them are:

– קדם (*qedem*), **East, antiquity, front, forward**. The word *qedem* is also the word for "*the past.*" In the ancient Hebrew mind, the past is in front of you while the future is behind you, the opposite way we think of the past and future. One can liken the Hebrew formulation of time to a man rowing a boat with his back to the bow; from that view, his past was before him (*qedem*); his future is behind his back (*aharit*). The earliest use of *qedem* anywhere in the Hebrew Bible is in Job where it means *forward*, as in a direction taken (Job 23:8). (Job is the earliest written book of the Bible, composed before the Exodus, in an unknown language that predated Hebrew, and read by the Israelites in their captivity in Egypt.)
– קדם (*qadam*), **meet, confront, go before**. **Also meaning, antiquity, and former estate.** This is the abstract noun form of the derived meaning (ancient, former) of *qedem*.
– קדם (*qeydem*), **East**. The "*East*" has both good and bad textual connotations – it is the direction towards the rising Sun and therefore represents the path toward God, but it is also the habitat of the men who built Babel (Gen. 11:2). When it refers to those who dwell to the East or northeast of Canaan, it is referring to tribes in conflict with Israel (Num. 23: 7).

Similarly, we find multiple meanings for the word, *Eden* (עדן). *Eden* not only refers to the Garden, but also as we have shown in this work, the entire Physical sphere. It also is used in ancient scripture to mean *pleasure* (Gen. 18:12).

In the verse shown above, context is the driving factor in establishing meaning, not the text itself. Of importance are the verses that lead up to it; in them we find Cain in direct contact with God, as a result of his murderous actions. In fact, he is in the presence of the LORD pleading for protection from those who would seek to do him harm. It is this direct contact (presence) with God that Cain "*went out from*" not *Eden*, or any location within the immortal Physical sphere. Simply put, Cain was a Temporal being, wholly an inhabitant of the Temporal sphere; his murderous, jealous actions could not occur in the Physical sphere or the Garden within it.

The verse details that Cain found a home, dwelling "*in the land of Nod, on the East of Eden.*" Nod has come to mean in Hebrew, *wanderer*, *exile* and *fugitive*. These meanings, however, are most probably derivative and not preceding, being associated with the proclamation of the LORD in Verse 12, "*a fugitive and a vagabond shalt thou be in the earth.*"

That Nod is "*East of Eden*" is not descriptive of geography, for within the Temporal sphere no location is *East* of any place within the Physical, but the phrase (in context) means that where Cain dwelt was *forward*, *beyond* his *former estate*, which was *pleasurable* to both himself and to God. Cain found no joy in his wanderings.

When Cain "*went out*" to the land of Nod, contrary to what the extant text suggests, he no doubt moved westward *away* from the East, which we have shown was associated with the precincts of the Father. In moving West from the precincts adjacent to Adam-Ondi-Ahman, he would have crossed along the Bearing land bridge(s) into Asia and continued on until he met, and interacted with, early man in South western Asia, the Middle East, southern Europe, and eastern North Africa. This in fact is exactly where we find paleontological and archeological evidence of Modern Man and earlier forms of man cohabitating (here actually and nowhere else).

Cain was no doubt followed by other antediluvian fathers and their families; evidence of this as fact can be seen in an examination of the Scriptures. D&C 107 provides a brief history of all of the antediluvian Patriarchs. In it, can be found information that tells us (when the City of Zion was taken up into Heaven) Enoch's father, grandfather, great grandfather and great-great grandfather, as well as his son and grandson all remained behind on Earth. No doubt many of these patriarchs and their families were no longer in the area of Adam-Ondi-Ahman. In all probability, many (if not all) had followed Cain westward. It is certain Noah was in the precincts of the *ha-a-dam* and the *hu-hal* as he preached repentance to the sons and daughters of God, who had interacted so intimately with them in the years before the Great Flood. This westward migration out of the Americas and into southwest Asia and beyond, is what places Noah, his family and the ark in the foothills of Ararat along the shores of the Black Sea at the time of the Bosphorus Breach and moves the history of the sons and daughters of God from the New World to the Old.

## Envisioning the Temporal Reality

What do we mean when we say the Temporal overlays the Physical, that it is a subtraction of a greater whole?

This concept can be difficult to comprehend, but perhaps a helpful analogy can be put forth that can help us visualize what is actually going on (if, however, imperfect our image will be).

Imagine a grand symphony orchestra playing a brilliantly beautiful and sweeping piece of music. Imagine the piece to be as emotionally moving as you can, full of woodwinds, strings and percussion, dulcet horns and bright brass. The orchestration sweeps you away and overwhelms you.

In the midst of the orchestra, all of your senses are involved; you *hear* the sound, *see* the movement of the musicians and *feel* the vibration in your frame – This is *MUSIC*, there is no doubt about it.

Now look at the image here, this is music too; in fact it could be the very piece you have imagined in your mind. This is what our Temporal sphere is like when compared to the Physical. The fullness, grandeur and beauty of the symphony collapses into just notes on a page. What we see and hear and feel here, in this Universe, what we know to be "*real,*" is but a constrained image, a simulacrum, of the full reality that is all around us. Our Universe, our Planet and our mortal bodies are but notes written on a page, they are not the music itself (the Physical sphere).

We understand here that what we experience as real is what is left over after most of what is, is withdrawn. What we are left with in Temporality are tiny bits of what really is. We push and move these "*things*" around, we study them, build with them, they are all we can see of all that God has created, but they are not all there is.

The sheet music can be found in the middle of the Orchestral chamber, the music it represents is all around it, filling the room even vibrating the very paper the notes are written on, yet if you were deaf and isolated from the orchestra itself within the room, the images on the page is all you would know of the music.

In Temporality this "*image*" is all that we know. Here we fall in love, have children with, live our lives and die with but Temporal images of our loved ones, what we can sense of them around us, what we understand them to be. But our Spirits within tell us there is more, so very much more; the Spirit can hear and feel, fully sensing the music. Unfortunately, few of us really listen; most of us only see the notes on the paper.

How can this limited view of a fuller picture be so? In a Temporal sense, in consideration of what we know of the Universe around us, "*space*" is a very empty place, exquisitely so (99.9999999999999% empty). Virtually all of what there is, is the Zero Point Field (miniscule fluctuations of energy, on scales so small and in time segments so short, as to be unmeasurable). Atoms we sense to be substantive and *solid*, if enlarged to a scale we can comprehend, will be found to be incredibly empty and void.

If the atomic nucleus were to be the size, *say, of a golf ball*, the electrons then would be the size of a poppy seed orbiting in the distance fully a mile away. Yet "*physically*" (in the reality we know) the entire structure, *nucleus and electron shell*, is sensed by us to be a singular *solid* object filling the space it occupies. If all of the empty space in all of the atoms comprising the Empire State building were to be removed (if the subatomic particles were gathered together as to be in direct contact with one another), all of the matter would fill a volume of space no larger than a single grain of rice while weighing over 100 million pounds.

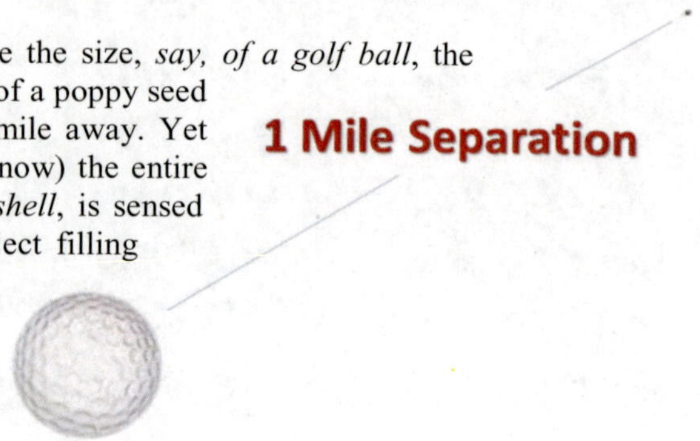


What we sense to be real and substantive, is really not at all what it seems. Like a symphony that can fill a room with sound is reduced to small notes on a page full of open white space, we understand that objects we know to be substantial are themselves comprised of very small things with much empty space in between. Now consider the structures we see before us: tall buildings, even larger planets, immense stars and tremendous galaxies are themselves just tiny pieces of a much larger whole as well. They (the things of the Temporal sphere) are just the notes representing a grander whole, a whole so beautiful, so magnificent and so large, as to be incomprehensible.

The analogy runs much deeper. Music, its freedom of expression, fluidity and creativity, cannot be fully fathomed within the rigidity of the written notes and the rules that must be attendant to ensure they can accurately present in a confined way the richness and intricacy of the piece they represent.

No matter how well the conductor understands the written music before him, it can only be fully comprehended when played and released into the air. – *Science and the scientist cannot fully comprehend the fullness of God's creation by studying only the Temporal.*

Regardless of how rich and full our understandings are, how beautiful and symmetric our formulas, equations and theories seem to be, they are incomplete, for they only address the Temporal. On paper the notes stand separate in space, black dense spots with white spaces in between, but the music they represent is a continuum, a broad spectrum. – *Temporality in all of its majesty, can only present a constrained, limited view of the whole of what there is.*

Those with experience, those who are aware, can hear and feel and sense the reality of the music, in all its fullness by simply looking at the notes on the paper. – *With the assistance of the Spirit, inspired men of God, and those who seek after His wisdom, can see in the Temporal, the things of the eternal Physical.*

# GLOSSARY OF IMPORTANT UNDERSTANDINGS

**Bara (Barau) (Hebrew, בָּרָא)** Rendered in the KJV as *"created,"* *bara* is *exclusively* used to describe the creative activity of God. By this distinction, it is understood that no other being can achieve the power inherent in this word, for it describes something only God can do.

**Be-ne- ha-e-lo-him (Hebrew, בְנֵי־הָאֱלֹהִים)** Rendered in the KJV as "the sons of God," *Be-ne- ha-e-lo-him* has been used in this work in its proper context to identify Adam and Eve and all of their offspring as *the sons and daughters of God.* We understand here these to be those who, after accepting the challenge presented in the Grand Council, enter into the Temporal Sphere fully aware of its challenges with the potential to progress and attain the fullness of the blessing of God. The *Be-ne- ha-e-lo-him* are those instilled with the *Neshama*.

**Bohu (Hebrew, בֹּהוּ)** *Bohu* is rendered in the KJV as "*void*" (meaning emptiness or even expanse). We have understood it to mean *filled with the building blocks of matter*, so we associate bohu with the very matter of the Temporal Universe, every particle and sub-particle that has a tangible existence and is subject to the rule of physical law. *Bohu* is a composite word, meaning, *bo* – in it and *hu* – there is. These building blocks are infused with the knowledge of God, and they have the potential to assemble and create everything there is in the Temporal Sphere.

**Ha-a-dam (Hebrew, הָאָדָם)** *Ha-a-dam* is rendered in the KJV as simply "*man.*" This translation misses the nuance we associate with it here to properly mean "*people of the clay,*" those whom we understand in this work to be pre-Adamites, or the *human animals* that populated the Earth prior to, and for some time after, the arrival of Adam and Eve. The *ha-a-dam* interbred with the sons and daughters of God, leading up to the Great Flood and other events associated with their elimination from the Earth.

**Hu-hal (Hebrew, הוּחַל)** *Hu-hal* is translated in the KJV as "*began men.*" We understand here in this work the *hu-hal* to be those descendants of Cain and the *ha-a-dam* who began to associate themselves with the religious understandings and convictions of the sons and daughters of God. This was problematic for they lacked the benefits of the *Neshama* and, therefore, lacked the ability to comprehend the things of God and the tools needed for the exercise of agency and the progression that it attends.

**Nefesh (Hebrew, נֶפֶשׁ)** *Nefesh* has many renderings in the KJV, among them are, "*soul, living being, life, self, person, desire, passion, appetite, emotion.*" The Nefesh is understood in Hebrew to be the *Godly force* that gives Animals life. *Nefesh* is most often associated with animal instinct and the source of self-identity within living creatures. We understand it here to be a form of the Spirit of God shared by Him through his "*Breathing*" (the *Ruach*) to all living things.

**Neshama (Hebrew, נְשָׁמָה)** *Neshama* is rendered in the KJV as "spirit, breath, breathes." We understand it here to be the second form of Spirit to be shared through the *Ruach* to living things, however, the Neshama only pertains to Adam, Eve and their offspring, only the sons and daughters of God. The Neshama is of a higher order than the *Nefesh.* The sons and daughters of God have both the *Nefesh* and the *Neshama*. In this work and in Hasidic Rabbinical thought, the *Neshama* is understood to be *an actual part of God* shared through his "*Breathing*" (the *Ruach*) to *bind* the Children of God to him during their sojourn in Temporality. The Neshama is the divine compass within us; it is our intellect, the source of our reason, and our sense of right and wrong, *our morality*. The *Neshama* is the *logos* within each of us; it is that which emanates from the Divine *Logos,* the Word of God.

**Ruach (Hebrew, רוּחַ)** The KJV translators render *ruach* as "*wind, spirit*." The Hebrew word *Ruach* has the common meaning of wind, however, as *Ruach* applies to God and as we use it here in this book, the word means very much more. It is the *Ruach* that is identified as the breath of God when He bestows *the breath of life* (the breathing) to both animals and man. Every living thing has a *Ruach.* Because of this, the classification of this word changes from a verb (taking a breath) to a noun (the breath); as it is understood that through the act of *breathing*, God *bestows* a tangible thing, *the breath of life* (*Ruach*) on us. The *Ruach* is the life-giving force that animates a living thing. We understand that force to be a bond between God and the living.

**Tohu (Hebrew, תֹּהוּ)** *Tohu* is rendered in the KJV as "*without form*" (meaning chaotic). In this work, we have understood it to mean much more. The Kabalah teaches that *tohu* is the solitary primordial substance (substrate) from which all that is material (tangible) was to be formed. We have here associated *tohu* with the random energy fluctuations in the Zero Point Field and at times with the Zero Point Field itself.

**Zero Point Field** (See Appendix)